MONOGRAPHS ON STATISTICS AND APPLIED PROBABILITY

General Editors

D.R. Cox, V. Isham, N. Keiding, T. Louis, N. Reid, R. Tibshirani, and H. Tong

Statistics
in the
21st Century

Edited by
Adrian E. Raftery
Martin A. Tanner
Martin T. Wells

American Statistical Association
(ASA)
Alexandria, Virginia

CHAPMAN & HALL/CRC

Boca Raton   London   New York   Washington, D.C.

Library of Congress Cataloging-in-Publication Data

Statistics in the 21st century / edited by Adrian E. Raftery, Martin A.
Tanner, Martin T. Wells.
        p. cm. -- (Monographs on statistics and applied probability ;
93)
    Includes bibliographical references and index.
    ISBN 1-58488-272-7 (alk. paper)
    1. Mathematical statistics. I. Title: Statistics in the
twenty-first century. II. Raftery, Adrian E. III. Tanner, Martin, Abba,
1957- IV. Wells, Martin. (Martin T.) V. Series
QA276.16. S84444 2001
519.5--dc21   2001028887

Co-published by

CRC Press LLC   and   American Statistical Association
2000 N.W. Corporate Blvd.   1429 Duke Street
Boca Raton, FL 33431   Alexandria, VA 22314-3415

Visit the CRC Press Web site at www.crcpress.com

© 2002 by American Statistical Association

No claim to original U.S. Government works
International Standard Book Number 1-58488-272-7
Library of Congress Card Number 2001028887
Printed in the United States of America   4 5 6 7 8 9 0
Printed on acid-free paper

Contents

Introduction
By Adrian E. Raftery, Martin A. Tanner, and Martin T. Wells

Chapter 3

Chapter 4
Theory and Methods of Statistics   273
Guest Edited By George Casella

Introduction

Where is statistics headed in the 21st century? What are the main themes in statistics as it emerges from the 20th century?

In this volume, over 70 leading statisticians and quantitative methodologists from other disciplines reflect on those questions in vignettes, or short review articles, each of which discusses an important area of statistics. The vignettes highlight some of the most important statistical advances and outline potentially fruitful areas of research. They are not exhaustive reviews, but rather selected "snapshots" of the world of statistics at the dawn of the 21st century. The purpose of this volume is to examine our statistical past, comment on our present, and speculate on our future.

A first major theme that emerges is that the development of statistics has been driven by the broader environment within which it operates: by applications in the sciences, the social sciences, medicine, engineering, and business, by the appearance of new types of data demanding interpretation, and by the rapid advance in computer technology. This is not a new theme, but what does seem without precedent is the range of applications and new data types that are pushing the discipline forward. In the 19th and early 20th centuries, statistical development was largely driven by applications in a small number of areas (astronomy, official statistics, agriculture). In the second half of the 20th century, statistics has come to be a central part of many disciplines that involve numerical data, and even nonnumerical data, and much of the research has been driven by the demand for new methods from the disciplines for which statistics has become an essential tool.

A second theme is that the computer revolution has transformed statistics. Statistics is largely built on a foundation of mathematics, but over the past 30 years, fast computing has become a cornerstone. This has made possible new kinds of analysis and modeling that previously were not only impossible but unthinkable. These range from the early interactive software such as GLIM in the 1970s, through the bootstrap and software such as S that allowed easy visual exploration of data in the 1980s, to the Bayesian revolution of the 1990s made possible by Markov chain Monte Carlo methods.

Because of this, we have organized this volume around major areas of application of statistics, leading up to a concluding group of vignettes that discuss the theory and methods of the discipline in their own right. The volume is divided into four main sections, each edited by a Guest Editor: Statistics in the Life and Medical Sciences, Statistics in Business and Social Science, Statistics in the Physical Sciences and Engineering, and Theory and Methods of Statistics.

Although the coverage of this volume is broad and the topics diverse, the same themes recur in different contexts, pointing to the underlying unity of the field of statistics. As one example, consider the analysis of point processes consisting of the times at which one or several events occur, such as death, divorce, or machine failure.

In the health sciences this is called survival analysis and is discussed by Oakes, in social science it is called event history analysis and is reviewed by Raftery and by Xie, and in engineering it is called reliability theory and is reviewed by Lawless. The underlying analysis strategy is the same in all these areas: the basic primitive is the hazard rate, and one develops models for this; the Cox proportional hazards model is influential everywhere. Applications of point processes in seismology are written about by Vere-Jones, and in the analysis of Internet data by Cleveland and Sun.

The analysis of multivariate discrete data is reviewed in general terms by Fienberg, Christensen, and McCulloch, and in the context of wildlife applications by Pollock and of sociology by Raftery. Causal analysis using counterfactuals is discussed for the health sciences by Greenland and for the social sciences by Sobel. Hierarchical models and related methods are discussed in general by Carlin and Louis and by Hobert, and in the context of epidemiology by Thomas, of receiver operating characteristic data by Pepe, of toxicology by Ryan, and of animal breeding by Gianola. Time series analysis is discussed from different perspectives by Tsay and by Solo. Coding and information theory are discussed by Rissanen and Yu, and by Soofi.

The development and application of a coherent and comprehensive set of methods for analyzing medical and public health data is perhaps the greatest collective achievement of the discipline of statistics in the second half of the 20th century. This has led to the development of biostatistics, which is a thriving subdiscipline in its own right, while remaining an integral part of the broader statistics profession. This seems like a model to follow for other areas where the penetration of statistics has not yet been as extensive. The first set of vignettes, Statistics in the Life and Medical Sciences, guest edited by Norman E. Breslow, bears witness to the extraordinary development of statistical methods in these areas, as well as the extent of collaborative work between statisticians and other scientists.

Several cross-cutting themes are apparent in this set of vignettes. They highlight three main methodologies: causal analysis, survival analysis, and hierarchical modeling, as well as a rich array of applications. Causal analysis using counterfactuals was pioneered by Neyman and Fisher and applied in medicine in the form of the randomized clinical trial (see Harrington's vignette); for more recent developments see Greenland's vignette. The basic tools of survival analysis have been the Kaplan-Meier estimator, the logrank test, and Cox's proportional hazards model; see the vignettes by Oakes, Thomas, Ryan, Pollock, and Gianola for review and recent developments. Hierarchical modeling and the related generalized estimating equations (GEE) approach are very important and are reviewed by Gianola, Thomas, and Pepe.

Perhaps the most active area of science at the moment is the study of the genome, and statistical aspects of this are reviewed by Weir and by Wong. Four areas are highlighted: the two more established areas of gene location and sequence analysis, and the two newer and rapidly expanding areas of protein structure prediction and gene expression data analysis. Other application areas in the life and medical sciences reviewed include the environment (Guttorp), wildlife population estimation (Pollock), animal breeding (Gianola), human fertility (Weinberg and Dunson), and toxicology (Ryan).

The Business and Social Science set of vignettes, guest edited by Mark P. Becker, reviews the state of statistics in a range of disciplines: finance (Lo), marketing (Rossi and Allenby), political science (Beck), sociology (Raftery), psychology (Browne), the law (Eisenberg), and demography (Xie).

The Physical Sciences and Engineering vignettes, guest edited by Diane Lambert, do likewise for disciplines within their scope: atmospheric science (Nychka), seismology (Vere-Jones), reliability (Lawless), process control (Stoubmos et al.), the pharmaceutical industry (Gunter and Holder), and manufacturing (Nair, Hansen, and Shi). The emerging field of the analysis of Internet data is discussed by Cleveland and Sun.

Bayesian statistics and Markov chain Monte Carlo have had a major impact on cutting edge statistical practice in the past ten years, and they are mentioned in many of the vignettes. The Theory and Methods vignettes, guest edited by George Casella, include several where they are the main focus (Berger, Cappé and Robert, Carlin and Louis, Gelfand, George). The bootstrap, which ignited the current explosion of computationally intensive methods in statistics, is reviewed by its inventor, Efron. Nonparametric and robust methods are reviewed by Fan, Hettmansperger, McKean and Sheather, Portnoy and He, missing data methods by Meng, and measurement error models by Stefanski. Likelihood and related foundational concepts are reviewed by Reid and by Robins and Wasserman, decision theory by Brown and W.E. Strawderman, and asymptotics by R.L. Strawderman.

In spite of the broad coverage of this volume, many statistical topics, some of them very important, have been omitted. We can only plead the impossibility of covering such a broad and dynamic discipline completely in one volume and mention some of the omissions that appear to us most glaring. Data collection methods generally get short shrift in this volume. The design of experiments, which played a major role in launching modern statistics early in the 20th century, is not represented here in spite of a recent spurt of interest and new applications. Similarly, survey sampling is not covered, although the proliferation of new ways of collecting social and behavioral data through the Internet seems likely to spark a revival of this field.

Exploratory data analysis and visualization do not have a separate vignette, although their influence is apparent in many of the vignettes. Multivariate analysis is not covered explicitly, although aspects are discussed in Browne's vignette on psychometrics and in some others. Graphical models, neural networks, and cluster analysis, in particular, are areas of multivariate analysis that are progressing rapidly at the moment. There is not a separate vignette about econometrics, but its influence is pervasive in several of the disciplines reviewed in the Business and Social Science chapter. And applications of statistics to the arts and the humanities are not mentioned, although there have been some, for example in history and music. Even topics that are treated may have been viewed from a limited perspective.

So where is the field of statistics going in the new millennium? While prediction is hard, especially about the future, it does seem safe to say that new developments will be driven by new kinds of data requiring analysis and by the development of computing to make them possible. Gene expression data is one current example of this, and this is a field where statisticians have rapidly become deeply involved. Datamining is another; this started life as the analysis of retail barcode data, and statisticians have become involved more slowly there. One area where statistics has been largely absent, but where new theory and computing power may allow it to make a contribution, is the analysis of simulation or mechanistic models, which are mostly deterministic and dominate scientific endeavor in many disciplines, often largely to the exclusion of more conventional statistical models. We encourage our successors to produce a sequel to this work to begin the 22nd century!

This volume shows statistics to be broad and diverse, but we feel that it also shows the essential intellectual unity of the field. Three basic ideas underlie a great deal of what statisticians do and are influential in almost every vignette: the representation of the phenomenon being studied by a probability model, the summarization of information in the data using the resulting likelihood function, and the basic principle put forward by Tukey in 1962 that one should look at the data as part of the model-building process. The methodology for implementing these principles can involve either mathematics, such as asymptotic approximations, or, increasingly, intensive Monte Carlo computing, such as the use of simulation to evaluate methods, the bootstrap, importance sampling, or Markov chain Monte Carlo.

We are very grateful to the many people who have contributed to making this millennial project a reality, primarily, of course, to the vignette authors themselves and to the Guest Editors. The vignettes in this volume were first published in the year 2000 in the *Journal of the American Statistical Association*, of which we were the editors in that year. We are very grateful to our editorial coordinators, Janet Wilt, Lisa Johnson, Mary Rogers, and Katherine Roberts, to Mary Fleming and Carol Edwards and their staff at the American Statistical Association, to Eric Sampson, and to Cathy Frey and her staff at Cadmus Press. We are also grateful to Jonas Ellenberg, Jim Landwehr, and Al Madansky for helping to make the publication of the vignettes a reality. And, finally, we thank Kirsty Stroud, Tom Louis, and Chapman & Hall for helping us to bring this book to publication.

Adrian E. Raftery, Seattle, WA
Martin A. Tanner, Evanston, IL
Martin T. Wells, Ithaca, NY
February 2001

Chapter 1

Statistics in the Life and Medical Sciences

Norman E. Breslow

One of the pleasures of working as an applied statistician is the awareness it brings of the wide diversity of scientific fields to which our profession contributes critical concepts and methods. My own awareness was enhanced by accepting the invitation from the editors of *JASA* to serve as guest editor for this section of vignettes celebrating the significant contributions made by statisticians to the life and medical sciences in the 20th century. The goal of the project was not an encyclopedic catalog of all the major developments, but rather a sampling of some of the most interesting work. Of the 12 vignettes, 10 focus on particular areas of application: environmetrics, wildlife populations, animal breeding, human fertility, toxicology, medical diagnosis, clinical trials, environmental epidemiology, statistical genetics, and molecular biology. The two vignettes that begin the series focus more on methods that have had, or promise to have, impact across a range of subject matter areas: survival analysis and causal analysis.

The concept of a counterfactual true treatment effect was introduced by Neyman for agricultural field experiments in the 1920s, and Fisher's method of randomization provided a physical basis for making causal inferences. Bradford Hill's advocacy of these principles for use in medicine led to the randomized, double-blind, placebo-controlled clinical trial. As Harrington points out, this was arguably the most important scientific advance in medicine during the 20th century. Greenland's vignette describes recent theory and methods developed from these same foundations for causal analysis of observational data that may help sort out some vexing public health issues.

The impact of survival analysis has been immense. Weinberg and Dunson discuss survival methods for population monitoring of fertility. Ryan describes how transition rate models for carcinogenicity underlie the analysis and interpretation of data from the lifetime rodent bioassay, which still strongly influence regulatory policy. Oakes mentions the importance of multivariate survival methods for genetic epidemiology, Pollock cites applications to wildlife studies, and Gianola notes increased use of survival models even in animal breeding. But these many applications still represent only a small sampling of the whole. Kaplan and Meier's product limit estimate, Mantel

Norman E. Breslow is Professor of Biostatistics, University of Washington, Seattle, WA 98195 (E-mail: norm@biostat.washington.edu).

and Peto's log-rank test and Cox's proportional hazards regression model are the indispensable tools of a large cadre of statisticians working on clinical trials in industry, government, and academia. The fact that Cox received the 1990 General Motors prize for clinical cancer research underscores the enormously beneficial impact of this work on clinical medicine.

Preventive medicine has been no less affected by the concepts and methods of survival analysis. The key epidemiologic measure of incidence rate is rooted firmly in the centuries-old tradition of the life table, whereas the more recent concept of relative risk is best understood as a ratio of such rates. The proportional hazards model provided the mathematical foundation for classical epidemiologic methods of relative risk estimation. It paved the way for modern developments by connecting the field to Fisher's likelihood inference and its semiparametric extensions. Particularly important are the new epidemiologic designs that have been stimulated by ideas from survival analysis: the nested case-control design, the case-cohort design, the case-crossover design, and two-phase stratified versions of all of these. The vignettes by Oakes and Thomas reference some of this work and cite recent, comprehensive reviews.

Hierarchical modeling is a cross-cutting development whose great importance is chronicled in several vignettes. Statisticians who have discovered its value in their own areas of application owe a great debt to the pioneering efforts of those working in the field of animal breeding, notably Henderson and Patterson and Thompson. Gianola argues that the mixed model equations and their best linear unbiased predictors (BLUPs) of genetic value are probably "the most important technological contribution of statistics to animal breeding." Analogous predictors of random effects in both linear and nonlinear mixed-effects models play no less a role in spatial statistics. Thomas, for example, notes their value for smoothing of small area disease rates prior to map construction.

Although hierarchical modeling can proceed using only the mixed model equations and restricted maximum likelihood (REML) estimation of variance components, the advantages of a full Bayes approach are increasingly apparent. Gianola argues that this provides the only satisfactory solution to assessing uncertainty in variance components and BLUPs. Markov chain Monte Carlo (MCMC) calculations, furthermore, are essential for fitting models with large (he cites a case with 700,000) numbers of random effects. Thomas calls attention to the importance of Bayes model averaging techniques in epidemiology. Guttorp cites several applications of MCMC for spatial prediction in environmental problems, and Wong notes the use of MCMC for multiple alignment of DNA sequence data. But Bayesians are not alone in their use of MCMC and other computationally intensive procedures. Efron's bootstrap has also dramatically impacted both the theory and practice of statistics. Pollock in particular notes its application to capture-recapture data.

Public health statisticians tend to favor marginal mean regression models over their hierarchical counterparts, because the parameters then have a desired interpretation in terms of population averages. The generalized estimating equation (GEE) approach with a specified "working" correlation matrix, as developed by Liang and Zeger, has revolutionized the analysis of longitudinal and other forms of clustered data. Ryan notes the impact of these methods on the analysis of data from reproductive toxicology studies, where the correlation of outcomes among littermates is of little intrinsic interest, and Thomas mentions their importance in epidemiology. Pepe cites both marginal and hierarchical approaches to the analysis of receiver operating characteristic data.

This series of short vignettes provides a sampling of the fascinating statistical problems that arise from the life and medical sciences, of the crucial contributions made by statisticians to those sciences, and of the statistical concepts and techniques that have led to this success. They confirm that the statistics of the 21st century will be heavily influenced by the revolutionary developments in technology, particularly in the information and biomedical sciences, and by the availability of vast new repositories of geographic and molecular data. The authors, referees, and editors who have contributed their hard work to this project will be amply rewarded if the series helps to attract students of statistical science into the fields that have so stimulated their own interest and productivity.

Survival Analysis

David Oakes

1. INTRODUCTION

Survival analysis concerns data on times T to some event; for example, death, relapse into active disease after a period of remission, failure of a machine component, or time to secure a job after a period of unemployment. Such data are often right-censored; that is, the actual survival time $T_i = t_i$ for the ith subject is observed only if $t_i < c_i$ for some potential censoring time c_i. Otherwise, the fact that $\{T_i \geq c_i\}$ is observed, but the actual value of T_i is not. For example, in a study of mortality following a heart attack, we will typically know the exact date of death for patients who died, but for those patients who survived, we will know only that they were alive on the date of their last follow-up. As an important but sometimes overlooked practical point, these event-free follow-up times must be recorded to allow any meaningful analysis of the data. Usually the c_i will vary from patient to patient, typically depending on when they entered the study. The paper of Kaplan and Meier (1958) in this journal brought the analysis of right-censored data to the attention of mathematical statisticians by formulating and solving this estimation problem via nonparametric maximum likelihood. Over the next few years, attention focused largely on extending nonparametric tests, such as logrank, Wilcoxon, and Kruskal–Wallis, to allow for possible right censoring. In this context, Efron (1967) introduced the notion of self-consistency ("to thine own self be true"), a key to the modern approach to missing-data problems via the EM algorithm (Dempster, Laird, and Rubin 1977). Breslow and Crowley (1974) proved the weak convergence of the normalized Kaplan–Meier estimator to Brownian motion.

2. COX'S PROPORTIONAL HAZARDS MODEL

Emphasis shifted from hypothesis testing to modeling effects of explanatory variables ("covariates") on survival following the introduction by Cox (1972) of the proportional hazards model. Cox's model includes the unknown baseline hazard as a nuisance function, but the effects of the covariates on the hazard are modeled via a simple multiplicative factor.

David Oakes is Professor and Chair, Department of Biostatistics, University of Rochester, Rochester, NY 14642.
This work was supported in part by National Cancer Institute grant R01 CA52572.

Specifically, the hazard function $\lim_{\Delta \to 0+} (1/\Delta)\mathrm{pr}(T_i \leq t + \Delta | T_i \geq t)$ for the survival time T_i of individual i is given by

$$h_i(t) = \exp(\boldsymbol{\beta}\mathbf{x}_i)h_0(t), \tag{1}$$

where $\mathbf{x}_i$ is the column vector of covariates, $\boldsymbol{\beta}$ is a corresponding row vector of regression coefficients to be estimated, and $h_0(t)$ is the baseline hazard function; that is, the hazard for an individual with $\mathbf{x} = 0$. Cox (1972, 1975) suggested an ingenious method of estimating $\boldsymbol{\beta}$ without knowledge of $h_0(t)$. The full likelihood, which involves both $h_0(t)$ and $\boldsymbol{\beta}$, is decomposed into a product of terms corresponding to the time to the first failure, the identity of the first failure, the gap time between the first and second failure, the identity of the second failure, and so on. Each term is taken conditionally on the information given by all previous terms of the sequence. Cox's insight was that the usual large-sample properties of the full likelihood function are inherited by the *partial likelihood*, consisting of the product of the alternate terms corresponding to the identities of the successive failures. This product has the form

$$\mathrm{lik} = \prod_{i \in \mathcal{D}} \frac{\exp(\boldsymbol{\beta}\mathbf{x}_i)}{\sum_{\mathcal{R}_i} \exp(\boldsymbol{\beta}\mathbf{x}_j)}, \tag{2}$$

where $\mathcal{D} = \{i; T_i \leq c_i\}$ is the set of observed event times and $\mathcal{R}_i = \{j : \min(T_j, c_j) \geq T_i\}$ is the set of individuals still at risk of an event at time T_i.

Alternative motivations of the partial likelihood estimate can be given from counting process theory (see, e.g., Andersen, Borgan, Gill, and Keiding 1993 for a thorough review) or from Johansen's (1983) representation as a semiparametric profile likelihood estimator. Johansen's representation allows many nonstandard problems in survival analysis to be treated by general techniques for maximizing likelihood functions, such as the EM algorithm. Rigorous proofs of the asymptotic normality of the partial likelihood estimator $\hat{\boldsymbol{\beta}}$ started to appear in preprints around 1977, with the first published proof given by Tsiatis (1981) based on a careful decomposition of the partial likelihood score function. He showed joint asymptotic normality of $\{\hat{\boldsymbol{\beta}}, \hat{H}_0(t)\}$, where

$$\hat{H}_0(t) = \sum_{T_i \leq t \wedge c_i} \left\{ \sum_{j \in R_i} \exp(\hat{\boldsymbol{\beta}}\mathbf{x}_j) \right\}^{-1}$$

is a natural estimator of the cumulative baseline hazard function $\int^t h_0(u)\,du$. Andersen and Gill (1982) gave a proof using counting process theory. Prentice and Self (1983) allowed hazard ratio functions of general form $r(\boldsymbol{\beta}\mathbf{x})$ to replace the $\exp(\boldsymbol{\beta}\mathbf{x})$ in (1). Numerous authors (e.g., Efron 1977; Kalbfleisch 1974; Oakes 1977) have shown that the partial likelihood approach has high efficiency with respect to a fully parametric approach (assuming, e.g., the Weibull distribution with power-law hazard function $h_0(t) = \rho t^\kappa$) as long as β is not too far from 0 and the censoring pattern is not too highly dependent on the covariates. Jeong and Oakes (1999) noted that this high efficiency for estimation of β does not extend to estimation of $H_0(t)$, for which there may be a substantial loss of information, as in the corresponding single-sample problem (Miller 1981). The relative computational simplicity of Cox's model has made it a useful case study for formal semiparametric estimation theory (Begun, Hall, Huang, and Wellner 1982; Bickel, Klaassen, Ritov, and Wellner 1993).

Following publication of the text by Kalbfleisch and Prentice (1980) and the incorporation of software for fitting proportional hazards models into packages such as BMDP and SAS, this model has, for better or worse, became standard for the analysis of survival data. But the assumption of proportional hazards has no compelling mathematical justification and is often found to be false in applications. Cox (1972) provided a simple test of fit using *time-dependent covariates*; that is, replacing the $\mathbf{x}_j$ in (1) by $\mathbf{x}_j(t)$ for some prespecified (or at least predictable; see Andersen et al. 1993) functions $\mathbf{x}_j(t)$ of time. The propriety of using such variables, and thus in effect subsuming nonproportional hazards models within the methodology, is a great strength of Cox's approach. Subsequent developments in goodness-of-fit testing emphasized martingale residuals based on the "observed minus (conditionally) expected" formula familiar in epidemiologic work (see, e.g., Barlow and Prentice 1988; Therneau, Grambsch, and Fleming 1990). Interestingly, an apparently different approach, via the so-called "generalized residuals" of Cox and Snell (1968), leads to essentially the same result.

3. OTHER REGRESSION MODELS

Many competitors to the proportional hazards model have been developed; however, none allows such a simple semiparametric analysis. In the *scale change* (*accelerated life*) model, the predictors act by speeding up or slowing down the time scale. Barring pathological examples involving periodicities (Doksum and Nabeya 1984), the accelerated life and proportional hazards models can both hold only if the survival functions follow the Weibull form. The regression coefficients in the two formulations are related by the factor κ in the index of the power law. Least squares methodology can be extended to censored data (Buckley and James 1979; Miller 1976), but, as for uncensored data, is inefficient when the error distribution is nonnormal. Robins and Tsiatis (1992) developed an efficient approach to semiparametric estimation in the accelerated life model that extends to a version proposed by Cox and Oakes (1984, chap. 5), allowing time-dependent explanatory variables that act via a monotonic but possibly nonlinear transformation of the time scale.

The proportional hazards model is not preserved when a covariate is added to or removed from the model, as is routinely done by "stepwise selection" procedures, even when the added term is statistically independent of the other covariates. For example, suppose that one covariate in a proportional hazards model has hazard ratio W that follows a gamma distribution and is independent of the other covariates, so that (1) becomes $h_i(t) = W_i \exp(\beta \mathbf{x}_i) h_0(t)$. The resulting model obtained by integrating over the distribution of W no longer has proportional hazards, and the effect of the observed covariates is attenuated. In particular, exponentially distributed W_i lead to the proportional odds model for the survivor function introduced by Bennett (1983). Murphy, Rossini, and van der Vaart (1997) justified Bennett's semiparametric estimation procedure. Hougaard (1986) noticed that the proportional hazards property is preserved if instead W has a positive stable distribution with index α. Here the effect of the observed covariates is attenuated by a factor α. An additive model, with (1) replaced by $h_i(t) = r(\beta \mathbf{x}_i) + h_0(t)$, has some attractions—although a partial likelihood is no longer available, estimating equations for β can be derived from counting process theory (Aalen 1978; Lin and Ying 1994).

4. MULTIPLE FAILURE TIMES

Often individuals may experience several different types of events, with possibly more than one event occurring to a subject. *Competing risks* models apply when at most one event can occur to a single subject. In competing risks models, the assumption that risks are independent can never be tested satisfactorily, so predicted survival curves calculated on this assumption can be seriously misleading. Pepe (1991) proposed that inferences be focused instead on observable quantities such as the cumulative incidence. When each individual can experience multiple events these can be viewed as transitions of a Markov or semi-Markov process among a variety of possible states (Prentice, Williams, and Peterson 1981). As long as these transitions are modeled in a way that respects the time ordering of the events (i.e., intensity functions at each time are modeled as a function of the entire history of the individual at that time), the full likelihood for the entire process factors into a product of terms for each possible transition, and standard asymptotic theory applies. This principle is violated by "marginal models" in which the intensity of an event of one type at time t is modeled as a function only of predictors available at $t = 0$. There may be compelling scientific reasons for the use of a marginal model; for example, to preserve the comparability between treatment groups in a randomized study, one typically measures the time to each possible event from the date of randomization rather than from a previous event. Wei, Lin, and Weissfeld (1989) suggested maximization of the product of the separate partial likelihood functions for the times to each type of event, ignoring any previous events. This product is no longer itself a partial likelihood and a "sandwich" formula must be used to estimate the variance of the estimator. Cai and Prentice (1995) considered weighting these estimators to achieve greater efficiency.

The partial likelihood argument extends to *nested case-control* studies in which the sum in the denominator of the ith term in (2) is taken over the set $\{i\} \cup \mathcal{C}_i$, where $\mathcal{C}_i$

are independent random samples from the set $\mathcal{R}_i \setminus \{i\}$ of event-free individuals in $\mathcal{R}_i$ (Oakes 1981). However, the simpler *case-cohort* design proposed by Prentice (1986), in which $\mathcal{R}_i$ is replaced by $\{i\} \cup \mathcal{C}$, where the single random sample $\mathcal{C}$ is chosen at the outset, does not yield a partial likelihood function. An estimating equation approach is needed that allows for correlation between the terms of the (pseudo) score function. Genetic studies involving ascertainment of disease incidence in probands and relatives of probands lead to still more complex designs and analyses. The AIDS epidemic has spurred development of methodology to handle more complicated data structures involving *truncation* (delayed entry) and/or *interval censoring* (Lagakos, Barraj, and DeGruttola 1988). Genetic studies provide a fertile area of application for multivariate survival analysis, where events to related individuals occur along separate unrelated time scales. Clayton (1978) introduced a bivariate model for assessing the dependence between risks of heart attacks in fathers and their sons. The random-effects (frailty) representation (Oakes 1989) provides a broad class of models for such data. Fully nonparametric estimation of a bivariate survival distribution under arbitrary patterns of censorship remains a challenging problem. Estimators have been proposed by Dabrowska (1988), Prentice and Cai (1992), and others, but whether these are asymptotically efficient is unknown. Simpler, inefficient estimators are available under restricted patterns of censorship (Lin and Ying 1993). The idea, familiar in sampling theory contexts, is akin to weighting each observation by the reciprocal of the estimated probability that it is uncensored. This approach has been developed extensively by Robins and coworkers to attack the problem of dependent risks in more complex models (see, e.g., Robins and Rotnitzky 1992).

5. FUTURE DEVELOPMENTS

The future evolution of survival analysis is far from a predictable process. The view has been expressed that survival analysis is but a special case of the analysis of longitudinal data (Diggle, Liang, and Zeger 1994) with incomplete follow-up, so that it may eventually disappear as a separate field of study. Alternatively, we may focus on the incompleteness aspect, especially when considering more complex mechanisms for incomplete observations than simple right censoring. From this viewpoint, correct modeling of the "missingness" mechanism then becomes the crucial element. The notion of *coarsening at random* (Gill, van der Laan, and Robins 1997) is one approach to developing this concept. Hastie and Tibshirani (1990, sec. 8.3) applied the techniques of generalized additive models to fit time-dependent coefficients in Cox's model. A related notion is the use of smoothing or using kernels or polynomial splines to smooth the underlying baseline hazard function (Kooperberg, Stone, and Truong 1995).

Survival analysis seems unlikely soon to become extinct (or to reach any other absorbing state!), because of the ubiquity of this type of data and the mathematical interest of the problems raised by the development of proper methods of analysis. As in other fields, the computational Bayesian approach provides an attractive tool for fitting models whose complexity would rule out other approaches. Sinha and Dey (1997) gave a comprehensive review of recent such work. Counting processes and their associated martingales will undoubtedly continue to play an important role in the theory. The notion of a martingale captures the fundamental characteristic of survival analysis, that information about events and associated covariates arrives over time, not all at once, and that information, inferences, and predictions at a specific time must involve only what is known at that time.

REFERENCES

Aalen, O. (1978), "Nonparametric Inference for a Family of Counting Processes," *The Annals of Statistics*, 6, 701–726.

Andersen, P. K., Borgan, O., Gill, R. D., and Keiding, N. (1993), *Statistical Models Based on Counting Processes* New York: Springer.

Andersen, P. K., and Gill, R. D. (1982), "Cox's Regression Model for Counting Processes, a Large-Sample Study," *The Annals of Statistics*, 10, 1100–1120.

Barlow, W. E., and Prentice, R. L. (1988), "Residuals for Relative Risk Regression," *Biometrika*, 75, 65–74.

Begun, J. M., Hall, W. J., Huang, W.-M., and Wellner, J. A. (1982), "Information and Asymptotic Efficiency in Parametric Nonparametric Models," *The Annals of Statistics*, 11, 432–452.

Bennett, S. (1983), "Analysis of Survival Data by the Proportional Odds Model," *Statistics in Medicine*, 2, 273–277.

Bickel, P. J., Klaassen, C. A., Ritov, Y., and Wellner, J. A. (1993), *Efficient and Adaptive Estimation in Semiparametric Models*, Baltimore: Johns Hopkins University Press.

Breslow, N. E., and Crowley, J. J., (1974), "A Large-Sample Study of the Life Table and Product Limit Estimates Under Random Censorship," *The Annals of Statistics*, 2, 437–453.

Buckley, J. D., and James, I. R. (1979), "Linear Regression With Censored Data," *Biometrika*, 66, 429–436.

Cai, J., and Prentice, R. L. (1995), "Estimating Equations for Hazard Ratio Parameters Based on Correlated Failure Time Data," *Biometrika*, 82, 151–164.

Clayton, D. G. (1978), "A Model for Association in Bivariate Life-Tables and Its Application in Epidemiological Studies of Familial Tendency in Chronic Disease Incidence," *Biometrika*, 65, 141–151.

Cox, D. R. (1972), "Regression Models and Life Tables" (with discussion), *Journal of the Royal Statistical Society*, Ser. B, 34, 187–220.

——— (1975), "Partial Likelihood," *Biometrika*, 62, 269–276.

Cox, D. R., and Oakes, D. (1984), *Analysis of Survival Data*, London: Chapman and Hall.

Cox, D. R., and Snell, E. J. (1968), "A General Definition of Residuals" (with discussion), *Journal of the Royal Statistical Society*, Ser. B, 30, 248–275.

Dabrowska, D. M. (1988), "Kaplan–Meier Estimate on the Plane," *The Annals of Statistics*, 16, 1475–1489.

Dempster, A. P., Laird, N. M., and Rubin, D. B. (1977), "Maximum Likelihood Estimation From Incomplete Data via the EM Algorithm" (with discussion), *Journal of the Royal Statistical Society*, Ser. B, 39, 1–38.

Diggle, P. J., Liang, K-Y., and Zeger, S. L. (1994), *Analysis of Longitudinal Data*, Oxford, U.K.: Oxford University Press.

Doksum, K. A., and Nabeya, S. (1984), "Estimation in Proportional Hazard and Log-Linear Models," *Journal of Statistical Planning and Inference*, 9, 297–303.

Efron, B. (1967), "The Two-Sample Problem With Censored Data," in *Proceedings of the Fifth Berkeley Symposium in Mathematical Statistics and Probability*, Vol. 4, New York: Prentice-Hall, pp. 831–853.

——— (1977), "The Efficiency of Cox's Likelihood Function for Censored Data," *Journal of the American Statistical Association*, 72, 557–565.

Gill, R. D., van der Laan, M. J., and Robins, J. M. (1997), "Coarsening at Random: Characterizations, Conjectures, Counterexamples," in *Proceedings of the First Seattle Symposium in Biostatistics: Survival Analysis*, eds. D. Y. Lin and G. R. Fleming, New York: Springer, pp. 255–294.

Hastie, T. J., and Tibshirani, R. J. (1990), *Generalized Additive Models*, London: Chapman and Hall.

Hougaard, P. (1986), "Survival Models for Heterogeneous Populations Derived From Stable Distributions," *Biometrika*, 73, 387–396.

Jeong, J-Y., and Oakes, D., (1999), "On the Asymptotic Relative Efficiency of Cumulative Hazard Estimates From Cox's Regression Model," technical report, University of Rochester.

Johansen, S. (1983), "An Extension of Cox's Regression Model," *International Statistical Review*, 51, 258–262.

Kalbfleisch, J. D. (1974), "Some Efficiency Calculations for Survival Distributions," *Biometrika*, 61, 31–38.

Kalbfleisch, J. D., and Prentice, R. L. (1980), *The Statistical Analysis of Failure Time Data*, New York: Wiley.

Kaplan, E. L. and Meier, P. (1958), Nonparametric Estimation From Incomplete Observations," *Journal of the American Statistical Association*, 53, 457–481.

Kooperberg, C., Stone, C. J., and Truong, Y. K. (1995), "Hazard Regression," *Journal of the American Statistical Association*, 90, 90–94.

Lagakos, S. W., Barraj, L. M., and DeGruttola, V. (1988), "Nonparametric Analysis of Truncated Survival Data, With Applications to AIDS," *Biometrika*, 75, 515–523.

Lin, D. Y., and Ying, Z. (1993), "A Simple Nonparametric Estimator of the Bivariate Survival Function Under Univariate Censoring," *Biometrika*, 80, 573–581.

——— (1994), "Semiparametric Analysis of the Additive Risks Model," *Biometrika*, 81, 61–74.

Miller, R. G. (1976), "Least Squares Regression With Censored Data," *Biometrika*, 63, 449–464.

——— (1981), "What Price Kaplan–Meier?" *Biometrics*, 39, 1077–1082.

Murphy, S., Rossini, A. J., and van der Vaart, A. W. (1997), "Maximum Likelihood Estimation in the Proportional Odds Model," *Journal of the American Statistical Association*, 92, 968–976.

Oakes, D. (1977), "The Asymptotic Information in Censored Survival Data," *Biometrika*, 59, 441–448.

——— (1981), "Survival Times: Aspects of Partial Likelihood" (with discussion), *International Statistical Review*, 49, 235–264.

——— (1989), "Bivariate Survival Models Induced by Frailties," *Journal of the American Statistical Association*, 84, 487–493.

Pepe, M. S. (1991), "Inference for Events With Dependent Risks in Multiple Endpoint Studies," *Journal of the American Statistical Association*, 86, 770–778.

Prentice, R. L. (1986), "A Case-Cohort Design for Epidemiologic Cohort Studies and Disease Prevention Trials," *Biometrika*, 73, 1–11.

Prentice, R. L., and Cai, J. (1992), "Covariance and Survivor Function Estimation Using Censored Multivariate Failure Time Data," *Biometrika*, 79, 495–512.

Prentice, R. L., and Self, S. G. (1983), "Asymptotic Distribution Theory for Cox-Type Regression Models With General Relative Risk Form," *The Annals of Statistics*, 11, 804–813.

Prentice, R. L., Williams, B. J., and Peterson, A. V. (1981), "On the Regression Analysis of Multivariate Failure-Time Data," *Biometrika*, 68, 373–379.

Robins, J., and Rotnitzky, A. (1992), "Recovery of Information and Adjustment for Dependent Censoring Using Surrogate Markers," in *AIDS Epidemiology—Methodological Issues*, eds. N. Jewell, K. Dietz, and V. Farewell, Boston: Burkhauser, pp. 297–331.

Robins, J., and Tsiatis, A. A. (1992), "Semiparametric Estimation of an Accelerated Failure Time Model," *Biometrika*, 79, 311–319.

Sinha, D., and Dey, D. K. (1997), "Semiparametric Bayesian Analysis of Survival Data," *Journal of the American Statistical Association*, 92, 1195–1212.

Therneau, T. M., Grambsch, P. M., and Fleming, T. R. (1990), "Martingale-Based Residuals for Survival Models," *Biometrika*, 77, 147–160.

Tsiatis, A. A. (1981), "A Large-Sample Study of Cox's Regression Model," *The Annals of Statistics*, 9, 93–108.

Wei, L. J., Lin, D. Y., and Weissfeld, L. (1989), "Regression Analysis of Multivariate Incomplete Failure Time Data by Modeling Marginal Distributions," *Journal of the American Statistical Association*, 84, 1065–1073.

Causal Analysis in the Health Sciences

Sander Greenland

1. INTRODUCTION

The final quarter of the 20th century witnessed a burgeoning of formal methods for the analysis of causal effects. Of the methods that appeared in the health sciences, most can be identified with approaches to causal analysis that originated much earlier in the century in other fields: counterfactual (potential outcomes) models, graphical models, and structural equations models. Connections among these approaches were elucidated during the 1990s, and the near future may bring a unified methodology for causal analysis. This vignette briefly reviews the counterfactual approach to causal analysis in the health sciences, its connections to graphical and structural equations approaches, its extension to longitudinal data analysis, and some areas needing further work. For deeper and more extensive reviews, I especially recommend Sobel's (1995) discussion of the connections among causal concepts in philosophy, statistics, and social sciences; Pearl's (2000) unified approach to counterfactual, graphical, and structural equations models; and Robins's (1997) review of causal analysis for longitudinal data.

2. CAUSAL ANALYSIS WITH COUNTERFACTUAL (POTENTIAL) OUTCOMES

The ideas behind counterfactual analysis of causation can be traced to philosophers in the 18th and 19th centuries (see Lewis 1973; Rubin 1990; Sobel 1995), but statistical implementations had to await a formal theory for inference from randomized experiments (Fisher 1935; Neyman 1923). In its simplest form, dealing with treatments applied at one point in time and their effect on a subsequent outcome Y, one imagines that observational unit i receives exactly one of the $J + 1$ treatments listed in the vector $\mathbf{x} = (x_0, \ldots, x_J)'$. One also imagines that at time of treatment unit i has an associated vector $\mathbf{y}_i = (y_{0i}, y_{1i}, \ldots, y_{Ji})'$ of *potential outcomes*, and that the actual outcome Y_i of unit i equals the potential outcome y_{ji} if x_j is administered to unit i; Y_i is thus a function of x_j. This notation assumes that Y_i depends only on the treatment assigned to unit i, called the no-interference assumption (Cox 1958). We

Sander Greenland is Professor, Department of Epidemiology, UCLA School of Public Health, Los Angeles, CA 90095. The author would like to thank Michael Sobel and the referees for their helpful comments on the initial draft.

can observe element y_{ji} of $\mathbf{y}_i$ only if unit i receives treatment x_j; the remainder of $\mathbf{y}_i$ then becomes unobserved, missing, or counterfactual. (The term "missing" suggests that a potential outcome y_{ji} exists as a well-defined characteristic of unit i even when x_j is not administered to unit i, whereas the term "counterfactual" is used to emphasize the hypothetical, "contrary-to-fact" nature of y_{ji} when x_j is not administered to unit i.)

Let $\mathbf{r}_i$ be the $J+1$ element treatment indicator vector for unit i, which has elements $r_{ji} = 1$ if unit i receives treatment x_j and 0 otherwise. Then the observed treatment and outcome of unit i are $X_i = \mathbf{r}_i'\mathbf{x}$ and $Y_i = \mathbf{r}_i'\mathbf{y}_i$. We say that receiving treatment x_j instead of x_0 (would have) had an effect on Y_i if $y_{ji} \neq y_{0i}$, for then Y_i differs according to whether unit i receives treatment x_j or x_0. Also, we say that $y_{ji} - y_{0i}$ is the effect on Y_i of receiving x_j instead of x_0. (If the potential outcomes are strictly positive, we may instead measure this causal effect by y_{ji}/y_{0i}, although this leads to identification problems (Greenland 1987)). As an example, suppose that x_1 is a new leukemia treatment, x_0 is a standard treatment, i indexes patients in a clinical trial, and Y_i is the survival time of patient i from treatment until death. We imagine that patient i will survive y_{1i} months under the new treatment and y_{0i} months under the standard treatment. If i is given the new treatment, $\mathbf{r}_i = (0, 1)'$, and we observe $Y_i = (0, 1)(y_{0i}, y_{1i})' = y_{1i}$; y_{0i} is then unobserved, and the effect of the new treatment on survival Y_i is $y_{1i} - y_{0i}$.

Given that, for every observed unit, only one element of $\mathbf{y}_i$ can be observed, how can inference about causal effects proceed? Suppose that our research goals would be advanced if we could just estimate the *average causal effect* of receiving treatment x_j instead of treatment x_0,

$$\mathrm{ACE}_j = \sum_{i=1}^{N}(y_{ji} - y_{0i})/N$$

$$= \sum_{i=1}^{N} y_{ji}/N - \sum_{i-1}^{N} y_{01}/N = \bar{y}_j - \bar{y}_0,$$

among the N units for which treatment X and outcome Y were observed. Suppose also that treatment is randomized (or, more generally, that treatment assignment $\mathbf{r}$ is independent of each of the potential outcomes y_j), and that $N_j > 0$ units are observed at treatment x_j, for a total of $J+1$ observed subsamples. Then the average outcomes of the subsamples,

$$\sum_{i:r_j=1} Y_i/N_j = \sum_{i:r_j=1} y_{ji}/N_j,$$

will be unbiased estimators of the unobserved total sample means $\bar{y}_j$, and statistical inferences about these means and their contrasts (such as the ACE_j) follow from randomization theory (see, e.g., Copas 1973; Cox 1958; Rosenbaum 1995; Rubin 1978). The unit-specific potential responses in $\mathbf{y}_i$ may be replaced by unit-specific distributional parameters, to better capture random variation in outcomes. For example, potential survival times $(y_{0i}, y_{1i})'$ may be replaced by expected survival times $\mu_i = (\mu_{0i}, \mu_{1i})'$ under a log-linear survival time model; the observed survival time Y_i is then drawn from the distribution with expectation $r_i'\mu_i$.

3. BENEFITS OF THE COUNTERFACTUAL APPROACH

Because counterfactual causal models have been resisted by some authors (e.g., Dawid 2000; Salmon 1998; Shafer 1996), it is worth noting the benefits that have accrued from their use. The following list is only partial, and as yet no alternative to counterfactuals has led to as many operational techniques (as opposed to informal guidelines) for causal analysis of health outcomes.

An important development in the final third of the 20th century was the extension of formal counterfactual models to settings in which the randomization assumption or the no-interference assumption may fail (Halloran and Struchiner 1995; Lewis 1973; Rubin 1974, 1978; Simon and Rescher 1966). By viewing r_{ji} as the "response observed" (nonmissing) indicator for y_{ji}, causal inference can be viewed as a missing-data problem, opening the way for extending concepts developed for survey response problems to causal analysis of observational data (Robins, Rotnitzky, and Scharfstein 1999; Rubin 1978). One causal analysis methodology arising from this link is propensity scoring (Joffe and Rosenbaum 1999; Rosenbaum 1995; Rosenbaum and Rubin 1983), in which one controls confounding via stratification on probability of treatment, in much the same manner as survey analyses control selection bias via stratification on probability of sampling. Another such methodology is marginal structural modeling, in which one controls confounding via weighting observations by inverse probability of treatment received, in much the same manner as survey analyses control selection bias via weighting by the inverse probability of sampling (Robins, Hernán, and Brumback 2000). Other applications of counterfactual models have led to new methods for causal inference in the face of noncompliance (Angrist, Imbens, and Rubin 1996; Balke and Pearl 1997; Robins 1994), and to formal methods for sensitivity analysis of unmeasured confounding (Copas and Li 1997; Robins et al. 1999; Rosenbaum 1995).

In addition to giving rise to many techniques for causal analysis, the counterfactual model also provides a conceptual clarity that can resolve issues left unsettled by other approaches (Holland 1986; Sobel 1995). For example, the importance of randomization for causal inference has at times been questioned in modern Bayesian writings (see, e.g., Howson and Urbach 1993, chap. 11). The counterfactual model shows the value of randomization in alleviating the profound sensitivity of Bayesian causal inferences to prior distributions (Rubin 1978). The counterfactual model also allows mathematically precise distinctions between key concepts in causal analysis and superficially similar concepts in noncausal (associational) modeling, which are often confused in observational research. For example, my colleagues and I have found the model invaluable in clarifying distinctions between direct and indirect effects (Robins and Greenland 1992), between causal and associational structures for event histories (Robins, Greenland, and Hu 1999), between causal confounding and associational collapsibility (Greenland, Robins, and Pearl 1999), between biological and statistical interaction (Greenland and Rothman 1998), between attributable fractions and the probability of causation (Greenland 1999), and between causal and associational conditional inference (Greenland 1991).

4. CONNECTIONS TO OTHER APPROACHES

Outside of the health sciences, two popular approaches to causal modeling are causal diagrams and structural equations models. Early versions of these two approaches appeared by the 1920s in the form of path analysis (see, e.g., Wright 1921), and connections between the two were appreciated from the start. Figure 1 gives an example of a causal diagram that encodes a set of qualitative assumptions about the causal relations among five variables $X_1, \ldots, X_5$. For example, presence of a single-headed arrow from X_1 to X_3 encodes the assumptions that X_1 "directly affects" X_3, whereas absence of such an arrow from X_2 to X_4 encodes the assumption that X has no "direct effect" on X_4. Causal descriptors such as "direct effect" are taken as primitives (i.e., undefined starting points) in the theory of causal diagrams; that is, unlike terms such as "causal effect" in counterfactual models, they have no formal definition within the theory. But the presence of these causal descriptors distinguishes

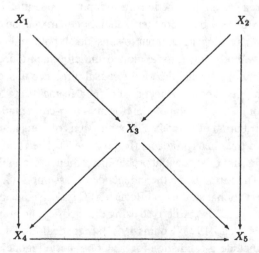

Figure 1.   An Example of a Causal Diagram (adapted from Pearl 1995).

the theory of causal diagrams from the more general theory of graphical probability models, and allows translation of results from the latter theory into identification theorems for causal effects (Pearl 1995, 2000; Spirtes, Glymour, and Scheines 1993).

Another way to formalize assumptions about causal relations is via a system of causal functional relations, often called a structural equations model. A nonparametric causal system that encodes the same assumptions as Figure 1 is $x_1 = f_1(\varepsilon_1)$, $x_2 = f_2(\varepsilon_2)$, $x_3 = f_3(x_1, x_2, \varepsilon_3)$, $x_4 = f_4(x_1, x_3, \varepsilon_4)$, $x_5 = f_5(x_2, x_3, x_4, \varepsilon_5)$, with the ε_j here assumed to be jointly independent but not otherwise specified. The presence of a variable on the right (input) side of an equation encodes the assumption that it directly affects the variable on the left (output) side, whereas its absence from the right encodes the assumption that it has no direct effect on the left. Structural equations are more often presented in a parametric form such as a system of linear equations, which, however, can encourage confusion with multivariate (noncausal) regression models. The structural equations format is intended to convey the notion that the functions f_i represent the behavior of causal mechanisms whose inputs are the function arguments, although some of the inputs (especially the ε_j) may only represent unmeasured and possibly random disturbances (Pearl 1995, 2000). Thus each function f_j in a system of structural equations is a hypothesized causal dependence of an output on inputs. The system comprises these hypotheses and the hypothesis that each dependency is autonomous; that is, each function is invariant to changes in the form of the other functions (Simon 1953).

Connections between counterfactual and structural equations models were not recognized until long after both approaches were established (Simon and Rescher 1966). The range of a structural function corresponds to the set of potential values for the output variable; these potential outcomes are thus completely (though not uniquely) indexed by the joint domain of the input variables. That is, a system of structural equations is a system of functions from inputs to potential outcomes. Structural equations thus extend the basic counterfactual model, in that the actual outcome from a given function may serve as an input to subsequent potential outcome functions.

Statistical methods that fully exploited the aforementioned connection had to await development of a formal theory for causal inference in sequential randomized experiments. The first complete theory emerged in the 1980s (Robins 1987). The basis of this theory is the g-computation formula or g-computation algorithm, which provides the distribution of outcomes in populations of units administered longitudinally randomized treatment protocols (Robins 1997). The formula decomposes this distribution into a sum of sequential probability products along the different possible event histories. The formula and the independence assumptions necessary for its application to observational studies of effects was first derived using causal tree graphs (Robins 1987), and can also be derived using directed acyclic graph models and counterfactual models (Pearl 1995; Robins 1997). Statistical models and methods based on the formula include g-estimation for structural nested models (Robins 1997) and inverse-probability-of-treatment-weighted estimation for marginal structural models (Robins et al. 2000). These methods allow causal inferences under much less restrictive independence assumptions than those required by standard methods, such as proportional hazards or generalized estimating equation (GEE) regression (Robins and Greenland 1994; Robins, Greenland, and Hu 1999); for example, these methods can provide valid effect estimates even when treatment and confounders vary over time and treatment affects the confounders.

5. SOME AREAS NEEDING DEVELOPMENT

The models discussed earlier are useful formalizations of causal structures, regardless of study design. From a health sciences viewpoint, however, one of the most pressing needs is for extension of the resulting statistical methods to case-control data. As of the time of this writing, I am unaware of formal extensions of propensity scoring or g-estimation to arbitrary case-control studies. Although they can be extended to case-control studies nested within an enumerated cohort and to case-cohort studies, most case-control studies arise from unenumerated source populations. Marginal structural modeling can, however, be applied to arbitrary case-control studies given a rare-disease assumption (Robins 1999).

There are a number of other research problems for which extensions are needed. For example, postmarketing surveillance for drug and device side effects typically involves many factors that affect the structure of the observed data, suggesting that fairly elaborate nonignorable missing-data models will be required for realistic causality assessments. In problems with multiple levels of unit definition, there is a need for elaboration of hierarchical (multilevel) counterfactual models to estimate the absolute and relative impacts of different levels of information. These issues arise, for example, in comparing the effectiveness of school-based versus community-based vaccination programs from observational data when both levels of intervention may be present.

There is also a need for performance evaluations and comparisons among specific causal modeling and fitting techniques. For example, there is as yet only limited published applications of g-estimation and marginal structural modeling. Issues such as model misspecification and model diagnostics are as yet largely unexplored. Both simulation studies and practical experiences would help identify problem areas. Practical experiences with the newer techniques would be facilitated by distribution of easily used software for the techniques; so, software is thus another pressing need.

Finally, real data applications of the methods discussed here often involve models with numerous parameters. Thus it would be of interest to examine the performance of methods such as propensity scoring, g-estimation, and marginal structural modeling when combined with techniques for fitting high-dimensional models (e.g., penalized likelihood and related hierarchical modeling or "shrinkage" techniques).

REFERENCES

Angrist, J. D., Imbens, G. W., and Rubin, D. B. (1996), "Identification of Causal Effects Using Instrumental Variables" (with discussion), *Journal of the American Statistical Association*, 91, 444–472.

Balke, A., and Pearl, J. (1997), "Bounds on Treatment Effects From Studies With Imperfect Compliance," *Journal of the American Statistical Association*, 92, 1171–1178.

Copas, J. B. (1973), "Randomization Models for Matched and Unmatched 2 × 2 Tables," *Biometrika*, 60, 467–476.

Copas, J. B., and Li, H. G. (1997), "Inference for Non-Random Samples" (with discussion), *Journal of the Royal Statistical Society*, Ser. B, 59, 55–95.

Cox, D. R. (1958), *The Planning of Experiments*, New York: Wiley, chap. 5.

Dawid, A. P. (2000), "Causal Inference Without Counterfactuals" (with discussion), *Journal of the American Statistical Association*, 95, in press.

Fisher, R. A. (1935), *The Design of Experiments*, Edinburgh: Oliver and Boyd.

Greenland, S. (1987), "Interpretation and Choice of Effect Measures in Epidemiologic Analyses," *American Journal of Epidemiology*, 125, 761–768.

——— (1991), "On the Logical Justification of Conditional Tests for Two-by-Two Contingency Tables," *The American Statistician*, 45, 248–251.

——— (1999), "Relation of the Probability of Causation to the Relative Risk and the Doubling Dose: A Methodologic Error that Has Become a Social Problem," *American Journal of Public Health*, 89, 1166–1169.

Greenland S., Robins, J. M., and Pearl, J. (1999), "Confounding and Collapsibility in Causal Inference," *Statistical Science,* 14, 29–46.

Greenland, S., and Rothman, K. J. (1998), "Concepts of Interaction," in *Modern Epidemiology*, eds. K. J. Rothman and S. Greenland, Philadelphia: Lippincott-Raven, chap. 18, pp. 329–342.

Halloran, M. E., and Struchiner, C. J. (1995), "Causal Inference for Infectious Diseases," *Epidemiology*, 6, 142–151.

Holland, P. W. (1986), "Statistics and Causal Inference" (with discussion), *Journal of the American Statistical Association*, 81, 945–970.

Howson, C., and Urbach, P. (1993), *Scientific Reasoning: The Bayesian Approach* (2nd ed.), LaSalle, IL: Open Court Press.

Joffe, M. M., and Rosenbaum, P. R. (1999), "Propensity Scores," *American Journal of Epidemiology*, 150, 327–333.

Lewis, D. (1973), "Causation," *Journal of Philosophy*, 70, 556–567.

Neyman, J. (1923), "On the Application of Probability Theory to Agricultural Experiments. Essay on Principles. Section 9." English translation of excerpts by D. Dabrowska and T. Speed (1990), *Statistical Science*, 5, 465–472.

Pearl, J. (1995), "Causal Diagrams for Empirical Research," *Biometrika*, 82, 669–710.

——— (2000), *Causality*, New York: Cambridge University Press.

Robins, J. M. (1987), "A Graphical Approach to the Identification and Estimation of Causal Parameters in Mortality Studies With Sustained Exposure Periods," *Journal of Chronic Diseases*, 40 (suppl. 2), 139S–161S.

——— (1994), "Correcting for Non-Compliance in Randomized Trials Using Structural Nested Mean Models," *Communications in Statistics*, 23, 2379–2412.

——— (1997), "Causal Inference From Complex Longitudinal Data," in *Latent Variable Modeling With*

Applications to Causality, ed. M. Berkane, New York: Springer-Verlag, pp. 69–117.

——— (1999), "Comment," *Statistical Science*, 14, 281–293.

Robins, J. M., and Greenland, S. (1992), "Identifiability and Exchangeability for Direct and Indirect Effects," *Epidemiology*, 3, 143–155.

——— (1994), "Adjusting for Differential Rates of Prophylaxis Therapy for PCP in High- Versus Low-Dose AZT Treatment Arms in an AIDS Randomized Trial," *Journal of the American Statistical Association*, 89, 737–749.

Robins, J. M., Greenland, S., and Hu, F. C. (1999), "Estimation of the Causal Effect of a Time-Varying Exposure on the Marginal Means of a Repeated Binary Outcome" (with discussion), *Journal of the American Statistical Association*, 94, 687–712.

Robins, J. M., Hernán, M. A., and Brumback, B. (2000), "Marginal Structural Models and Causal Inference in Epidemiology," *Epidemiology*, 11, 550–560.

Robins, J. M., Rotnitzky, A., and Scharfstein, D. O. (1999), "Sensitivity Analysis for Selection Bias and Unmeasured Confounding in Missing Data and Causal Inference Models," in *Statistical Models in Epidemiology*, ed. M. E. Halloran, New York: Springer-Verlag.

Rosenbaum, P. R. (1995), *Observational Studies*, New York, Springer-Verlag.

Rosenbaum, P. R., and Rubin, D. B. (1983), "The Central Role of the Propensity Score in Observational Studies of Causal Effects," *Biometrika*, 70, 41–55.

Rubin, D. B. (1974), "Estimating Causal Effects of Treatments in Randomized and Nonrandomized Studies," *Journal of Educational Psychology*, 66, 688–701.

——— (1978), "Bayesian Inference for Causal Effects: The Role of Randomization," *The Annals of Statistics*, 7, 34–58.

——— (1990), "Comment: Neyman (1923) and Causal Inference in Experiments and Observational Studies," *Statistical Science*, 5, 472–480.

Salmon, W. C. (1998), "Causality Without Counterfactuals," in *Causality and Explanation*, ed. W. C. Salmon, New York: Oxford University Press, chap. 16, pp. 248–260.

Shafer, G. (1996), *The Art of Causal Conjecture*, Cambridge, MA: MIT Press.

Simon, H. A. (1953), "Causal Ordering and Identifiability," in *Studies in Econometric Method*, eds. W. C. Hood and T. C. Koopmans, New York: Wiley.

Simon, H. A., and Rescher, N. (1966), "Cause and Counterfactual," *Philosophy and Science*, 33, 323–340.

Sobel, M. E. (1995), "Causal Inference in the Social and Behavioral Sciences," in *Handbook of Statistical Modeling for the Social and Behavioral Sciences*, eds. G. Arminger, C. C. Clogg, and M. E. Sobel, New York: Plenum Press.

Spirtes, P., Glymour, C., and Scheines, R. (1993), *Causation, Prediction, and Search*, New York: Springer-Verlag.

Wright, S. (1921), "Correlation and Causation," *Journal of Agricultural Research*, 20, 557–585.

Environmental Statistics

Peter Guttorp

1. INTRODUCTION

The field of environmental statistics is relatively young. The term "environmetrics" was apparently introduced in a National Science Foundation proposal by Philip Cox in 1971 (Hunter 1994). During the last decade, the field has achieved some recognition, in that there now are three journals wholly or partially devoted to the field (*Environmetrics* published by the International Environmetrics Society and Wiley; *Ecological and Environmental Statistics* published by Kluwer, and *Journal of Agricultural, Biological and Environmental Statistics* published by the American Statistical Association). The ASA has a section on Statistics and the Environment, and the International Statistical Institute is currently discussing such a section. Volume 12 of the series *Handbook of Statistics* (Patil and Rao 1994) was devoted to the topic of environmental statistics. Its 28 chapters constitute an interesting overview over the field.

In this vignette I present some of the current areas of research in environmental statistics. This is of course by no means an overview of the field as it stands; rather, it is a list of areas in which I can see the need for, and largely also the tools for, methodological developments.

2. ENVIRONMENTAL MONITORING

Environmental monitoring design deals mainly with two quite different sorts of design problems: monitoring for trend, where spatial and temporal dependence is of importance, and monitoring for "hot spots," or regions of local high intensity, which is often used for monitoring compliance with pollution regulations. The basic theory of optimal design for spatial random fields was outlined by Ripley (1981, chap. 3). Among the popular designs are systematic random sampling designs, in which a

Peter Guttorp is Director, National Research Center for Statistics and the Environment, University of Washington, Seattle, WA 98195 (E-mail: peter@stat.washington.edu). The assistance of numerous colleagues in working groups at the National Research Center for Statistics and the Environment is gratefully acknowledged. Although the research described in this article had partial support from the U.S. Environmental Protection Agency under cooperative agreement CR 825173-01-0 with the University of Washington, it has not been subjected to the Agency's required peer and policy review and thus does not necessarily reflect the views of the Agency, and no official endorsement should be inferred.

point is chosen uniformly over the study area, and a regular design (consisting of squares, triangles, or hexagons) is put down starting at the chosen point. When the sample mean is used to estimate the spatial mean of an isotropic random field over a region, the regular sampling plans are most efficient (Matérn 1960, chap. 5). The hexagonal design requires fewer sampling sites than a square or triangular design to cover the same area, but does not take into account spatial covariance heterogeneity or temporal nonstationarity. The covariance mapping technique mentioned in Section 3 can be used to deal with spatial heterogeneity, by implementing a spatial design in the transformed space.

Zidek and coworkers (e.g., Caselton, Kan, and Zidek 1992; Guttorp, Le, Sampson, and Zidek 1993) have developed an approach to network design that can deal with heterogeneous random fields. The basic idea is to consider a number of potential monitoring sites, some of which are gauged and some ungauged. In a multivariate normal setting, the design maximizes the amount of information (of the Kullback–Leibler type) about the ungauged sites that can be obtained from the gauged sites. This can be particularly useful when trying to redesign a current network, by adding and removing stations.

It is frequently the case that data from a monitoring network will serve more than one purpose. For example, in analyzing trends in tropospheric ozone (Reynolds, Das, Sampson, and Guttorp 1998), the data were collected by the state of Washington to monitor compliance with the Clean Air Act. Consequently, the network was aimed at finding areas of high air pollution and was changing over time. Statistical methods for analyzing data from a network adaptively designed to find the extremes of a random field need to be developed.

In 1989, the U.S. Environmental Protection Agency (EPA) started an ambitious monitoring program called the Environmental Monitoring and Assessment Program (EMAP). This was intended to create a "report card" for the state of the U.S. environment. The basic design of the EMAP study (Overton, White, and Stevens 1990) is a hexagonal grid, with a random starting point and a side of 27 km, resulting in 12,600 grid points over the continental United States, of which 25 fall in the Delaware Bay on the eastern U.S. coast, where EMAP has an ongoing study of benthic invertebrates. The EMAP protocol required revisiting some of the sites on a 3-year rotating basis. The measurements made at each site (three times each summer) included a bottom grab sample of benthic organisms, together with measurements of covariates such as temperature, depth, and salinity.

The basic biological tenet behind this sampling scheme is that environmental insults affect the distribution of organisms, in that pollution-tolerant species tend to get a larger proportion of the sample than do pollution-sensitive species. To deal with species composition data, Aitchison (1986) developed a methodology based on transforming the proportions from the unit simplex to Euclidean space. The proportions are then treated as multivariate normal data in the transformed space. Billheimer

(1995) extended this model to space–time data, showed how to estimate parameters using Markov chain Monte Carlo (MCMC) techniques, and how by backtransforming to the simplex the parameters can be given a natural interpretation as proportions (Billheimer, Guttorp, and Fagan 1999). In fact, it is possible to develop an algebra of proportions, which allows the statement of common models, such as regression, in terms of proportions. To account for counts of species from samples, the proportions are thought of as hidden state variables, and the counts are, for example, conditionally multinomial given the (unobserved) proportions. Billheimer et al. (1997) analyzed the spatial distribution of EMAP data from the Delaware Bay, and Silkey (1998) looked at changes over time.

Another example of compositional data in environmental statistics deals with particulate matter air pollution. Here the chemical analysis determines the distribution of chemical species among the particles. Regression on known pollution profiles enables identification of sources (Park, Spiegelman, and Henry 1999), but the aforementioned compositional analysis approach may yield additional insight, particularly into seasonal patterns.

3. SPATIAL PREDICTION

Environmental monitoring data are often used to develop regional summaries of pollution fields. To do so, values at unobserved sites have to be predicted. Geostatistical methods, such as kriging, were originally developed to do spatial prediction from a single observation of a network of sites. The main difference in the environmental context is that we generally have a time series of observations. Where ordinary geostatistical methods are forced to make strong assumptions on the spatial covariance structure, such as isotropy, these are not needed (and often not warranted) in the environmental context. Methods are available to study spatially heterogeneous covariance structures (Guttorp and Sampson 1994). Such methods are needed, for example, when hydrology or meteorology determines the covariance structure.

Our preferred approach is to use the class of covariance functions of the form $c(x, y) = c_0(\|f(x) - f(y)\|)$, where c_0 is an isotropic covariance function and f is a smooth mapping taking the geographic coordinates (x, y) into a different space in which covariances are isotropic. (Some facts regarding this class of covariance functions have been presented by Perrin and Meiring 1999.) The mapping f can be estimated nonparametrically, and current work involves implementing the fitting procedure using MCMC techniques.

Given a covariance model, spatial prediction traditionally proceeds as a generalized least squares problem. The standard error of the least squares prediction has three components: one due to the uncertainty about the random field, one due to the uncertainty in the covariance estimation, and one due to the choice of c_0 and f. Traditional geostatistical work ignores the second and third components. Use of MCMC estimation of the covariance function allows direct estimation of the second component. Model uncertainty calculations (e.g., Clyde 1999) can be used to estimate the overall uncertainty by estimating the support of the data for each of several potential covariance models c_0.

4. RISK ASSESSMENT

The EPA is committed to assessing environmental problems using risk analysis. Traditionally, this has been done by putting down a deterministic model of the relationship between level of pollutant and effect. The typical risk function is a differential equation, with parameters that are determined from a variety of sources, such as laboratory experiments, measurements on exposed individuals, and scientific consensus. When the model has to do with human health effects, the basis for the risk function is more often than not experiments on animals, which are then rescaled to provide a risk function for humans using a fairly arbitrary scaling factor. (For further discussion of health effects estimation, see the vignette by Thomas 2000.)

Recently, much emphasis has been put on uncertainty analysis of these risk assessments. Primarily, it has been noted that the values of the parameters in the model are subject to uncertainty, which then propagates through the whole assessment and results in uncertainty about the final risk. The method of probabilistic risk analysis (Cullen and Frey 1999) assigns what a statistician would call a prior distribution to each of the parameters. Typically, the parameters are treated as independent a priori, with simple marginal distributions such as uniform or normal. The analysis is done by simulating values from the prior distributions and summarized by producing simulated confidence intervals for quantiles from the resulting risk distribution. Current work aims at assessing the uncertainty more accurately by looking at the entire model uncertainty (e.g., Givens, Raftery, and Zeh 1995; Poole and Raftery 1999). This includes, in addition to the uncertainty of the parameters mentioned earlier, uncertainty of the data used to fit and/or assess the model, and uncertainty of the model itself.

5. ENVIRONMENTAL STANDARDS

The detailed understanding of the health effects of a pollutant is one of the tools needed for setting scientifically valid standards for environmental compliance. As an example, the U.S. standard for ozone requires that all sites in a region have an expected number of annual maximum daily 1-hour exceedances of 120 ppb of not more than one. Such a standard is not enforceable, because the expected number of exceedances is not directly measurable, and measurements cannot be taken everywhere in the region. Rather, it describes an ideal of compliance and may be termed an ideal standard. The standard is implemented by requiring that each site in an approved monitoring network have no more than three exceedances in 3 years. In effect, this rule applies the law of large numbers to $n = 3$.

Barnett and O'Hagan (1997) introduced the concept of statistically realizable ideal standards. Their idea is to combine an ideal standard with a statistically based rule of implementation. A simple approach to the problem of setting scientifically defensible environmental standards uses very traditional statistical tools, namely the Neyman–Pearson approach to hypothesis testing. The basic null hypothesis to be tested is that the region is in violation of the regulation; that is, in the ozone case, that the expected number of exceedances is more than one per year. Type I errors are more serious, as they indicate unacceptable health risks to the population, whereas type II errors can have serious consequences for the state environmental administrators in having to develop control strategies that are not strictly speaking needed. When viewing the EPA regulations from this standpoint, they entail type I error probabilities that would be viewed as unacceptable by statisticians (Carbonez, El-Shaarawi, and Teugels 1999; Cox, Guttorp, Sampson, Caccia, and Thompson 1999). In addition, a statistical approach to testing the basic null hypothesis would use test statistics different from the number of exceedances.

In air quality data, measurements are generally made on multiple pollutants. Standards are, however, set on individual pollutants. (In the U.S. these are the *criteria pollutants*: carbon monoxide, ozone, particulate matter, sulfur dioxide, and nitrous oxides.) How to set multivariate standards, taking into account the joint health effect of several correlated primary and secondary pollutants, is an open problem.

6. GRAPHICAL METHODS

An area of considerable importance in all of modern statistics is the management, display, and analysis of massive datasets. Land use data from satellite-based sensors, automated air quality sensors, and continuous water flow meters are among a variety of new measurement devices producing vast amounts of data. We are lacking tools for displaying spatially expressed data with uncertainty measures (see, however, Lindgren and Rychlik 1995 and Polfeldt 1999 for two approaches). Recent advances in three-dimensional visualization (virtual reality) allow a viewer to immerse herself in spatially expressed multivariate data (Cook et al. 1998).

As in all visualization of multivariate data, the tools of linked plots and brushing are extremely useful. There are promising developments in multi-platform graphical systems design using Java-based tools (e.g., the ORCA system; Sutherland, Cook, Lumley, and Rossini 1999). In particular, views of projections of multiple multivariate time series can yield valuable insights into the temporal structure of values that are multivariate outliers originating in a particular temporal part of the data, something that may not be visible in a rotating scatter cloud and even less so in a bivariate scatterplot matrix.

7. THE FUTURE OF ENVIRONMENTAL STATISTICS

Many important environmental problems directly involve multidimensional, spatially heterogeneous, and temporally nonstationary random fields. My personal belief is that the development of statistical research tools for classes of such processes may prove to be the most useful development in the field of environmental statistics. The multivariate aspect in particular is very important, in that there are few symmetries in space and time that can be used in setting up models for realistic situations. As an example, if we are studying the joint distribution of SO_2 and SO_4 during situations of similar meteorology, then we will find different space–time correlations for positive and negative time lags, because most of the SO_4 is produced from SO_2 emissions. As pointed out earlier, tools for looking at the joint behavior of several pollutants and for developing control strategies for their behavior are currently the focus of intensive research.

This vignette has focused on examples from the air quality arena. There are equally important, and often more complex, issues in water quality and more generally in ecological assessment of natural resources. In the long run, a battery of tools for describing, analyzing, and controlling the state of ecological systems must be developed. Significant challenges lie ahead for environmental statisticians.

REFERENCES

Aitchison, J. (1986), *The Statistical Analysis of Compositional Data*, London: Chapman and Hall.

Barnett, V., and O'Hagan, A. (1997), *Setting Environmental Standards*, London: Chapman and Hall.

Billheimer, D. (1995), "Statistical Analysis Of Biological Monitoring Data: State–Space Models For Species Composition," Ph.D. dissertation, University of Washington, Dept. of Statistics.

Billheimer, D., Cardoso, T., Freeman, E., Guttorp, P., Ko, H., and Silkey, M. (1997), "Natural Variability of Benthic Species Composition in the Delaware Bay," *Environmental and Ecological Statistics*, 4, 95–115.

Billheimer, D., Guttorp, P., and Fagan, W. F. (1999), "Statistical Analysis and Interpretation of Discrete Compositional Data," NRCSE Technical Report Series, 11, University of Washington. Available at http://www.nrcse.washington.edu/research/reports/papers/trs11_interp/trs11_interp.pdf.

Carbonez, A., El-Shaarawi, A. H., and Teugels, J. L. (1999), "Maximum Microbiological Contaminant Levels," *Environmetrics*, 10, 79–86.

Caselton, W. F., Kan, L., and Zidek, J. V. (1992), "Quality Data Networks That Minimize Entropy," in *Statistics in the Environmental & Earth Sciences*, eds. A. T. Walden and P. Guttorp, London: Edward Arnold, pp. 10–38.

Clyde, M. (1999), "Bayesian Model Averaging and Model Search Strategies (with discussion)," in *Bayesian Statistics*, 6, eds. J. M. Bernardo, A. P. Dawid, J. O. Berger, and A. F. M. Smith, Oxford, U.K.: Oxford University Press, pp. 157–185.

Cook, D., Cruz-Neira, C., Kohlmeyer, B. D., Lechner, U., Lewin, N., Nelson, L., Olsen, A., Pierson, S., and Symanzik, J. (1998), "Exploring Environmental Data in a Highly Immersive Virtual Reality Environment," *Environmental Monitoring and Assessment*, 51, 441–450.

Cox, L. H., Guttorp, P., Sampson, P. D., Caccia, D. C., and Thompson, M. L. (1999), "A Preliminary Statistical Examination of the Effects of Uncertainty and Variability on Environmental Regulatory

Criteria for Ozone," in *Environmental Statistics: Analyzing Data for Environmental Policy, Novartis Foundation Symposium 220*, Chichester, U.K.: Wiley, pp. 122–143.

Cullen, A. C., and Frey, H. C. (1999), *Probabilistic Techniques in Exposure Assessment*, New York: Plenum Press.

Givens, G. H., Raftery, A. E., and Zeh, J. E. (1995), "Inference From a Deterministic Population Dynamics Model for Bowhead Whales" (with discussion), *Journal of the American Statistical Association*, 90, 402–430.

Guttorp, P., Le, N. D., Sampson, P. D., and Zidek, J. V. (1993), "Using Entropy in the Redesign of an Environmental Monitoring Network," in *Multivariate Environmental Statistics*, eds. G. P. Patil and C. R. Rao, Amsterdam: North-Holland, pp. 175–202.

Guttorp, P., and Sampson, P. D. (1994), "Methods for Estimating Heterogeneous Spatial Covariance Functions With Environmental Applications," in *Handbook of Statistics, Vol. XII: Environmental Statistics*, eds. G. P. Patil and C. R. Rao, Amsterdam: North-Holland, pp. 661–690.

Hunter, S. (1994), "Environmetrics: An Emerging Science," in *Handbook of Statistics, Vol. XII: Environmental Statistics*, eds. G. P. Patil, C. R. Patil, and C. R. Rao, Amsterdam: North-Holland, pp. 1–8.

Lindgren, G., and Rychlik, I. (1995), "How Reliable are Contour Curves? Confidence Sets for Level Contours," *Bernoulli*, 1, 301–319.

Matérn, B. (1960), "Spatial Variation, Meddelanden fran Statens Skogsforskningsinstitut," 49, vol. 5. Republished in Lecture Notes in Statistics, Vol. 36, New York: Springer.

Overton, W. S., White, D., and Stevens, D. K. (1990), "Design Report for EMAP: Environmental Monitoring and Assessment Program," EPA/600/3-91/053, Washington, DC: U.S. Environmental Protection Agency.

Park, E. A., Spiegelman, C. H., and Henry, R. C. (1999), "Bilinear Estimation of Pollution Source Profiles in Receptor Models," NRCSE Technical Report Series 19, University of Washington. Available at http://www.nrcse.washington.edu/research/reports/papers/ trs19_vertex/trs19_vertex.pdf.

Patil, G. P., and Rao, C. R. (Eds.) (1994), *Handbook of Statistics, Vol. XII: Environmental Statistics*, Amsterdam: North-Holland.

Perrin, O., and Meiring, W. (1999), "Identifiability for Non-Stationary Spatial Structure," *Journal of Applied Probability*, 36, 1244–1250.

Polfeldt, T. (1999), "On the Quality of Contour Lines," *Environmetrics*, in press.

Poole, D., and Raftery, A. E. (1999), "Inference for Deterministic Simulation Models: The Bayesian Melding Approach," Technical Report 346, University of Washington, Dept. of Statistics. Available at http://www.stat.washington.edu/tech.reports/tr346.ps.

Reynolds, J. H., Das, B., Sampson, P. D., and Guttorp, P. (1998), "Meteorological Adjustment of Western Washington and Northwest Oregon Surface Ozone Observations With Investigation of Trends," NRCSE Technical Report Series 15, University of Washington. Available at http://www.nrcse.washington.edu/research/reports/papers/trs15_doe/ trs15_doe.pdf.

Ripley, B. D. (1981), *Spatial Statistics*, New York: Wiley.

Silkey, M. (1998), "Evaluation of a Model of the Benthic Macro Invertebrate Distribution of Delaware Bay," master's thesis, graduate program in Quantitative Ecology and Resource Management, University of Washington.

Sutherland, P., Cook, D., Lumley, T., and Rossini, T. (1999), "ORCA-A Toolkit for Statistical Visualization," available at http://pyrite.cfas.washington.edu/orca/.

Thomas, D. C. (2000), "Some Contributions of Statistics to Environmental Epidemiology," *Journal of the American Statistical Association*, 95, 315–319.

Capture–Recapture Models

Kenneth H. Pollock

1. INTRODUCTION

Here I briefly review capture–recapture models as they apply to estimation of demographic parameters (e.g., population size, survival, recruitment, emigration, and immigration) for wild animal populations. These models are now also widely used in a variety of other applications, such as the census undercount, incidence of disease, criminality, homelessness, and computer bugs (see Pollock 1991 for many references). Although they have their historical roots in the sixteenth century, capture–recapture models are basically a twentieth century phenomenon. The papers by Petersen and Lincoln (Seber 1982) from late last century and early this century represent early attempts by biologists to use capture–recapture methods. Later, as statistical inference took its modern form and provided powerful tools such as maximum likelihood methods, biometricians became involved. There has been an explosion of research that still seems to be accelerating at the century's end. Fortunately, most of the research is still rooted in the need to solve biological questions. Section 2 reviews closed models; Section 3, open models; and Section 4, combined methods. I conclude the article with my views on fruitful current and future research thrusts and how the pace of change is affecting them.

2. CLOSED MODELS

The most basic model is the Lincoln–Petersen model for a closed population of size N (an unknown parameter). In one sample some animals are marked (M), and in a second later sample of size $n = (m + n)$, both marked (m) and unmarked (u) animals are captured. Simple intuition suggests equating the sample proportion marked with the population proportion marked $m/n = M/N$. Thus our estimate is $\hat{N} = nM/m$. This estimator is maximum likelihood under a model with the following assumptions: (a) the population is closed to additions and deletions; (b) all animals are equally likely to be captured in each sample; and (c) marks are not lost and not overlooked by the observer (Seber 1982). The second assumption may be violated in two ways: (a) heterogeneity, which occurs when different animals have

Kenneth H. Pollock is Professor, North Carolina State University, Raleigh, North Carolina 27695.

inherently different capture probabilities; and (b) trap response when the probability of capture depends on the animals' prior capture history. One often samples the population more than twice. Each time, every unmarked animal caught is uniquely marked, previously marked animals have their captures recorded, and usually, all animals are released back into the population. This more extensive sampling enables sophisticated modeling that permits unequal catchability due to heterogeneity and trap response. Otis, Burnham, White, and Anderson (1978) considered a set of eight models where capture probabilities vary due to time, heterogeneity, and trap response in all possible combinations. They also provided a computer program, CAPTURE, to compute the estimates and select between models. The heterogeneity models that use a distribution of capture probabilities have caused theoretical difficulty for statisticians. An early ad hoc approach based on the "jackknife" method (Burnham and Overton 1978) proved quite useful. Later, Chao and colleagues used a method based on sample coverage (e.g., Lee and Chao 1994). Others have suggested using log-linear models (Cormack 1989). Maximum likelihood estimation, where the heterogeneity is modeled as a finite mixture distribution, usually with two or three support points, is a recent development (Norris and Pollock 1996a; Pledger 2000). Another approach to modeling heterogeneity uses covariates (Alho 1990; Huggins 1989). Current research is integrating the covariate approach with the distribution approach. Lloyd and Yip (1991) and others have applied martingales to various closed capture–recapture models. Another difficulty has been in finding a good method of model selection for this series of models. The original approach of Otis et al. (1978) does not work well. I suspect that the finite mixture approaches will aid in solving this problem. There is also current ongoing research exploring Bayesian methods of comparing models (S. Ghosh, personal communication). Model averaging may also be used. An interesting type of removal and recapture combined model, called Markov recapture, was proposed by Wileyto, Ewens, and Mullen (1994), who applied the model to insect pests in stored grain.

3. OPEN MODELS

Often capture–recapture studies have a long duration, rendering the closed models impractical. Thus there has been a need for the development of models that allow for additions (recruits and immigrants) and deletions (deaths and emigrants). The first general open model was developed independently by Jolly (1965) and by Seber (1965). Their model, which requires equal catchability and equal survival rates of all animals at each sampling time, enables one to estimate population sizes, survival rates, and birth numbers for almost all samples. Detailed treatments have been given by Pollock, Nichols, Brownie, and Hines (1990) and Seber (1982). Recently there has been an emphasis on integration of recruitment in the likelihood (Pradel 1996; Schwarz and Arnason 1996). Since the original Jolly and Seber papers, there has been much research on modeling survival rates allowing for multiple strata (e.g., sex, age, location). An important work is that of Lebreton, Burnham, Clobert, and Anderson (1992). An interesting development has been the shift to fitting large numbers of models, which has necessitated the development of model selection criteria based on the Akaike information criterion. The alternative approach of averaging over some reasonable models is also being considered. The recent book by Burnham and Anderson (1998) on model selection is very important. Related models based on band recoveries of exploited animals (Brownie, Anderson, Burnham, and Robson 1985) have been developed and widely used to analyze band return data on migratory waterfowl and other animals. These models led to important work on the compensatory versus additive mortality hypotheses for hunting mortality. Recently, these models have been used in fisheries studies in a slightly different form using instantaneous rates of fishing and natural mortality (Hoenig, Barrowman, Hearn, and Pollock 1998). Biologists began to use radio tags as a method of studying animal movements. As the technology improved and sample sizes increased, this approach has also been used to estimate survival rates. Early works were by Trent and Rongstad (1974) and Heisey and Fuller (1985). Pollock, Bunck, and Curtis (1989) used and modified the Kaplan–Meier method widely used in medical applications of survival analysis for wildlife telemetry. There have been many articles in the wildlife journals using these methods. In some cases, the Cox proportional hazards model for relating auxiliary variables to survival was used. The early survival modeling work considered fixed strata, like sex, or strata where the transition to the next stage was automatic, as in successive age classes. Recent works have looked at transitions between stages in a probabilistic manner. For example, animals may be marked in two different locations and, besides surviving between periods, may move to another location. Stage-structured models allow estimation of both the survival and the movement probabilities. Some important works that outline important theoretical results and provide interesting examples are those of Brownie, Hines, Nichols, Pollock, and Hestbeck (1993), Hestbeck, Nichols, and Malecki (1991), Nichols, Sauer, Pollock, and Hestbeck (1992), and Schwarz, Schweigert, and Arnason (1993). Research in this important area is continuing.

4. COMBINED METHODS

Another focus of research has been to develop combinations of different sampling methods. One early example used a "robust" design that combines both open and closed models in one analysis (Pollock 1982). Since then, many articles have used this design for many reasons; for example, to allow for unequal catchability, to separate recruitment from immigration, and to estimate temporary emigration. Burnham (1993), Barker (1997), and others have developed models for combining tag returns from dead animals with live recoveries. This research is continuing and expanding to include radio telemetry tagging. Many works have combined line transects methodology with capture–recapture (e.g., Borchers, Zucchini, and Fewster 1998). It is also possible to combine change-in-ratio methods and capture–recapture models. Currently, many large, important monitoring studies rely on indices of relative population abundance. I believe that there will be the need to include internal validation in these studies. This will combine the sampling of a small number of the plots more intensively using mark and recapture or other methods of estimating absolute abundance with the more extensive but crude sampling for relative abundance.

5. INFLUENCE OF CHANGE ON THE FIELD

5.1 Changing Ecological Paradigms

Early in this century, ecologists emphasized obtaining a population size estimate at a single point in time (Lincoln–Petersen model). Later, emphasis switched to estimating survival and recruitment for a single population (Jolly–Seber model). Concern about the assumption of equal catchability led to more general models for closed populations and to the robust design for open populations. Growing interest in comparing many different ages, sexes, and populations necessitated computer programs capable of fitting and selecting among many similar, but distinct models. Separation of emigration estimates from survival estimates cannot be achieved by the Jolly–Seber model. However, it is feasible in more modern analyses using radio tags or the combination of dead recoveries and live recaptures. A recent emphasis on habitat fragmentation has led to metapopulation theory (Hanski and Gilpin 1991). Metapopulations consist of discrete patches linked by potential movement between the patches. Stage-structured capture–recapture models that I discussed earlier have arisen and are evolving to meet the need to estimate patch-specific population sizes, survival, and movement probabilities. Habitat fragmentation and the explosion of endangered species of plants and animals has brought a growing interest in estimation of species richness and related parameters. Burnham and Overton (1978) first recognized that capture–recapture methods could be used to estimate the number of species as well as the numbers of the species. Research will continue on this topic (Nichols, Boulinier, Hines, Pollock, and Sauer 1998).

5.2 Changing Field Technology

An example of how field technology has changed the models is through the nature of the mark itself. A simple nonunique mark is all that is needed to fit the Lincoln–Petersen model for a closed population with two samples. However, individually identifiable marks are necessary to fit the more general closed models that allow unequal catchability of animals. Open models that allow additions and deletions to the population also require unique marks. Use of unique individual marks allows the monitoring of animals that remain alive and have not emigrated. Radio transmitter tags, however, allow biologists to also relocate newly dead animals and animals that have emigrated out of the study area provided that a search, often using a radio receiver mounted on a plane, is feasible. "Smart" radio tags can provide even more information, such as continuous location and animal activity summaries. These tags are likely to be quite common in the future. Another use of technology is use of radioisotopes for identifying animals so that feces can be examined for marks. In the future, DNA "fingerprints" will be used to identify animals, which will give rise to more exciting new modeling efforts.

5.3   Changing Computer Technology

Increasing computer power and user-friendly interfaces have revolutionized the way statisticians analyze data and capture–recapture is no exception. Now it is common to fit many related models to a single dataset. Modern computer packages like MARK (White and Burnham 1999; see also www.cnr.colostate.edu/~gwhite/software.html) make this easy with their high power and interactive interfaces. Bootstrapping and other resampling methods are now commonly used to obtain variances of estimates (Norris and Pollock 1996b). Computer intensive modern hierarchical Bayesian methods of analysis are beginning to be used for capture–recapture modeling, and I expect this to increase in the future (e.g., Vounastou and Smith 1995; see also the BUGS software at www.mrc-bsu.cam.ac.uk/bugs/mainpage.html).

5.4   Changing Statistical Theory

Early models used method-of-moments estimators, whereas later, maximum likelihood estimators became widely used. Profile likelihood methods of constructing confidence intervals and likelihood ratio tests for comparing models gained favor when computer capabilities had increased. Goodness-of-fit tests based on distributions that condition on minimal statistics have been widely used along with the traditional chi-squared goodness-of-fit test (Pollock, Nichols, Brownie, and Hines 1990). I suspect to see an increase in model pictorial diagnostics. Classical resampling methods like the jackknife (Burnham and Overton 1978) and the bootstrap (Norris and Pollock 1996b) are now widely used. Generalized linear models (Cormack 1989) and linear logistic models (Huggins 1989) with fixed and random effects (Pledger 2000) are now being used, as are Markov chain Monte Carlo (MCMC) and other modern Bayesian methods (Vounsatu and Smith 1995).

6. CONCLUSION

The continual interaction of animal ecologists and statisticians has been crucial to the development of sound capture–recapture models. In fact, I believe that all of the important statistical innovations of this century have depended on the interaction of subject matter scientists and statisticians. For example, R. A. Fisher's work on experimental design arose out of his position at Rothamsted Agricultural Experiment Station. Currently, although I believe that more theoretical innovations are possible, I think a major limitation is the nature and quality of the capture–recapture data that can be collected in the field. For an example from the recent past, about 20 years ago, radio tags became widely available to biologists, which has had a profound influence on model development. Similarly, I suspect that new methods of data collection will arise, opening up many new and challenging areas of model development.

REFERENCES

Alho, J. M. (1990), "Logistic Regression in Capture–Recapture Models," *Biometrics*, 46, 623–635.

Barker, R. J. (1997), "Joint Modelling of Live Recapture, Tag-Resight and Tag-Recovery Data," *Biometrics*, 53, 666–677.

Borchers, D. L., Zucchini, W., and Fewster, R. M. (1998), "Mark-Recapture Models for Line Transect Surveys," *Biometrics*, 54, 1207–1220.

Brownie, C., Anderson, D. R., Burnham, K. P., and Robson, D. S. (1985), *Statistical Inference From Band-Recovery Data: A Handbook* (2nd ed.), Washington, DC: Fish and Wildlife Service, U.S. Department of the Interior.

Brownie, C., Hines, J. E., Nichols, J. D., Pollock, K. H., and Hestbeck, J. B. (1993), "Capture–Recapture Studies for Multiple Strata Including Non-Markovian Transitions," *Biometrics*, 49, 1173–1187.

Burnham, K. P. (1993), "A Theory for Combined Analysis of Ring Recovery and Recapture Data," In *Marked Individuals in the Study of Bird Population*, eds. J.-D. Lebreton and P. M. North, Verlag, Basel, Switzerland: Birkhauser, pp. 199–213.

Burnham, K. P., and Anderson, D. R. (1998), *Model Selection and Inference: A Practical Information Theoretic Approach*, New York: Springer-Verlag.

Burnham, K. P., and Overton, W. S. (1978), "Estimation of the Size of a Closed Population When Capture Probabilities Vary Among Animals," *Biometrika*, 65, 625–633.

Cormack, R. M. (1989), "Loglinear Models for Capture–Recapture," *Biometrics*, 45, 395–413.

Hanski, L. A., and Gilpin, M. E. (1991), "Metapopulation Dynamics: Brief History and Conceptual Domain," *Biological Journal of the Linnean Society*, 42, 3–16.

Heisey, D. M., and Fuller, T. K. (1985), "Evaluation of Survival and Cause Specific Mortality Rates Using Telemetry Data," *Journal of Wildlife Management*, 49, 668–674.

Hestbeck, J. G., Nichols, J. D., and Malecki, R. A. (1991), "Estimates of Movement and Site Fidelity Using Mark-Resight Data of Wintering Canada Geese," *Ecology*, 72, 523–533.

Hoenig, J. M., Barrowman, N. J., Hearn, W. S., and Pollock, H. (1998), "Multiyear Tagging Studies Incorporating Fishing Effort Data," *Canadian Journal of Fisheries and Aquatic Science*, 55, 1466–1476.

Huggins, R. M. (1989), "On the Statistical Analysis of Capture Experiments," *Biometrika*, 76, 133–140.

Jolly, G. M. (1965), "Explicit Estimates From Capture–Recapture Data With Both Death and Immigration—Stochastic Model," *Biometrika*, 52, 225–247.

Lebreton, J-D., Burnham, K. P., Clobert, J., and Anderson, D. R. (1992), "Modelling Survival and Testing Biological Hypotheses Using Marked Animals: A Unified Approach with Case Studies," *Ecological Monographs*, 62, 67–118.

Lee, S-M., and Chao, A. (1994), "Estimating Population Size via Sample Coverage for Closed Capture–Recapture Models," *Biometrics*, 50, 88–97.

Lloyd, C. J., and Yip, P. (1991), "A Unification of Inference From Capture–Recapture Studies Through Martingale Estimating Equations," In *Estimating Functions*, ed. V. P. Godambe, Oxford, U.K.: Oxford University Press, pp. 65–88.

Nichols, J. D., Boulinier, T., Hines, J. E., Pollock, K. H., and Sauer, J. R. (1998), "Estimating Rates of Local Species Extinction, Colonization, and Turnovers in Animal Communities," *Ecological Applications*, 8, 1213–1225.

Nichols, J. D., Sauer, J. R., Pollock, K. H., and Hestbeck, J. B. (1992), "Estimating Transition Probabilities for Stage Based Population Projection Matrices Using Capture–Recapture Data," *Ecology*, 73, 306–312.

Norris, J. I., and Pollock, K. H. (1996a), "Nonparametric MLE Under Two Closed Capture–Recapture Models with Heterogeneity," *Biometrics*, 52, 639–649.

——— (1996b), "Including Model Uncertainty in Estimating Variances in Multiple Capture Studies," *Journal of Environmental and Ecological Statistics*, 3, 235–244.

Otis, D. L., Burnham, K. P., White, G. C., and Anderson, D. R. (1978), "Statistical Inference from Capture Data on Closed Animal Populations," *Wildlife Monograph*, 62, 135 pp.

Pledger, S. (2000), "Unified Maximum Likelihood Estimates for Closed Capture–Recapture Models Using Mixtures," *Biometrics*, 56, 434–442.

Pollock, K. H. (1982), "A Capture–Recapture Design Robust to Unequal Probability Capture," *Journal of Wildlife Management*, 46, 752–757.

——— (1991), "Modeling Capture, Recapture, and Removal Statistics for Estimation of Demographic Parameters for Fish and Wildlife Populations: Past, Present, and Future," *Journal of the American Statistical Association*, 86, 225–238.

Pollock, K. H., Bunck, C. M., and Curtis, P. D. (1989), "Survival Analysis in Telemetry Studies: The Staggered Entry Design," *Journal of Wildlife Management*, 53, 7–15.

Pollock, K. H., Nichols, J. D., Brownie, C., and Hines, J. E. (1990), "Statistical Inference for Capture–Recapture Experiments," *Wildlife Society Monographs*, 107, 1–97.

Pradel, R. (1996), "Utilization of Capture–Mark–Recapture for the Study of Recruitment and Population Growth Rate," *Biometrics*, 52, 703–709.

Schwarz, C. J., and Arnason, A. N. (1996), "A General Methodology for the Analysis of Capture–Recapture Experiments in Open Populations," *Biometrics*, 52, 860–873.

Schwarz, C. J., Schweigert, J. F., and Arnason, A. N. (1993), "Estimating Migration Rates Using Tag-Recovery Data," *Biometrics*, 49, 177–193.

Seber, G. A. F. (1965), "A Note on the Multiple Recapture Census," *Biometrika*, 52, 249–259.

——— (1982), "The Estimation of Animal Abundance and Related Parameters," (2nd ed.), London: Charles W. Griffin.

Trent, T. T., and Rongstad, O. J. (1974), "Home Range and Survival of Cottontail Rabbits in Southwestern Wisconsin," *Journal of Wildlife Management*, 38, 459–472.

Vounastou, P., and Smith, A. F. M. (1995), "Bayesian Analysis of Ring-Recovery Data via Markov Chain Monte Carlo Simulation," *Biometrics*, 51, 687–728.

White, G. C., and Burnham, K. P. (1999), "Survival Estimation from Populations of Marked Animals," *Bird Study*, 46, 120–138.

Wileyto, E. P., Ewens, W. J., and Mullen, M. A. (1994), "Markov-Recapture Population Estimates: A Tool for Improving Interpretation of Trapping Experiments," *Ecology*, 75, 1109–1117.

Statistics in Animal Breeding

Daniel Gianola

1. INTRODUCTION

Genetic improvement programs for livestock aim to maximize the rate of increase of some merit function expected to have a genetic basis. Animals producing progeny with the highest expected merit are kept as parents of subsequent generations, and those with the lowest merit are discarded. Merit can be a linear or nonlinear combination of genetic values for several traits that are economically important. Genetic merit cannot be observed, so it must be inferred from data. Relevant statistical problems are (a) assessing whether traits have a genetic basis, (b) developing accurate methods for inferring merit ("genetic evaluation"), and (c) designing mating plans. I do not deal with (c) here. The data consist of records of performance, such as growth rate, milk yield, and composition, records on diseases, egg production in layers, litter size in pigs, calving difficulty, survival, and length of productive life. Recently, data on molecular markers have become available, but its use is still at an early stage.

Many traits (continuous or discrete) seem to have a polygenic mode of inheritance and are subject to strong environmental influences. There are sex-limited traits as well (e.g., milk production), observable in females only. Often it is more important to infer the genetic merit of males because of their higher impact on the rate of improvement. For example, artificial insemination and frozen semen allow dairy bulls to sire thousands of daughters in several countries, thus generating opportunities for international sire evaluation, although this requires complex statistical models.

Animal breeding datasets are large (e.g., millions of lactation milk yields in dairy cattle), multivariate (several traits modeled simultaneously), and seemingly Gaussian in some instances and nonnormal in others. Data structure can be cross-sectional or longitudinal (as in a study of growth curves), can be very unbalanced, and with nonrandomly missing observations. For example, not all first-lactation cows produce a second lactation, because of sequential selection for low production, reproductive failure, or disease. Some sires are used more intensively than others, because of differences in estimated or perceived genetic value. Hence it is not surprising that statistical science has been important in animal breeding. A review of statistical methods and

Daniel Gianola is Professor, Department of Animal Sciences, Department of Biostatistics and Medical Informatics, and Department of Dairy Science, University of Wisconsin, Madison, WI 53706 (E-mail: gianola@calshp.cals.wisc.edu).
The author thanks A. Blasco, William Hill, Luc Janss, Lawrence Schaeffer, Daniel Sorensen, and three anonymous reviewers for their useful comments.

problems discussed over the last 25 years was given by Gianola and Hammond (1990). Our objective is to describe main developments in statistical methods in animal breeding over the last decades, concentrate on landmarks (Sec. 2), and speculate on some future issues (Sec. 3).

2. LANDMARKS

2.1 Statistical Genetic Models

Statistical models in animal breeding consist of (a) a mathematical function relating observations to location parameters and random effects (Bayesians view all unknowns as random); (b) genetic and environmental dispersion parameters, such as components of variance and covariance; and (c) assumptions about the joint distribution of the observations and of the random effects. The latter include genetic components, such as additive genetic values, dominance and epistatic deviations, and permanent environmental effects; all of these create correlations between cross-sectional and longitudinal records of performance.

The most widely used (and abused) assumption has been that of normality. A reason is that many traits seem to be inherited in a multifactorial manner, with a large number of genes acting, and it is believed that the effects of many gene substitutions are additive and infinitesimally small. Molecular information suggests that this assumption is often reasonable. Some recent evidence in dairy cattle, using molecular markers, suggests the presence of "quantitative trait loci" affecting fat percentage in milk in five chromosomes; this type of investigation is preliminary. If the alleles ("genes") at these loci act additively and have small effects, then their sum produces a normal process right away.

Fisher (1918) gave foundations for the infinitesimal model, describing the statistical implications of Mendelian inheritance. He posited that "observation = genetic value + residual" and gave a precursor of his analysis of variance, by proposing a partitioning of genetic variance into additive and dominance components. From this, the expected correlations between different types of relatives follow. The additive model was the statistical genetic point of departure for the development of predictors of breeding value, leading to fairly precise evaluation of dairy sires, and it continues to be used extensively, although in a sophisticated, vectorial manner. Wright (1921) arrived at the same results for the additive part using "path coefficients." This involves describing a system of correlations via standardized random-effects linear models. The procedure faded away in animal breeding because of its inability to take into account interactions and nuisance parameters in a flexible and computationally attractive manner.

Kempthorne (1954) partitioned the genetic variance further by including interactions called "epistatic" components. He used probability of identity by descent and described the epistatic variance in terms of several components, depending on the number of loci involved. This permitted him to express the covariance between traits measured in relatives in a random mating population as a function of several genetic components of variance and covariance. Additional extensions are accommodated; for example, maternal effects and cloning and cytoplasmic inheritance.

2.2 Best Linear Unbiased Prediction

"Predicting" or "estimating" genetic merit in candidates for selection is central. Lush (1931) gave formulas for assessing the genetic merit of dairy sires and found that regression to the mean (shrinkage) was needed. He assumed that means and genetic and environmental components of variance were known. Robertson (1955) showed that the statistic also stems from a weighted average between "population" information and data, anticipating a Bayesian interpretation. Henderson (1973) formulated the problem in a more general framework, and derived what later became known as best linear unbiased prediction (BLUP). Henderson, Searle, Kempthorne, and vonKrosigk (1959) posited the linear (univariate or multivariate) mixed-effects model $y = X\beta + Zu + e$, where β is a fixed (over repeated sampling) vector and $u \sim N(0, G)$ and $e \sim N(0, R)$ are independent random vectors; X and Z are known incidence matrices; and G and R are variance-covariance matrices, these being a function of (assumed known) dispersion parameters. Maximization of the joint density of u and y with respect to β and u leads to Henderson's mixed model equations (MME):

$$\begin{bmatrix} X'R^{-1}X & X'R^{-1}Z \\ Z'R^{-1}X & Z'R^{-1}Z + G^{-1} \end{bmatrix} \begin{bmatrix} \hat{\beta} \\ \hat{u} \end{bmatrix} = \begin{bmatrix} X'R^{-1}y \\ Z'R^{-1}y \end{bmatrix}.$$

Henderson thought that he had maximized a likelihood function, and he viewed $\hat{\beta}$ and $\hat{u}$ as maximum likelihood estimators (MLEs) of β (which it is, under normality) and of u (which it is not, as this vector is random). In fact, the objective function is a joint posterior density, in a certain Bayesian setting, or a "penalized" or "extended" likelihood. This error had a happy ending: Henderson and Searle showed in later works that even without normality, $\hat{\beta}$ is the generalized least squares estimator of β and $\hat{u}$ is the BLUP of u. The inverse of the coefficient matrix yields the covariance matrices of $\hat{\beta}$ and of $\hat{u} - u$. This holds for multivariate and univariate settings. When β is known, $\hat{u}$ is the best linear predictor of u and, under normality, $\hat{u}$ is the best predictor in the mean squared error sense. With known β, the predictor gives the "selection index" derived by Smith (1936) and Hazel (1943) in less general settings.

The MME algorithm has been used world-wide for genetic evaluation of livestock, using models where components of u have a genetic meaning. Arguably, it may be the most important technological contribution of statistics to animal breeding. The order of the system can be in the million of equations, especially for models (univariate or multivariate) where a random additive genetic effect is fitted for each animal with a record of production, as well as for animals that appear in the genealogy and need to be included. Hence iterative methods must be used and approximations are needed for assessing uncertainty. The MME can be used to advantage in computing algorithms for several methods of variance component estimation in linear and generalized mixed-effects linear models.

A question is to how to invert G when the order of u is in the millions, as in routine genetic evaluation of dairy cows in the United States. Here Henderson (1976) made a remarkable breakthrough. Suppose that $G = G_0 \otimes A$, where G_0 has order equal to the number of traits and A is a matrix of "additive genetic relationships." He discovered that A^{-1} could be written directly from a list of parents of the animals, enabling the use of all available relationships in genetic evaluation. This gave more precise inferences about genetic values and allowed accounting for biases due to ignoring many relationships in naive variance component analyses.

2.3   Variance and Covariance Component Estimation

Application of animal breeding theory depends on knowledge of variance and covariance parameters. Hofer (1998) reviewed a large body of literature on estimation methods. Because the datasets are large and unbalanced, and the models have a large number of nuisance parameters, simple analysis of variance (ANOVA) type estimation methods seldom work well. Henderson (1953) described three methods for unbalanced data. The more general, method 3, uses quadratic forms based on least squares and yields unbiased estimators. Harvey (1960) implemented method 3 in software for variance and covariance component estimation in animal breeding. Then minimum norm quadratic unbiased estimation and its minimum variance version (under normality) entered into the picture. These estimators can be formulated in terms of the MME, but optimality requires knowledge of the true parameters. Animal breeders leaned toward maximum likelihood (ML) instead, assuming normality; works by Hartley and Rao (1967) and Harville (1977) were influential. Many algorithms for ML can be derived using the MME (Harville 1977). The bias of the MLE of the residual variance led to great interest in a method called "restricted" maximum likelihood (REML). Patterson and Thompson (1971) aimed to account for the "loss of degrees of freedom" incurred in estimating fixed effects, and noted that maximization of the location invariant part of the likelihood led to estimating equations similar to those in ANOVA in balanced layouts. The idea was to reduce bias, but was this perhaps at the expense of precision? A strong argument favoring REML comes from its Bayesian interpretation. Harville (1974) showed that REML is the mode of the posterior distribution of the variance parameters after integrating the fixed effects (with respect to an improper uniform prior) out of the joint posterior distribution; hence the probability calculus takes uncertainty about β into account. Gianola, Foulley, and Fernando (1986) observed that under Gaussian assumptions, best linear unbiased estimation (BLUE) and BLUP with the unknown (co)variance parameters evaluated at the REML estimates corresponds to an approximate integration of the dispersion parameters. This gives an approximate Bayesian solution to inferences about breeding values when dispersion components are unknown, provided that the restricted likelihood is sharp enough. This problem lacks a simple frequentist solution.

2.4   Bayesian Procedures

Animal breeders did not remain insensitive to the Bayesian revival that occurred in the mid-1960s. Lindley and Smith (1972) provided a link between mixed models and Bayesian approaches. Dempfle (1977) and Rönningen (1971) investigated connections between BLUP and Bayesian ideas. Subsequently, Gianola and Fernando (1986) suggested the Bayesian approach as a general framework for solving a large number of animal breeding problems. It was not until the advent of Markov chain Monte Carlo (MCMC) methods, however, that Bayesian methods were taken seriously. Early applications of Gibbs sampling in animal breeding were discussed by Wang, Rutledge, and Gianola (1993). An important development was using Bayesian measures for assessing uncertainty in response to genetic selection (Sorensen, Wang, Jensen, and Gianola 1994); this is another problem in animal breeding without a satisfactory frequentist solution.

2.5   Nonlinear and Generalized Linear Models and Longitudinal Responses

Dempster and Lerner (1950) laid out foundations for quantitative genetic analysis of binary data. Gianola and Foulley (1983) and Harville and Mee (1984) addressed inferences about fixed and random effects in generalized mixed linear models for ordered categorical responses. The two approaches give the same answer. This has been extended to models with Gaussian and categorical responses, to multivariate binary responses, and to hierarchical models where a categorical response variable depends on a count with a Poisson conditional distribution. Foulley, Im, Gianola, and Höschele (1987), Harville and Mee (1984), and Tempelman and Gianola (1996) discussed estimation of variance components in such settings. For categorical responses, Gilmour, Anderson, and Rae (1985) described a procedure based on quasi-likelihood. Their variance component estimators and predictors of random effects lack formal justification, but reduce to REML and BLUP with Gaussian responses. Sorensen, Andersen, Gianola, and Korsgaard (1995) presented a fully Bayesian solution for ordered polychotomies based on Gibbs sampling.

Animal breeders also developed empirical Bayes and REML-type estimators of breeding values and (co)variance components for linear and nonlinear functions describing longitudinal trajectories (milk yield, wool growth). These are "random regression" models. MCMC implementations have been presented as well.

There has been recent interest in modeling individual cow milk yield trajectories;

so random regressions (linear or nonlinear) are natural candidates here. A seemingly different approach has been that of "covariance" functions, where the covariance between records of individuals is a continuous function of time. Another area receiving increased attention has been survival analysis (e.g., Ducrocq and Casella 1996).

2.6 Statistical Computing

Much effort has been devoted to making BLUP and REML computationally feasible, even in multivariate settings. A comparison of packages was given by Misztal (1998). Janss and de Jong (1999) fitted an univariate mixed-effects model with about 1.4 million location effects, including almost 700,000 additive genetic values (with a relationship matrix, **A**, of corresponding order), to milk yield of Dutch cattle, and used Gibbs sampling. REML is unfeasible here; so this is a case where MCMC allows estimating an entire distribution, whereas deterministic likelihood computations break down.

2.7 Effects of Selection on Inferences

Animal breeding data seldom arise from a purely random mechanism, because of selection and assortative mating. Except in designed experiments, the history of the selection process is known incompletely. How robust are inferences if selection and assortative mating are ignored? Important contributions are those of Kempthorne and von Krosigk (in Henderson et al. 1959), Curnow (1961), and, notably, Henderson (1975). When selection is not ignorable, it is essential to model the "missing-data" process.

3. FUTURE DEVELOPMENTS

With continued growth in computer power and with better algorithms, there will be more flexibility for fitting more realistic functional forms and distributions, and for challenging models—an area that has received scant attention. There has been work on fitting heavy-tailed distributions and on using splines. An area of explosive interest is the use of molecular information in inferences about genetic values and in gene mapping. Major challenges reside in multipoint linkage analysis and fine mapping of genes affecting quantitative traits. Animal breeders have assumed multivariate normality in high-dimensional analyses and at several levels of hierarchical models. Violation of normality may not have serious consequences in point prediction of breeding values, but it may affect entire probabilistic inference adversely. Hence a search for robust methods of inference (especially considering that some animals receive undeclared preferential treatment) should be in the research agenda. Model selection issues will surely become more important than in the past.

REFERENCES

Curnow, R. N. (1961), "The Estimation of Repeatability and Heritability From Records Subject to Culling," *Biometrics*, 17, 553–566.

Dempfle, L. (1977), "Relation Entre BLUP (Best Linear Unbiased Prediction) et Estimateurs Bayesians," *Annales de Génetique et de Séléction Animale*, 9, 27–32.

Dempster, E. R., and Lerner, I. M. (1950), "Heritability of Threshold Characters," *Genetics*, 35, 212–236.

Ducrocq, V., and Casella, G. (1996), "Bayesian Analysis of Mixed Survival Models," *Genetics, Selection, Evolution*, 28, 505–529.

Fisher, R. A. (1918), "The Correlation Between Relatives on the Supposition of Mendelian Inheritance," *Royal Society (Edinburgh) Transactions*, 52, 399–433.

Foulley, J. L., Im, S., Gianola, D., and Höschele, I. (1987), "Empirical Bayes Estimation of Parameters for n Polygenic Binary Traits," *Genetics, Selection, Evolution*, 19, 197–224.

Gianola, D., and Fernando, R. L. (1986), "Bayesian Methods in Animal Breeding Theory," *Journal of Animal Science*, 63, 217–244.

Gianola, D., and Foulley, J. L. (1983), "Sire Evaluation for Ordered Categorical Data With a Threshold Model," *Genetics, Selection, Evolution*, 15, 201–224.

Gianola, D., Foulley, J. L., and Fernando, R. L. (1986), "Prediction of Breeding Values When Variances are not Known," *Proceedings of the Third World Congress on Genetics Applied to Livestock Production*, XII, Lincoln, NE: Agricultural Communications, University of Nebraska, pp. 356–370.

Gianola, D., and Hammond, K. (Eds.) (1990), *Advances in Statistical Methods for Genetic Improvement of Livestock*, Heidelberg: Springer-Verlag.

Gilmour, A. R., Anderson, R. D., and Rae, A. L. (1985), "The Analysis of Binomial Data by a Generalized Linear Mixed Model," *Biometrika*, 72, 593–599.

Hartley, H. O., and Rao, J. N. K. (1967), "Maximum Likelihood Estimation for the Mixed Analysis of Variance Model," *Biometrika*, 54, 93–108.

Harvey, W. R. (1960), *Least-Squares Analysis of Data With Unequal Subclass Numbers*, Bulletin 20-8, Washington, DC: United States Department of Agriculture, Agricultural Research Service.

Harville, D. A. (1974), "Bayesian Inference for Variance Components Using Only Error Contrasts," *Biometrika*, 61, 383–385.

——— (1977), "Maximum Likelihood Approaches to Variance Component Estimation and to Related Problems," *Journal of the American Statistical Association*, 72, 320–340.

Harville, D. A., and Mee, R. W. (1984), "A Mixed Model Procedure for Analyzing Ordered Categorical Data," *Biometrics*, 40, 393–408.

Hazel, L. N. (1943), "The Genetic Basis for Constructing Selection Indexes," *Genetics*, 28, 476–490.

Henderson, C. R. (1953), "Estimation of Variance and Covariance Components," *Biometrics*, 9, 226–252.

——— (1973), "Sire Evaluation and Genetic Trends," *Proceedings of the Animal Breeding and Genetics Symposium in Honor of Dr. Jay L. Lush*, 10–41, Champaign: American Society of Animal Science and the American Dairy Science Association.

——— (1975), "Best Linear Unbiased Estimation and Prediction Under a Selection Model," *Biometrics*, 31, 423–449.

——— (1976), "A Simple Method for Computing the Inverse of a Numerator Relationship Matrix Used in Prediction of Breeding Values," *Biometrics*, 32, 69–83.

Henderson, C. R., Searle, S. R., Kempthorne, O., and vonKrosigk (1959), "Estimation of Environmental and Genetic Trends from Records Subject to Culling," *Biometrics*, 15, 192–218.

Hofer, A. (1998), "Variance Component Estimation in Animal Breeding: A Review," *Journal of Animal Breeding and Genetics*, 115, 247–265.

Janss, L. L. G., and de Jong, G. (1999), "MCMC Based Estimation of Variance Components in a Very Large Dairy Cattle Data Set," *Computational Cattle Breeding 99*, Helsinki: MTT.

Kempthorne, O. (1954), "The Correlation Between Relatives in a Random Mating Population," *Royal Society (London) Proceedings*, Ser. B, 143, 103–113.

Lindley, D. V., and Smith, A. F. M. (1972), "Bayes Estimates for the Linear Model" (with discussion), *Journal of the Royal Statistical Society*, Ser. B, 34, 1–41.

Lush, J. L. (1931), "The Number of Daughters Necessary to Prove a Sire," *Journal of Dairy Science*, 14, 209–220.

Misztal, I. (1998), "Comparison of Software Packages in Animal Breeding," *Proceedings of the Sixth World Congress on Genetics Applied to Livestock Production*, 22, 3–10.

Patterson, H. D., and Thompson, R. (1971), "Recovery of Inter-Block Information When Block Sizes are Unequal," *Biometrika*, 58, 545–554.

Robertson, A. (1955), "Prediction Equations in Quantitative Genetics," *Biometrics*, 11, 95–98.

Rönningen, K. (1971), "Some Properties of the Selection Index Derived by Henderson's Mixed Model Method," *Zeitschrift für Tierzuchtung und Züchtungsbiologie*, 8, 186–193.

Smith, F. H. (1936), "A Discriminant Function for Plant Selection," *Annals of Eugenics*, 7, 240–250.

Sorensen, D. A., Andersen, S., Gianola, D., and Korsgaard, I. (1995), "Bayesian Inference in Threshold Models Using Gibbs Sampling," *Genetics, Selection, Evolution*, 27, 229–249.

Sorensen, D. A., Wang, C. S., Jensen, J., and Gianola, D. (1994), "Bayesian Analysis of Genetic Change due to Selection Using Gibbs Sampling," *Genetics, Selection, Evolution*, 26, 333–360.

Tempelman, R. J., and Gianola, D. (1996), "A Mixed Effects Model for Overdispersed Count Data in Animal Breeding," *Biometrics*, 52, 265–279.

Wang, C. S., Rutledge, J. J., and Gianola, D. (1993), "Marginal Inference About Variance Components in a Mixed Linear Model Using Gibbs Sampling," *Genetics, Selection, Evolution*, 25, 41–62.

Wright, S. (1921), "Systems of Mating. I. The Biometric Relations Between Parent and Offspring," *Genetics*, 6, 111–123.

Some Issues in Assessing Human Fertility

Clarice R. Weinberg and David B. Dunson

1. INTRODUCTION

While the human population continues to grow, depleting natural resources and reducing biodiversity, scientists have become concerned about our continued capacity to reproduce (perhaps a testament to the enduring value that we place on procreation). Several lines of evidence have contributed to this concern. Ecologists have documented reproductive abnormalities, including malformations and effects on sexual dimorphism and behavior, in certain species exposed to polluted waters (Burkhart et al. 1998; Guillette et al. 1994). Some reports describe declines in the concentrations and quality of human sperm over the past several decades (Swan, Elkin, and Fenster 1997), and increases in rates of testicular cancer and the birth defect cryptorchidism (undescended testicles). Laboratory studies document that certain chemicals can mimic and/or disrupt reproductive hormones, and research on endocrine disruption is flourishing, both epidemiologically and in the laboratory (Daston et al. 1997).

Interest in statistical assessment of reproduction dates back to Galton (1869) and Fisher (1958). Trying to understand why the families of the English aristocracy suffered from subfertility, Galton concluded that subfertility confers a selective aristocratic advantage by tending to concentrate a family's wealth in the estates of fewer descendents. Fisher studied families and inferred (correctly) that fertility must be highly heterogeneous across couples, an observation with important implications for models.

In this article we discuss statistical methods for identifying and characterizing factors that can modify human fertility, through either unintended effects, as with reproductively toxic exposures, or through intended interventions, such as contraceptive methods or clinical treatments for infertility.

2. BACKGROUND

Reproductive systems are highly variable across species. Human females do not have an estrous cycle. We are not reflex ovulators, like rabbits, that release ova in

Clarice R. Weinberg is Chief and David B. Dunson is Staff Fellow, Biostatistics Branch, National Institute of Environmental Health Sciences, Research Triangle Park, NC 27709 (E-mail: weinberg@niehs.nih.gov; dunson1@niehs.nih.gov).

response to coitus. We do not normally give birth to litters. In fact, species display such a diversity of reproductive strategies that there is no good animal model for human fertility; we must study humans.

Under regulation by the hypothalamus and pituitary, one of a woman's two ovaries releases a mature egg once each menstrual cycle, an event called ovulation. If the egg is properly swept into the fallopian tube, encounters healthy sperm, is successfully fertilized, implants in a receptive uterine lining, and avoids rejection by the maternal immune system, then pregnancy may ensue; otherwise, the uterine lining is menstrually sloughed, and the cycle must begin again.

Human fertility is inherently hard to study. First, the unit of study is the couple (a complication in itself), and couples vary in their fertility. Empirically, the conception rate among couples followed after discontinuing contraception decreases markedly over time. This decline reflects sorting, where the more fertile couples conceive rapidly and are absent from later risk sets. Determining the sources of such heterogeneity among couples has long been of interest to demographers (Sheps and Mencken 1973) and remains a primary challenge for fertility research. A second complicating factor is that human couples exercise a great deal of control over their reproduction. A third complication is that humans get multiple opportunities (some intended, some not) to evidence their fertility, and self-selection plays an important role.

The interplay of these phenomena creates novel forms of confounding. For example, among couples recruited for prospective study, an exposure under study may be correlated with past use of contraception. If smokers have historically taken more risks than nonsmokers, then the smokers with higher fertility may have had all of their desired pregnancies through unintended conceptions, leaving only the relatively subfertile smokers to be at risk of a self-identified planned conception. In this way, sorting can induce spurious associations. Improved statistical tools for studying fertility are needed for monitoring the reproductive health of a population, for time-to-pregnancy studies, and for highly detailed prospective studies, where intercourse records are kept for couples attempting conception and the time of ovulation is identified for each menstrual cycle at risk. Innovative methods are also needed to model biologic markers, such as menstrual cycle characteristics, hormone patterns, and properties of semen.

3. POPULATION MONITORING

In much of the world, fertility is under considerable voluntary control, and our capacity to reproduce could deteriorate markedly before the usual indices would pick up the decline. Methods are needed for monitoring a population to establish a baseline against which future trends can be compared. Unfortunately, there are currently no methods in place that allow us to determine whether fertility has declined over recent decades, say, in the United States. A recent idea is based on case-cohort methods (Olsen and Andersen 1999). Briefly, one samples a subcohort from a defined population, identifies couples in the subcohort who are trying to conceive, and follows them for a certain number of months. Pregnancies that occur over follow-up in the larger population are also identified. Thus, as in a case-cohort design, all "cases" are identified, and control information comes from a randomly sampled subcohort. The combined data allow estimation of the survival function (cumulative nonconception rate) across time.

4. TIME TO PREGNANCY

Studies of the time required for noncontracepting, sexually active couples to achieve conception can provide estimates of fertility parameters and can permit comparative studies of infertility treatments or potentially toxic exposures. Time to pregnancy (TTP) is best measured on a discrete scale, because the natural biologic time unit is the menstrual cycle. TTP can be studied either prospectively or retrospectively, but accidental pregnancies are usually excluded, because a meaningful TTP cannot be reconstructed retrospectively and contracepting couples are usually not efficient to study prospectively. Retrospective studies are either based on women's recall of past planned pregnancies (e.g., her first or her most recent), or based on women identified through an obstetrics service and asked about their current pregnancy. When sampling is based on the pregnancy, sterile couples are excluded by design, subfertile couples are underrepresented, and parameter estimates must be interpreted accordingly. In prospective studies of TTP, the sampling unit is the attempt, not the pregnancy. Such studies do include couples who fail to conceive, but can be difficult to carry out. This is particularly true in an occupational setting, where few couples may be at risk of conception at any given point in time. One can enroll couples who have already been trying for some time, provided that the left censoring is appropriately accounted for in modeling. In such a study, one must ascertain the prestudy attempt time (in cycles), so that couples recruited mid-attempt can be entered into the appropriate cycle-based risk set for analysis.

Modeling can then proceed according to any of several approaches. One can simply fit a proportional probabilities model, which is analogous to a proportional hazards model, except that time is discrete. The baseline conception rate can decline across menstrual cycles to account for couple-to-couple heterogeneity, and the model allows estimation of a fecundability ratio (analogous to a hazard ratio) as a summary measure of effect of an exposure (Weinberg and Wilcox 1998). A beta-geometric model has also been proposed (Sheps and Mencken 1973) and is simple to fit as a generalized linear model with an inverse link. For prospective data, one can allow for a proportion of sterile couples; that is, a probability mass at zero fecundability. An extension confers robustness against digit preference seen in retrospective studies (Ridout and Morgan 1991). The beta-based models do not, however, provide a readily interpretable summary measure of covariate effect, although Crouchley and Dassios (1998) developed a quantile ratio that has an individual-level interpretation. Heckman and Walker (1990) outlined a general mixture of geometrics model and developed goodness-of-fit tests.

One problem with the aforementioned approaches is that although time-dependent covariates can be formally included, the resulting models are not plausible because the time-specific conditional fecundability distribution among couples with heterogeneous fecundability needs to depend on both the number of prior failures, the exposures, and the couples' exposure histories. Scheike and Jensen (1997) proposed a discrete-time random-effects model that accommodates time-dependent covariates and also allows for multiple TTPs per couple. Dunson and Zhou (2000) later developed a probit model with similar advantages, which can additionally account for a sterile subpopulation of couples.

With studies that allow recruitment mid-attempt, one must deal with left censoring of the survival time. Methods that are fundamentally life table-based (e.g., the proportional probabilities model and the beta-geometric) are valid under minimal assumptions, provided that couples are entered into the appropriate risk sets. The random-effects models (Dunson and Zhou 2000; Scheike and Jensen 1997) can also handle left censoring, provided again that couples are not credited with pre-enrollment failures.

For retrospective studies of TTP in relation to potentially toxic exposures, trends in prevalence of the exposure through historical time can complicate the analysis. Even if such exposures are accurately ascertained on a cycle by cycle basis, with proper methods used to treat them as time dependent, bias will result. This problem was discovered in the context of a study of dental assistants, where wearing latex gloves was found to confer markedly increased fertility. This was recognized to be an artifact induced by the fact that when the TTP had been short, the attempt spanned recent, post-AIDS epidemic time, when glove use was common; when the TTP had been long, the attempt had begun prior to the AIDS epidemic (Weinberg, Baird, and Rowland 1993), an era when glove use was rare. Artifacts can also arise due to trends in initiation of pregnancy attempts over time; for example, from increasingly delayed child bearing in recent decades in the industrialized world. Such trends over time can cause older women to appear to be more fertile than young women, based on retrospective TTP data. Analytic methods that are robust against such distortions remain to be developed.

5. TIMING OF INTERCOURSE MODELS

It is possible to design a more informative study by prospectively collecting more detailed data, including dates of menses, ovulation, and unprotected intercourse, for each couple (Barrett and Marshall 1969; Masarotto and Romualdi 1997). Because of the large number of possible patterns of intercourse around ovulation, a model is required to relate the cycle-specific coital histories to the probability of (detectable) conception. A model advanced by Peter Armitage (personal communication) asserted simply that batches of sperm that arrive in the female reproductive tract on different days commingle and then compete independently in fertilizing the egg. If k indexes the day in relation to ovulation, and variables X_k indicate that there was (1) or was not (0) intercourse on day k, then the model specifies that

$$\Pr(\text{conception}|\mathbf{X}) = 1 - \prod_k (1 - p_k)^{X_k}. \tag{1}$$

The parameters p_k are interpretable as the probability that conception would occur in a cycle in which there was intercourse only on day k, and independence permits aggregation of the contributions of multiple days in the same cycle. The model is not biologically plausible, however, because it ignores factors other than the timing of intercourse. The egg must be viable, the uterine lining must be favorable to implantation, and so on. A later modification (Schwartz, MacDonald, and Heuchel 1980) includes a susceptibility multiplier that represents the probability that for a menstrual cycle, all events not related to the timing of intercourse are favorable for initiation of a detectable conception:

$$\Pr(\text{conception}|\mathbf{X}) = A \left[1 - \prod_k (1 - p_k)^{X_k} \right]. \tag{2}$$

Further extensions allow A and the p_k's to depend on covariates (Royston 1982; Weinberg, Gladen, and Wilcox 1994), and A to be heterogeneous among couples (Dunson and Zhou 2000; Zhou, Weinberg, Wilcox, and Baird 1996). Such extensions can be applied to studies of contraceptive efficacy, even studies that include couples whose use of a method is intermittent (Weinberg and Zhou 1997). More recently, attention has focused on problems that arise due to unreported intercourse (Dunson and Weinberg 2000) and errors in the identification of the day of ovulation. The latter problem produces frame-shift errors where the measured X_k is really X_{k+i} for all k.

6. MALE BIOMARKERS

Studies of conception are difficult to carry out, and intermediate markers, such as those based on semen quality, can provide valuable endpoints for identifying and characterizing toxic effects. Men can be studied whether or not they and their partner are attempting pregnancy, effects detected based on semen are unambiguously assignable to the man (rather than the couple), and a large sample of gametes is readily available from each man willing to be studied. Computer-assisted semen analysis (CASA) has made available an array of continuous descriptors of sperm movement for each sperm studied, based on the X-Y coordinates of the head position at repeated time points (Perreault 1998), creating almost too much data. Recently, methods have been developed that are robust to CASA machine settings and enable a reduction in the dimensionality of the data. Mortimer, Swan, and Mortimer (1996) assessed sperm activation using the fractal dimension of each sperm's trajectory. Dunson, Weinberg, Perreault, and Chapin (1999) later proposed an exponential model, inspired by complexity theory, that describes each sperm's movement using measures related to displacement velocity, linearity, and predictability of the path. The proposed measures are currently being evaluated using rodent reproductive toxicity data from a series of studies conducted by the National Toxicology Program. Videotaped sperm data are subject to errors that occur as a result of the intersection of sperm paths and loss of data from sperm that move too quickly off the field. Novel approaches are needed that avoid the resulting biases and preserve information about the individual movement patterns.

7. FEMALE BIOMARKERS

Women are more difficult to study than men in that their gametes cannot be sampled except in special clinical settings, such as in vitro fertilization (IVF) protocols, and one must rely on less direct markers, such as menstrual cycle hormone patterns or the characteristics of the cycle itself. The menstrual cycle, the time from the beginning of one menstrual period to the next, can be studied noninvasively, and disturbance of its length or increases in its variability may indicate reproductive dysfunction. Hormone assays of saliva or urine collected daily can provide additional detail, including confirmation that ovulation occurred, its timing, and the preovulatory and postovulatory patterns of hormone excretion (Waller, Swan, Windham, Elkin, and Lasley 1998). Comparisons across women are complicated by the existence of metabolic variability in the fate of hormones from woman to woman. Nonetheless, a recent method that combines smoothing splines and mixed-effects models has been usefully applied to hormone data (Brumback and Rice 1998).

This is an area in need of more research. We need methods to enable us to summarize the important features of the multivariate hormone profile, so that those optimally predictive features can be studied in relation to outcomes such as conception or early pregnancy loss. Such summary measures may need to account for frame shifts, where the event relative to which the hormones are measured (e.g., ovulation) is measured with some error.

8. DISCUSSION

Fertility has been a fruitful area of research, and interesting statistical problems persist. Our treatment in this article has been necessarily selective. In particular, we have focused on methods related to reproductive epidemiology and have limited our discussion of demographic models. However, the gap between epidemiologic and demographic approaches to fertility analysis has narrowed substantially in recent years. Demographers have long been interested in biological sources of variability in fertility. Bongaarts (1978) suggested seven factors through which social, economic, and cultural conditions affect fertility: frequency of intercourse, duration of the fertile period, contraception, induced abortion, lactational infecundability, spontaneous intrauterine mortality, and sterility. Until recently, models to account for these factors were overly simplified. However, hazards modeling has emerged as a promising approach. Frailty models have been developed that account for heterogeneity, for changing fecundability with age and parity (Larsen and Vaupel 1993), for sterility (Wood et al. 1994), and for hidden physiological processes (Yashin, Iachine, Andreev, and Larsen 1998).

Realistic models are needed that allow for unintended conceptions and that accommodate general covariate effects, including changes across time in exposures and in the demographics of couples attempting pregnancy. Improved methods for fitting these models and for including supplemental information from biological studies are also needed. Areas of considerable interest are the correlation in fertility within families and variation in fecundability across regions. Such dependency can potentially be accommodated using a multilevel model and may provide important insight into genetic and environmental influences. Other areas needing statistical development include the progression through puberty, recovery of fertility following childbirth or lactation, onset of menopause, evaluation of clinical treatments for infertility, and assessment of data from couples undergoing IVF.

REFERENCES

Barrett, J. C., and Marshall, J. (1969), "The Risk of Conception on Different Days of the Menstrual Cycle," *Population Studies*, 23, 455–461.

Bongaarts, J. (1978), "A Framework for Analyzing the Proximate Determinants of Fertility," *Population and Development Review*, 4, 105–132.

Brumback, B. A., and Rice, J. A. (1998), "Smoothing Spline Models for the Analysis of Nested and Crossed Samples of Curves," *Journal of the American Statistical Association*, 93, 961–976.

Burkhart, J., Helgen, J., Fort, D., Gallagher, K., Bowers, D., Propst, T., Gernes, M., Magner, J., Shelby, M., and Lucier, G. (1998), "Induction of Mortality and Malformation in Xenopus Laevis Embryos by Water Sources Associated With Field Frog Deformities," *Environmental Health Perspectives*, 106, 841–848.

Crouchley, R., and Dassios, A. (1998), "Interpreting the Beta Geometric in Comparative Fecundability Studies," *Biometrics*, 54, 161–167.

Daston, G., Gooch, J., Breslin, W., Shuey, D., Nikiforov, A., Fico, T., and Gorsuch, J. (1997), "Environmental Estrogens and Reproductive Health: A Discussion of the Human and Environmental Data," *Reproductive Toxicology*, 11, 465–481.

Dunson, D. B., and Weinberg, C. R. (2000), "Accounting for Unreported and Missing Intercourse in Human Fertility Studies," *Statistics in Medicine*, 19, 665–679.

Dunson, D. B., Weinberg, C. R., Perreault, S. D., and Chapin, R. E. (1999), "Summarizing the Motion of Self-Propelled Cells: Applications to Sperm Motility," *Biometrics*, 55.

Dunson, D. B., and Zhou, H. (2000), "A Bayesian Model for Fecundability and Sterility," *Journal of the American Statistical Association*, 95, 1054–1062.

Fisher, R. A. (1958), *The Genetical Theory of Natural Selection*, New York: Dover Press.

Galton, F. (1869), *Hereditary Genius: An Inquiry Into Its Laws and Consequences*, London: Macmillan.

Guillette, L. J., Gross, T., Masson, G., Matter, J., Percival, H., and Woodward, A. (1994), "Developmental Abnormalities of the Gonad and Abnormal Sex Hormone Concentrations in Juvenile Alligators From Contaminated and Control Lakes in Florida," *Environmental Health Perspectives*, 102, 680–688.

Heckman, J. J., and Walker, J. R. (1990), "Estimating Fecundability From Data on Waiting Times to First Conception," *Journal of the American Statistical Association*, 85, 283–294.

Larsen, U., and Vaupel, J. W. (1993), "Hutterite Fecundability by Age and Parity: Strategies for Frailty Modeling of Event Histories," *Demography*, 30, 81–102.

Masarotto, G., and Romualdi, C. (1997), "Probability of Conception on Different Days of the Menstrual Cycle: An Ongoing Exercise," *Advances in Contraception*, 13, 105–115.

Mortimer, S., Swan, M., and Mortimer, D. (1996), "Fractal Analysis of Capacitating Human Spermatozoa," *Human Reproduction*, 11, 1049–1054.

Olsen, J., and Andersen, P. (1999), "We Should Monitor Human Fecundity, But How?" *Epidemiology*, 10, 419–421.

Perreault, S. D. (1998), "Gamete Toxicology: The Impact of New Technologies," in *Reproductive and Developmental Toxicology*, ed. K. Korach, New York: Marcel Dekker, pp. 637–656.

Ridout, M., and Morgan, B. (1991), "Modeling Digit Preference in Fecundability Studies," *Biometrics*, 47, 1423–1433.

Royston, J. P. (1982), "Basal Body Temperature, Ovulation and the Risk of Conception, With Special Reference to the Lifetimes of Sperm and Egg," *Biometrics*, 38, 397–406.

Schcike, T. H., and Jensen, T. K. (1997), "A Discrete Survival Model With Random Effects: An Application to Time to Pregnancy," *Biometrics*, 53, 318–329.

Schwartz, D., MacDonald, P. D. M., and Heuchel, V. (1980), "Basal Body Temperature, Ovulation and the Risk of Conception, With Special Reference to the Lifetime of Sperm and Egg," *Population Studies*, 34, 397–400.

Sheps, M. C., and Mencken, J. A. (1973), *Mathematical Models of Conception and Birth*, Chicago: University of Chicago Press.

Swan, S. H., Elkin, E., and Fenster, L. (1997), "Have Sperm Densities Declined? A Reanalysis of Global Trend Data," *Environmental Health Perspectives*, 105, 1228–1232.

Waller, K., Swan, S. H., Windham, G. C., Elkin, E. P., and Lasley, B. L. (1998), "Use of Urine Biomarkers to Evaluate Menstrual Function in Healthy Premenopausal Women," *American Journal of Epidemiology*, 147, 1071–1080.

Weinberg, C. R., Baird, D. D., and Rowland, A. (1993), "Pitfalls Inherent in Retrospective Time-to-Event Data: The Example of Time to Pregnancy," *Statistics in Medicine*, 12, 867–879.

Weinberg, C. R., Gladen, B. C., and Wilcox, A. J. (1994), "Models Relating the Timing of Intercourse to the Probability of Conception and the Sex of the Baby," *Biometrics*, 50, 358–367.

Weinberg, C. R., and Wilcox, A. J. (1998), "Chapter 29: Reproductive Epidemiology," in *Modern Epidemiology*, 2nd edition, eds. K. Rothman and S. Greenland, Philadelphia: Lippincott-Raven.

Weinberg, C. R., and Zhou, H. (1997), "Model-Based Approaches to Studying Fertility and Contraceptive Efficacy," *Advances in Contraception*, 13, 97–103.

Wood, J. W., Holman, D. J., Yashin, A. I., Peterson, R. S., Weinstein, M., and Chang, M-C. (1994), "A Multistate Model of Fecundability and Sterility," *Demography*, 31, 403–426.

Yashin, A. I., Iachine, I. A., Andreev, K. F., and Larsen, U. (1998), "Multistate Models of Postpartum Infecundity, Fecundability and Sterility by Age and Parity: Methodological Issues," *Mathematical Population Studies*, 7, 51–78.

Zhou, H., Weinberg, C. R., Wilcox, A. J., and Baird, D. D. (1996), "A Random Effects Model for Cycle Viability in Fertility Studies," *Journal of the American Statistical Association*, 91, 1413–1422.

Statistical Issues in Toxicology

1. INTRODUCTION

Toxicology is "the study of the nature and mechanism of toxic effects of substances on living organisms and other biologic systems" (Lu 1996). Sometimes, data from human populations serve as the sentinel event indicating adverse health effects associated with environmental exposures. For example, the serious developmental effects associated with prenatal methyl mercury exposure were discovered only after some rural Japanese women ate fish contaminated by effluent from a nearby factory. Usually, however, the complexity and inherent variability of human populations complicates the evaluation of adverse environmental effects. For this reason, the field of toxicology has traditionally relied heavily on controlled studies in laboratory animals. Conducting studies in animals also allows researchers to explore questions that are difficult or unethical to address in a human population. For example, the pharmaceutical industry assesses safety in controlled animal experiments before products are used in humans.

This article reviews just a few of the interesting statistical problems in toxicology. After presenting a short history, I turn to some problems of current interest, followed by some emerging problems in the field. Many important topics (e.g., the use of toxicological information to inform risk assessment decisions and policy making) are touched on only briefly or not at all.

2. A BRIEF HISTORY

Much of modern toxicology and risk assessment have their roots in laws and regulations designed to ensure food purity and safety. King John of England proclaimed the first English food law in 1202, prohibiting the adulteration of bread with chickpeas and beans! After establishing the Food and Drug Administration (FDA) in 1930, Congress passed the Food, Drug, and Cosmetics Act in 1938, setting the standards for many of the regulatory practices that protect us today. Among a variety of amendments to this act is the now-infamous Delaney Clause of 1958, which prohibits the approval of any food additive shown to cause cancer in humans or animals. Public

Louise M. Ryan is Professor, Harvard School of Public Health and Dana-Farber Cancer Institute, Boston, MA 02115.

support for the regulatory role of the FDA grew further with events such as the thalido-mide tragedy of the 1960s. Not until the first half of the 20th century did regulators begin to consider the broader question of how chemical exposures might affect the environment in general. At first, the environmental movement focused on ecological concerns related to fish and wildlife. However, with the provocative 1962 publication of *The Silent Spring* (Carson 1962), the public began to realize that environmental contamination could also affect human health. The environmental movement gained momentum throughout the 1960s, culminating in 1970 with the first "Earth Day" and the establishment of the Environmental Protection Agency (EPA). (See the EPA's history page at www.epa.gov and the FDA's at www fda.gov.)

Statisticians have always played an important role in toxicology and regulatory science, and this involvement has led to many new statistical innovations. Much of the earlier work was motivated by pharmacological evaluations, rather than regulatory science. For example, Bliss (1934) provided one of the earliest applications of probit regression to fit a dose–response model and calculate the LD_{50}, or the dose that kills 50% of the test animals. Berkson (1944) suggested the logistic model as an alternative. Feiller's theorem was suggested as a means to finding a better confidence interval for the LD_{50}. Finney and others contributed many articles regarding the "bioassay" prob-lem of comparing the potency of two or more different pharmacological preparations (see, e.g., Finney 1965).

In the late 1960s and 1970s, statisticians began to address some of the broader issues raised by the emergence of modern regulatory science. For example, there was a need to quantify regulatory concepts, such as "acceptable daily intakes" (World Health Organization 1962), aimed at identifying dose or exposure levels corresponding to low or zero risk. By this time, toxicologists had put in place the idea of estimating safe doses by the "no observed adverse effect level" (NOEL or NOAEL), which corresponds to the highest observed experimental dose level that is not significantly different from controls with respect to a suitable adverse outcome. Clearly, there are considerable philosophical problems with this approach, as failure to reject a null hypothesis of no difference does not mean no difference in reality. In the context of cancer risk assessment, alternative strategies based on dose–response models have been accepted for some years now (see Krewski and Brown 1981 for a guide to some of the earlier literature on this topic). The NOEL approach is still used in the noncancer settings, although there seems to be widespread support for using a dose–response approach there as well, based on the "benchmark dose" suggested by Crump (1984).

During the 1980s and 1990s, statistical research in toxicology has specialized into a number of areas. Toxicology is an enormous field, with scientists interested in health-related issues ranging from cancer to reproduction, skin irritation, neurotoxic-ity, nephrotoxicity, and respiratory problems (see Lu 1996 for an interesting review). Other branches of toxicology (e.g., aquatic toxicology) were developed to study envi-ronmental effects on the ecosystem. Each of these specialty areas involves a different study design, leading to unique and interesting statistical problems. I touch on a few topics that have been popular with statisticians over the past decade or two: three-state models for carcinogenicity and the analysis of clustered data and multiple outcomes from teratology experiments. More detailed discussion on these and other topics can be found in several excellent texts (see, e.g., Piegorsch and Bailer 1997).

3. LONG-TERM CARCINOGENICITY STUDIES

In a long-term cancer bioassay or carcinogenicity study, control and exposed mice or rats, usually 50 or 60 per dose group, are observed over a typical lifetime (24 months) and are examined at death or at the end of the experiment for a variety of different tumors. Because the high doses given to exposed animals often shorten their lifetimes due to general toxicity, age-adjusted statistical analyses are needed to ensure a valid test for carcinogenicity. Prevalence tests (Hoel and Walburg 1972) and time-to-event analyses (Peto 1974) have been suggested, but these approaches make strong and unrealistic assumptions about tumor lethality. To address this limitation, Kodell and Nelson (1980) formulated the problem as the three-state illness-death model shown in Figure 1.

In the figure, $\lambda(t)$ represents the instantaneous rate of tumor onset at time t, $\beta(t)$ the instantaneous death rate at time t, and $\alpha(t, x)$ the instantaneous death rate at time t for an animal that developed a tumor at time x. Although the goal is simply to characterize the effect of exposure on the tumor onset rate, $\lambda(t)$, the problem is challenging because tumor onset is unobserved. Kodell and Nelson's creative suggestion sparked much activity (see, e.g., McKnight and Crowley 1984). Many of these approaches proved infeasible in practice because they assumed the availability of extensive interim sacrifice, wherein randomly selected animals are killed at prechosen times during the experiment and examined for tumors. Dinse (1991) and Lindsey and Ryan (1993) advocated using semiparametric alternatives that assume either an additive $(\alpha(t, x) = \beta(t) + \Delta)$ or multiplicative relationship $(\alpha(t, x) = \beta(t)e^{\Delta})$ be-

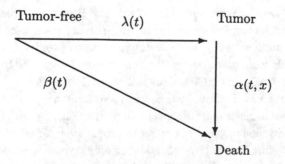

Figure 1.   *Three-State Model for Carcinogenicity.*

tween the hazards for death with and without tumor. These approaches are appealing from a biological perspective and can be applied to a standard-sized experiment with only a single terminal sacrifice. The semiparametric approach is also appealing in that it reframes the problem to a form more closely related to other familiar statistical problems. For instance, when Δ equals 0, we have a current status problem (Keiding, Begtrup, Scheike, and Hasibeder 1996). Methods for interval censored data are also relevant. Nonparametric smoothing of the baseline hazards is likely to prove fruitful in such settings (Betensky, Lindsey, Ryan, and Wand 1999; Kooperberg and Stone 1992). Although statistical research related to the three-state model continues, these models are still not widely used in practice. Indeed, the method favored of late by the National Toxicology Program is the poly-k method proposed by Portier and Bailer (1989).

Carcinogenicity studies have also motivated much interesting work on the use of historical control data, which can help in the interpretation of a current study, particularly when control tumor rates are low and when marginal significance levels are obtained in a test for dose effects (Haseman, Huff, and Boorman 1984). Many authors have developed statistical methods to incorporate historical data on tumor incidence rates into the analysis of a current study (e.g., Fung, Krewski, and Smythe 1996). However, these methods are not widely used in practice because they are not age adjusted. Ibrahim, Ryan, and Chen (1998) developed methods to incorporate historical information into age-adjusted tests, but these methods require strong assumptions about tumor lethality. Methods to incorporate historical control information directly into analyses based on the three-state model would be useful.

An issue of recent interest is whether and how to adjust for animal weight in the analysis of a carcinogenicity experiment. Ignoring weight in the analysis of carcinogenicity data may lead to bias because test compounds often cause exposed animals to weigh less, and lighter animals tend to have a lower incidence of certain tumors (see Seilkop 1995). The relatively low tumor incidence rate in most experiments means that there is seldom enough information within any one experiment to support a regression-based adjustment, and hence historical control data are needed (see also Ibrahim et al. 1998).

Ensuring good small-sample properties of statistical testing and analysis procedures and appropriately managing multiple testing represent two additional areas where statistical research has played an important role. Resource limitations and the general principle of minimizing animal usage mean that toxicity experiments are often relatively small. In the absence of strong effects, event rates (e.g., numbers of tumors) will generally be very low. Sometimes, this can be beneficial. For example, Haseman (1984) argued that inflation of type I errors due to multiple testing is unlikely to be a problem in this setting, due to the natural conservativeness of statistical tests in such settings. The exact approach suggested by Westfall and Young (1993) has had a strong influence in this area, which continues to generate considerable interest.

4. TERATOLOGY, DEVELOPMENTAL, AND REPRODUCTIVE TOXICITY STUDIES

A variety of different study designs are available for assessing toxic effects on reproduction and development in animals. A reproductive toxicity study is designed to assess whether chemical exposures interfere with successful fertilization or implantation of fertilized ova in the uterus. Much statistical work has been motivated by developmental toxicity studies concerned with the effects of exposures during gestation on the offspring themselves. In a typical study, 20–30 pregnant animals are randomized to each dose group. Typical control litter sizes (number of live-born offspring) range from around 8–10 in rabbits to 12–15 in mice and rats. Outcomes measured on the live offspring include presence or absence of malformations, body weight and size (e.g., crown to rump length), and sometimes organ weights. Accounting for the clustering of offspring within a litter and the multiplicity of outcomes has led to interesting statistical research. Williams (1975) suggested using a beta-binomial model for dose–response modeling of teratology data. Many people began to advocate using quasi-likelihood as an alternative to likelihood-based methods for clustered data (see, for example, Williams 1982). The emerging popularity of generalized estimating equations (GEE) and particularly the availability of GEE software (Liang and Zeger 1986) sparked numerous works motivated by teratology data. Some of the more interesting and challenging problems are those related to multiple outcomes. Unlike the carcinogenicity setting, where chemicals usually cause only one or two different types of cancer, developmental toxicants typically affect various organ systems in the developing fetus. Correlated multinomial models (Chen, Kodell, Howe, and Gaylor 1991) were suggested for handling hierarchically related outcomes such as death and malformation. Lefkopoulou, Moore, and Ryan (1989) suggested an approach based on GEEs for analyzing multiple binary outcomes measured on the same fetus.

Developmental toxicity also generated interesting work on the analysis of multiple outcome data, particularly of mixed type, for example binary and continuous outcomes such as weight and malformation from a teratology study. Catalano and Ryan (1992) and Fitzmaurice and Laird (1995) both proposed solutions that link a marginal model for one outcome with a conditional model for the other. These ideas are linked to more general theory related to the general location model of Liu and Rubin (1998), as well as the chain graph models of Cox and Wermuth (1996).

Aside from the basic challenges of developing good multivariate models, particularly in the clustered data setting, additional technical challenges arise in applying these models in real data settings. For example, correlation structures are likely to change with exposure (Kupper, Portier, Hogan, and Yamamoto 1986). Ignoring this can lead to bias in the case of likelihood-based models and loss of efficiency for GEE-based methods. Determining how to use a complex model for risk assessment purposes—for example, calculating benchmark doses (Crump 1984)—is another area of interest. Likelihood-based methods may have some advantages for that purpose, because they more naturally facilitate combining endpoints to calculate an overall probability of effect (Regan and Catalano 1999).

5. SOME OTHER DESIGNS AND ISSUES

The last two sections touched on statistical problems in the analysis of data from two common toxicological study designs. Challenging issues arise in various other settings as well. For example, the EPA is interested in test systems to evaluate neurotoxicity. Methyl mercury and PCBs are examples of chemicals thought to affect the central nervous system in ways that may be devastating for the developing fetus and young children (Lu 1996, chap. 16). Quantifying neurotoxicity is complicated, as effects tend to manifest in relatively subtle behavioral and performance changes. Current recommended toxicity testing is based on the so-called "functional observational battery" or FOB (United States Environmental Protection Agency 1995), which is a set of 20 to 30 items measured on laboratory animals. Statistical methods for analyzing the FOB are relatively undeveloped at this time. Whereas Blackwell and Catalano (2001) proposed using latent variable models in this setting, these methods tend to work well only for fairly large experiments.

The multigenerational toxicity study design also raises a number of interesting statistical challenges, though there do not appear to be any papers in the statistical literature on this topic. The design involves exposing male and female animals for 10 weeks prior to breeding. Exposure then continues for the females throughout gestation and lactation. To evaluate effects on the reproductive functioning of the offspring, a second generation is bred and reared to reproductive maturity. Endpoints of interest include the numbers of successful pregnancies and number of live offspring in the first and second generations, and survival. Opportunities abound for development of innovative survival analysis models in this context.

6. THE FUTURE?

Advances in molecular biology have led to increased emphasis on mechanisms of toxicity. Indeed, the EPA is in the process of revising its carcinogen risk assessment guidelines to accommodate mechanistic considerations. As an alternative to "statistical" models, the toxicological literature increasingly refers to "biologically based dose response models" (see Moolgavkar, Leubeck, de Gunst, Port, and Schwetz 1990 for an example). Broadly speaking, such models aim to characterize the relationship between exposure and response by specifying some or all of the biological steps in between, including dose at the target organ and mechanism of action. Many toxicologists believe that such models are the key to explaining variability between species and hence to making effective extrapolation from animals to humans. Despite active research on the development of biologically based models, they have not been widely used in real-life risk assessment settings. It is generally difficult to determine the appropriate biological mechanisms to include in the model, and often there will be a wide range of expert opinions on this topic. Even when appropriate mechanisms have been identified, the corresponding model-fitting procedures are complex, requiring the solution to systems of differential equations characterizing transitions to and from multiple compartments. In practice, data need to be drawn from several studies or from the literature, and a relatively advanced knowledge of toxicology is needed. For all of these reasons, biologically based models tend to be of more interest to biomathematicians than to statisticians, and almost all of the literature on this topic appears in the toxicological and biomathematics literature. Given that mechanistic models are most likely here to stay, however, statisticians interested in toxicological applications will need to invest time learning about basic physiology and the mechanisms of disease. The development of statistical models to incorporate biomarkers may represent a middle ground between the two extremes just described and represents a promising avenue for further research.

The identification of biomarkers such as DNA adducts and genetic markers of susceptibility and cellular damage have opened new avenues of toxicological research in human populations. This, along with advances in environmental science leading to better methods of exposure assessment, has made it possible in some cases (e.g., arsenic, passive smoking, butadiene) to use human data for risk assessment. Although issues of confounding and measurement error are always complications in that setting, risk assessment based on human data has the major advantage of avoiding the need for interspecies extrapolation. As the field of environmental epidemiology continues to evolve, human data likely will play an even greater role in risk assessment. Statistical methods to adjust for measurement error and to incorporate biomarkers will become more important than ever.

Today's rapid changes in information technology are also having an impact on risk assessment and regulatory science. Society as a whole is more informed about the potential of drugs and chemicals to affect health and the environment. For example, anyone can use the EPA website to access information about toxic waste sites within their own community. Regulatory agencies are under pressure not only to move more quickly in establishing regulatory standards, but also to address issues that more directly affect people's day-to-day lives. Dealing with complex mixtures is a good example. Because regulatory toxicology has traditionally focused on characterizing the risks associated with a lifetime exposure to a single chemical, it often falls short in real-world settings involving mixtures of different chemicals with nonconstant exposure patterns. Diesel exhaust, for instance, is a complex mixture of benzene, 1,3 butadiene, ethylene dibromide, and many other chemicals. Furthermore, diesel exposure patterns vary dramatically according to traffic patterns. As a result of these complications, the EPA and other regulatory agencies have struggled for years to set diesel standards. Tackling a problem like this requires a multidisciplinary effort to problem solving that in many ways goes against traditional scientific training that specializes in one narrow field. In some ways, our training as statisticians prepares us well for the kind of wide-ranging, cross-disciplinary thinking needed in modern toxicology and risk assessment. As we move into the 21st century, this kind of perspective likely will become even more valuable. However, some new statistical tools may be needed. For example, there likely will be increased demand for ways to synthesize information from many different sources. Although traditional meta-analysis techniques accomplish this to some extent, a broader approach is needed that can effectively combine different kinds of information, including subjective expert opinion, to help inform decision making.

Recently, I had the pleasure of interviewing Professor Fred Mosteller for a newsletter article, and asked how he thought our profession was doing and what changes he would like to see. He commented that not enough statisticians were being trained in policy. While writing this vignette, his comments have come to my mind several times. There is no doubt that toxicology is a fascinating science that will continue to yield lots of interesting statistical problems for statisticians to tackle. However, its real value lies in its potential to inform regulatory decisions and policy making related to the chemical and drug exposures that we encounter in our daily lives. Statisticians who choose to work in this interesting area have the potential to make a real impact as well, though to do so may involve moving beyond the usual disciplinary boundaries. Although statistical research over the past several decades has made valuable contributions to the field of toxicology, many of us shy away from the really difficult problems, such as biologically based models and exposure assessment, that require sophisticated biological as well as statistical knowledge. Many are predicting that this will change over the next several decades. Our field will need to become even more interdisciplinary to stay competitive with emerging disciplines such as computational biology (see Waterman 1998).

REFERENCES

Berkson, J. (1944), "Application of the Logistic Function to Bioassay," *Journal of the American Statistical Association*, 39, 357–365.

Betensky, R. A., Lindsey, J. C., Ryan, L. M., and Wand, M. P. (1999), "Local EM Estimation of the Hazard Function for Interval-Censored Data," *Biometrics*, 55, 238–245.

Blackwell, B., and Catalano, P. J. (2001), "A Random Effects Latent Variable Model for Multivariate Ordinal Data," in press.

Bliss, C. I. (1934), "The Method of Probits," *Science*, 79, 38–39.

Carson, R. (1962), *The Silent Spring*, New York: Houghton Mifflin.

Catalano, P. J., and Ryan, L. (1992), "Bivariate Latent Variable Models for Clustered Discrete and Continuous Outcomes," *Journal of the American Statistical Association*, 87, 651–658.

Chen, J. J., Kodell, R. L., Howe, R. B., and Gaylor, D. W. (1991), "Analysis of Trinomial Responses from Reproductive and Developmental Toxicity Experiments," *Biometrics*, 47, 1049–1058.

Cox, D. R., and Wermuth, N. (1996), *Multivariate Dependencies: Models, Analysis and Interpretation*, London: Chapman and Hall.

Crump, K. S. (1984), "A New Method for Determining Allowable Daily Intakes," *Fundamental and Applied Toxicology*, 4, 854–871.

Dinse, G. E. (1991), "Constant Risk Differences in the Analysis of Animal Tumorigenicity Data," *Biometrics*, 47, 681–700.

Finney, D. J. (1965), "The Meaning of Bioassay," *Biometrics*, 21, 785–798.

Fitzmaurice, G. M., and Laird, N. M. (1995), "Regression Models for a Bivariate Discrete and Continuous Outcome with Clustering," *Journal of the American Statistical Association*, 90, 845–852.

Fung, K. Y., Krewski, D., and Smythe, R. T. (1996), "A Comparison of Tests for Trend with Historical Controls in Carcinogen Bioassay," *The Canadian Journal of Statistics*, 24, 431–454.

Haseman, J. K. (1984), "Statistical Issues in the Design, Analysis, and Interpretation of Animal Carcinogenicity Studies," *Environmental Health Perspectives*, 58, 385–392.

Haseman, J. K., Huff, J., and Boorman, G. A. (1984), "Use of Historical Control Data in Carcinogenicity Studies in Rodents," *Toxicologic Pathology*, 12, 126–135.

Hoel, D. G., and Walburg, H. E. (1972), "Statistical Analysis of Survival Experiments," *Journal of the National Cancer Institute*, 49, 361–372.

Ibrahim, J. G., Ryan, L. M., and Chen, M. H. (1998), "Using Historical Controls to Adjust for Covariates in Trend Tests for Binary Data," *Journal of the American Statistical Association*, 93, 1282–1293.

Keiding, N., Begtrup, K., Scheike, T. H., and Hasibeder, G. (1996), "Estimation from Current-Status Data in Continuous Time," *Lifetime Data Analysis*, 2, 119–129.

Kodell, R. L., and Nelson, C. J. (1980), "An Illness-Death Model for the Study of the Carcinogenic Process Using Survival/Sacrifice Data," *Biometrics*, 36, 267–277. Corr: 37, 875.

Kooperberg, C., and Stone, C. J. (1992), "Logspline Density Estimation for Censored Data," *Journal of Computational and Graphical Statistics*, 1, 301–328.

Krewski, D., and Brown, C. (1981), "Carcinogenic Risk Assessment: A Guide to the Literature," *Biometrics*, 37, 353–366.

Kupper, L. L., Portier, C., Hogan, M. D., and Yamamoto, E. (1986), "The Impact of Litter Effects on Dose-Response Modeling in Teratology (C/R: V44 p 305–309)." *Biometrics*, 42, 85–98.

Lefkopoulou, M., Moore, D., and Ryan, L. (1989), "The Analysis of Multiple Correlated Binary Outcomes: Application to Rodent Teratology Experiments," *Journal of the American Statistical Association*, 84, 810–815.

Liang, K. Y., and Zeger, S. L. (1986), "Longitudinal Data Analysis Using Generalized Linear Models," *Biometrika*, 73, 13–22.

Lindsey, J. C., and Ryan, L. M. (1993), "A Three-State Multiplicative Model for Rodent Tumorigenicity Experiments," *Applied Statistics*, 42, 283–300.

Liu, C. and Rubin, D. R. (1998), "Ellipsoidally Symmetric Extensions of the General Location Model for Mixed Categorical and Continuous Data," *Biometrika*, 85, 673–688.

Lu, F. C. (1996), *Basic Toxicology*, London: Taylor and Francis.

McKnight, B., and Crowley, J. (1984), "Tests for Differences in Tumor Incidence Based on Animal Carcinogenesis Experiments," *Journal of the American Statistical Association*, 79, 639–648.

Moolgavkar, S. H., Leubeck, E. G., de Gunst, M., Port, R. E., and Schwetz, M. (1990), "Quantitative Analysis of Enzyme-Altered Foci in Rat Hepatocarcinogenesis Experiments 1: Single Agent Regimen," *Carcinogenesis*, 11, 1271–1278.

Peto, R. (1974), "Guidelines on the Analysis of Tumor Rates and Death Rates in Experimental Animals," *British Journal of Cancer*, 29, 101–105.

Piegorsch, W. W., and Bailer, A. J. (1997), *Statistics for Environmental Biology and Toxicology*, Boca Raton, FL: Chapman & Hall.

Portier, C. J., and Bailer, A. J. (1989), "Testing for Increased Carcinogenicity Using a Survival-Adjusted Quantal Response Test," *Fundamental and Applied Toxicology*, 12, 731–737.

Regan, M., and Catalano, P. J. (1999), "Likelihood Models for Clustered Binary and Continuous Outcomes: Application to Developmental Toxicology," *Biometrics*, 55, 760–768.

Seilkop, S. K. (1995), "The Effect of Body Weight on Tumor Incidence and Carcinogenicity Testing in B6C3F1 Mice and F344 Rats," *Fundamental and Applied Toxicology*, 24, 247–259.

United States Environmental Protection Agency (1995), "Proposed Guidelines for Neurotoxicity Testing," *Federal Register*, 60, 52032–52036.

Waterman, M. (1998), *Introduction to Computational Biology: Maps, Sequences, and Genomes*, Chapman & Hall.

Westfall, P. H., and Young, S. S. (1993), "On Adjusting *p*-values for Multiplicity (Disc: p944–945)," *Biometrics*, 49, 941–944.

Williams, D. A. (1975), "The Analysis of Binary Responses from Toxicological Experiments Involving Reproduction and Teratogenicity," *Biometrics*, 31, 949–952.

—— (1982), "Extra-Binomial Variation in Logistic Linear Models," *Applied Statistics*, 31, 144–148.

World Health Organization (1962), *Sixth Report of the Joint FAD/WHO Expert Committee on Food Additives*, Geneva: WHO.

Receiver Operating Characteristic Methodology

Margaret Sullivan Pepe

1. INTRODUCTION

Diagnostic medicine has progressed tremendously in the last several decades, and the trend promises to continue well into the next millennium. Advances in technology provide new methods for detecting disease or physical impairment. Some examples include the use of biochemical serum markers such as prostate-specific antigen for prostate cancer and CA-125 for ovarian cancer, of radiographic imaging procedures such as mammography for breast cancer, and of electrophysical procedures such as brain stem response testing for hearing impairment. Research studies to assess the operating characteristics of diagnostic tests are clearly important to ensure that accurate and cost-effective procedures are selected for widespread use. Development of appropriate statistical methods for designing such studies and for analyzing data from them will be key to their success.

A statistical tool that is becoming popular for describing diagnostic accuracy is the receiver operating characteristic (ROC) curve. To define an ROC curve, first consider diagnostic tests with dichotomous outcomes, with positive outcomes suggesting presence of disease. For dichotomous tests, there are two potential types of error. A false-positive error occurs when a nondiseased individual has a positive test result, and conversely, a false-negative error occurs when a diseased individual has a negative test result. The rates with which these errors occur, termed the false-positive and false-negative rates, together constitute the operating characteristics of the dichotomous diagnostic test. Statisticians are already familiar with these concepts in the context of statistical hypothesis testing. ROC curves generalize these notions to nonbinary tests in the following fashion: Let D be a binary indicator of true disease status with $D = 1$ for diseased subjects. Let X denote the test result with the convention that larger values of X are more indicative of disease. For any chosen threshold value c, one can define a dichotomous test by the positivity criterion $X \geq c$, and calculate the associated error rates. A plot of 1 minus the false-negative rate (or true positive rate) versus the false-positive rate for all possible choices of c is the ROC curve for X. By definition, this is a monotone increasing function from

Margaret Sullivan Pepe is Member, Division of Public Health Sciences, Fred Hutchinson Cancer Research Center, Seattle, WA 98109 (E-mail: mspepe@u.washington.edu).

[0, 1] to [0, 1], with higher curves associated with better tests.

The ROC curve is primarily a descriptive device displaying the range of trade-offs between true-positive and false-positive rates possible with the test. It transforms the test results to a scale that pertains to accuracy for detecting disease. This is particularly useful for comparing tests with numerical results that are on different measurement scales and for which no meaningful comparisons can be based on the raw data. Mathematically, the ROC curve can be written as $\text{ROC}(t) = F_D(F_{\bar{D}}^{-1}(t))$ for $t \in (0, 1)$, where F_D and $F_{\bar{D}}$ denote the "survivor" functions for X in the diseased and nondiseased populations. It follows that the ROC curve is invariant to monotone increasing transformations of the data.

2. ORDINAL RATING DATA

Although ROC analysis had its roots in the electronic signal detection theory developed in the 1950s (Green and Swets 1966), it was not until the early 1980s that it started to be used in biomedical applications. It became especially popular in radiology for characterizing the accuracy of diagnostic imaging modalities. Indeed, most of the statistical methodologic work on ROC analysis has been done in this context (see Hanley 1998 for a review). Why did it fit particularly well with the needs in radiology? Because image assessments are made subjectively by radiologists, and implicit criteria for assessments vary among radiologists. To see this, consider the classic setting where each reader assesses an image on an ordinal scale, $k = 1, \ldots, K$, with the lowest category labeled perhaps as "definitely no disease present" to the highest labeled as "disease definitely present." It is assumed (Metz 1986) that with each reading, Y, there is a continuous latent decision variable, X, and that the reader classifies the image in category k if the decision variable falls within the interval (τ_{k-1}, τ_k), $k = 1, \ldots, K$ with $\tau_0 = -\infty$ and $\tau_K = +\infty$. That is, $Y = k$ if $\tau_{k-1} < X < \tau_k$. Although one cannot know the value of X, K points on its ROC curve are identifiable, and the whole curve is identifiable under parametric modeling assumptions. Points for readers using different decision criteria (i.e., different threshold values $\{\tau_1, \ldots, \tau_{K-1}\}$) will simply fall at different locations on the ROC curve for X. In this way, ROC analysis accommodates variation among readers in their decision criteria and purports to disentangle this variation from the inherent discriminatory capacity of the diagnostic test.

The classical approach for estimating an ROC curve in this context is to assume the binormal model; that is, that some monotone-increasing transformation of X has a standard normal distribution in the nondiseased population and a normal distribution with (mean, SD) $= (a, b^{-1})$ in the diseased population. The ROC curve for X then has the form $\text{ROC}(t) = \Phi(a + b\Phi^{-1}(t))$ for $t \in (0, 1)$. Using readings of images from a set of diseased subjects, $\{Y_1^D, \ldots, Y_{n_D}^D\}$, and from a set of nondiseased subjects, $\{Y_1^{\bar{D}}, \ldots, Y_{n_{\bar{D}}}^{\bar{D}}\}$, the parameters (a, b) are estimated via maximum likelihood (Dorfman and Alf 1969). If different readers or different imaging modalities give rise to different ROC curves, then the data are stratified accordingly and stratum-specific parameters (a_s, b_s) are estimated. As described later, classical comparisons between curves are based on the area under the ROC curve, which for the binomial curve is given by $\Phi(a/(1 + b^2)^{1/2})$.

Tosteson and Begg (1988) proposed that ordinal regression modeling methods could be applied to radiology rating data to make inference about ROC curves. This opened up avenues for much more sophisticated and flexible ROC analysis than had been available previously. In the simplest setting, a location ordinal model is postulated,

$$P[X < \tau_k] = P[Y \leq k] = g(C_k - \alpha_1 D - \alpha_2 \mathbf{Z} - \alpha_3 \mathbf{Z} \cdot D),$$

where $\mathbf{Z}$ is a vector of covariates and g is a cumulative distribution function. The ROC curve corresponding to X conditional on $\mathbf{Z}$ is then $\mathrm{ROC}_{\mathbf{Z}}(t) = 1 - g(g^{-1}(1 - t) - \alpha_1 - \alpha_3 \mathbf{Z})$, which reduces to $\Phi(\Phi^{-1}(t) + \alpha_1 + \alpha_3 \mathbf{Z})$, for example, when g is the standard normal cdf. Thus covariates that have no interaction with disease status do not affect the ROC curve, but are only associated with shifting the operating points on the curve.

The ability to explore effects of covariates on test accuracy was a huge step forward in the analysis of radiology rating data. Important types of covariates include characteristics of subjects from which images are taken, characteristics of procedures used to process images, and characteristics of the readers such as experience or institutional affiliation. Other indicators of disease status such as clinical signs or symptoms, or indeed the results of another diagnostic test, could also be considered as covariates in the model. This strategy allows one to assess the incremental value of the imaging test over and above this information.

The general model proposed by Tosteson and Begg included the possibility for covariates to affect scale parameters as well and can be written as

$$P[Y \leq k] = g(\{C_k - \alpha_1 D - \alpha_2 \mathbf{Z} - \alpha_3 D \cdot \mathbf{Z}\}/\exp\{\beta_1 D + \beta_2 \mathbf{Z} + \beta_3 D \cdot \mathbf{Z}\}).$$

Synthesis of covariate effects on the ROC curve is not as straightforward as in the location-only model. However, it does allow for estimation of covariate adjusted ROC curves. Note that the classic binormal model of Dorfman and Alf (1969) is a special case of the Tosteson and Begg model, when a probit link function is used and no covariates are included.

By embedding ROC methodology in the framework of ordinal regression, issues of correlated data that had previously been difficult to deal with now became much easier because of the growth in methods for handling clustered data that also occurred in the 1980s. Toledano and Gatsonis (1995) chose to model marginal probabilities and account for correlation with GEE, whereas Gatsonis (1995) suggested incorporating random reader effects into the ordinal regression framework. An advantage of the latter approach is that it allows for assessment of inter-rater variability in decision criteria and in accuracy, which is of considerable interest in diagnostic medicine (Beam, Layde, and Sullivan 1996).

Studies of diagnostic or screening tests often suffer from so-called verification bias. This occurs when subjects with certain test results, X, or other characteristics, Z, indicative of disease are more likely to be assessed for disease, D, with the gold standard than are other subjects. This sort of selection makes practical clinical sense when the gold standard is invasive or costly, such as requiring surgery for cancer detection. However, it introduces bias into estimates of test accuracy unless adjustments are made in the analysis. Although some attempts at ROC adjustments had been made previously, embedding ROC analysis into the ordinal regression framework again provided access to a rich variety of missing-data techniques that were already available for generalized linear regression analysis. These include inverse probability weighting and EM-based maximum likelihood methods.

3. CONTINUOUS DATA

Concepts of ROC have more recently gained popularity in biomedical applications involving tests with results on non-ordinal scales. Many biochemical measurements—for example, including serum antigen or enzyme concentrations (Zweig and Campbell 1993)—are continuous in nature. At first glance, one might expect that ROC analysis for continuous data would be more straightforward than it is for ordinal data, because in this setting the decision variable itself is available rather than just a categorized version. However, new challenges present themselves with continuous data.

Consider, for example, estimation of the ROC curve from data $\{Y_i^D, i = 1, \ldots, n_D, : Y_j^{\bar{D}}, j = 1, \ldots, n_{\bar{D}}\}$, where Y_i^D denotes the test result for the ith diseased study unit and $Y_j^{\bar{D}}$ that for the jth nondiseased unit. The curve $\widehat{\mathrm{ROC}}(t) = \hat{F}_D(\hat{F}_{\bar{D}}^{-1}(t))$, where $\hat{F}_D$ and $\hat{F}_{\bar{D}}$ are empirical estimators of F_D and $F_{\bar{D}}$ based on $\{Y_i^D, i = 1, \ldots, n_D\}$ and $\{Y_j^{\bar{D}}, j = 1, \ldots, n_{\bar{D}}\}$, is a jagged curve. Curves that incorporate reasonable smoothness assumptions can be based on parametric or smoothed estimators of F_D and $F_{\bar{D}}$. However, these estimated ROC curves do not enjoy a fundamental property of ROC curves—namely, invariance to monotone-increasing transformations of the data. Metz, Herman, and Shen (1998) noted that classical smooth (parametric) ROC curve estimators for ordinal data, on the other hand, do have this property. Indeed, they proposed that smooth ROC curve estimators for continuous data be based on application of ordinal data methods to categorized versions of the continuous data.

The fact that decision criteria are explicit with continuous data rather than implicit also impacts on analysis. One has the capacity in practice to specify the decision criterion and thus to control the false-positive rate. Because higher ranges of false-positive rates may be of no interest for future application of the test, it may be appropriate to focus the analysis on a restricted range, $\{\text{ROC}(t), t \in [0, t_0]\}$, for some $t_0 < 1$. Wieand, Gail, James, and James (1989) proposed test statistics for comparing ROC curves over restricted intervals.

For regression modeling of covariate effects on ROC curves, the analog of Tosteson and Begg's approach is to model the test outcome with location and scale components that are functions of disease status and covariates. This approach again suffers from lack of invariance to monotone data transformations. Pepe (1997) proposed a fundamentally different approach to regression. She suggested that instead of modeling the test result and indirectly ascertaining induced covariate effects on the ROC curve, one could directly model the ROC curve itself. The general model takes the form

$$\text{ROC}_{\mathbf{Z}}(t) = g(\mathbf{Z}\beta, h(\gamma, t))$$

for some specified functions g and h and unknown parameters (β, γ). A special case of this is the generalized linear model

$$\text{ROC}_{\mathbf{Z}}(t) = g(\Sigma \gamma_l h_l(t) + \mathbf{Z}\beta),$$

where $h_1, \ldots, h_L$ are "basis" functions of t and g is a link function. Parameter estimation as proposed by Pepe (1997) is cumbersome, although newer approaches to inference using standard binary regression methods appear promising. Advantages of direct modeling of ROC curves over modeling the test results have been summarized (Pepe 1998) and include the ability to (a) restrict inference about ROCs to restricted ranges of false-positive rates; (b) incorporate interactions between covariate effects and t, so that covariates have different effects over different ranges of t; and, most important, (c) compare ROCs for tests with results of numerically different form and thus cannot be modeled sensibly in a single regression model for test results. Finally, by omitting the covariate component from the general model, it can be seen that this framework provides an avenue for developing smooth estimators of the ROC curve that are invariant to monotone data transformations.

4. THE AREA UNDER THE RECEIVER OPERATING CHARACTERISTIC CURVE

Traditionally, the area under the ROC curve (AUC) has been used as a summary index of test accuracy (Hanley 1989). Indeed, comparisons between two diagnostic tests are classically based on differences between estimated AUC's. This has long been the standard of practice in radiology where AUC is estimated with the binormal model (Swets and Pickett 1982) and for continuous data where nonparametric estimation of the AUC is possible with the Mann–Whitney U-statistic (Hanley and McNeil 1982). Regression analysis based on AUC statistics has been proposed. Here an AUC statistic is calculated as a derived variable from each subset of data that is homogeneous in regards to covariates, and then a regression model is fit to the AUC's (Obuchowski 1995). This regression framework, however, is more restrictive than others. It cannot, for example, accommodate continuous covariates.

The AUC statistic can be interpreted as the probability that the test result from a randomly chosen diseased individual is more indicative of disease than that from a randomly chosen nondiseased individual, $AUC = P(X_i \geq X_j | D_i = 1, D_j = 0)$. Thus it can be thought of as simply a nonparametric measure of the distance between the distributions of test results for diseased and nondiseased individuals. Despite these interpretations, however, many investigators find the index unappealing, because it has no clinically relevant meaning. Another valid concern is that a large part of the area comes from the rightmost part of the curve that includes false-positive rates unlikely to be used in practice. If the curves for two tests cross, moreover, a meaningful difference between the tests over a range of interest might not be picked up by the AUCs. These considerations lead to consideration of a partial AUC, the area under the ROC curve in a restricted range of false-positive rates (Thompson and Zucchini 1989; Wieand et al. 1989).

5. CONCLUDING REMARKS

Methods for evaluating diagnostic tests have not received the same level of attention from biostatisticians as have, say, methods for evaluating new therapeutic strategies (Begg 1987). With impetus from regulatory agencies and from public health administrators, it appears that there will be an increasing demand for standards to be set for the design and analysis of studies evaluating new tests. ROC methodology is a popular statistical approach in this context. I consider ROC analysis most useful in the development phase of diagnostic testing. Its purpose is to determine whether a test has the capacity to effectively discriminate between diseased and nondiseased states. This is an important property, but it does not necessarily indicate its practical value for patient care (Zweig and Campbell 1993). Issues of cost, disease prevalence, and consequences of misdiagnosis will enter into the ultimate evaluation of test usefulness.

I mention in closing some of the major statistical challenges for evaluating diagnostic tests in general and for applying ROC methodology in particular. First, in many settings a definitive gold standard assessment of disease status, D, is not available. Infection with *Chlamydia trachomis*, for example, can be assessed only imprecisely with standard bacterial culture techniques. How can inference for an ROC curve be accomplished in this setting? Second, the statistical literature on diagnostic testing typically assumes that the test result is a simple numeric value. However, test results may be much more complicated, involving several components. Do ROC curves have a role to play in determining how to combine different sources of information to optimize diagnostic accuracy? Third, disease status is often not a fixed entity, but rather can evolve over time. How can the time aspect be incorporated sensibly into ROC analysis? Finally, are there alternatives to the ROC curve for describing test accuracy? For binary outcomes, two ways of describing test accuracy are to report (a) true- and false-positive rates, and (b) positive and negative predictive values. ROC curves can be thought of as generalizing the former to continuous tests; that is, ROC curves generalize the binary test notions of true-positive and false-positive rates to continuous tests. Are there analogs of ROC curves that similarly generalize the notions of predictive values to continuous tests?

REFERENCES

Beam, C. A., Layde, P. M., and Sullivan, D. C. (1996), "Variability in the Interpretation of Screening Mammograms by U.S. Radiologists," *Archives of Internal Medicine*, 156, 209–213.

Begg, C. E. (1987), "Biases in the Assessment of Diagnostic Tests," *Statistics in Medicine*, 6, 411–423.

Dorfman, D. D., and Alf, E. (1969), "Maximum Likelihood Estimation of Parameters of Signal Detection Theory and Determination of Confidence Intervals—Rating Method Data," *Journal of Mathematical Psychology*, 6, 487–496.

Gatsonis, C. A. (1995), "Random Effects Models for Diagnostic Accuracy Data," *Academic Radiology*, 2, 514–521.

Green, D. M., and Swets, J. A. (1996), *Signal Detection Theory and Psychophysics*, New York: Wiley.

Hanley, J. A. (1989), "Receiver Operating Characteristic (ROC) Methodology: The State of the Art," *Clinical Reviews in Diagnostic Imaging*, 29, 307–335.

——— (1998), "Receiver Operating Characteristic (ROC) Curves," *Encyclopedia of Biostatistics*, 5, 3738–3745.

Hanley, J. A., and McNeil, B. J. (1982), "The Meaning and Use of the Area Under a Receiver Operating Characteristic (ROC) Curve," *Radiology*, 143, 29–36.

Metz, C. E. (1986), "ROC Methodology in Radiologic Imaging," *Investigative Radiology*, 21, 720–733.

Metz, C. E., Herman, B. A., and Shen, J-H. (1998), "Maximum Likelihood Estimation of Receiver Operating Characteristic (ROC) Curves From Continuously Distributed Data," *Statistics in Medicine*, 17, 1033–1053.

Obuchowski, N. A. (1995), "Multireader, Multimodality Receiver Operating Characteristic Studies: Hypothesis Testing and Sample Size Estimation Using an Analysis of Variance Approach With Dependent Observations," *Academic Radiology*, 2, 522–529.

Pepe, M. S. (1997), "A Regression Modelling Framework for Receiver Operating Characteristic Curves in Medical Diagnostic Testing," *Biometrika*, 84(3), 595–608.

——— (1998), "Three Approaches to Regression Analysis of Receiver Operating Characteristic Curves for Continuous Test Results," *Biometrics*, 54, 124–135.

Swets, J. A., and Pickett, R. M. (1982), *Evaluation of Diagnostic Systems. Methods From Signal Detection Theory*, New York: Academic Press.

Thompson, M. L., and Zucchini, W. (1989), "On the Statistical Analysis of ROC Curves," *Statistics in Medicine*, 8, 1277–1290.

Toledano, A., and Gatsonis, C. A. (1995), "Regression Analysis of Correlated Receiver Operating Characteristic Data," *Academic Radiology*, 2, S30–S36.

Tosteson, A., and Begg, C. B. (1988), "A General Regression Methodology for ROC Curve Estimation," *Medical Decision Making*, 8, 204–215.

Wieand, S., Gail, M. H., James, B. R., and James, K. L. (1989), "A Family of Nonparametric Statistics for Comparing Diagnostic Markers With Paired or Unpaired Data," *Biometrika*, 76, 585–592.

Zweig, M. H., and Campbell, G. (1993), "Receiver-Operating Characteristic (ROC) Plots: A Fundamental Evaluation Tool in Clinical Medicine," *Clinical Chemistry*, 39, 561–577.

The Randomized Clinical Trial

David P. Harrington

1. INTRODUCTION

The randomized clinical trial is among the most important methodological tools in biostatistics. Some have conjectured that it could be the most significant advance in scientific medicine in the 20th century (Smith 1998). In the statistical community, it is the method of choice for controlling confounding in medical studies where two or more interventions are compared and where the choice of intervention by either the subject or the study investigator could lead to selection bias. Among physicians, it is central to evidence-based medicine and is the major step along the way from treatment choice based on opinion and experience to that based on scientific finding.

Randomization in experimental designs is, of course, not limited to studies of human subjects. Fisher (1926, 1935) advocated it in the study of biological problems, and randomization became widely used in agricultural experiments. But randomized clinical trials in medicine have special features that set them apart from other experimental designs: the experimental units are human subjects, sometimes diagnosed with a potentially fatal disease, and the study subjects and their physicians cede treatment choice to a random allocation scheme. A methodologic device for statisticians supplants one of the central aspects of the patient–physician relationship—personal choice of treatment. This intrusion into the caregiving role in medicine will always spark controversy (Hellman and Hellman 1991).

It is impossible to estimate the frequency with which randomized trials have been used in medicine, but some estimate that the number of randomized trials runs into the hundreds of thousands (Chalmers 1998). The Cochrane Collaboration (1999) lists review teams in more than 50 diseases that maintain libraries of reviews of published trials, and that archive now indexes more than 250,000 trials.

Close relatives of the randomized clinical trial were conducted as early as 1747 (Lind 1753), when James Lind studied six strategies for treating scurvy on the British ship HMS Salisbury. Although Lind appears to not have assigned his treatments randomly, he recognized that bias can result when responses are correlated more with disease characteristics or environmental features than with a putatively successful treatment. He surely sought to avoid the shoals described by his contemporary,

David Harrington is Professor of Biostatistics, Dana-Farber Cancer Institute and Harvard School of Public Health, Boston, MA 02115. This research was partly supported by National Cancer Institute grant CA-39929.

Thomas Jefferson, in an 1807 letter to Dr. Caspar Wistar: "The patient, treated on the fashionable theory, sometimes gets well in spite of the medicine. The medicine therefore restored him, and the young doctor receives new courage to proceed in his bold experiments on the lives of his fellow creatures" (Jefferson 1984). Lind's study design attempted to hold constant the factors within the control of the investigator (other aspects of diet, the environment of the patient) while varying only the treatment. By today's standards, Lind's study was flawed for two reasons: He assigned only two patients to each treatment (he would not be the last investigator to conduct an underpowered study), and he apparently failed to recognize that unmeasured patient characteristics beyond his observation and control could influence results.

The modern randomized trial emerged in the mid-20th century. Some believe that the 1948 report in the *British Medical Journal* (Medical Research Council 1948) of a randomized trial comparing streptomycin with a no-treatment control is the first detailed published account of randomized treatment assignments. The use of randomization in that trial was due in large part to the pioneering work of Bradford Hill. In addition to controlling possible confounding, randomization apparently also provided an acceptable way to ration the small supply of streptomycin (Doll 1998). Within 20 years, randomization in medical research had become widely adopted.

In one of the delightful ironies of modern science, the randomized trial "adjusts" for both observed and unobserved heterogeneity in a controlled experiment by introducing chance variation into the study design. If interventions for patients are chosen by chance, then the law of large numbers implies that the average values of patient characteristics should be roughly equal in the intervention groups. Under the null hypothesis of no differential treatment effect, the assigned interventions should be unimportant labels attached to patients, and the null distribution of a statistic that compares any two of the groups should, in most cases, cluster around 0. It is a disarmingly simple but effective construction.

2. ADAPTIVE RANDOMIZATION AND SEQUENTIAL DESIGNS

The widespread use of the randomized trial has inspired methodologic research into the design and analysis of these trials, as well as into the ethical quandaries that arise. Not surprisingly, each line of research has influenced the other. Specific randomization strategies became an area of active research in the 1960s. Because an investigator cannot control the order in which patients are enrolled in a trial, the sophisticated randomization schemes used in agricultural experiments did not adapt well to clinical trials. Simple randomization, however, can lead to a lack of balance in the number of subjects assigned to, say, two treatments and, even more severely, to the distributions of important prognostic factors between the two treatment groups. Many remedies have been proposed. The simplest may be Zelen's (1974) use of randomly permuted blocks within predefined subgroups (strata) of patients. More complicated adaptive schemes have been proposed that use biased coin designs to adjust adaptively randomization fractions to nudge the distributions of treatments overall and within subgroups toward the desired allocation ratio (usually .5). Efron's (1971) design seems to have been the earliest of these; the method of minimization proposed by Pocock and Simon (1975) has become the most popular of these covariate-adaptive designs. Other, more controversial designs have addressed some of the ethical issues of randomization by adapting randomization probabilities to the observed history of patient responses. "Play the winner" designs (Wei and Durham 1983; Zelen 1969) favor the treatment arm showing the better outcome at the time of randomization, with the intent of providing participating physicians partial relief from the dilemma of continued randomization when therapeutic effects may be emerging. These response-adaptive designs have had limited practical impact largely because a physician with knowledge of the history of a trial has some information with which to predict the next treatment assignment and selectively enroll or withhold patients. A spirited discussion of a play the winner design in a controversial trial in the treatment of a respiratory disorder in infants was given by Ware (1989). A useful summary of randomization methods was given by Kalish and Begg (1985) and more prescriptive advice was provided by Lachin, Matts, and Wei (1988).

Unlike response-adaptive randomization methods, sequential designs that allow for early stopping of trials have become widely used. Armitage, McPherson, and Rowe (1969) provided conceptual and computational results for easy computation of experiment-wise significance levels when repeated significance tests are used to monitor a trial as groups of responses become available (see Armitage 1975 for a more complete account). Several versions of these group-sequential designs that maintain the specified experiment-wise type I error probability are now commonly used (Lan and DeMets 1983; O'Brien and Fleming 1979; Pocock 1977; Whitehead 1983). These designs have been extended to allow for early stopping to accept, as well as to reject, the null hypothesis (Emerson and Fleming 1989; Pampallona and Tsiatis 1994; Whitehead 1983). Jennison and Turnbull (1989) advocated the use of repeated confidence intervals with confidence coefficients adjusted for multiple "looks" at the data. Many phase III trials now have sequential designs used by data monitoring committees that have exclusive access to outcome data, so that treating physicians and study investigators are no longer aware of developing trends.

To date, sequential methods for clinical trials have been dominated by the frequentist perspective, though some have argued strongly that the Bayesian perspective should see more use (see, e.g., Freedman, Spiegelhalter, and Parmar 1994).

3. ANALYSIS ISSUES

Many methods of analysis are used with data from clinical trials, but one question is especially pertinent to randomized trials: how should a trial be analyzed when some subjects do not receive their assigned therapy? "Noncompliance" to randomized treatment assignment received widespread attention during the early trials of HIV disease (Robins and Greenland 1994), when substantial fractions of patients on those trials elected either additional self-medication or stopped therapy altogether, but the issue arises in nearly all diseases. In many cases the reason for treatment deviations may be beyond the control of a patient (e.g., inadequate drug supply, an intolerable side effect), but noncompliance has nevertheless become the technical term of choice. These discrepancies between the treatment assigned and that actually received challenge the very foundation of a randomized trial; they begin to "undo" the randomization when the reasons for those changes are correlated with response. More directly, randomization allows an investigator to conclude that differences in outcome between treatment groups were caused by the treatments themselves; noncompliance threatens that causal link. The data analyst faces a difficult choice: an analysis of all eligible randomized subjects according to treatment assigned (the intent-to-treat principle) provides an unbiased test of the null hypothesis of no treatment differences but produces a potentially biased estimate of treatment effects. An analysis of patients according to treatment received will very likely produce a test of the null hypothesis with distorted type I error and often still does not accurately estimate treatment differences. Robins and Tsiatis (1991) and others (Fischer-Lapp and Goetghebeur 1999) have studied structural statistical models, which, under some conditions, restore the causal interpretation of treatment differences in the presence of noncompliance and maintain type I error rates.

Problems encountered in the design and analysis of clinical trials have inspired a rich history of methodologic research in many areas of statistics, and there is certainly not the space here to discuss any of those results in detail. Survival analysis is widely used in clinical trials, and much of what is known about the proportional hazards regression model for right-censored data was originally investigated because of, and later applied to, data from clinical trials. Exact methods for large but sparse contingency tables are now routinely used in trials with outcomes measured on a categorical scale. The analysis of quality-of-life data presents difficult problems in nonignorable missingness that are still under active study. Recent results in the computation of Bayes procedures are just now spurring increased use of those methods. Vaccine and screening trials present special issues of design and inference that are far from solved.

4. ETHICAL ISSUES

Many difficult issues intersect the ethical and sociological problems that arise in medical research. When is it ethical to randomize? The classical ethical theory of clinical trials presupposes "clinical equipoise" on the question of which treatment in a trial is preferable, but it is far from clear how to operationalize that abstract notion of balance in clinical opinion. The ECMO trial in infants (Ware 1989) involved a therapy that many believed unethical to withhold; others felt just as strongly that the procedure's potential benefit had not been established in a properly controlled study. Peto's working definition (Peto and Baigent 1998) is perhaps the most practical: "A patient can be entered if, and only if, the responsible clinician is substantially uncertain which of the trial treatments would be most appropriate for that particular patient."

As scientific and commercial organizations expand internationally, clinical judgments are influenced by differences in culture or access to medical care, and equipoise does not always cross the membrane of national boundaries. Brennan's (1999) account (with a rejoinder in Levine 1999) of the debate on the use of a no-treatment control in a study of perinatal treatment for HIV in Africa, after an effective drug had been established in United States trials but before that treatment was widely used in developing countries, provides a fascinating case study and a platform for discussing pending changes to the Declaration of Helsinki (World Medical Association 1964).

Many randomized trials are too small, so that moderate treatment differences cannot be distinguished, at least with near certainty, from random fluctuation. Small trials result from a complex web of constraints, including physician or patient reluctance toward randomization, the increased cost of medical care on a clinical trial, and/or the lack of an available trial at a patient's treatment site. The U.S. National Cancer Institute estimates that approximately 5% of adults with cancer participate in clinical trials. These small trials lack precision, are of questionable value in generalizing estimates of treatment differences to the intended target population (Harrington 1999), and violate the implicit contract with study subjects, who believe that their participation will contribute to the answer of an important pending medical question. Important statistical research into methods for meta-analyses has helped but not cured this malady.

Successful trials of new therapeutic interventions may have substantial financial implications for pharmaceutical companies. Although many countries have codified regulations on good clinical practice, the specter of a conflict of interest between profit and the protection of patient rights still exists. A pharmaceutical company benefits greatly when a drug can be marketed as soon as possible after discovery (patients reap this benefit as well for effective treatments), but the quest for efficiency and speed may crowd patient interests, especially when treating physicians are reimbursed for the costs of placing patients on study.

Investigators and some medical journals have a propensity to publish trials that demonstrate treatment benefits, leaving the data from equally valuable negative trials in study archives. The failure to publish may itself be a violation of the implicit contract with a trial participant, and publication bias may distort even the most careful literature review by the conscientious clinician.

5. CONCLUSIONS

Despite its blemishes, the arguments that the randomized trial is the best experimental design for controlling for selection bias are compelling. In the Eastern Cooperative Oncology Group, where this author works as a statistician, some of the largest observed "treatment" differences have been between identical treatments in hypothetically identical populations treated on different trials, and no statistical models could explain those differences. To paraphrase Peto and Baigent (1998), large therapeutic advances are unlikely in major diseases such as cancer, HIV infection, and cardiovascular disease, and randomized trials are the only way to reliably confirm important incremental advances that can be distorted or lost altogether in nonrandomized studies. In cancer and HIV alone, randomized trials have established that for small enough breast cancer tumors, the far less disfiguring lumpectomy operation is as effective as complete surgical removal of the affected breast, that most operable breast cancers are best treated with postoperative chemotherapy, that many forms of pediatric leukemia are treatable, that the perinatal administration of antiretroviral agents in HIV-positive pregnant women reduces the risk of transmission of HIV disease to the newborn, and that combinations of antiretroviral drugs increase the latency period of the AIDS virus. Similar examples exist in diabetes, cardiovascular disease, and psychiatric disorders that have changed the practice of medicine dramatically.

There are several excellent texts covering methodologic, regulatory, and ethical issues in clinical trials. The best places to start discovering the richness of this subject are the books by Friedman, Furberg, and DeMets (1998), Piantadosi (1997), and Pocock (1983).

REFERENCES

Armitage, P. (1975), *Sequential Medical Trials* (2nd ed.), Oxford, U.K.: Blackwell.

Armitage, P., McPherson, K., and Rowe, B. C. (1969), "Repeated Significance Tests on Accumulating Data," *Journal of the Royal Statistical Society A*, 132, 235–244.

Brennan, T. A. (1999), "Proposed Revisions to the Declaration of Helsinki—Will They Weaken the Ethical Principles Underlying Human Research," *New England Journal of Medicine*, 341, 527–531.

Chalmers, I. (1998), "Unbiased, Relevant, and Reliable Assessments in Health Care," *British Journal of Medicine*, 317, 1167–1168.

Cochrane Collaboration (1999), http://www.update.software.com/ ccweb/cochrane/cc-broch.htm#CRG.

Doll, R. (1998), "Controlled Trials: the 1948 Watershed," *British Medical Journal*, 317, 1217–1220.

Efron, B. (1971), "Forcing a Sequential Experiment to be Balanced," *Biometrika*, 58, 403–417.

Emerson, S., and Fleming, T. R. (1989), "Symmetric Group Sequential Test Designs," *Biometrics*, 45, 905–923.

Fischer-Lapp, K., and Goetghebeur, E. (1999), "Practical Properties of Some Structural Mean Analyses of the Effect of Compliance in Randomized Trials," *Controlled Clinical Trials*, 20, 531–546.

Fisher, R. A. (1926), "The Arrangement of Field Experiments," *Journal of the Ministry of Agriculture*, 33, 503–513.

—— (1935), *The Design of Experiments*, Edinburgh: Oliver and Boyd.

Freedman, L. S., Spiegelhalter, D. J., and Parmar, M. K. B. (1994), "The What, Why and How of Bayesian Clinical Trials Monitoring," *Statistics in Medicine*, 1371–1384.

Friedman, L. M., Furberg, C. D., and DeMets, D. L. (1998), *Fundamentals of Clinical Trials* (3rd ed.), New York: Springer-Verlag.

Harrington, D. (1999), "The Tea Leaves of Small Trials," *Journal of Clinical Oncology*, 17, 1336–1338.

Hellman, S., and Hellman, D. S. (1991), "Of Mice But Not Men: Problems of the Randomized Clinical Trial," *New England Journal of Medicine*, 324, 1585–1589.

Jefferson, T. (1984), *Letters, 1760–1826*, ed. M. D. Peterson, New York: Library of America

Jennison, C., and Turnbull, B. W. (1989), "Interim Analyses: The Repeated Confidence Interval Approach," *Journal of the Royal Statistical Society*, Ser. B, 51, 305–361.

Kalish, L. A., and Begg, C. B. (1985), "Treatment Allocation Methods in Clinical Trials: A Review," *Statistics in Medicine*, 4, 129–144.

Lachin, J. M., Matts, J. P., and Wei, L. J. (1988), "Randomization in Clinical Trials: Conclusions and Recommendations," *Statistics in Medicine*, 9, 365–374.

Lan, K. K. G., and DeMets, D. L. (1983), "Discrete Sequential Boundaries for Clinical Trials," *Biometrika*, 70, 659–663.

Levine, R. J. (1999), "The Need to Revise the Declaration of Helsinki," *New England Journal of Medicine*, 341, 531–534.

Lind, J. (1753), "A Treatise of the Scurvy. Of Three Parts Containing an Inquiry into the Nature, Causes and Cure of that Disease," Edinburgh: Sands, Murray and Cochran.

Medical Research Council (1948), "Streptomycin Treatment of Pulmonary Tuberculosis," *British Medical Journal*, 2, 769–782.

O'Brien, P. C., and Fleming, T. R. (1979), "A Multiple Testing Procedure for Clinical Trials," *Biometrics*, 35, 549–556.

Pampallona, S., and Tsiatis, A. A. (1994), "Group Sequential Designs for One-Sided and Two-Sided Hypothesis Testing, With Provision for Early Stopping in Favor of the Null Hypothesis," *Journal of Statistical Planning and Inference*, 42, 19–35.

Peto, R., and Baigent, C. (1998), "Trials: The Next 50 Years," *British Medical Journal*, 317, 1170–1171.

Piantadosi, S. (1997), *Clinical Trials: A Methodologic Perspective*, New York: John Wiley & Sons.

Pocock, S. (1977), "Group Sequential Methods in the Design and Analysis of Clinical Trials," *Biometrika*, 64, 191–199.

—— (1983), *Clinical Trials: A Practical Approach*, New York: John Wiley & Sons.

Pocock, S., and Simon, R. (1975), "Sequential Treatment Assignment with Balancing for Prognostic Factors in the Controlled Clinical Trial," *Biometrics*, 31, 103–115.

Robins, J. M., and Greenland, S. (1994), "Adjusting for Differential Rates of Prophylaxis Therapy for PCP in High- Versus Low-dose AZT Arms in an AIDS Randomized Trial," *Journal of the American Statistical Association*, 89, 737–749.

Robins, J. M., and Tsiatis, A. A. (1991), "Correcting for Non-Compliance in Randomized Trials Using Structural Nested Failure-Time Models," *Communications in Statistics A—Theory and Methods*, 20, 2609–2631.

Smith, R. (1998), "Fifty Years of Randomised Controlled Trials," *British Medical Journal*, 317, 1166.

Ware, J. H. (1989), "Investigating Therapies of Potentially Great Benefit: ECMO (with discussion)," *Statistical Science*, 4, 298–340.

Wei, L. J., and Durham, S. (1978), "The Randomized Play-the-Winner Rule in Medical Trials," *Journal of the American Statistical Association*, 73, 840–843.

Whitehead, J. (1983), *The Design and Analysis of Sequential Medical Trials*, New York: Halstead Press.

World Medical Association (1964), "Declaration of Helsinki: Recommendations for Guiding Medical Doctors in Biomedical Research Involving Human Subjects," (revised 1975, 1983, 1989), Helsinki: World Medical Association.

Zelen, M. (1969), "Play the Winner Rule and the Controlled Clinical Trial," *Journal of the American Statistical Association*, 64, 131–146.

——— (1974), "The Randomization and Stratification of Patients to Clinical Trials," *Journal of Chronic Diseases*, 27, 365–375.

Some Contributions of Statistics to Environmental Epidemiology

Duncan C. Thomas

1. INTRODUCTION

The field of epidemiology has come to rely particularly heavily on statistical methods because of its observational nature and the widespread acceptance of a complex "web of causation" as its conceptual basis. As much of modern chronic disease epidemiology is oriented to the study of disease incidence data, regression models for binary, Poisson, and survival time data have figured prominently in epidemiologic applications. Thus logistic regression for case-control studies and Poisson and Cox regression models for cohort data have become standard tools in the epidemiologist's armamentarium (Breslow and Day 1980, 1987). In other areas, methods of longitudinal data analysis (Liang and Zeger 1986) have provided a similar unified framework for modeling changes in continuous outcomes over time, such as lung function measurements. Without these important statistical contributions, it is arguable that epidemiology would not have been able to progress so far in understanding diseases of complex etiology.

These tools are broadly applicable to the study of many different risk factors, including genetic and environmental factors and their interactions, although specialized techniques have been developed in these two subdisciplines. Environmental epidemiology can in turn be broadly defined as comprising everything that is not genetic (e.g., diet, lifestyle factors like smoking, the social milieu, medical exposures, infectious agents) or narrowly defined as focusing on the exogenous environment (e.g., air and water pollution, indoor radon, fallout from nuclear weapons development and testing, electromagnetic fields, pesticides). In this vignette I touch only briefly on statistical methods in standard risk factor epidemiology (see Thomas 1988, 1998 for more extensive treatments of that topic) and focus instead on some of the unique problems that arise in the latter case, owing to the geographically defined nature of many environmental exposures: (a) evaluations of disease clusters and the spatial distribution of disease in relation to possible environmental causes using Geographic Information Systems (GIS); (b) the design of studies involving geographical and temporal comparisons; (c) the control of measurement error; and (d) problems of multicollinearity.

Duncan C. Thomas is Professor and Verna R. Richter Chair in Cancer Research, Department of Preventive Medicine, University of Southern California, Los Angeles, CA 90033.

There has been an explosion in the availability of geographically defined data on health and exposures, and statistical methods to exploit such resources are still in their infancy.

2. HYPOTHESIS GENERATION USING GEOGRAPHIC DATA

Public health agencies often find themselves functioning in a reactive mode to public concerns about disease clusters or point source exposures. Although such observations are difficult to put in any statistical framework for formal inference because of their anecdotal nature, this is nonetheless one way by which hypotheses in environmental health are sometimes generated. Other hypotheses are generated by concerns about low-dose human exposures to substances that are toxic to animals (Ryan 2000) or by extrapolation to the general population of risks previously studied in highly exposed groups, such as workers exposed to pesticides or vinyl chloride. Although the investigation of specific disease clusters has seldom yielded clearcut or generalizable insights (Rothman 1990), mapping disease rates and ecologic regression techniques for explanatory variables have proven useful. Different environmental exposures are manifest over quite different scales of spatial variability, but in general one would expect disease rates to vary smoothly over space and to reflect the aggregate effect of many different factors, racial, and socioeconomic as well as environmental. This leads to a classic trade-off between geographic resolution and statistical stability. On the one hand, fine-scale mapping is desirable to adequately capture the effects of the many different factors that can influence disease rates, but on the other hand, the reliability of fine-scale mapping is limited by the frequency of the disease under study. Many different statistical measures, including the observed disease rates, their statistical significance, and various combinations thereof, have been used for constructing disease maps (Devine, Annest, Kirk, Holmgreen, and Emrich 1991; Pickle, Mason, Howard, Hoover, and Fraumeni 1987). The most promising development has been the use of empirical Bayes techniques for smoothing maps of local disease rates (Clayton and Kaldor 1987; Marshall 1991; Mollie and Richardson 1991), borrowing strength from nearby areas that presumably share at least some of the same determinants. However, the extent of spatial correlation may also vary locally due to omitted, misspecified, or imperfectly measured spatial covariates, so that joint modeling of the means and covariances would be helpful. Euclidean distance may not be the best metric for drawing such maps, so that factors such as meteorology, topography, and population density might suggest a more natural metric. For example, Le and Zidek (1992) described techniques for modeling air pollution based on a transformation of space to one with uniform spatial correlation, followed by back transformation for presentation of the final map.

Although an appropriately smoothed map can be useful in itself as a hypothesis-generating tool, even more useful are spatial regression techniques that can incorporate geographically defined explanatory data (Mugglin and Carlin 1998; Waller, Louis, and Carlin 1997). Such techniques were used by Richardson, Monfort, Green, Draper, and Muirhead (1995), for example, for testing a hypothesis suggested by Henshaw, Eatough, and Richardson (1990) concerning the risk of leukemia from indoor radon. They found consistent evidence of spatial clustering of leukemia rates over the 459 districts of England, Scotland, and Wales but no correlation of these rates with either radon or gamma radiation. Guttorp (2000) gave a review of spatial monitoring and prediction techniques.

3. ISSUES OF STUDY DESIGN

Because so many environmental exposures like air pollution are geographically determined, standard epidemiologic designs based on comparisons between individuals may have limited utility. For example, a case-control design with neighborhood controls (which might be the design of choice for studying individual risk factors) would be useless for studying outdoor air pollution, because there would be little, if any, variation in exposure within matched sets. Instead, environmental epidemiologists tend to rely on geographical and temporal comparisons, or on combinations of the two. For example, in air pollution research, chronic effects generally have been studied using "ecologic" designs that relate mean health indices for different cities to ambient pollution levels averaged over time. The major concern with such comparisons is the so-called "ecologic fallacy" (Greenland and Robins 1994) that effects seen at the group level may not reflect individual-level associations. In large part, this fallacy arises from uncontrolled confounding and can be addressed by measuring the relevant confounders at the individual level. This suggests a multilevel design (Navidi, Thomas, Stram, and Peters 1994; Prentice and Sheppard 1990), in which individuals are enrolled in several different communities, data on potential confounders and outcomes are obtained and controlled at the individual level, and the adjusted community means are then regressed on community ambient exposures at the ecologic level. This analysis then correctly allows for uncontrolled between-community residual variation, with the effective sample size being the number of communities, not the number of persons. In a variant of this approach (Prentice, personal communication), cross-sectional surveys are conducted in a large number of communities to obtain information on the joint distribution of exposures and confounders—but not disease outcomes—and the resulting means and covariances are then used as predictors in an ecologic regression analysis of population disease rates. Such designs have great potential for correlating separate data bases on health (e.g., SEER cancer registries) and exposure (e.g., the AIRS air pollution data base).

Acute effects have commonly been studied using a time-series design, in which short-term fluctuations in population disease rates have been correlated with short-term fluctuations in ambient pollution levels, after filtering out long-term cycles and trends (Samet, Zeger, and Berhane 1995; Schwartz 1994). A very promising extension is a parallel implementation of this approach in a nationwide sample of many U.S. cities (Dominici, Samet, Zeger, and Xu 2000), which is taking the results of separate time-series analyses in each city to develop an aggregate model for the variability in pollution mortality coefficients in relation to city-specific predictors, using spatial smoothing techniques.

In some cases, it may be possible to exploit all three levels of comparison— between communities, between individuals, and between times—in a single unified study design and analysis. This is currently being done, for example, in a longitudinal study of air pollution in school children from 12 southern California communities, where rates of change in annual lung function measurements are being related to time-varying and fixed individual covariates and community ambient pollution measurements using a three-level hierarchical mixed model (Gauderman et al. 2000). Similar analyses of acute effects on school absences are being conducted using multilevel mixed models for binary time series data. The latter are closely related to the case-crossover design (Maclure 1991), which compares the covariate values for an individual at the time of failure to the same person's covariate values at a sampled control time. However, subtleties are involved in applying this approach to environmental exposures where the exposure series is seen as fixed rather than random, which requires careful consideration of the choice of crossover times, particularly for events that lead to censoring (Lumley and Levy 1999; Navidi 1998). The major advantage of the case-crossover design over the time-series approach is the ability to incorporate individual time-varying covariates either as exposures or as confounders and modifiers of an ambient exposure effect.

4. EXPOSURE MEASUREMENT ERROR

Environmental exposures are notoriously difficult to measure accurately at the individual level. Personal dosimeters may not be available or may be expensive or cumbersome. Past exposures may have to be reconstructed using pathway models (Till and Meyer 1983). Exposures may be geographically defined, and local or personal variability may be difficult to assess. Even if measurements are available, they may be subject to great uncertainty. For all of these reasons, it is important to consider the influence of exposure errors on the epidemiologic associations.

Available methods are based largely on two different approaches to the problem, depending upon whether direct measurements are available for each person or whether exposures are estimated using some prediction model. Again, air pollution provides a good example to illustrate the issues. Although personal monitors are available for some air pollutants, it generally is not feasible to obtain direct measurements of personal exposure over an extended period for thousands of study subjects.

Central site monitors have been deployed in many communities to measure ambient pollution levels over time, but of course provide no information about personal exposures, which are influenced by such factors as time spent outdoors and residential characteristics like air exchange rates and indoor sources. Personal and "microenvironmental" measurements on subsamples can be used to build models for individual exposures that can then be evaluated using questionnaire data for all subjects in the main study. As a result, each subject can be characterized by a predicted exposure and a model-based estimate of the uncertainty of that prediction (Navidi and Lurman 1995). Such data might be expected to have a Berkson error structure (Berkson 1950), in which the individual's true exposures are distributed around their model predictions with some variability. On the other hand, direct measurements of personal exposures would be expected to have a classical error structure; that is, distributed around the true values with some measurement error. The implications of these two error structures is rather different (Thomas, Stram, and Dwyer 1993), with classical error having a general tendency to bias exposure-response relationships towards the null, whereas Berkson error does not (at least in linear models) but leads to some loss of power essentially because the exposure categories begin to overlap. There is an extensive literature on methods for correcting for exposure measurement error (Carroll, Ruppert, and Stefanski 1995; Fuller 1987). Dose-reconstruction models often contain elements of both types of error, since source measurements might have a classical error component while unmeasured factors would introduce Berkson error; as a result, individuals with similar histories might have correlated errors, a problem that has not received adequate attention in the statistical literature. Some of the most promising approaches involve a combination of exposure measurements and model predictions in the same analysis. For example, in an analysis of childhood leukemia, Bowman, Thomas, Jiang, and Peters (1999) used a physically based model to predict magnetic fields from detailed power line wiring configuration information, fitted to actual measurements in the children's homes, and then combined the predicted and measured fields to assign exposures to individuals. The predicted fields were generally found to be more strongly related to leukemia than the measurements themselves.

5. STATISTICAL ANALYSIS ISSUES

At the outset, I noted the profound impact on epidemiology generally of the statistical techniques for modeling event data (Breslow and Day 1980, 1987). A recurring theme in epidemiology, and especially environmental epidemiology, is the time-dependent nature of most exposures, requiring consideration of complex exposure time–response relationships, incorporating such features as latency and modification by age at exposure and age at risk, duration of exposure, and dose-rate effects. For example, very sophisticated models incorporating such time-dependent modifiers have been built for lung cancer risk from radon exposure in miners with extended and time-varying exposure histories that can be used for projecting risks from lifetime exposure to domestic radon (Lubin, Boice, and Edling 1994; NAS 1988). Similar considerations arise in constructing an exposure metric for time-dependent environmental exposures like air pollution, taking into account possible threshold and interactive effects. For example, Davidian and Gallant (1992) discussed nonlinear regression techniques for dose–response models with individual random-effects parameters estimated nonparametrically. (See also Thomas 1988, 1998 for reviews of descriptive approaches to this problem and Moolgavkar 1986 for a discussion of biologically based approaches in the context of cancer epidemiology.)

Many environmental exposures occur as complex mixtures whose component effects may be difficult or impossible to separate, some of whose components may have synergistic effects. Again taking air pollution as the example, the major pollutants of concern often tend to be highly correlated in time and in space, due in part to their common sources and in part to atmospheric chemical reactions that convert one pollutant into another. Furthermore, some pollutants like particulate matter comprise many constituents with distinct size and chemical composition, which may have different health effects but are too highly correlated to be separated statistically. Simply putting all of the relevant pollutants (each of which may be measured in various ways or over different time periods) into a multiple regression model is unlikely to be rewarding, as the resulting coefficients will be too unstable. Single-pollutant models, on the other hand, are likely to be misleading because of uncontrolled confounding by other pollutants. To the extent that this problem of multicollinearity derives from common sources, and hence the relevant agent is the common source rather than specific pollutants, source apportionment methods (Schauer et al. 1996) may be helpful for deriving a set of less highly correlated predictors. Alternatively, one might need to consider a range of alternative two- or three-pollutant models that will illustrate the effects of a particular pollutant, adjusted for some of the other pollutants. Greenland (1993) provided a critique of variable selection methods for model building in epidemiology with multiple exposure factors and advocates empirical Bayes methods. Bayes model selection and model averaging techniques (George and Mc-Culloch 1993; Madigan and Raftery 1994; Raftery, Madigan, and Hoeting 1997) may prove useful for summarizing such analyses, taking into account our uncertainty as to which are the relevant causal factors. For example, Clyde and DeSimone-Sasinowska (1997) used Bayes model averaging to examine the association between particulate air pollution and daily mortality, averaging over the space of all possible models for adjusting for weather and other confounders. Shaddick, Wakefield, and Elliott (1998) applied similar methods to multipollutant data. In some cases it may not be possible to tease apart the effects of different constituents of a complex mixture—it may be the mixture itself that is the relevant risk factor. On the other hand, if the same component appears consistently in different studies with different levels of or correlations with co-pollutants, then that would tend to implicate it as the causal factor, as has been argued for particulate matter air pollution (Schwartz 1994).

6. CONCLUSIONS AND PRIORITIES FOR FUTURE DEVELOPMENT

This brief vignette has tried to touch on the key distinguishing aspects of environmental epidemiology—its reliance on ecologic and other nontraditional epidemiologic designs, the problem of exposure measurement error, and the problem of multicollinearity. Future exploratory studies are likely to rely ever more heavily on combinations of large data bases, possibly supplemented with specialized studies to fill in gaps. Methodologic research on the challenges such studies pose, both computational and conceptual, is badly needed. There are certainly other statistical problems in this field, many of which are shared with other epidemiologic specialties. However, greater attention by statisticians to these problems in particular would certainly help advance the field.

REFERENCES

Berkson, J. (1950), "Are There Two Regressions?," *Journal of the American Statistical Association*, 45, 164–180.

Bowman, J., Thomas, D., Jiang, F., and Peters, J. (1999), "Residential Magnetic Fields Predicted From Wiring Configurations: I. Exposure Model," *Bioelectromagnetics*, 20, 399–413.

Breslow, N. E., and Day, N. E. (1980), *Statistical Methods in Cancer Research: I. The Analysis of Case-Control Studies*, Lyon: IARC.

——— (1987), *Statistical Methods in Cancer Research: II. The Design and Analysis of Cohort Studies*, Lyon: IARC.

Carroll, R. J., Ruppert, D., and Stefanski, L. A. (1995), "Measurement Error in Nonlinear Models," London: Chapman and Hall.

Clayton, D., and Kaldor, J. (1987), "Empirical Bayes Estimates of Age-Standardized Relative Risks for Use in Disease Mapping," *Biometrics*, 43, 671–681.

Clyde, M., and DeSimone-Sasinowska, H. (1997), "Accounting for Model Uncertainty in Regression Models: Particulate Matter and Mortality in Birmingham, Alabama," Institute of Statistical and Decision Sciences, Duke University.

Davidian, M., and Gallant, A. R. (1992), "Smooth Nonparametric Maximum Likelihood Estimation for Population Pharmacokinetics, With Application to Quinidine," *Journal of Pharmacokinetics and Biopharmaceutics*, 20, 529–556.

Devine, O. J., Annest, J. L., Kirk, M. L., Holmgreen, P., and Emrich, S. S. (1991), *Injury Mortality Atlas of the United States 1979–1987*, Atlanta: Centers for Disease Control.

Dominici, F., Samet, J., and Zeger, S. G. (2000), "Combining Evidence on Air Pollution and Daily Mortality From the Largest 20 U.S. Cities: A Hierarchical Modeling Strategy," *Journal of the Royal Statistical Society*, Ser. A, 163, 263–284.

Fuller, W. (1987), *Measurement Error Models*, New York: Wiley.

Gauderman, W. J., McConnell, R., Gilliland, F., London, S., Thomas, D., Avol, E., Berhane, K., Rappaport, E. B., Lurman, F., Margolis, H. G., and Peters, J. (2000), "Association Between Air Pollution and Lung Function Growth in Southern California Children," *American Journal of Respiratory and Critical Care Medicine*, 162, 1383–1390.

George, E. I., and McCulloch, R. E. (1993), "Variable Selection via Gibbs Sampling," *Journal of the American Statistical Association*, 88, 881–889.

Greenland, S. (1993), "Methods for Epidemiologic Analyses of Multiple Exposures: A Review and Comparative Study of Maximum Likelihood, Preliminary Testing, and Empirical-Bayes Regression," *Statistics in Medicine*, 12, 717–736.

Greenland, S., and Robins, J. (1994), "Invited Commentary: Ecologic Studies—Biases, Misconceptions, and Counterexamples," *American Journal of Epidemiology*, 139, 747–760.

Guttorp, P. (2000), "Environmental Statistics," *Journal of the American Statistical Association*, 95, 289–292.

Henshaw, D. L., Eatough, J. P., and Richardson, R. B. (1990), "Radon as a Causative Factor in Induction of Myeloid Leukaemia and Other Cancers," *Lancet*, 335, 1008–1012.

Le, N. D., and Zidek, J. V. (1992), "Interpolation With Uncertain Spatial Covariances: A Bayesian Alternative to Kriging," *Journal of Multivariate Analysis*, 43, 351–374.

Liang, K. Y., and Zeger, S. L. (1986), "Longitudinal Data Analysis Using Generalized Linear Models," *Biometrika*, 73, 13–22.

Lubin, J. H., Boice, J. D. J., and Edling, C. (1994), *Radon and Lung Cancer Risk: A Joint Analysis of 11 Underground Miners Studies*, Bethesda, MD: U.S. Department of Health and Human Services.

Lumley, T., and Levy, D. (1999), "Bias in the Case-Crossover Design: Implications for Studies of Air Pollution," unpublished manuscript.

Maclure, M. (1991), "The Case-Crossover Design: A Method for Studying Transient Effects on the Risk of Acute Events," *American Journal of Epidemiology*, 133, 144–153.

Madigan, D., and Raftery, A. E. (1994), "Model Selection and Accounting for Model Uncertainty in Graphical Models Using Occam's Window," *Journal of the American Statistical Association*, 89, 1335–1346.

Marshall, R. (1991), "Mapping Disease and Mortality Rates Using Empirical Bayes Estimators," *Statistics in Medicine*, 40, 283–294.

Mollie, A., and Richardson, S. (1991), "Empirical Bayes Estimates of Cancer Mortality Rates Using Spatial Models," *Statistics in Medicine*, 10, 95–112.

Moolgavkar, S. H. (1986), "Carcinogenesis Modeling: From Molecular Biology to Epidemiology," *Annual Review of Public Health*, 7, 151–169.

Mugglin, A., and Carlin, B. P. (1998), "Hierarchical Modeling in Geographic Information Systems: Population Interpolation Over Incompatible Zones," *Journal of Agricultural, Biological, and Environmental Statistics*, 3, 111–130.

NAS, Committee on the Biological Effects of Ionizing Radiations, (1988), *Health Risks of Radon and Other Internally Deposited Alpha-Emitters (BEIR IV)*, Washington, DC: National Academy Press.

Navidi, W. (1998), "Bidirectional Case-Crossover Designs for Exposures With Time Trends," *Biometrics*, 54, 596–605.

Navidi, W., and Lurman, F. (1995), "Measurement Error in Air Pollution Exposure Assessment," *Journal of Exposure Analysis and Environmental Epidemiology*, 5, 111–124.

Navidi, W., Thomas, D., Stram, D., and Peters, J. (1994), "Design and Analysis of Multilevel Analytic Studies With Applications to a Study of Air Pollution," *Environmental Health Perspectives*, 102, 25–32.

Pickle, L. W., Mason, T. J., Howard, N., Hoover, R., and Fraumeni, J. F. J. (1987), *Atlas of Cancer Mortality Among Whites, 1950–1980*, Washington, DC: U.S. Government Printing Office.

Prentice, R. L., and Sheppard, L. (1990), "Dietary Fat and Cancer: Consistency of the Epidemiologic Data, and Disease Prevention That May Follow from a Practical Reduction in Fat Consumption," *Cancer Causes & Control*, 1, 81–97.

Raftery, A. E., Madigan, D., and Hoeting, J. A. (1997), "Bayesian Model Averaging for Linear Regression Models," *Journal of the American Statistical Association*, 92, 179–191.

Richardson, S., Monfort, C., Green, M., Draper, G., and Muirhead, C. (1995), "Spatial Variation of Natural Radiation and Childhood Leukaemia Incidence in Britain," *Statistics in Medicine*, 14, 2487–2501.

Rothman, K. J. (1990), "A Sobering Start for the Cluster Busters Conference," *American Journal of Epidemiology*, 132 (Suppl), S6–13.

Ryan, L. (2000), "Statistical Issues in Toxicology," *Journal of the American Statistical Association*, 95, 304–308.

Samet, J., Zeger, S., and Berhane, K. (1995), *The Association of Mortality and Particulate Air Pollution: The Phase I Report of the Particulate Epidemiology Evaluation Project*, Boston: Health Effects Institute.

Schauer, J. J., Rogge, W. F., Hildemann, L. M., Mazurek, M. A., Cass, G. R., and Simoneit, B. R. T. (1996), "Source Apportionment of Airborne Particulate Matter Using Organic Compounds as Tracers," *Atmospheric Environment*, 30, 3837–3855.

Schwartz, J. (1994), "Air Pollution and Daily Mortality: A Review and Meta-Analysis," *Environmental Research*, 64, 36–52.

Shaddick, G., Wakefield, J., and Elliott, P. (1998), "Estimating the Association Between Air Pollution and Daily Mortality Using Bayesian Model Averaging (abstract)," *Epidemiology*, 9, S-60.

Thomas, D. C. (1988), "Exposure Time–Response Relationships With Applications to Cancer Epidemiology," *Annual Review of Public Health*, 9, 451–482.

—— (1998), "New Approaches to the Analysis of Cohort Studies," *Epidemiologic Reviews*, 14, 122–134.

Thomas, D., Stram, D., and Dwyer, J. (1993), "Exposure Measurement Error: Influence on Exposure–Disease Relationships and Methods of Correction," *Annual Review of Public Health*, 14, 69–93.

Till, J. E., and Meyer, H. R. (1983), *Radiological Assessment*, Washington, DC: U.S. Nuclear Regulatory Commission.

Waller, L. A., Louis, T. A., and Carlin, B. P. (1997), "Bayes Methods for Combining Disease and Exposure Data in Assessing Environmental Justice," *Environmental and Ecological Statistics*, 4, 267–281.

Challenges Facing Statistical Genetics

B. S. Weir

1. INTRODUCTION

The fields of statistics and genetics grew together during the 20th century, and each faced a period of tremendous growth as the century ended. At the beginning of the 21st century, the need for statistical interpretation of genetic data is greater than ever, and there is a corresponding opportunity for the development of new statistical methods. This vignette reflects a personal viewpoint and is colored by the recent increased interest in statistical genetics experienced at North Carolina State University. C. Clark Cockerham, a giant in the field who led the N.C. State program for 40 years, died 3 years before the century ended.

2. THE GENOME REVOLUTION

The former editor of *Science* (Abelson 1998) considered that the genome revolution will have as great an impact on society as the industrial and computer revolutions before it. Readers of this journal will be familiar with the impact of widely available computing power on statistical analyses. Students may now be called on to handle large datasets, replete with missing values, where earlier they had to be taught with examples of unrealistically small and balanced sets. They can perform computationally intensive procedures of resampling or permutation to avoid invoking asymptotic theories, and can generate complex posterior distributions by simulation.

The impact of the genome revolution, however, will affect us all. As a consequence of new technologies supporting the Human Genome Project, it is becoming possible to study all of the genes in an organism at once. The science of genetics in the last century dealt with a few genes at a time, and indeed the first geneticist looked at seven single-gene traits in the garden pea. Although Mendel did perform some three-factor analyses on his data, only very recently have complete catalogs of genes become available and both the possibility and the necessity for many-gene analyses arisen. Once the yeast *Saccharomyces cerevissiae* was sequenced, all 6,200 genes (or at least the potential coding regions) could be determined by searching the 12 million base-pair (bp) DNA sequence for features characteristic of genes. DNA subsequences specific for each of these genes were then constructed on a silicon chip and used to

B. S. Weir is Professor, Program in Statistical Genetics, North Carolina State University, Raleigh, NC 27695-8203. This research was supported in part by National Institutes of Health grant GM 45344.

determine which genes were being expressed (producing products) in yeast colonies subject to different environments. Parallel activity is now projected for the human genome, where maybe 100,000 genes are encoded by about 10% of the 3×10^9 bp genome. Only a fraction of these genes is expressed in any specific cell of the body, and that fraction will vary with disease status and environment. Mendel published data on seven genes with a few thousand observations from 84 crosses of peas. His human genetics successors will collect data from 100,000 genes in different tissues at different times from different people in different populations. These expression data, with their spatial and temporal components, will be vastly more extensive than the DNA sequence data that are now emerging. Although we can anticipate a summary of a person's DNA sequence being encrypted on a credit card-sized medium, at least that information will not vary from conception until death and will not vary within the person.

Determining the genetic basis for any character, with the eventual potential for improving economic traits or correcting human defects, will be a major task facing 21st century geneticists. Extracting signals from very noisy data and accounting for the substantial interactions will pose severe statistical problems. Although there are many other tasks assigned to statistical geneticists, that of relating genotype to phenotype provides a convenient path for describing past successes and predicting future problems.

3. LOCATING GENES FOR DISCRETE CHARACTERS

Assigning genes to chromosomal locations is ultimately a physical exercise, but much can done with statistical analysis. A very successful approach has been the detection and estimation of genetic linkage (Ott 1999). Consider two genes **A** and **B** that exist in a population with alternative allelic forms A, a and B, b. Each individual receives two copies of each gene, one from each parent, so some individuals are doubly heterozygous AB/ab because one parent transmitted the alleles A, B and the other transmitted the alleles a, b. The individual in turn may transmit a parental pair AB or ab or a recombinant pair Ab or aB. The extent to which these two pair types are not equally probable reflects the degree of linkage between the genes. Completely linked genes do not allow the production of recombinant pairs of alleles, and unlinked genes have equal probabilities for parental and recombinant pairs. Genes on different chromosomes are unlinked. Although the relationship between the degree of linkage and amount of physical separation on a chromosome is complex, it is generally monotonic, so that evidence for strong linkage suggests close proximity. Linkage studies are conducted within family pedigrees, and the analysis is made complicated by the many dependencies among family members. The elegant algorithm of Elston and Stewart (1971) provided a method for the recursive computation of likelihoods on pedigrees and gave the basis for computer programs and many subsequent successful linkage analyses. The use of likelihood methods for genetic data has been advanced substantially by the many works of E. A. Thompson, whose 1986 book concentrated on human pedigree analyses.

Linkage analysis rests on the availability of genetic markers; that is, easily scored heritable units with known chromosomal locations. Statisticians have been substantially involved in developing methods for detecting and estimating linkage from data collected in a pedigree framework. The task is not a trivial one, however, as the case of Huntington's disease shows. Initial evidence for linkage to a marker on human chromosome 4 was published in 1983 (Gusella et al. 1983), but not until 1993 was a more precise localization of the gene made (Huntington's Disease Collaborative Research Group 1993).

Part of the problem with linkage studies for fine-scale mapping of human disease genes is that the proportion of recombinant allele pairs for genes closer together than about 1 million bp is less than 1%, so that these pairs are unlikely to be seen in family pedigrees of a few hundred people or less. Although there will be strong evidence for linkage, it will not be possible to provide reliable estimates for recombination frequencies lower than .01. A population-based sampling framework has thus been adopted as an alternative to a pedigree-based one. In essence, the data are examined for evidence of the effects of recombination during all of the previous generations in which the disease has been present in the population. As with all statistical genetic endeavors, this approach requires specification of an evolutionary model.

Population-association approaches to gene location have attracted attention from the pharmaceutical industry. In April 1999, a consortium was announced to develop 300,000 genetic markers for mapping human disease genes. This means that a marker will be sought about every 10,000 bp, and adjacent markers will have an average recombination rate of about .0001. The technology to enable scoring of so many markers is not yet completely in place, and the statistical problems inherent in addressing such data are formidable. Certainly, data-mining techniques will be used in place of traditional single-marker analyses.

4. LOCATING GENES FOR CONTINUOUS CHARACTERS

As difficult as locating genes affecting discrete genes is, at least the dependent variable, disease status, is assumed known without error. With some exceptions, such as disease-resistance genes, most economic traits are continuous and have an error component. Similarly, susceptibility or risk factors for human disease are generally continuous. These characters may be thought to have a component influenced by a small number of genes whose location is sought, but also to be influenced by many other genes and by environmental factors. It is no longer possible to infer the trait genotype from the phenotype of an individual. This is unlike the case for Huntington's disease, where even before the gene was located, it was known that someone with the disease must have inherited a "disease" allele from at least one parent.

The first published effort to locate genes affecting continuous traits was by Sax in 1923. He used seed coat color as a marker for seed size in beans. In work that anticipated much current activity, Thoday (1961) exploited the simplicities following the crosses between inbred lines of *Drosophila* selected in opposite directions for the trait of interest. For most of the 20th century, however, the statistical genetic analysis of continuous traits was less concerned with identifying the genes than with describing components of trait variance. Similarity in trait values of individuals known to be unrelated would reflect environmental factors, whereas similarity of trait values in unilineal relatives (e.g., half sibs) would reflect the effects of alleles acting singly: the additive effects. Trait values for bilinear relatives (e.g., full sibs) allowed inferences to be made for the dominance effects of alleles acting in pairs. Between-gene interactions, known as epistasis, could also be estimated from the covariances of relatives (Cockerham 1954). The genetic architecture of complex traits thus could be expressed in terms of components of variance even though the underlying physical basis was not known. This rich theory now available for quantitative genetics stems from the monumental paper of Fisher (1918) in which he laid out the partitioning of variance in work that "presumably led to the analysis of variance" (Moran and Smith 1966). Fisher's paper also was crucial in reconciling the opposing views of the Mendelians and biometricians that featured prominently in the statistical genetic literature at the beginning of the 20th century. Quantitative genetic theory has been sufficient to allow prediction of genetic gain from selection schemes proposed for domestic species and was also useful in describing natural populations. The most comprehensive review is that of Lynch and Walsh (1998).

Genetic engineering is based on the concept of altering genes by direct manipulation, instead of by the gradual accumulation of desirable genotypes through choice of the best individuals in selection schemes, and this requires detailed knowledge of gene location. Early work in such quantitative trait locus (QTL) location simply compared trait means in classes distinguished by genetic markers. Differences in means implied an association between marker and trait genes, and possible physical proximity. The confounding of the magnitudes of trait gene effects and marker gene location could be removed by considering intervals between pairs of markers, and the effects of QTLs outside a specific interval could then be removed by partial regression on other markers. This interval mapping methodology has now progressed to allow the location of pairs of QTLs and estimation of their effects and interactions (Kao, Zeng, and Teasdale 1999). Extension to many QTLs and to a wide range of individual sampling schemes lies ahead. Determining the direct and epistatic effects of all genes affecting a continuous trait is beyond current statistical methodology, and substantial experimental design issues remain to be faced.

Design issues also arise in the interpretation of expression array data. The possibility of screening all genes to determine which ones are being expressed has been mentioned, but current technology produces data that are continuous in nature, and levels of expression are revealed by light intensities in hybridization reactions. Local variation in intensities within an array needs to be factored into comparisons between arrays (i.e., across environments or genomes). The attention beginning to be paid to how and when genes act (i.e., functional genomics) has emerged as a new data-rich field.

5. SEQUENCE ANALYSES

The initial focus of the Human Genome Project has been the production of DNA sequence information. Prior to this coordinated effort to determine the complete sequence of man and five model organisms (a bacterium, *Escherichia coli*; a yeast, *Saccharomyces cerevisiae*; a worm, *Cenorhabditis elegans*; a fruit fly, *Drosophila melanogaster*; and the laboratory mouse *Mus musculus*), DNA sequences tended to be determined for specific regions or genes in various species. The field of comparative genomics is building on the availability of homologous sequences—those with shared ancestry—in different species. Statistical comparisons have already been made to demonstrate the variation in rates of change in different evolutionary lineages and to examine rates for pairs of genes (e.g., Muse and Gaut 1997), and more extensive analyses will be needed as sequence data accumulate.

Although DNA sequences have been termed the "blueprint of life," it is the physical structure of proteins that determines function. A great deal of statistical modeling has been undertaken to relate DNA sequence features to protein structure. There appear to be selective constraints to preserve protein structure, and this must affect mutational changes in the underlying DNA sequences. It is not easy to incorporate structural features into evolutionary models, and statistical work (e.g., Thorne, Goldman, and Jones 1996) will also be challenged by more data.

6. CONCLUSION

Genetics is having an impact on almost all aspects of life, with obvious examples in medicine, agriculture, and forensic science. Along with the burst in the science of genetics has come a burst in the volume of genetic data, and a consequent need for new statistical methods. The whole-genome nature of genetic data in the 21st century means that the single-gene analyses of the twentieth century (e.g., Weir 1998) will need to be extended.

REFERENCES

Abelson, P. H. (1998), "A Third Technological Revolution," *Science*, 279, 2019.

Cockerham, C. C. (1954), "An Extension of the Concept of Partitioning Hereditary Variance for Analysis of Covariances Among Relatives When Epistasis is Present," *Genetics*, 39, 859–882.

Elston, R. C., and Stewart, J. (1971), "A General Model for the Analysis of Pedigree Data," *Human Heredity*, 21, 523–542.

Fisher, R. A. (1918), "The Correlation Between Relatives on the Supposition of Mendelian Inheritance," *Transactions of the Royal Society of Edinburgh*, 52, 399–433.

Gusella, J., Wexler, N. S., Conneally, P. M., Naylor, S. L., Anderson, M. A., Tanzi, R. E., Watkins, P. C., Ottina, K., Wallace, M. R., Sakaguchi, A. Y., Young, A. B., Shoulson, I., Bonilla, E., and Martin, J. B. (1983), "A Polymorphic DNA Marker Genetically Linked to Huntington's Disease," *Nature*, 306, 234–238.

Huntington's Disease Collaborative Research Group (1993), "A Novel Gene Containing a Trinucleotide Repeat That is Expanded and Unstable on Huntington's Disease Chromosomes," *Cell*, 72, 971–983.

Kao, C. H., Zeng, Z. B., and Teasdale, R. D. (1999), "Multiple Interval Mapping for Quantitative Trait Loci," *Genetics*, 152, 1203–1216.

Lynch, M., and Walsh, B. (1998), *Genetics and Analysis of Quantitative Traits*, Sunderland, MA: Sinauer.

Moran, P. A. P., and Smith, C. A. B. (1966), Commentary on "Correlation Between Relatives on the Supposition of Mendelian Inheritance," by R. A. Fisher, Galton Laboratory, University College London, Eugenics Laboratory Memoirs 41.

Muse, S. V., and Gaut, B. S. (1997), "Comparing Patterns of Nucleotide Substitution Rates Among Chloroplast Loci Using the Relative Ratio Test," *Genetics*, 146, 393–399.

Ott, J. (1999), *Analysis of Human Genetic Linkage* (3rd ed.), Baltimore: Johns Hopkins University Press.

Sax, K. (1923), "The Association of Size Differences With Seed-Coat Patterns and Pigmentation in *Phaseolus Vulgaris*," *Genetics*, 8, 552–560.

Thoday, J. M. (1961), "Location of Polygenes," *Nature*, 191, 368–370.

Thompson, E. A. (1986), *Pedigree Analysis in Human Genetics*, Baltimore: Johns Hopkins University Press.

Thorne, J. L., Goldman, N., and Jones, D. T. (1996), "Combining Protein Evolution and Secondary Structure," *Molecular Biology and Evolution*, 13, 666–673.

Weir, B. S. (1998), *Genetic Data Analysis II*, Sunderland, MA: Sinauer.

Computational Molecular Biology

Wing Hung Wong

1. INTRODUCTION

Molecular biology is one of the most important scientific frontiers in the second half of the 20th century. During this period, the basic principles of how genetic information is encoded in the DNA and how this information is used to direct the function of a cell were worked out at the molecular level, and methods were developed to clone, to sequence, and to amplify DNA. As a result, a large amount of biological sequence information has been generated and deposited into publicly accessible data bases. Starting from fewer than 1 million base pairs (bp) in 1982, the amount of DNA sequence data in GenBank has been growing exponentially and had reached 3.4 billion bp in August 1999. By 2003 there should be close to 20 billion bp in GenBank, including the complete DNA sequence of the human genome. Parallel to the growth of DNA sequence data is the rapid accumulation of data on the three-dimensional structure of biopolymers such as proteins and RNAs. Such structural information, obtained through time-consuming X-ray crystallography or nuclear magnetic resonance spectroscopy experiments, is key to our understanding of how proteins perform their functions in the various cellular processes.

The phenomenal growth of DNA sequence data is underpinned by a fundamental shift in the way such data are produced. Although individual investigators are still cloning and sequencing specific genes related to particular biological questions, the bulk of sequence data is currently coming from a number of high-throughput genome centers. Thus the production of data is decoupled from the biological questions under investigation. The data are produced and deposited immediately for use by researchers worldwide to answer a multitude of questions. In the next few years, such high-throughput data production will spread from genome research to other areas of biological research. The availability of massive amounts of fundamental data and the need to extract the embedded information by computational and analytical means has spurred the development of computational molecular biology and bioinformatics. Here I briefly review selected recent progress and discuss some opportunities for statistical researchers.

Wing Hung Wong is Professor of Statistics, University of California, Los Angeles, CA. This work is partially supported by National Science Foundation grant DMS-9703918. The author is grateful to David Haussler and to Jun Liu for useful comments.

2. SEQUENCE ALIGNMENT AND ANALYSIS

Many components of basic cellular processes are highly conserved, although their uses and regulation have diverged greatly among extant organisms. Sequence alignment is the basic tool that allows us to detect these conserved components. To align two protein sequences, similarity scores are assigned to all possible pairs of amino acids, and the sequences are aligned to each other so as to maximize the sum total of scores in the sequence of pairings induced by the alignment. Table 1 displays a part of the alignment of a ribosomal protein from a group of anciently divergent organisms. The alignment not only establishes the evolutionary linkage of these proteins, it also suggests a candidate RNA-binding motif in these proteins. The dashes indicate gaps in the alignment that are allowed at the cost of an additive penalty to the total score. Because of the gaps, the space of all possible alignments of two sequences is very large. The beginning of computational molecular biology can be traced to the development of dynamic programming-based algorithms for the solution of the optimal global and local alignment problem (Needleman and Wunsch 1970; Smith and Waterman 1981). Later, "word-based" alignment algorithms that may give suboptimal solutions but are extremely fast were introduced for large database searches (FASTA by Pearson and Lipman 1988; BLAST by Altschul, Gish, Miller, Myers, and Lipman 1990). The interpretation of alignment scores was aided by the derivation of the asymptotic distribution of ungapped alignment scores (Karlin and Altschul 1990). These classic results have by now become indispensable tools for all biomedical researchers in their dealing with molecular sequence data. For the computational biologist, sequence alignment often serves as a key step in the analysis of protein families, genomes, and phylogenies.

Table 1. A Part of the Alignment of Four Ribosomal Proteins

```
VITPEEIDELAKDKRRAKKLANSIDFFIAEAPLMPEIGRKLGPVLGPRGKIPQP-IP-P LADPKPFIDRLRN-SVK

VGMEDLADQIKKG-------EMNFDVVIASPDAMRVVG-QLGQVLGPRGLMPNPKVGTV TPNVAEAVKNAKAGQVR

VGGDELINKILKEE------WTDFDVAIATPEMMPKVA-KLGRILGPRGLMPSPKTGTV TTNVEQAIKDAKRGRVE

MSVDDLKKLNKNKK-LIKKLSKKYNAFIASEVLIKQVPRLLGPQLSKAGKFPTPVSH-- NDDLYGKVTDVRS-TIK

                            **************
```

NOTE: First row: protein RPL1p from the archaeon *A. fulgidus*. Second and third rows: RPL1 from the bacteria *E. coli* and *A. aeolicus*. Fourth row: RPL1B from the eukaryote budding yeast. The *'s mark a candidate RNA binding motif identified from this alignment.

Probabilistic and statistical reasoning has played a central role in recent advances in this area. In profile analysis (Gribskov, McLachlan, and Eisenberg 1987), a statistical model was constructed based on prealigned multiple sequences to characterize the regularity of protein sequence family and to increase the sensitivity of searches. Later, Lawrence et al. (1993) used a block-based product multinomial model and Bayesian Monte Carlo computation to perform multiple alignment and to sample and detect subtle sequence motifs, and Krogh, Brown, Mian, Sjolander, and Haussler (1994) developed hidden Markov models (HMMs) for protein family with multiple sequences. (HMM models were used earlier for DNA sequence analysis in Churchill 1992.) A notable feature of the recent approaches is that the multiple alignment information is regarded as missing data to be inferred together with other parameters of the statistical model. In other words, there is a parametric statistical model (e.g., a Markov chain) that specifies a distribution on the space of multiply aligned sequences. However, only the sequences are available as data; the information on how they should be aligned to each other is not available. Advanced statistical modeling and computation techniques such as the EM algorithm and Markov chain Monte Carlo are typically used for simultaneous model estimation and multiple alignment (for further review see Durbin, Eddy, Krogh, and Mitchison 1998; Liu, Neuwald, and Lawrence 1999; Waterman 1995). The statistical model obtained can be used to score how well a new query sequence fits the model. It can also be used for database searches. In recent comparisons of the ability of alignment and search tools to detect distantly related proteins, statistical model-based search methods are found to provide substantial improvement over PSI-Blast, which is currently the most powerful non-model-based method (Park et al. 1998). In these comparisons, a database of structural classification of proteins provides a means for testing the performance of homology detection (Brenner, Chothia, and Hubbard 1998). These studies also found that sensitivity is greatly improved if one searches the database based on a family of close homologs to the query sequence rather than using just the query sequence itself. Such a family of homologs can be built up from iterative searches (Neuwald, Liu, Lipman, and Lawrence 1997). Clearly, the improvement is due to the fact that the initial multiply aligned sequences provide statistical information on the relative importance of the various parts of the query sequence.

3. PROTEIN STRUCTURE PREDICTION

Most genes code for protein products. Because the function of a protein is determined by its three-dimensional structure, the ability to predict the structure of a protein from its amino acid sequence is a fundamental and long-standing problem. The solution to this problem becomes more urgent with the advent of gene cloning technology that allows numerous genes in numerous organisms to be cloned and sequenced before their gene products have been studied. As an example, consider the recently sequenced genome of the nematode *C. elegans*. There are about 20,000 predicted coding regions in this 100 million bp genome. The structure or function of some of these can be predicted by homology modeling using the search tools discussed in Section 2. However, more than 50% of them turn out to have no sequence similarity outside of Nematoda. To predict protein structure from such "orphan" sequences, computational biologists have in recent years developed an "inverse folding" approach. This is based on the assumption that there is a small collection of "folds," perhaps several hundreds in number, that can be used to model the majority of protein domains in all organisms (Holm and Sander 1996). A significant fraction of these folds may already have been sampled in the current data base of experimentally determined structures. The problem of structure prediction is then reduced to the task of classifying, based on its primary sequence, the protein in question into one of these folding classes. Statistical reasoning is key to several approaches to this classification problem. For example, in the "threading" method (Byrant and Lawrence 1993; Jones, Taylor, and Thornton 1992; Sippl and Weitckus 1992), the statistical models are Gibbs-type models based on sets of "potentials" that parameterize the propensity for pairs of amino acid residues to be in close contact or to be separated by a given distance. These parameters were estimated by fitting to data derived from known structures. To access the compatibility of a query sequence to such a model, the sequence is aligned (threaded) to the positions in the structure so that its probability is maximized. In the method of three-dimensional profiling (Bowie, Luthy, and Eisenberg 1991), the local environmental features, such as degree of exposure to solvent molecule, are computed for each residue in the structure. Any residue in a structure is classified according to these environmental features into a number of classes. In this way, a three-dimensional structure is reduced to a one-dimensional sequence of environment classes (called a three-dimensional profile). The relative frequencies of the amino acids in a given environment class were estimated using a training set of structure data, and these were used to construct scores for the alignment of a query sequence to the three-dimensional profile. An attractive feature of this method is that the alignment step can be implemented efficiently using standard dynamic programming. Through these and other related algorithms, fold recognition methods have produced some successful predictions even for proteins that fall in the "twilight zone" of detectability by sequence comparison (Marchler-Bauer and Byrant 1997).

Instead of fold recognition, one may try to directly compute a protein's structure from its sequence based on biophysical considerations. This "protein folding" problem has been a grand challenge ever since Anfinsen's experiments more than 30 years ago demonstrated that, at least for some proteins, the sequence of a protein determines its folded conformation. Typically, one sets up an energy function based on considerations of known covalent bonding geometry, electrostatic, and van der Waals forces. Conformation of the protein can then be sampled either by numerically solving the corresponding Newton equation, or by Monte Carlo sampling of the corresponding Boltzmann distribution (for a review, see Frenkel and Smit 1996). This problem is of interest to computational statisticians interested in simulation and global optimization methodology, as the energy landscape here is much harder than those we are likely to encounter in most statistical models. If successful, direct folding certainly would give deeper insight than the inverse approach. Currently, however, best results are of only modest precision (3–4 angstrom) and are obtained by using drastically simplified energy functions to reduce the energy and computational complexity. For small proteins of sizes 30–70 residues, some promising prediction results have been reported recently (Cui, Chen, and Wong 1997; Srinivasan and Rose 1995; Yue and Dill 1996). Further progress on this fundamental problem may be realized through improvement of the energy functions, multi-level formulation and search strategy, and efficient usage of commodity cluster computing resources.

4. GENE EXPRESSION STUDIES

Although different cell types in a multicellular organism have the same genome, they can have drastically different shapes and structures because the expression levels of their genes can be very different. Within the same cell, tightly regulated gene expression is also essential for various processes such as proper response to intercellular signals, cell division, and cell differentiation. Traditionally, gene expression studies were carried out in a gene-by-gene manner, and the understanding of expression profile on a global (genome-wide) scale was often lacking. During the past few years, the development of DNA-array technology has provided the means to monitor the expression levels of a large number of genes simultaneously. In such "global" expression studies, messenger RNAs are extracted from the cell culture. Complementary DNAs (cDNA) are generated from the RNAs and are amplified, labeled, and then hybridized to a large array of DNA probes immobilized on a solid surface. The array is then scanned by a laser to obtain the fluorescent signal for each probe region. From the signal strengths of the probes from a particular gene, one can infer the expression level of the gene in the cell type under study. Currently, microarray technology (Schena, Shalon, Davis, and Brown 1995) can print thousands of cDNA spots on a microscope slide, and light-directed array technology can synthesize up to 400,000 oligonucleotide probes in a small glass chip, with each gene probed by a set of 20–40 probes (Lockhart et al. 1996).

These arrays will be invaluable in diverse areas ranging from yeast genetics to cancer research. In one example, the temporal program of gene expression accompanying the cellular response to an external condition, such as induced metabolic shift from fermentation to respiration in yeast (DeRisi, Iyer, and Brown 1997), is studied by expression profiling of the cells at multiple time points during the experimentally induced shift. In another example, genes involved in complex phenotypes may be identified by comparing expression profiles of animals with extreme phenotypes in experimental crosses. Numerous statistical issues arise in connection with these studies, including

- What is the intrinsic noise characteristics of the data generated by the various array technologies?
- How to compare signals corresponding to the same gene (or probe) across different experiments?
- How to cluster genes that show similar expression levels across different time points, or to identify genes that show differential expression in normal and cancerous cells?

Careful statistical investigation will be needed to establish the sensitivity of the techniques in measuring differences in expression levels, to estimate sample sizes in the planning of large scale experiments, and to design ways to account for the effects of data-mining and selection from the large number of genes being profiled.

The complete sequences of the yeast and nematode genomes, and the availability of arrays capable of monitoring gene expression on a genomic scale, offer powerful tools for dissecting the complex genetic control of early eukaryotic and metazoan life. There is an unusually rich opportunity to combine recently developed bioinformatic methods with the new experimental approaches. For example, recently Roth, Hughes, Estep, and Church (1998) proposed a novel strategy for finding DNA binding motifs in noncoding regions. Based on expression profile data, they identified clusters of yeast genes that may be coregulated in induced responses of the yeast cell to various treatments. For the genes in each cluster, they extracted the corresponding upstream regions (about 600 bp) from the complete yeast genome and then applied the Bayesian motif sampler (see Sec. 2) to identify candidate binding sites for gene-regulatory protein factors. Again, statistical reasoning will play a key role in this and other novel approaches to the investigation of eukaryotic gene regulation.

5. CONCLUSION

Despite its youth, computational molecular biology is now firmly established as an important part of cutting-edge biological research. This article has reviewed some recent developments, but leaves many other important topics untouched. Examples include physical mapping, sequence assembly, genome annotation, phylogenetic algorithms, and the analysis of EST sequences. There is also a vast literature on statistical genetic mapping that is reviewed separately by Weir in this series of vignettes. It is safe to predict that this field will continue its rapid growth and will provide a rich source of new scientific challenges. Statistical researchers who have core expertise in data analysis, modeling, and computation and who have acquired a good understanding of molecular genetics and cell biology will be well positioned to contribute to this exciting frontier of research.

REFERENCES

Altschul, S. F., Gish, W., Miller, W., Myers, E. W., and Lipman, D. (1990), "Basic Local Alignment Search Tool," *Journal of Molecular Biology*, 215, 403–410.

Bowie, J. U., Luthy, R., and Eisenberg, D. (1991), "A Method to Identify Protein Sequences That Fold into a Known Three-Dimensional Structure," *Science*, 253, 164–170.

Brenner, S. E., Chothia, C., and Hubbard, T. (1998), "Assessing Sequence Comparison Methods With Reliable Structurally Identified Distant Evolutionary Relationships," *Proceedings of the National Academy of Science*, 95, 6073–6078.

Bryant, S., and Lawrence, C. L. (1993), "An Empirical Energy Function for Threading Protein Sequence Through the Folding Motif," *Proteins*, 16, 92–112.

Churchill, G. A. (1992), "Hidden Markov Chains and the Analysis of Genome Structure," *Computers and Chemistry*, 16, 107–115.

Cui, Y., Chen, R. S., and Wong, W. H. (1997), "Protein Folding Simulation With Genetic Algorithm and Super-Secondary Structure Constraints," *Proteins*, 31, 247–257.

DeRisi, J. L., Iyer, V. R., and Brown, P. O. (1997), "Exploring the Metabolic and Genetic Control of Gene Expression on a Genomic Scale," *Science*, 278, 680–686.

Durbin, R., Eddy, S., Krogh, A., and Mitchison, G. (1998), *Biological Sequence Analysis*, Cambridge, U.K.: Cambridge University Press.

Frenkel, D., and Smit, B. (1996), *Understanding Molecular Simulation*, San Diego: Academic Press.

Gribskov, M., McLachlan, A. D., and Eisenberg, D. (1987), "Profile Analysis: Detection of Distantly Related Proteins," *Proceedings of the National Academy of Science*, 84, 4355–4358.

Holm, L., and Sander, C. (1996), "Mapping the Protein Universe," *Science*, 273, 595–602.

Jones, D. T., Taylor, W. R., and Thornton, J. M. (1992), "A New Approach to Protein Fold Recognition," *Nature*, 358, 86–89.

Karlin, S., and Altschul, S. F. (1990), "Methods for Assessing the Statistical Significance of Molecular Sequence Features by Using General Scoring Schemes," *Proceedings of the National Academy of Science*, 87, 2264–2268.

Krogh, A., Brown, M., Mian, I. S., Sjolander, K., and Haussler, D. (1994), "Hidden Markov Models in Computational Biology: Applications to Protein Modeling," *Journal of Molecular Biology*, 235, 1501–1531.

Lawrence, C. E., Altschul, S. F., Boguski, M. S., Liu, J. S., Neuwald, A. F., and Wootton, J. C. (1993), "Detecting Subtle Sequence Signals: A Gibbs Sampling Strategy for Multiple Alignment," *Science*, 262, 208–214.

Liu, J., Neuwald, A., and Lawrence, C. (1999), "Markovian Structures in Biological Sequence Alignments," *Journal of the American Statistical Association*, 94, 1–15.

Lockhart, D. L., Dong, H., Byrne, M. C., Follettie, M. T., Gallo, M. V., Chee, M. S., Mittmann, M., Wang, C., Kobayashi, M., Horton, H., and Brown, E. L. (1996), "Expression Monitoring by Hybridization to High-Density Oligonucleotide Arrays," *Nature Biotechnology*, 14, 1675–1680.

Marchler-Bauer, A., and Bryant, S. H. (1997), "A Measure of Success in Fold Recognition," *Trends in Biochemical Sciences*, 22, 236–240.

Needleman, S. B., and Wunsch, C. D. (1970), "A General Method Applicable to the Search for Similarities in the Amino Acid Sequence of Two Proteins," *Journal of Molecular Biology*, 48, 443–453.

Neuwald, A., Liu, J. S., Lipman, D. J., and Lawrence, C. E. (1997), "Extracting Protein Alignment Models From Sequence Databases," *Nucleic Acids Research*, 25, 1665–1677.

Park, J., Karplus, K., Barrett, C., Hughey, R., Haussler, D., Hubbard, T., and Chothia, C. (1998), "Sequence Comparisons Using Multiple Sequences Detect Three Times as Many Remote Homologues as Pairwise Methods," *Journal of Molecular Biology*, 284, 1201–1210.

Pearson, W. R., and Lipman, D. J. (1988), "Improved Tools for Biological Sequence Comparison," *Proceedings of the National Academy of Science*, 85, 2444–2448.

Roth, F. P., Hughes, J. D., Estep, P. W., and Church, G. W. (1998), "Finding DNA Regulatory Motifs Within Unaligned Noncoding Sequences Clustered by Whole-Genome mRNA Quantitation," *Nature Biotechnology*, 16, 939–945.

Schena, M., Shalon, D., Davis, R. W., and Brown, P. O. (1995), "Quantitative Monitoring of Gene Expression Patterns With a Complementary DNA Microarray," *Science*, 270, 467–470.

Sippl, M. J., and Weitckus, S. (1992), "Detection of Native Like Models for Amino Acid Sequences of Unknown Three-Dimensional Structure in a Data Base of Known Protein Conformations," *Proteins*, 13, 258–271.

Smith, T. H., and Waterman, M. S. (1981), "Identification of Common Subsequences," *Journal of Molecular Biology*, 147, 195–197.

Srinivasan, R., and Rose, G. (1995), "LINUS: A Hierarchic Procedure to Predict the Fold of a Protein," *Proteins*, 22, 81–99.

Waterman, M. S. (1995), *Introduction to Computational Biology: Sequences, Maps and Genomes*, London: Chapman and Hall.

Yue, K., and Dill, K. A. (1996), "Folding Proteins With a Simple Energy Function and Extensive Conformational Searching," *Protein Science*, 5, 254–261.

Chapter 2

Statistics in Business and Social Science

Mark P. Becker

This collection of vignettes, the second in a series of four collections to appear in the *Journal of the American Statistical Association* in the year 2000, explores the interdigitation of statistics with business and social science. The 10 vignettes in the collection cover a broad range of methodologies and applications. The collection opens with vignettes on finance and marketing, followed by a series of methodologically focused vignettes on times series and forecasting, contingency tables, and causal inference. Attention then moves to disciplines, with coverage of political science, psychology, sociology, demography, and the law. The reader of this collection should gain an appreciation for the history and role of statistical thinking and methodology in the evolution of studies in business and social science, and for the considerable promise that exists for continued innovations at the interstices of statistics and these fields of inquiry.

This collection of vignettes is not meant to be encyclopedic in coverage, and certainly many additional topics could have been included were there sufficient space. That said, the authors are to be commended for the quality and scope of their respective contributions, and one will indeed find discussion of many topics not explicitly covered by separate or free-standing vignettes, including event-history analysis, structural equation models, factor analysis, spatial analysis, and the field of economics.

It may seem odd that there is not a separate vignette covering economics, but the interdigitation of economics and statistics is so complete that explicating the role of statistics in economics would essentially necessitate an almost complete accounting of the field of statistics itself. Rather, interlaced through much of the collection are numerous references to the roles of economics and economists in the development of statistical methodology in business and social science. The role of economics in finance is inescapable, and Lo places the principle of supply and demand front and center in his vignette. Likewise, it is not surprising that much of the methodological development for empirical marketing studies cited by Rossi and Allenby has some-

Mark P. Becker is Professor of Biostatistics, School of Public Health, University of Michigan, Ann Arbor, MI 48109. He is also a Faculty Associate in the University of Michigan Institute for Social Research, and co-editor (with Michael E. Sobel) of *Sociological Methodology*.

what of an econometric flavor, or that the econometric literature figures prominently in Tsay's accounting of time series and forecasting. Economists have long been interested in principles of selection, identifiability, and causation, and this is evident in Sobel's contribution. Eisenberg points out that issues of causality also arise in the application of statistics to legal issues, and econometric methods and language are prominent in the vignette. Xie takes the viewpoint that demography provides an empirical foundation for social science, with the economic perspective one of several that shape the field. Beck places the influence of the econometric tradition at the center of his explication of statistical methodology in political science. Beck's point is that political scientists find the econometric approach to methodological developments more appealing than the statistical approach, because it is in keeping with their primary goal of evaluating substantive theories, rather than in the more statistical tradition of exploratory data analysis. I like to couch such discussion in the following terms: If the model and the data are not congruent, the statistician questions the model, whereas the economist questions the data. It has been my experience that both sides agree with this, and neither takes offense at the portrayal. The econometric perspective is not prominent in Raftery's account of sociology, but suffice it to say that the viewpoint taken by Beck is also pervasive in sociology and in the training of sociologists.

Technological innovations and the opportunities that they present for obtaining new types or massive quantities of data figure prominently in many of the vignettes. Raftery devotes an entire section to the topic of new data, new challenges, and new methods, focusing on social networks and spatial data, textual data, narrative and sequence analysis, (deterministic) simulation models, and macrosociology. Rossi and Allenby open their vignette by highlighting the influence of different types of data in the development of statistical methodology in marketing, drawing attention to innovations such as scanners and web-browsers and the new opportunities and challenges these present. Browne notes that computer-generated adaptive tests create new challenges in psychometrics. Eisenberg points out that new sources of data are making it easier to study how our legal system functions, and identifies the emergence of DNA evidence as an area where statistical methodology has played a significant role in determining the outcomes of legal proceedings. The quantity issue is highlighted by Tsay and by Fienberg. Tsay points out that advances in data acquisition technologies for financial markets and communications networks are important driving forces for future time series research. Fienberg notes that even with major advances in computing power and storage technology, the need remains for innovative approaches to model selection when many variables are to be considered.

Another important theme woven through the collection is the increasing importance of the Bayesian paradigm across business and social science. It is indeed the case that the Bayesian paradigm has figured prominently in business applications for quite some time, but in this collection one sees that recent computational advances, particularly Markov chain Monte Carlo methodology, are facilitating a more rapid and broader adoption of Bayesian methodology. The importance of Bayesian thought and methodology is manifest in the vignettes of Beck, Eisenberg, Lo, Raftery, Rossi and Allenby, Sobel, and Tsay.

One of the real pleasures of pulling this collection of vignettes together was to see a large number of themes recur through many of the vignettes. In addition to those just mentioned, latent variables, ecological regression, nonlinearity, long-range dependence, graphical models, and networks (social and neural) are among the topics that reemerge. Latent variables arise in a variety of contexts, including mixture models and latent trait models, and they figure into most of the vignettes. Beck highlights the role of political scientists in developing approaches to addressing the ecological inference problem, and Eisenberg notes the importance of this methodology in voting rights cases as well as the controversy that surrounds its use. Nonlinearity is a recurrent theme in various modeling contexts throughout the collection. For example, it arises in Browne's vignette in the context of generalizations of structural equation models and in Tsay's vignette in the context of nonlinear processes. Both authors point to these as important areas for future work. Tsay discusses long-range dependence in the context of data from communications networks and from financial markets, and Lo goes into some depth on this issue in his discussion of the stochastic nature of financial asset prices. Fienberg discusses graphical models in the context of log-linear model theory, and Sobel discusses them in the context of their usage for drawing causal inferences. Both Fienberg and Raftery note the importance of recent work on social networks, as well as the importance of work that remains to be done. The vignettes of Lo and of Rossi and Allenby make references to ways that neural networks are being used in business applications.

In summary, this collection points to the excitement of past and future developments arising from the interdigitation of statistics with business and social science. Though the types of questions arising in the various fields and the motivation behind them vary to some extent, it is clear that statistical thought and methodology is central to advancement of our understanding of human behavior and interactions. The opportunities presented by new and evolving technologies for collecting more and better data are abundant, and these will no doubt continue to motivate new statistical research and applications for many years to come. It is hoped that these vignettes will stimulate statistical scientists to become more deeply engaged in the challenges and problems of business and social science. The authors have pointed the way to a wide array of interesting challenges arising at the interstices of statistics with the economic, behavioral, and social sciences, and there are suggestions that we stand to profit by also bringing the biological and physical sciences to bear on some of these challenges. An attraction of the field of statistics has always been its broad applicability to interesting and important problems, and this collection demonstrates the numerous and intellectually challenging opportunities for making valuable contributions in various areas.

Finance: A Selective Survey

Andrew W. Lo

1. INTRODUCTION

Ever since the publication in 1565 of Girolamo Cardano's treatise on gambling, *Liber de Ludo Aleae* (*The Book of Games of Chance*), statistics and financial markets have been inextricably linked. Over the past few decades, many of these links have become part of the canon of modern finance, and it is now impossible to fully appreciate the workings of financial markets without them. In this brief survey, I hope to illustrate the enormous research opportunities at the intersection of finance and statistics by reviewing three of the most important ideas of modern finance: efficient markets, the random walk hypothesis, and derivative pricing models. Although it is impossible to provide a thorough exposition of any of these ideas in this brief essay, my less ambitious goal is to communicate the excitement of financial research to statisticians and to stimulate further collaboration between these two highly complementary disciplines. It is also impossible to provide an exhaustive bibliography for each of these topics—that would exceed the page limit of this entire article—and hence my citations are selective, focusing on more recent and most relevant developments for the readers of this journal. (For a highly readable and entertaining account of the recent history of modern finance, see Bernstein 1992.)

To develop some context for the three topics that I have chosen, consider one of the most fundamental ideas of economics, the principle of supply and demand. This principle states that the price of any commodity and the quantity traded are determined by the intersection of supply and demand curves, where the demand curve represents the schedule of quantities desired by consumers at various prices and the supply curve represents the schedule of quantities that producers are willing to supply at various prices. The intersection of these two curves determines an "equilibrium," a price–quantity pair that satisfies both consumers and producers simultaneously. Any other price–quantity pair may serve one group's interests, but not the other's.

Even in this simple description of a market, all the elements of modern finance are present. The demand curve is the aggregation of many individual consumers' desires,

Andrew W. Lo is Harris & Harris Group Professor and director of the MIT Laboratory for Financial Engineering, MIT Sloan School of Management, Cambridge, MA 02142 (E-mail: alo@mit.edu). The author is grateful to Samantha Arrington and Mark Becker for helpful comments. This research was partially supported by MIT Laboratory for Financial Engineering and the National Science Foundation grant SBR-9709976.

each derived from optimizing an individual's preferences subject to a budget constraint that depends on prices and other factors (e.g., income, savings requirements, borrowing costs). Similarly, the supply curve is the aggregation of many individual producers' outputs, each derived from optimizing an entrepreneur's preferences subject to a resource constraint that also depends on prices and other factors (e.g., costs of materials, wages, trade credit). Probabilities affect both consumers and producers as they formulate their consumption and production plans through time and in the face of uncertainty—uncertain income, uncertain costs, and uncertain business conditions.

It is the interaction between prices, preferences, and probabilities—sometimes called the "three p's of total risk management" (see Lo 1999)—that gives finance its richness and depth. Formal models of financial asset prices such as those of Breeden (1979), Lucas (1978), and Merton (1973a) show precisely how the three p's simultaneously determine a "general equilibrium" in which demand equals supply across *all* markets in an uncertain world where individuals and corporations act rationally to optimize their own welfare. Typically, these models imply that a security's price is equal to the present value of all future cashflows to which the security's owner is entitled. Several aspects make this calculation unusually challenging: individual preferences must be modeled quantitatively, future cashflows are uncertain, and so are discount rates. Pricing equations that account for such aspects are often of the form

$$P_t = E_t \left[\sum_{k=1}^{\infty} \gamma_{t,t+k} D_{t+k} \right],$$ (1)

and their intuition is straightforward; today's price must equal the expected sum of all future payments D_{t+k} multiplied by discount factors $\gamma_{t,t+k}$ that act as "exchange rates" between dollars today and dollars at future dates. If prices do not satisfy this condition, this implies a misallocation of resources between today and some future date, not unlike a situation in which two commodities sell for different prices in two countries even after exchange rates and shipping costs have been taken into account (a happy situation for some enterprising arbitrageurs, but not likely to last very long).

What determines the discount factors $\gamma_{t,t+k}$? They are determined through the equalization of supply and demand, which in turn is driven by the preferences, resources, and expectations of all market participants; that is, they are determined in general equilibrium. It is this notion of equilibrium, and all of the corresponding ingredients on which it is based, that lies at the heart of financial modeling.

2. EFFICIENT MARKETS

There is an old joke, widely told among economists, about an economist strolling down the street with a companion when they come upon a $100 bill lying on the ground. As the companion reaches down to pick it up, the economist says "Don't bother—if it were a real $100 bill, someone else would have already picked it up."

This humorous example of economic logic gone awry strikes dangerously close to home for proponents of the efficient markets hypotheses, one of the most controversial and well-studied propositions in all the social sciences. It is disarmingly simple to state, has far-reaching consequences for academic pursuits and business practice, and yet is surprisingly resilient to empirical proof or refutation. Even after three decades of research and literally hundreds of journal articles, economists have not yet reached a consensus about whether markets—particularly financial markets—are efficient or not.

As with so many of the ideas of modern economics, the origins of the efficient markets hypothesis can be traced back to Paul Samuelson (1965), whose contribution is neatly summarized by the title of his article, "Proof that Properly Anticipated Prices Fluctuate Randomly." In an informationally efficient market, price changes must be unforecastable if they are properly anticipated; that is, if they fully incorporate the expectations and information of all market participants. In the context of the basic pricing equation (1), the conditional expectation operator $E_t[\cdot] \equiv E[\cdot | \Omega_t]$ is defined with respect to a certain set of information Ω_t; hence elements of this set cannot be used to forecast future price changes, because they have already been impounded into current prices. Fama (1970) operationalized this hypothesis—summarized in his well-known expression "Prices fully reflect all available information"—by specifying the elements of the information set Ω_t available to market participants; for example, past prices, or all publicly available information, or all public and private information.

This concept of informational efficiency has a wonderfully counterintuitive and "Zen-like" quality to it: The more efficient the market, the more random the sequence of price changes generated by such a market, and the most efficient market of all is one in which price changes are completely random and unpredictable. In contrast to the passive motivation that inspires randomness in physical and biological systems, randomness in financial systems is not an implication of the principle of insufficient reason, but instead is the outcome of many active participants attempting to profit from their information. Motivated by unbridled greed, speculators aggressively pounce on even the smallest informational advantages at their disposal, and in doing so they incorporate their information into market prices and quickly eliminate the profit opportunities that gave rise to their speculation. If this occurs instantaneously, which it must in an idealized world of "frictionless" markets and costless trading, then prices must always fully reflect all available information, and no profits can be garnered from information-based trading (because such profits have already been captured).

Such compelling motivation for randomness is unique among the social sciences and is reminiscent of the role that uncertainty plays in quantum mechanics. Just as Heisenberg's uncertainty principle places a limit on what we can know about an electron's position and momentum if quantum mechanics holds, this version of the efficient markets hypothesis places a limit on what we can know about future price changes if the forces of financial self-interest are at work.

However, one of the central tenets of modern finance is the necessity of some trade-off between risk and expected returns, and whether or not predictability in security prices is inefficient can be answered only by weighing it against the risks inherent in exploiting such predictabilities. In particular, if a security's price changes are predictable to some degree, then this may be just the reward needed to attract investors to hold the asset and bear the associated risks (see, e.g., Lucas 1978). Indeed, if an investor is sufficiently risk averse, then he might gladly *pay* to avoid holding a security that has unforecastable returns.

Despite the eminent plausibility of such a trade-off—after all, investors must be rewarded to induce them to bear more risk—operationalizing it has proven a formidable challenge to both finance academics and investment professionals. Defining the appropriate measures of risk and reward, determining how they might be linked through fundamental principles of economics and psychology, and then estimating such links empirically using historical data and performing proper statistical inference are issues that have occupied much of the finance literature for the past half-century, beginning with Markowitz's (1952) development of portfolio theory and including Sharpe's (1964) capital asset pricing model (CAPM), Merton's (1973a) intertemporal CAPM, Ross's (1976) arbitrage pricing theory, and the many empirical tests of these models. Moreover, recent advances in methods of statistical inference, coupled with corresponding advances in computational power and availability of large amounts of data, have created an exciting renaissance in the empirical analysis of efficient markets, both inside and outside the halls of academia; in earlier work (Lo 1997) I provided an overview and a more complete bibliography of this literature.

3. THE RANDOM WALK

Quite apart from whether or not financial markets are efficient, one of the most enduring questions of modern finance is whether financial asset price changes are forecastable. Perhaps because of the obvious analogy between financial investments and games of chance, mathematical models of financial markets have an unusually rich history that predates virtually every other aspect of economic analysis. The vast number of prominent mathematicians, statisticians, and other scientists who have applied their considerable skills to forecasting financial security prices is a testament to the fascination and the challenges that this problem poses.

Much of the early finance literature revolved around the random walk hypothesis and the martingale model, two statistical descriptions of unforecastable price changes that were (incorrectly) taken to be implications of efficient markets. One of the first tests of the random walk was devised by Cowles and Jones (1937), who compared the frequency of *sequences* and *reversals* in historical stock returns, where the former are pairs of consecutive returns with the same sign and the latter are pairs of consecutive returns with opposite sign. Many others performed similar tests of the random walk (see Lo 1997 and Lo and MacKinlay 1999 for a survey of this literature), and with the exception of Cowles and Jones (who subsequently acknowledged an error in their analysis), all reported general support for the random walk using historical stock price data.

However, some recent research has sharply contradicted these findings. Using a statistical comparison of variances across different investment horizons applied to the weekly returns of a portfolio of stocks from 1962 to 1985, Lo and MacKinlay (1988) found that the random walk hypothesis can be rejected with great statistical confidence (well in excess of .999). In fact, the weekly returns of a portfolio containing an equal dollar amount invested in each security traded on the New York and American Stock Exchanges (called an *equal-weighted* portfolio) exhibit a striking relation from one week to the next: a first-order autocorrelation coefficient of .30.

An autocorrelation of .30 implies that approximately 9% of the variability of next week's return is explained by this week's return. An equally weighted portfolio containing only the stocks of "smaller" companies, companies with market capitalization in the lowest quintile, has a autocorrelation coefficient of .42 during the 1962–1985 sample period, implying that about 18% of the variability in next week's return can be explained by this week's return. Although numbers such as 9% and 18% may seem small, it should be kept in mind that 100% predictability yields astronomically large investment returns; a very tiny fraction of such returns can still be economically meaningful.

These findings surprise many economists, because a violation of the random walk necessarily implies that price changes are forecastable to some degree. But because forecasts of price changes are also subject to random fluctuations, riskless profit opportunities are not an immediate consequence of forecastability. Nevertheless, economists still cannot completely explain why weekly returns are not a "fair game." Two other empirical facts add to this puzzle:

1. Weekly portfolio returns are strongly positively autocorrelated, but the returns to individual securities generally are not; in fact, the average autocorrelation—averaged across individual securities—is negative (and statistically insignificant).

2. The predictability of returns is quite sensitive to the holding period; serial dependence is strong positive for daily and weekly returns but is virtually zero for returns over a month, a quarter, or a year.

For holding periods much longer than 1 week (e.g., 3–5 years), Fama and French (1988) and Poterba and Summers (1988) found negative serial correlation in U.S. stock returns indexes using data from 1926 to 1986. Although their estimates of serial correlation coefficients seem large in magnitude, there are insufficient data to reject the random walk hypothesis at the usual levels of significance. Moreover, a number of statistical biases documented by Kim, Nelson, and Startz (1991) and Richardson

(1993) cast serious doubt on the reliability of these longer-horizon inferences.

Despite these concerns, models of long-term memory have been a part of the finance literature ever since Mandelbrot (1971) applied Hurst's (1951) rescaled range statistic to financial data. Time series with long-term memory exhibit an unusually high degree of persistence, so that observations in the remote past are nontrivially correlated with observations in the distant future, even as the time span between the two observations increases. Nature's predilection toward long-term memory has been well documented in the natural sciences such as hydrology, meteorology, and geophysics, and some have argued that economic time series thus must also have this property.

But, using recently developed asymptotic approximations based on functional central limit theory, I (Lo 1991) constructed a test for long-term memory that is robust to short-term correlations of the sort uncovered by Lo and MacKinlay (1988, 1999), and concluded that despite earlier evidence to the contrary, there is little support for long-term memory in stock market prices. Departures from the random walk hypothesis can be fully explained by conventional models of short-term dependence for most financial time series. However, new data are being generated each day, and the characteristics of financial time series are unlikely to be stationary over time as financial institutions evolve. Perhaps some of the newly developed techniques for detecting long-term memory—borrowed from the statistical physics literature—will shed more light on this issue (see, e.g., Mandelbrot 1997; Pilgram and Kaplan 1998).

More recent investigations have focused on a number of other aspects of predictability in financial markets: stochastic volatility models (Gallant, Hsieh, and Tauchen 1997), estimation of tail probabilities and "rare" events (Jansen and de Vries 1991), applications of "chaos theory" and nonlinear dynamical systems (Hsieh 1991), Markov-switching models (Gray 1996), and mixed jump-diffusion models (Bates 1996). This research area is one of the most active in the finance literature, with as many researchers in industry as in academia developing tools to detect and exploit all forms of predictabilities in financial markets.

Finally, in contrast to the random walk literature, which focuses on the conditional distribution of security returns, another strand of the early finance literature has focused on the *marginal* distribution of returns, and specifically on the notion of "stability," the preservation of the parametric form of the marginal distribution under addition. This is an especially important property for security returns, which are summed over various holding periods to yield cumulative investment returns. For example, if P_t denotes the end-of-month-t price of a security, then its monthly continuously compounded return x_t is defined as $\log(P_t/P_{t-1})$, and hence its annual return is $\log(P_t/P_{t-12}) = x_t + x_{t-1} + \cdots + x_{t-11}$. The normal distribution is a member of the class of stable distributions, but the nonnormal stable distributions have a distinguishing feature not shared by the normal: they exhibit leptokurtosis or "fat tails," which seems to accord well with higher-frequency financial data, such as daily and weekly stock returns. Indeed, the fact that the historical returns of most securities have many more outliers than predicted by the normal distribution has rekindled interest in this literature, which has recently become part of a much larger endeavor known as "risk management."

Of course, stable distributions have played a prominent role in the early development of modern probability theory (see, e.g., Lévy 1937), but their application to economic and financial modeling is relatively recent. Mandelbrot (1960, 1963) pioneered such applications, using stable distributions to describe the cross-sectional distributions of personal income and of commodity prices. Fama (1965) and Samuelson (1967) developed the theory of portfolio selection for securities with stably distributed returns, and Fama and Roll (1971) estimated the parameters of the stable distribution using historical stock returns. Since then, many others have considered stable distributions in a variety of financial applications; McCulloch (1996) has provided an excellent and comprehensive survey.

More recent contributions include the application of invariance principles of statistical physics to deduce scaling properties in tail probabilities (Mandelbrot 1997; Mantegna and Stanley 1999), the use of large-deviation theory and extreme-value theory to estimate loss probabilities (Embrechts, Kluppelberg, and Mikosch 1997), and the derivation of option-pricing formulas for stocks with stable distributions (McCulloch 1996).

4. DERIVATIVE PRICING MODELS

One of the most important breakthroughs in modern finance is the pricing and hedging of "derivative" securities, securities with payoffs that depend on the prices of other securities. The most common example of a derivative security is a call option on common stock, a security that gives its owner the right (but not the obligation, hence the term "option") to purchase a share of the stock at a prespecified price K (the "strike price") on or before a certain date T (the "expiration date"). For example, a 3-month call option on General Motors (GM) stock with a $90 strike price gives its owner the right to purchase a share of GM stock for $90 any time during the next 3 months. If GM is currently trading at $85, is the option worthless? Not if there is some probability that GM's share price will exceed the $90 strike price some time during the next 3 months. It seems, therefore, that the price of the option should be determined in equilibrium by a combination of the statistical properties of GM's price dynamics and the preferences of investors buying and selling this type of security, as in the pricing equation (1).

However, Black and Scholes (1973) and Merton (1973b) provided a compelling alternative to (1), a pricing model based only on arbitrage arguments and not on general equilibrium. [In fact, the Black and Scholes (1973) framework does rely on equilibrium arguments—it was Merton's (1973b) application of continuous-time stochastic processes that eliminated the need for equilibrium altogether (see Merton 1992 for further discussion).] This alternative is best illustrated through the simple *binomial option-pricing model* of Cox, Ross, and Rubinstein (1979), a model in which there are two dates, 0 and 1, and the goal is to derive the date-0 price of a call option with strike price K that expires at date 1. In this simple economy, two other financial securities are assumed to exist: a riskless bond that pays a gross rate of return of r (e.g., if the bond yields a 5% return, then $r = 1.05$) and a risky security with date-0 price P_0 and date-1 price P_1 that is assumed to be a Bernoulli random variable:

$$P_1 = \begin{cases} uP_0 & \text{with probability } \pi \\ dP_0 & \text{with probability } 1 - \pi, \end{cases} \tag{2}$$

where $0 < d < u$. Because the stock price takes on only two values at date 1, the option price takes on only two values at date 1 as well:

$$C_1 = \begin{cases} C_u \equiv \text{Max}[uP_0 - K, 0] & \text{with probability } \pi \\ C_d \equiv \text{Max}[dP_0 - K, 0] & \text{with probability } 1 - \pi. \end{cases} \tag{3}$$

Given the simple structure that has been assumed so far, can one uniquely determine the date-0 option price C_0? It seems unlikely, as we have said nothing about investors' preferences nor the supply of the security. Yet C_0 is indeed completely and uniquely determined and is a function of K, r, P_0, d, and u. Surprisingly, C_0 is not a function of π!

To see how and why, consider constructing a portfolio of Δ shares of stock and $\$B$ of bonds at date 0, at a total cost of $X_0 = P_0\Delta + B$. The payoff X_1 of this portfolio at date 1 is simply:

$$X_1 = \begin{cases} uP_0\Delta + rB & \text{with probability } \pi \\ dP_0\Delta + rB & \text{with probability } 1 - \pi. \end{cases} \tag{4}$$

Now choose Δ and B so that the following two linear equations are satisfied simultaneously:

$$uP_0\Delta + rB = C_u, \qquad dP_0\Delta + rB = C_d \tag{5}$$

which is always feasible as long as the two equations are linearly independent. This is assured if $u \neq d$, in which case we have

$$\Delta^* = \frac{C_u - C_d}{(u - d)P_0}, \qquad B^* = \frac{uC_d - dC_u}{(u - d)r}. \tag{6}$$

Because the portfolio payoff X_1 under (6) is identical to the payoff of the call option C_1 in both states, the total cost X_0 of the portfolio must equal the option price C_0; otherwise, it is possible to construct an *arbitrage*, a trading strategy that yields riskless profits. For example, suppose that $X_0 > C_0$. By purchasing the option and selling the portfolio at date 0, a cash inflow of $X_0 - C_0$ is generated, and at date 1 the obligation X_1 created by the sale of the portfolio is exactly offset by the payoff of the option C_1. A similar argument rules out the case where $X_0 < C_0$. Thus the following pricing equation holds:

$$C_0 = P_0\Delta^* + B^* = \frac{1}{r}\left[\left(\frac{r-d}{u-d}\right)C_u + \left(\frac{u-r}{u-d}\right)C_d\right] \tag{7}$$

$$= \frac{1}{r}[\pi^* C_u + (1 - \pi^*)C_d], \qquad \pi^* \equiv \frac{r-d}{u-d}. \tag{8}$$

This pricing equation is remarkable in several respects. First, it does not seem to depend on investors' attitudes toward risk, but merely requires that investors prefer more money to less (in which case arbitrage opportunities are ruled out). Second, nowhere in (8) does the probability π appear, which implies that two investors with very different opinions about π will nevertheless agree on the price C_0 of the option. Finally, (8) shows that C_0 can be viewed as an expected present value of the option's payoff, but where the expectation is computed not with respect to the original probability π, but with respect to a "pseudoprobability" π^*, often called a *risk-neutral* probability or *equivalent martingale measure*. [Contrast (8) with the pricing equation (1) in which the discount factors $\gamma_{t,t+k}$ are also present.]

That π^* is a probability is not immediately apparent and requires further argument. A necessary and sufficient condition for $\pi^* \in [0, 1]$ is the inequality $d \le r \le u$. But this inequality follows from the assumption of the coexistence of stocks and riskless bonds in our economy. Suppose, for example, that $r < d \le u$; in this case, no investor will hold bonds, because even in the worst case, stocks will yield a higher return than r. Hence bonds cannot exist; that is, they will have zero price. Alternatively, if $d \le u < r$, then no investor will hold stocks, and hence stocks cannot exist. Therefore, $d \le r \le u$ must hold, in which case π^* can be interpreted as a probability. The fact that the option price is determined not by the original probability π, but rather by the equivalent martingale measure π^*, is a deep and subtle insight that has led to an enormous body of research in which the theory of martingales plays an unexpectedly profound role in the pricing of complex financial securities.

In particular, Merton's (1973b) derivation of the celebrated Black–Scholes formula for the price of a call option makes use of the Itô calculus, a sophisticated theory of continuous-time stochastic processes based on Brownian motion. Perhaps the most important insight of Merton's (1973b) seminal paper—for which he shared the Nobel prize in economics with Myron Scholes—is the fact that under certain conditions, the frequent trading of a small number of long-lived securities (stocks and riskless bonds) can create new investment opportunities (options and other derivative securities) that otherwise would be unavailable to investors. These conditions—now known collectively as *dynamic spanning* or *dynamically complete markets*—and the

corresponding financial models on which they are based have generated a rich literature and a multitrillion-dollar derivatives industry in which exotic financial securities
such as caps, collars, swaptions, and knock-out and rainbow options are synthet-
ically replicated by sophisticated trading strategies involving considerably simpler
securities.

This framework has also led to a number of statistical applications. Perhaps the
most obvious is the estimation of the parameters of Itô processes that are the inputs
to derivative pricing formulas. This task is complicated by the fact that Itô processes
are continuous-time processes, whereas the data are discretely sampled. The most
obvious method, maximum likelihood estimation, is practical for only a handful of
Itô processes—those for which the conditional density functions are available in
closed form; for example, processes with linear drift and diffusion coefficients. In
most other cases, the conditional density cannot be obtained analytically but can only
be characterized implicitly as the solution to a particular partial differential equation, the Fokker–Planck or "forward" equation (see Lo 1988 for further discussion).
Therefore, other alternatives have been developed, including generalized method-of-
moments estimators (Hansen and Scheinkman 1995), simulation estimators (Duffie
and Singleton 1993), and nonparametric estimators (Aït-Sahalia 1996).

Because the prices of options and most other derivative securities can be expressed
as expected values with respect to the risk-neutral measure [as in (8)], efficient Monte
Carlo methods have also been developed for computing the prices of these securities
(see Boyle, Broadie, and Glasserman 1997 for an excellent review). Moreover, option
prices contain an enormous amount of information about the statistical properties of
stock prices and the preferences of investors, and several methods have been developed recently to extract such information parametrically and nonparametrically (e.g.,
Aït-Sahalia and Lo 1998, 2000; Jackwerth and Rubinstein 1996; Longstaff 1995;
Rubinstein 1994; Shimko 1993).

Finally, the use of continuous-time stochastic processes in modeling financial
markets has led, directly and indirectly, to a number of statistical applications in
which functional central limit theory and the notion of *weak convergence* (see, e.g.,
Billingsley 1968) are used to deduce the asymptotic properties of various estimators,
such as long-horizon return regressions (Richardson and Stock 1990), long-range
dependence in stock returns (Lo 1991), and the approximation errors of continuous-
time dynamic hedging strategies (Bertsimas, Kogan, and Lo 2000).

5. CONCLUSIONS

The three ideas described here should convince even the most hardened skeptic
that finance and statistics have much in common. There are, however, many other examples in which statistics has become indispensable to financial analysis (see Campbell, Lo, and MacKinlay 1997 and Lo and MacKinlay 1999 for specific references and

a more complete survey). Multivariate analysis, especially factor analysis and principal components analysis, are important aspects of mean-variance models of portfolio selection and performance attribution. Entropy and other information-theoretic concepts have been used to construct portfolios with certain asymptotic optimality properties. Nonparametric methods such as kernel regression, local smoothing, and bootstrap resampling algorithms are now commonplace in estimating and evaluating many financial models, most of which are highly nonlinear and based on large datasets. Neural networks, wavelets, support vector machines, and other nonlinear time series models have also been applied to financial forecasting and risk management. There is renewed interest in the foundations of probability theory and notions of subjective probability because of mounting psychological evidence regarding behavioral biases in individual decisions involving financial risks and rewards. And Bayesian analysis has made inroads into virtually all aspects of financial modeling, especially with the advent of computational techniques such as Markov chain Monte Carlo methods and the Gibbs sampler.

With these developments in mind, can there be any doubt that the intersection between finance and statistics will become even greater and more active over the next few decades, with both fields benefiting enormously from the association?

REFERENCES

Aït-Sahalia, Y. (1996), "Nonparametric Pricing of Interest Rate Derivative Securities," *Econometrica*, 64, 527–560.

Aït-Sahalia, Y., and Lo, A. (1998), "Nonparametric Estimation of State-Price Densities Implicit in Financial Asset Prices," *Journal of Finance*, 53, 499–548.

———— (2000), "Nonparametric Risk Management and Implied Risk Aversion," *Journal of Econometrics*, 94, 9–51.

Bates, D. (1996), "Jumps and Stochastic Volatility: Exchange Rate Processes Implicit in Deutsche Mark Options," *Review of Financial Studies*, 9, 69–107.

Bernstein, P. (1992), *Capital Ideas*, New York: Free Press.

Bertsimas, D., Kogan, L., and Lo, A. (2000), "When Is Time Continuous?," *Journal of Financial Economics*, 55, 173–204.

Billingsley, P. (1968), *Convergence of Probability Measures*, New York: Wiley.

Black, F., and Scholes, M. (1973), "Pricing of Options and Corporate Liabilities," *Journal of Political Economy*, 81, 637–654.

Boyle, P., Broadie, M., and Glasserman, P. (1997), "Monte Carlo Methods for Security Pricing," *Journal of Economic Dynamics and Control*, 21, 1267–1321.

Breeden, D. (1979), "An Intertemporal Capital Pricing Model With Stochastic Investment Opportunities," *Journal of Financial Economics*, 7, 265–296.

Campbell, J., Lo, A., and MacKinlay, C. (1997), *The Econometrics of Financial Markets*, Princeton, NJ: Princeton University Press.

Cowles, A., and Jones, H. (1937), "Some A Posteriori Probabilities in Stock Market Action," *Econometrica*, 5, 280–294.

Cox, J., Ross, S., and Rubinstein, M. (1979), "Option Pricing: A Simplified Approach," *Journal of Financial Economics*, 7, 229–264.

Duffie, D., and Singleton, K. (1993), "Simulated Moments Estimation of Markov Models of Asset Prices," *Econometrica*, 61, 929–952.

Embrechts, P., Kluppelberg, C., and Mikosch, T. (1997), *Modelling Extremal Events for Insurance and Finance*, New York: Springer-Verlag.

Fama, E. (1965), "Portfolio Analysis in a Stable Paretian Market," *Management Science*, 11, 404–419.

—— (1970), "Efficient Capital Markets: A Review of Theory and Empirical Work," *Journal of Finance*, 25, 383–417.

Fama, E., and French, K. (1988), "Permanent And Temporary Components of Stock Prices," *Journal of Political Economy*, 96, 246–273.

Fama, E., and Roll, R. (1971), "Parameter Estimates for Symmetric Stable Distributions," *Journal of the American Statistical Association*, 66, 331–338.

Gallant, R., Hsieh, D., and Tauchen, G. (1997), "Estimation of Stochastic Volatility Models With Diagnostics," *Journal of Econometrics*, 81, 159–92.

Gray, S. (1996), "Modeling the Conditional Distribution of Interest Rates as a Regime-Switching Process," *Journal of Financial Economics*, 42, 27–62.

Hansen, L., and Scheinkman, J. (1995), "Back to the Future: Generating Moment Implications for Continuous-Time Markov Processes," *Econometrica*, 63, 767–804.

Hsieh, D. (1991), "Chaos and Nonlinear Dynamics: Application to Financial Markets," *Journal of Finance*, 46, 1839–1877.

Hurst, H. (1951), "Long-Term Storage Capacity of Reservoirs," *Transactions of the American Society of Civil Engineers*, 116, 770–799.

Jackwerth, J., and Rubinstein, M. (1996), "Recovering Probability Distributions From Contemporary Security Prices," *Journal of Finance*, 51, 1611–1631.

Jansen, D., and de Vries, C. (1991), "On the Frequency of Large Stock Returns: Putting Booms and Busts Into Perspective," *Review of Economics and Statistics*, 73, 18–24.

Kim, M., Nelson, C., and Startz, R. (1991), "Mean Reversion In Stock Prices? A Reappraisal of the Empirical Evidence," *Review of Economic Studies*, 58, 515–528.

Lévy, P. (1937), *Théorie de l'Addition des Variables Aléatoires*, Paris: Gauthier-Villars.

Lo, A. (1988), "Maximum Likelihood Estimation of Generalized Itô Processes With Discretely Sampled Data," *Econometric Theory*, 4, 231–247.

—— (1991), "Long-Term Memory in Stock Market Prices," *Econometrica*, 59, 1279–1313.

—— (Ed.) (1997), *Market Efficiency: Stock Market Behaviour in Theory and Practice*, Vols. I and II, Cheltenham, U.K.: Edward Elgar.

—— (1999), "The Three P's of Total Risk Management," *Financial Analysts Journal*, 55, 13–26.

Lo, A., and MacKinlay, C. (1988), "Stock Market Prices do not Follow Random Walks: Evidence From a Simple Specification Test," *Review of Financial Studies*, 1, 41–66.

—— (1999), *A Non-Random Walk Down Wall Street*, Princeton, NJ: Princeton University Press.

Longstaff, F. (1995), "Option Pricing and the Martingale Restriction," *Review of Financial Studies*, 8, 1091–1124.

Lucas, R. (1978), "Asset Prices in an Exchange Economy," *Econometrica*, 46, 1429–1446.

Mandelbrot, B. (1960), "The Pareto-Lévy Law and the Distribution of Income," *International Economic Review*, 1, 79–106.

—— (1963), "The Variation of Certain Speculative Prices," *Journal of Business*, 36, 394–419.

—— (1971), "When Can Price be Arbitraged Efficiently? A Limit to the Validity of the Random Walk and Martingale Models," *Review of Economics and Statistics*, 53, 225–236.

——— (1997), *Fractals and Scaling in Finance*, New York: Springer-Verlag.

Mantegna, R., and Stanley, E. (1999), *Scaling Approach to Finance*, Cambridge, U.K.: Cambridge University Press.

Markowitz, H. (1952), "Portfolio Selection," *Journal of Finance*, 7, 77–91.

McCulloch, H. (1996), "Financial Applications of Stable Distributions," in *Handbook of Statistics, Volume 14: Statistical Methods in Finance*, eds. G. Maddala and C. Rao, Amsterdam: Elsevier Science.

Merton, R. (1973a), "An Intertemporal Capital Asset Pricing Model," *Econometrica*, 41, 867–887.

——— (1973b), "Rational Theory of Option Pricing," *Bell Journal of Economics and Management Science*, 4, 141–183.

——— (1992), *Continuous-Time Finance* (rev. ed.), Oxford, U.K.: Basil Blackwell.

Pilgram, B., and Kaplan, D. (1998), "A Comparison of Estimators for $1/f$ Noise," *Physica D*, 114, 108.

Poterba, J., and Summers, L. (1988), "Mean Reversion in Stock Returns: Evidence and Implications," *Journal of Financial Economics*, 22, 27–60.

Richardson, M. (1993), "Temporary Components of Stock Prices: A Skeptic's View," *Journal of Business and Economics Statistics*, 11, 199–207.

Richardson, M., and Stock, J. (1990), "Drawing Inferences From Statistics Based on Multiyear Asset Returns," *Journal of Financial Economics*, 25, 323–348.

Ross, S. (1976), "The Arbitrage Theory of Capital Asset Pricing," *Journal of Economic Theory*, 13, 341–360.

Rubinstein, M. (1994), "Implied Binomial Trees," *Journal of Finance*, 49, 771–818.

Samuelson, P. (1965), "Proof that Properly Anticipated Prices Fluctuate Randomly," *Industrial Management Review*, 6, 41–49.

——— (1967), "Efficient Portfolio Selection for Pareto–Lévy Investments," *Journal of Financial and Quantitative Analysis*, 2, 107–122.

Sharpe, W. (1964), "Capital Asset Prices: A Theory of Market Equilibrium Under Conditions of Risk," *Journal of Finance*, 19, 425–442.

Shimko, D. (1993), "Bounds of Probability," *RISK*, 6, 33–37.

Statistics and Marketing

Peter E. Rossi and Greg M. Allenby

Statistical research in marketing is heavily influenced by the availability of different types of data. The last 10 years have seen an explosion in the amount and variety of data available to market researchers. Demand data from scanning equipment have now become routinely available in the packaged goods industries. Data from e-commerce and direct marketing are growing at an exponential rate and provide coverage to a wide assortment of different products. Web-based technology has dramatically lowered the cost of survey research. Web-browsing data provide an important new source of information about consumer tastes and preferences, which is becoming available for a large fraction of the total consumer population. In this vignette we explore some of the implications of this data explosion for the development of statistical methodology in marketing, with primary emphasis on the explosion in demand data.

Scanning equipment has provided the market researcher with a national panel of stores in addition to panels of households, altering the focus of marketing research. These data have stimulated a large literature on applied demand and discrete choice modeling. Demand models at the store level typically take the form of multivariate regression models in which demand for a vector of products is related to marketing variables such as prices, displays, and various forms of advertising. At the household level, demand is discrete, and a wide variety of multinomial logit and probit models have been applied to the data.

Early experience with scanner data revealed that households have very different patterns of buying behavior that cannot be explained simply by differences in the marketing environment. Some households, for example, exhibit strong brand loyalties, whereas other households readily switch brands when prices are lowered. Even at the store level, large differences have been detected in price and local advertising sensitivity. Initial observations of store and consumer heterogeneity created considerable interest in models of observed and unobservable heterogeneity, primarily of the random-effects form. The development and application of random-effects models in marketing has been dictated to a large degree by the available inference technology.

The first paper in this area by Kamakura and Russell (1989) used a finite mixture model of heterogeneity in a logit framework. Kamakura and Russell postulated a

Peter E. Rossi is Joseph T. Lewis Professor of Marketing and Statistics, Graduate School of Business, University of Chicago, IL 60637 (E-mail: peter.rossi@gsb.uchicago.edu). Greg M. Allenby is Helen C. Kurtz Chair in Marketing and Professor of Marketing and Statistics, Fisher College of Business, Ohio State University, Columbus, OH 43210 (E-mail: allenby.1@osu.edu).

discrete multivariate distribution for the intercepts and marketing mix variable coefficients in a multinomial logit model. The ease of computation of finite mixture models produced a stream of papers applying this idea in a wide variety of other model contexts. Until quite recently, it was not computationally practical to investigate multivariate continuous models of heterogeneity in a discrete choice setting. Simulation-based maximum likelihood (Geweke 1989) and Markov chain Monte Carlo (MCMC) Bayesian inference methods (Gelfand and Smith 1990) have now made this possible. Recent work (Allenby and Rossi 1999; Lenk, DeSarbo, Green, and Young 1996) has documented that finite mixtures provide poor approximations to the high-dimensional distributions necessary to capture heterogeneity not only in the intercepts, but also in the slopes of the multinomial choice models. Rossi, McCulloch, and Allenby (1996) used a multivariate normal distribution of random effects in a Bayesian hierarchical setting. Observable household characteristics such as demographics are included in the model by allowing the mean of the random-effects distribution to depend on these variables. These observable characteristics account for only a small amount of the total household heterogeneity. Brand preferences and marketing mix sensitivities thus are revealed primarily by household purchase behavior.

A distinguishing characteristic of random-effects applications in marketing is the interest not only in the hyperparameters of the random-effects distribution, but also in making inferences about household parameters. Marketers can use household-level parameters to target households for customized promotions or to cross-sell other products. Even at the store level, there is considerable interest in the allocation of marketing efforts across different geographic areas, which requires the building of store- or market-level models.

The sampling-theoretic approaches to random-effects models typically average the model likelihood over the random-effects distribution and provide no natural way to make household-level inferences. In contrast, Bayesian hierarchical models provide a natural setting in which inferences can be made concerning both the parameters of the random-effects distribution and the household- or store-level parameters. The combination of ability to make inferences about the random parameter draws and computational tractability for even very-high-dimensional problems has made Bayesian hierarchical methods very popular in the empirical marketing literature (Ainslie and Rossi 1998; Allenby, Arora, and Ginter 1998; Allenby and Lenk 1994; Allenby, Leone, and Jen 1999; Boatwright and Rossi 1999; Bradlow and Schmittlein 1999; Montgomery 1997; Neelemegham and Chintagunta 1999).

Just as the current statistical literature in marketing was stimulated by the availability of panel data, we expect the future course of research to be influenced by the growth in customer transaction databases. Direct marketers, credit card and financial service marketers, and retailers are now compiling large databases comprised of customer transactions and marketing contacts. These databases range from a complete record of all customer purchases along with information on other alternatives considered to much more incomplete data in which only the purchases are recorded (as in most frequent shopper or loyalty programs) or where only indirect evidence of product preferences is available (as in web-browsing data).

One of the most basic uses of customer transaction data is to find customers who are likely to express interest in or purchase a specific product. For example, a marketer could use credit usage data to find card holders likely to be interested in an upscale Italian restaurant and then target promotion activities at this subset. Another common example is the use of house file information in catalog purchases to identify customers likely to respond to a catalog mailing. The challenge is essentially a classification task. Can we construct a set of variables summarizing past transaction data and choose a model form that effectively locates the customers we want to reach? The industry uses various standard regression and logit models for the most part. Given the extremely large set of possible variables and functional forms possible, many of the classification problems are associated with a huge variable selection problem. Also, there is no particular theoretical reason to believe that standard linear regression or logistic regression models have the correct functional form. Recent developments in the statistics literature on variable selection and CART models might find very fruitful application here (Chipman, George, and McCulloch 1998; George and McCulloch 1995). It may also be useful for firms to build on the current investment in regression models by considering regression models with coefficients which change depending on the location in the explanatory variable space. Neural networks models (Ripley 1996) have already found application in predictive model building with marketing data.

In many instances, firms need to predict the response to a new or modified offering, ranging from new prices for old brands to entirely new offerings. A direct mail merchant who wants to predict the effects of a price change, for example, needs information about prices of competing brands. Although this information is often not available in a firm's transaction database, it can be imputed by augmenting the transaction data with data from other sources. For example, a firm could obtain the data for a subset of customers by recruiting a panel and having them record their purchases and prices for the other brands that were considered at the time. Once a joint model of all causal variables is available for a subset of customers, estimates of price sensitivity and brand preference can be generated from transaction data by constructing the conditional distribution of these variables given the available information.

Even when data are available to identify the parameters in a customer demand model, the resulting estimates are sometimes insufficient for marketing decisions. The brand intercepts, for example, reflect consumer preference for brands net the effects of price and the other variables that change through time. These preferences are driven by product attributes, the perceived performance of the brand on these attributes, and the importance that consumers attach to them. Furthermore, the importance of product attributes to a consumer can be traced to the concerns, interests, and motives that the consumer brings to the problem for which the brand is relevant. Understanding these more primitive constructs is important in many marketing decisions, such as market segmentation, advertising, and product development. A valuable area for future research is the merging of survey data with transaction data. For example, in choice models the brand intercepts reflect preference net the effect of price and other variable that change over time. Marketers want to relate these preferences to product attributes. For many products, these attributes are defined subjectively by the consumer. Survey techniques can be used to assess these attribute levels, which can then be entered into choice models to parameterize the intercept.

As more demand data become available, demand models can be used to assess the impact of marketing actions such as changes in price and advertising. Researchers are becoming increasingly aware that standard regression and choice models are not always adequate for optimal policy determination as the parameters of these models are not policy invariant. For example, the price sensitivity parameter in a choice model can give misleading predictions if the marketing environment is altered significantly. Dynamic models of consumer choice that take into account price expectations and inventory decisions are required for policy evaluation (Gonul and Srinivasan 1996). For example, the effects of a temporary price cut (sale) can be very different in a market environment with frequent, predictable sales than in an environment with infrequent sales.

Dynamic considerations also affect the interpretation of the intercept parameters in choice models. In general, the consumer may dynamically update his or her views about product quality. For some products, past consumption and/or advertising exposure may lead to "learning" or updates of the intercept values (Erdem and Keane 1996). In models with consumption- or advertising-based updating of brand preference, marketing interventions such as temporary price cuts and advertising can have long-run effects that require a dynamic structural model for policy evaluation.

In addition to dynamic considerations, we must also recognize the fact that the consumer may not be fully informed about all of the marketing mix variables when making a purchase decision. Typically, the choice models used in marketing applications assume that the consumer is aware of the prices of all products in the choice set. Studies have shown that consumers are not fully aware of prices and engage in a price search process in which they become selectively informed. This is best modeled in a search theoretic framework in which the consumer decides to sample or not sample a price of an item based on the expected return to further search. For example, if the consumer finds a relatively low price, as judged by the prior, on a preferred item, then further search may have a negative expected payoff. Estimating and postulating nontrivial choice models with price search presents an important challenge for the marketing literature (see Mehta, Rajiv, and Srinivasan 1999 for a pioneering effort). A search theoretic framework can also shed light on the role of information-style advertising such as in-store displays and feature ads in local newspapers. A major role of such advertising is to inform consumers of prices without incurring the cost of search. Finally, marketers have long debated the possibility of long-term negative effects of price promotions (frequent sales). In a search model, greater price variability gives rise to a greater return on search and heightens price sensitivity. A formal search model can provide marketers with a sense of the trade-off between the long-term and short-term effects of price promotions.

Reduction in the cost of acquiring survey data will result in increased interest in statistical and psychometric methods. For example, web-based consumer satisfaction surveys present methodological challenges, including the problem of respondent-specific scale usage patterns (Rossi, Gilula, and Allenby 1999; Greenleaf 1992). More generally, survey data provide unique statistical challenges that stem from measurement error problems (see Bagozzi, Yi, and Nassen 1999 for an excellent discussion of measurement error models), as well as the multivariate aspects of the data (see DeSarbo, Munrai, and Munrai 1994 for an overview of developments in multidimensional scaling). Low response rates to many marketing surveys also present a modeling and measurement challenge (see Bradlow and Zaslavsky 1999 for a recent approach to the problem of "no answer" responses).

In summary, the rapid growth in consumer data has caused a major change in the sorts of statistical models used by marketing researchers. Many standard models of demand data are often not sufficient to provide insights into consumer behavior, nor to deal with the often incomplete nature of demand data available to firms. The challenge to researchers working in marketing is to develop new statistical tools appropriate for this data environment, while using models that can shed meaningful insight on consumer behavior.

REFERENCES

Ainslie, A., and Rossi, P. E. (1998), "Similarities in Choice Behavior Across Product Categories," *Marketing Science*, 17, 91–106.

Allenby, G. M., Arora, N., and Ginter, J. L. (1998), "On The Heterogeneity of Demand," *Journal of Marketing Research*, 35, 384–389.

Allenby, G. M., and Ginter, J. L. (1995), "Using Extremes to Design Products and Segment Markets," *Journal of Marketing Research*, 32, 392–403.

Allenby, G. M., and Lenk, P. J. (1994), "Modeling Household Purchase Behavior With Logistic Normal Regression," *Journal of the American Statistical Association*, 89, 1218–1231.

Allenby, G. M., Leone, R. P., and Jen, L. (1999), "A Dynamic Model of Purchase Timing With Application to Direct Marketing," *Journal of the American Statistical Association*, 94, 365–374.

Allenby, G. M., and Rossi, P. E. (1999), "Marketing Models of Consumer Heterogeneity," *Journal of Econometrics*, 89, 57–78.

Bagozzi, R. P., Yi, Y., and Nassen, K. (1999), "Representation of Measurement Error in Marketing Variables: Review of Approaches and Extension to Three-Facet Designs," *Journal of Econometrics*, 89, 393–421.

Boatwright, P., and Rossi, P. E. (1999), "Estimating Price Elasticities With Theory-Based Priors," *Journal of Marketing Research*,

Bradlow, E. T., and Schmittlein, D. C. (1999), "The Little Engines That Could: Modeling the Performance of World Wide Web Search Engines," *Marketing Science*, 19, 43–62.

Bradlow, E. T., and Zaslavsky, A. M. (1999), "A Hierarchical Latent Variable Model for Ordinal Data From a Customer Satisfaction Survey With 'No Answer' Response," *Journal of the American Statistical Association*, 94, 43–52.

Chipman, H., George, E., and McCulloch, R. (1998), "Bayesian CART Model Search," *Journal of the American Statistical Association*, 93, 935–960.

DeSarbo, W. S., Munrai, A. K., and Munrai, L. A. (1994), "Latent Class Multi-Dimensional Scaling: A Review of Recent Developments in the Marketing and Psychometric Literature," in *Advanced Methods for Marketing Research*, ed. R. P. Bagozzi, Cambridge, U.K.: Blackwell.

Erdem, T., and Keane, M. P. (1996), "Decision-Making Under Uncertainty: Capturing Dynamic Brand Choice Processes in Turbulent Consumer Goods Markets," *Marketing Science*, 15, 1–20.

Gelfand, A. E., and Smith, A. F. M. (1990), "Sampling-Based Approaches to Calculating Marginal Densities," *Journal of the American Statistical Association*, 85, 398–409.

George, E., and McCulloch, R. (1995), "Variable Selection via Gibbs Sampling," *Journal of the American Statistical Association*, 88, 881–889.

Geweke, J. (1989), "Simulation Estimation Methods for Limited Dependent Variable Models," in *Handbook of Statistics*, Vol. 11, eds. G. S. Maddala, C. R. Rao, and H. D. Vinod, Amsterdam: North-Holland.

Gonul, F., and Srinivasan, K. (1996), "Estimating the Impact of Consumer Expectations of Coupons on Purchase Behavior: A Dynamic Structural Model," *Marketing Science*, 15, 262–279.

Greenleaf, E. (1992), "Improving Ratings Scale Measures by Detecting and Correcting Bias Components in Some Response Styles," *Journal of Marketing Research*, 29, 176–188.

Kamakura, W., and Russell, G. (1989), "A Probabilistic Choice Model for Market Segmentation and Elasticity Structure," *Journal of Marketing Research*, 26, 379–90.

Lenk, P., DeSarbo, W., Green, P., and Young, M. (1996), "Hierarchical Bayes Conjoint Analysis: Recovery of Partworth Heterogeneity from Reduced Experimental Designs," *Marketing Science*, 15, 173–191.

Mehta, N., Rajiv, S., and Srinivasan, K. (1999), "Active Versus Passive Loyalty: A Structural Model of Consideration Set Formation," working paper, Carnegie-Mellon University.

Montgomery, A. L. (1997), "Creating Micro-Marketing Pricing Strategies Using Supermarket Scanner Data," *Marketing Science*, 16, 315–337.

Neelemegham, R., and Chintagunta, P. (1999), "A Bayesian Model to Forecast New Product Performance in Domestic and International Markets," *Marketing Science*, 18, 115–136.

Ripley, B. D. (1996), *Pattern Recognition and Neural Networks*, Cambridge, U.K.: Cambridge University Press.

Rossi, P. E., Gilula, Z., and Allenby, G. M. (1999), "Overcoming Scale Usage Heterogeneity: A Bayesian Hierarchical Approach," working paper, University of Chicago, Graduate School of Business.

Rossi, P. E., McCulloch, R. E., and Allenby, G. M. (1996), "The Value of Purchase History Data in Target Marketing," *Marketing Science*, 15, 321–340.

Time Series and Forecasting: Brief History and Future Research

Ruey S. Tsay

1. A BRIEF HISTORY

Statistical analysis of time series data started a long time ago (Yule 1927), and forecasting has an even longer history. Objectives of the two studies may differ in some situations, but forecasting is often the goal of a time series analysis. In this article I focus on time series analysis with an understanding that the theory and methods considered are foundations and tools important in forecasting.

Applications played a key role in the development of time series methodology. In business and economics, time series analysis is used, among other purposes, (a) to study the dynamic structure of a process, (b) to investigate the dynamic relationship between variables, (c) to perform seasonal adjustment of economic data such as the gross domestic product and unemployment rate, (d) to improve regression analysis when the errors are serially correlated, and (e) to produce point and interval forecasts for both level and volatility series.

To facilitate discussion, I denote a time series at time index l by z_t and let Ψ_{t-1} be the information set available at time $t - 1$. It is often assumed, but not necessarily so, that Ψ_{t-1} is the σ field generated by the past values of z_t. A model for z_t can be written as

$$z_t = f(\Psi_{t-1}) + a_t \tag{1}$$

where a_t is a sequence of iid random variables with mean 0 and finite variance σ_a^2. It is evident from the equation that a_t is the one-step-ahead forecast error of z_t at time origin $t - 1$ and, hence, it is often referred to as the innovation or shock of the series at time t. The history of time series analysis is concerned with the evolution of the function $f(\Psi_{t-1})$ and the innovation $\{a_t\}$.

The series z_t is said to be (weakly) stationary if its first two moments are time-invariant under translation. That is, the expectation $E(z_t) = \mu$ is a constant, and the lag covariance function $\gamma_l = \text{cov}(z_t, z_{t-l})$ is a function of l only. The autocorrelation

Ruey S. Tsay is H. G. B. Alexander Professor of Econometrics and Statistics, Graduate School of Business, University of Chicago, IL 60637 (E-mail: ruey.tsay@gsb.uchicago.edu). This work was supported by the National Science Foundation, the Chiang Ching-Kuo Foundation, and the Graduate School of Business, University of Chicago.

function of a stationary z_t is simply $\rho_l = \gamma_l/\gamma_0$. For a linear series, understanding the behavior of ρ_l is the key to time series analysis.

1.1   Unification in Research

The publication of *Time Series Analysis: Forecasting and Control* by Box and Jenkins in 1970 was an important milestone for time series analysis. It provided a systematic approach that enables practitioners to apply time series methods in forecasting. It popularized the autoregressive integrated moving average (ARIMA) model by using an iterative modeling procedure consisting of identification, estimation, and model checking. In the framework of (1), an ARIMA(p, d, q) model assumes the form

$$w_t = (1 - B)^d z_t,$$

$$f(\Psi_{t-1}) = c + \sum_{i=1}^{p} \phi_i w_{t-i} - \sum_{j=1}^{q} \theta_j a_{t-j}, \qquad (2)$$

where p, d, and q are nonnegative integers; c is a constant; and B is the backshift operator such that $Bz_t = z_{t-1}$. The series w_t is referred to as the dth differenced series of z_t. Using polynomials, one can write the ARIMA model in a compact form, $\phi(B)(1 - B)^d z_t = c + \theta(B)a_t$, where $\phi(B) = 1 - \sum_{i=1}^{p} \phi_i B^i$ and $\theta(B) = 1 - \sum_{i=1}^{q} \theta_i B^i$ are two polynomials in B. The two polynomials $\phi(B)$ and $\theta(B)$ have no common factors, and their 0's are assumed to be outside the unit circle. In practice, it is common to assume further that a_t is Gaussian. The foregoing assumptions imply that z_t is stationary if $d = 0$. When $d \neq 0$, z_t is said to contain a unit root or to be unit-root nonstationary.

Once an ARIMA model is built and judged to be adequate, forecasts of future values are simply the conditional expectations of the model if one uses the minimum mean squared error as the criterion. The success of ARIMA models generated substantial research in time series analysis. However, the history of time series analysis was not as smooth as one might think. To begin with, time series analysis was originally divided into *frequency domain* and *time domain* approaches. Proponents of the two approaches did not necessarily see eye to eye, and there were heated debates and criticisms between the two schools. The time domain approach uses autocorrelation function ρ_l of the data and parametric models, such as the ARIMA models, to describe the dynamic dependence of the series (see Box, Jenkins, and Reinsel 1994 and references therein). The frequency domain approach focuses on spectral analysis or power distribution over frequency to study theory and applications of time series analysis. A power spectrum of a stationary z_t is the Fourier transform of the autocorrelation function ρ_l (see Brillinger 1975 and Priestley 1981 and references therein). Cooley and Tukey (1965) made an important advance in frequency-domain analysis by making spectral estimation efficient.

Times change, and it is fair to say that the sharp separation between the two approaches is gone. Now, the objective of an analysis and experience of the analyst are the determining factors between which approach to use. Likewise, the difference between Bayesian and non-Bayesian time series analyses is also diminishing. There remain some differences between Bayesian and non-Bayesians in our profession, but the issue has been shifted to those of practicality rather than philosophy (see Kitagawa and Gersch 1996 and West and Harrison 1989 and the references therein for Bayesian time series analysis.) Durbin and Koopman (2000) provided both classical and Bayesian perspectives in time series analysis.

1.2 Technical Developments

The advances in computing facilities and methods have profound impacts on time series analysis. Within the so called "traditional analysis" (i.e., linear Gaussian processes with parametric models) there are many important developments. Outlier analysis and detecting structural breaks have become an integral part of model diagnostics (see, e.g., Chang, Tiao, and Chen 1988 for outlier detection and Martin and Yohai 1986 for influential functionals). Outlier analysis in time series is concerned with aberrant observations in z_t and a_t and the changes in the mean of z_t and the variance of a_t. Many model selection criteria have been proposed to help model selection (e.g., Akaike 1974, Hannan 1980), and some important advances in pattern identification methods have also been developed; for example, the R- and S-array of Gray, Kelley, and McIntire (1978) and the extended autocorrelation function of Tsay and Tiao (1984) that is capable of handling both stationary and unit-root nonstationary series. Indeed, there have been many developments in ARMA model identification (see Choi 1992). The exact likelihood method now becomes the standard method of estimation (Ansley 1979; Hillmer and Tiao 1979). The foregoing developments are not in isolation with other developments in the area, and their impacts are not limited to linear Gaussian time series models.

Generally speaking, two important technical advances in the recent history of time series analysis have generated much interest on the topic. The first advance is the use of state-space parameterization and Kalman filtering (Harrison and Stevens 1976; Kalman and Bucy 1961 and the references therein). This happened largely in the 1980s, as evidenced by the explosion in the papers published in statistical journals that have "state-space" or "Kalman filter" in their titles. A simple state-space model for z_t can be written as

$$z_t = \mathbf{H}\mathbf{S}_t, \qquad \mathbf{S}_{t+1} = \mathbf{F}\mathbf{S}_t + \mathbf{R}_t, \tag{3}$$

where $\mathbf{S}_t$ is the state vector at time t, $\mathbf{H}$ is a row vector relating the observation z_t to the state vector, $\mathbf{F}$ is a state transition matrix, and $\mathbf{R}_t$ is a state innovation with mean 0 and a fixed nonnegative definite covariance matrix. The first equation in (3) is called the observation equation; the second, the state equation. An ARIMA model in (2) can be put into a state-space model in (3). Similarly, a state-space model in (3) implies an ARIMA model for z_t. However, the correspondence between state-space models in (3) and ARIMA models in (2) is not one-to-one (see Aoki 1987).

The original purpose of introducing Kalman filter into time series analysis was mainly to evaluate efficiently the exact Gaussian likelihood function of a model and to handle missing observations (Jones 1980). The exact likelihood function of a model can be written as a product of consecutive conditional distributions, $p(z_1, \ldots, z_n) = \prod_{i=1}^{n} p(z_i | \Psi_{i-1})$, and the Kalman filter provides closed-form formulas for the evolutions of the conditional mean and conditional variance of z_t. The usefulness of the technique was extended beyond estimation, however. It led to developments of new methods for signal extraction (Kohn and Ansley 1989; Wecker and Ansley 1983), for smoothing and seasonal adjustment (Kitagawa and Gersch 1996), and for renewal interest in structural models (Harvey 1989), among many others.

The second technical advance in recent time series analysis is the use of Markov chain Monte Carlo (MCMC) methods, especially Gibbs sampling (Gelfand and Smith 1990), and the idea of data augmentation (Tanner and Wong 1987). The applicability of MCMC methods to time series analysis is widespread, and indeed the technique has also led to various new developments in time series analysis. (See, e.g., Carlin, Polson, and Stoffer 1992 for nonnormal and nonlinear state-space modeling and McCulloch and Tsay 1993 for inference and prediction of autoregressive models with random mean and variance shifts, including using explanatory variables to estimate transition probabilities in mean and variance.) The MCMC methodology also led to increasing use of simulation methods in time series analysis, especially in tackling complicated problems that were impossible to handle a few years ago.

1.3   Methodological Developments

The past several decades also brought many important advances in time series methodology.

Nonlinearity and Nonnormality.   Theory and methods have been developed for many nonlinear and non-Gaussian models, marking a generalization of the functional form $f(\Psi_{t-1})$ in (1). In the statistical literature, Tong (1990) provided an excellent summary of recent developments in nonlinear models, such as bilinear models and threshold autoregressive (TAR) models. For a bilinear model, $f(\Psi_{t-1})$ involves some cross-product terms between z_{t-i} and a_{t-j}, where $i, j > 0$. A TAR model, on the other hand, uses a piecewise linear model for $f(\Psi_{t-1})$ over a threshold space. For example, the model

$$f(\Psi_{t-1}) = \begin{cases} \phi_1 z_{t-1} & \text{if } z_{t-1} \geq r \\ \phi_2 z_{t-1} & \text{if } z_{t-1} < r, \end{cases}$$

where $\phi_1 \neq \phi_2$, is called a two-regime TAR model of order 1 with threshold variable z_{t-1} and threshold r. This simple TAR model is piecewise linear in the space of z_{t-1}, not in time. It can produce several nonlinear characteristics commonly observed in practice, such as asymmetry between increasing and declining patterns in a periodic time series. Many tests now are available to detect nonlinearity in a series, and some studies show that nonlinear models can indeed improve forecasting in certain situations (see, e.g., Montgomery, Zarnowitz, Tsay, and Tiao 1998).

In the econometric literature, some important developments in nonlinear models have also emerged. The advance, however, focuses on the evolution of the conditional variance of a_t in (1). Let $h_t = E(a_t^2|\Psi_{t-1})$ be the conditional variance of a_t given Ψ_{t-1}. This quantity is referred to as the volatility of a_t in econometrics and finance. The autoregressive conditional heteroscedastic (ARCH) model of Engle (1982) assumes that h_t is a positive deterministic function of Ψ_{t-1}. For the simplest ARCH(1) model, h_t becomes $h_t = \alpha_0 + \alpha_1 a_{t-1}^2$, where $\alpha > 0$ and $1 > \alpha_1 \geq 0$. A special feature of the ARCH model is that under certain conditions the innovation a_t is heavy-tailed distributed and hence is unconditionally non-Gaussian even though it is conditionally Gaussian. Another feature of ARCH models is volatility clustering. Because volatility plays an important role in options pricing, the ARCH model has attracted much attention. Its also leads to the introduction of generalized ARCH (GARCH) models and stochastic volatility models. In general, the ARCH-family models use a positive deterministic (typically quadratic) equation to govern the evolution of h_t over time, whereas stochastic volatility models allow h_t to have its own innovational series. For example, a GARCH(1, 1) model assumes the form

$$h_t = \alpha_0 + \alpha_1 a_{t-1}^2 + \beta_1 h_{t-1}, \tag{4}$$

where $\alpha_0 > 0, \alpha_1 \geq 0, \beta_1 \geq 0$ and $1 > \alpha_1 + \beta_1$. A simple stochastic volatility model may assume the form $\ln(h_t) = \alpha_0 + \ln(h_{t-1}) + v_t$, where v_t is a white noise sequence. The added innovation v_t increases the flexibility of the model as well as complications. A GARCH model has an ARMA-type representation, so that many of its properties are similar to those of ARMA models. For example, let $\eta_t = a_t^2 - h_t$, which is the squared innovation without its conditional expectation. Then the GARCH(1, 1) model in (4) can be written as

$$a_t^2 = \alpha_0 + (\alpha_1 + \beta_1)a_{t-1}^2 + \eta_t - \beta_1\eta_{t-1}, \tag{5}$$

where it is easy to see that η_t is a martingale difference. The stationarity condition $\alpha_1 + \beta_1 < 1$ of a GARCH(1, 1) becomes apparent from (5). The econometric literature also concerns the evolution of $f(\Psi_{t-1})$. The Markov switching model of Hamilton (1989) uses a state variable to govern the choice of a linear model for $f(\Psi_{t-1})$. The state variable then evolves over time based on a transition probability matrix.

Multivariate Process. Methods for analyzing multivariate series have been developed, especially in structural specification of a vector system. The usefulness and need of considering jointly several related time series were recognized a long time ago (see Quenouille 1957). However, multivariate analysis is often confined to vector autoregressive (VAR) models. One can think of many reasons for this lack of progress, but two reasons stand out:

a. The generalization of univariate ARMA models to vector ARMA models encounters the problem of identifiability. As a simple illustration, the following bivariate AR(1) and MA(1) models are identical:

$$\begin{bmatrix} z_{1t} \\ z_{2t} \end{bmatrix} = \begin{bmatrix} 0 & 0 \\ 2 & 0 \end{bmatrix} \begin{bmatrix} z_{1,t-1} \\ z_{2,t-1} \end{bmatrix} + \begin{bmatrix} a_{1t} \\ a_{2t} \end{bmatrix},$$

$$\begin{bmatrix} z_{1t} \\ z_{2t} \end{bmatrix} = \begin{bmatrix} a_{1t} \\ a_{2t} \end{bmatrix} - \begin{bmatrix} 0 & 0 \\ -2 & 0 \end{bmatrix} \begin{bmatrix} a_{1,t-1} \\ a_{2,t-1} \end{bmatrix},$$

where z_{1t} and z_{2t} are two time series and $a_t = (a_{1t}, a_{2t})'$ is a sequence of independent bivariate Gaussian random vectors with mean 0 and a constant positive-definite covariance matrix. From the model, $z_{1t} = a_{1t}$, so that $z_{1,t-1} = a_{1,t-1}$. The equality of the two models then results from $z_{2t} = 2z_{1,t-1} + a_{2t} = 2a_{1,t-1} + a_{2t}$. Such exchangeable forms between AR and MA models cannot occur in the univariate case.

b. Multivariate models are much harder to estimate and to understand, and there is a propensity to use perceived simpler models.

These difficulties have been largely overcome. The identifiability problem can be solved by using either the Kronecker indices (Akaike 1976) or the scalar component model (SCM) of Tiao and Tsay (1989). The relationship between Kronecker indices and orders of SCM was given by Tsay (1991). Both the Kronecker index and SCM can easily identify the existence of exchangeable models; in fact, both methods identify the underlying structure of a linear system.

A related development in multivariate time series analysis is the cointegration of Engle and Granger (1987). Roughly speaking, cointegration means that a linear combination of marginally unit-root nonstationary series becomes a stationary series. It has become popular in econometrics because cointegration is often thought of as the existence of some long-term relationship between variables. In the statistical literature, the idea of a linear combination of unit-root nonstationary series becoming stationary was studied by Box and Tiao (1977). Associated with cointegration is the development of various test statistics to test for the number of cointegrations in a linear system. Despite the huge literature on cointegration, its practical importance is yet to be judged (see Tiao, Tsay, and Wang 1993). This is due primarily to the fact that cointegration is a long-term concept that overlooks the practical effects of scaling factors of marginal series.

Long-Range Dependence.   To model the fact that sample autocorrelation functions in some real-world time series tend to decay much slower than those of an ARMA model, Granger and Joyeux (1980) and Hosking (1981) introduced the idea of fractional difference to handle what is called long-range dependence in a series. This led to the development of autoregressive fractionally integrated moving average (ARFIMA) (p, d, q) model where d is a positive but noninteger real number. The long-range dependence occurs in many fields (see Beran 1994). In finance, it was found that the squared or absolute returns of an asset typically exhibit long-range dependence (e.g., Ding, Granger, and Engle 1993).

Finally, models for time series that assume only discrete values are also available; see a recent review article by Berchtold and Raftery (1999), which contains many references. For forecasting, there is a trend toward using the predictive distribution rather than the mean squared error to evaluate the accuracy of forecasts.

1.4   Theoretical Developments

The most recent widespread theoretical development in time series analysis is the theory of unit-root. In its simplest case, the theory is concerned with asymptotic properties of statistics of a random walk process, $z_t = z_{t-1} + a_t$, where $z_0 = 0$ and a_t is a martingale difference satisfying $E(a_t|\Psi_{t-1}) = 0$ and $E(|a_t|^{2+\delta}) < \infty$, where δ is a positive real number. Consider, for example, the ordinary least squares estimate $\hat{\phi} = (\sum_{i=1}^{n} z_{t-1}^2)^{-1}(\sum_{i=1}^{n} z_t z_{t-1})$ of the autoregression $z_t = \phi z_{t-1} + \varepsilon_t$, where ε_t denotes the error term. Unlike in the stationary case, here the limiting distribution of $\hat{\phi}$ is a functional of a standard Brownian motion. (See Chan and Wei 1988 for a comprehensive treatment of least squares estimates with various characteristic roots on the unit circle and Phillips 1987 for a nice treatment of time series regression with a unit-root.) This newly established asymptotic result is interesting in several ways. First, it is a case in which the convergence rate of $\hat{\phi}$ is n^{-1} not the usual $n^{-.5}$, where n is the sample size. Second, the mathematics involved are elegant. Third, the nonstandard limiting distribution is by itself exciting, opening a new set of testing problems that are of interest to many econometricians and statisticians. In particular, the unit-root test is to consider the null hypothesis $H_0 : \phi = 1$ versus the alternative hypothesis $H_a : \phi \neq 1$ (see Dickey and Fuller 1979).

The unit-root problem has attracted much attention because (a) it provides a formal test to determine the order of differencing in using ARIMA models, (b) it opens an area of testing in which the proper test statistic depends on the existence of a nonzero constant term c in (1) and the multiplicity of the unit root (i.e., d) in (2), and (c) the other AR and MA parameters, if they exist, become nuisance parameters that disappear asymptotically, but might have serious implications in finite-sample cases. A multivariate extension of Chan and Wei (1988) was given by Tsay and Tiao (1990) and that of the unit-root test is the so-called cointegration test. The unit-root problem has also been extended to the MA case (see Davis and Dunsmuir 1996).

Another area of theoretical development is the theory of nonlinear models, especially the issue of geometrical ergodicity and stationarity of nonlinear processes. For the simple TAR(1) model, Chen and Tsay (1991) and Petruccelli and Woolford (1984) provided some interesting results. In general, useful theorems for studying stability of a nonlinear process were given by Meyn and Tweedie (1993) and Tong (1990, appendices).

The theory of long-range dependent processes and that of processes with heavy-tailed distributions also mark substantial progress; see Beran (1994), Resnick (1997), Samorodnitsky and Taqqu (1994), and the references therein.

2. FUTURE RESEARCH

An important driving force of future research in time series analysis is the advance in high-volume data acquisition. Consider, for instance, transaction-by-transaction data common in financial markets, or communications networks, and in e-commerce on the internet. These data are available now and must be processed properly and efficiently in a globally competitive business environment. But the special features of the data, such as large sample sizes, heavy tails, unequally-spaced observations, and mixtures of multivariate discrete and continuous variables, can easily render existing methods inadequate. Analysis of these types of data will certainly influence the directions of future time series research.

In my personal opinion, the following topics will attract substantial interest of time series researchers. First, the use of multivariate models either in a vector ARMA framework or a state-space form will increase, partly because of the need to study the dynamic relationships between variables and partly because of the advances in computing facilities. Part of the work here will be to identify common characteristics between the marginal series. Second, theory and applications of nonlinear and non-Gaussian models will continue to flourish. The development here is likely to move closely with nonparametric methods or computing intensive methods. Within the parametric framework, we can expect to see new models that use different equations to govern the evolutions of lower-order conditional moments. Third, heavy-tail modeling and extreme value analysis will become a necessity in some areas of application, such as telecommunication and high-frequency financial data analysis. Fourth, there will be a trend to study not only the usual time series data, but also the time duration between observations. In other words, times of occurrence will play an increasingly important role in time series analysis and forecasting. This line of research also marks a marriage between theories of time series analysis and point processes and between researchers in econometrics and statistics. Fifth, methods will be developed to efficiently and effectively process large-scale datasets.

In terms of technical developments, we will continue to see mixtures of Bayesian and non-Bayesian analyses. The ideas of data augmentation and MCMC methods are here to stay and will flourish. Data mining will become part of time series data analysis, and we need to make good use of it.

REFERENCES

Akaike, H. (1974), "A New Look at the Statistical Model Identification," *IEEE Transactions on Automatic Control*, AC-19, 716–723.

——— (1976), "Canonical Correlation Analysis of Time Series and the Use of an Information Criterion," in *Systems Identification: Advances and Case Studies*, eds. R. K. Methra and D. G. Lainiotis, New York: Academic Press, pp. 27–99.

Ansley, C. F. (1979), "An Algorithm for the Exact Likelihood of a Mixed Autoregressive Moving Average Process," *Biometrika*, 66, 59–65.

Aoki, M. (1987), *State-Space Modeling of Time Series*, New York: Springer-Verlag.

Beran, J. (1994), *Statistics for Long-Memory Processes*, New York: Chapman and Hall.

Berchtold, A., and Raftery, A. E. (1999), "The Mixture Transition Distribution Model for High-Order Markov Chains and Non-Gaussian Time Series," Technical Report 360, University of Washington, Dept. of Statistics.

Box, G. E. P., Jenkins, G. M., and Reinsel, G. C. (1994), *Time Series Analysis: Forecasting and Control* (3rd ed.), Englewood Cliffs, NJ: Prentice-Hall.

Box, G. E. P., and Tiao, G. C. (1977), "A Canonical Analysis of Multiple Time Series," *Biometrika*, 64, 355–365.

Brillinger, D. R. (1975), *Time Series: Data Analysis and Theory*, New York: Holt, Rhinehart and Winston.

Carlin, B. P., Polson, N. G., and Stoffer, D. S. (1992), "A Monte Carlo Approach to Nonnormal and Nonlinear State-Space Modeling," *Journal of the American Statistical Association*, 87, 493–500.

Chan, N. H., and Wei, C. Z. (1988), "Limiting Distributions of Least Squares Estimates of Unstable Autoregressive Processes," *The Annals of Statistics*, 16, 367–401.

Chang, I., Tiao, G. C., and Chen, C. (1988), "Estimation of Time Series Parameters in the Presence of Outliers," *Technometrics*, 30, 193–204.

Chen, R., and Tsay, R. S. (1991), "On the Ergodicity of TAR(1) Processes," *Annals of Applied Probability*, 1, 613–634.

Choi, B. (1992), *ARMA Model Identification*, New York: Springer-Verlag.

Cooley, J. W., and Tukey, J. W. (1965), "An Algorithm for the Machine Calculation of Complex Fourier Series," *Mathematics of Computation*, 19, 297–301.

Davis, R., and Dunsmuir, W. (1996), "Maximum Likelihood Estimation of MA(1) Processes With a Root on or Near the Unit Circle," *Econometric Theory*, 12, 1–29.

Dickey, D. A., and Fuller, W. A. (1979), "Distribution of the Estimates for Autoregressive Time Series With a Unit Root," *Journal of the American Statistical Association*, 427–431.

Ding, Z., Granger, C. W. J., and Engle, R. F. (1993), "A Long Memory Property of Stock Returns and a New Model," *Journal of Empirical Finance*, 1, 83–106.

Durbin, J., and Koopman, S. J. (2000), "Time Series Analysis of Non-Gaussian Observations Based on State-Space Models From Both Classical and Bayesian Perspectives" (with discussion), *Journal of the Royal Statistical Society*, Ser. B, 62, 3–56.

Engle, R. F. (1982), "Autoregressive Conditional Heteroscedasticity With Estimates of the Variance of United Kingdom Inflations," *Econometrica*, 50, 987–1007.

Engle, R. F., and Granger, C. W. J. (1987), "Co-Integration and Error Correction: Representation, Estimation and Testing," *Econometrica*, 55, 251–276.

Gelfand, A. E., and Smith, A. F. M. (1990), "Sampling-Based Approaches to Calculating Marginal Densities," *Journal of the American Statistical Association*, 85, 398–409.

Granger, C. W. J., and Joyeux, R. (1980), "An Introduction to Long-Memory Time Series Models and Fractional Differencing," *Journal of Time Series Analysis*, 1, 15–29.

Gray, H. L., Kelley, G. D., and McIntire, D. D. (1978), "A New Approach to ARMA Modeling," *Communications in Statistics*, Part B, 7, 1–77.

Hamilton, J. D. (1989), "A New Approach to the Economic Analysis of Nonstationary Time Series and the Business Cycle," *Econometrica*, 57, 357–384.

Hannan, E. J. (1980), "The Estimation of the Order of an ARMA Process," *The Annals of Statistics*, 8, 1071–1081.

Harrison, P. J., and Stevens, C. F. (1976), "Bayesian Forecasting" (with discussion), *Journal of the Royal Statistical Society*, Ser. B, 38, 205–247.

Harvey, A. C. (1989), *Forecasting, Structural Time Series Models and the Kalman Filter*, Victoria, Australia: Cambridge University Press.

Hillmer, S. C., and Tiao, G. C. (1979), "Likelihood Function of Stationary Multiple Autoregressive Moving Average Models," *Journal of the American Statistical Association*, 74, 652–660.

Hosking, J. R. M. (1981), "Fractional Differencing," *Biometrika*, 68, 165–176.

Jones, R. H. (1980), "Maximum Likelihood Fitting of ARMA Models to Time Series With Missing Observations," *Technometrics*, 22, 389–395.

Kalman, R. E., and Bucy, R. S. (1961), "New Results in Linear Filtering and Prediction Theory," *Transactions of ASME, Journal of Basic Engineering*, 83D, 95–108.

Kitagawa, G., and Gersch, W. (1996), *Smoothness Priors Analysis of Time Series*, New York: Springer-Verlag.

Kohn, R., and Ansley, C. F. (1989), "A Fast Algorithm for Signal Extraction Influence and Cross-Validating in State-Space Models," *Biometrika*, 76, 65–79.

Martin, R. D., and Yohai, V. J. (1986), "Influence Functionals for Time Series," *The Annals of Statistics*, 14, 781–818.

McCulloch, R. E., and Tsay, R. S. (1993), "Bayesian Inference and Prediction for Mean and Variance Shifts in Autoregressive Time Series," *Journal of the American Statistical Association*, 88, 968–978.

Meyn, S., and Tweedie, R. L. (1993), *Markov Chains and Stochastic Stability*, New York: Springer-Verlag.

Montgomery, A. L., Zarnowitz, V., Tsay, R. S., and Tiao, G. C. (1998), "Forecasting the U.S. Unemployment Rate," *Journal of the American Statistical Association*, 93, 478–493.

Petruccelli, J., and Woolford, S. W. (1984), "A Threshold AR(1) Model," *Journal of Applied Probability*, 21, 270–286.

Phillips, P. C. B. (1987), "Time Series Regression With a Unit Root," *Econometrica*, 55, 277–301.

Priestley, M. B. (1981), *Spectral Analysis and Time Series*, Vols. I and II, London: Academic Press.

Quenouille, M. H. (1957), *The Analysis of Multiple Time Series*, London: Griffin.

Resnick, S. I. (1997), "Heavy Tail Modeling and Teletraffic Data," (with discussion), *The Annals of Statistics*, 25, 1805–1869.

Samorodnitsky, G., and Taqqu, M. S. (1994), *Stable Non-Gaussian Random Processes*, New York: Chapman and Hall.

Tanner, M., and Wong, W. H. (1987), "The Calculation of Posterior Distributions," (with discussion), *Journal of the American Statistical Association*, 82, 528–550.

Tiao, G. C., and Tsay, R. S. (1989), "Model Specification in Multivariate Time Series," (with discussion),

Journal of the Royal Statistical Society, Ser. B, 51, 157–213.

Tiao, G. C., Tsay, R. S., and Wang, T. (1993), "Usefulness of Linear Transformation in Multivariate Time Series Analysis," *Empirical Economics*, 18, 567–593.

Tong, H. (1990), *Non-Linear Time Series: A Dynamical System Approach*, Oxford, U.K.: Oxford University Press.

Tsay, R. S. (1991), "Two Canonical Forms for Vector ARMA Processes," *Statistica Sinica*, 1, 247–269.

Tsay, R. S., and Tiao, G. C. (1984), "Consistent Estimates of Autoregressive Parameters and Extended Sample Autocorrelation Function for Stationary and Nonstationary ARMA Models," *Journal of the American Statistical Association*, 79, 84–96.

——— (1990), "Asymptotic Properties of Multivariate Nonstationary Processes With Applications to Autoregressions," *The Annals of Statistics*, 18, 220–250.

Weeker, W. E., and Ansley, C. F. (1983), "The Signal Extraction Approach to Nonlinear Regression and Spline Smoothing," *Journal of the American Statistical Association*, 78, 81–89.

West, M., and Harrison, P. J. (1989), *Bayesian Forecasting and Dynamic Models*, Berlin: Springer-Verlag.

Yule, G. U. (1927), "On a Method of Investigating Periodicities in Disturbed Series With Special Reference to Wölfer's Sunspot Numbers," *Philosophical Transactions of the Royal Society London*, Ser. A, 226, 267–298.

Contingency Tables and Log-Linear Models: Basic Results and New Developments

Stephen E. Fienberg

1. HISTORICAL REMARKS ON CONTINGENCY TABLE ANALYSIS

Contingency table analysis is rooted in the turn-of-the-century work of Karl Pearson and George Udny Yule, who introduced the cross-product, or odds ratio, as a formal statistical tool. The subsequent contributions by R. A. Fisher linked their methods to basic statistical methodology and theory, but it was not until 1935 that Maurice Bartlett, as a result of a suggestion by Fisher, utilized Yule's cross-product ratio to define the notion of second-order interaction in a $2 \times 2 \times 2$ table and to develop an appropriate test for the absence of such an interaction (Bartlett 1935). The multivariate generalizations of Bartlett's work, beginning with a 1956 article by Roy and Kastenbaum, form the basis of the log-linear model approach to contingency tables, which is largely the focus of this vignette. Key articles in the 1960s by M. W. Birch (1963), Yvonne Bishop (1975), John Darroch (1962), I. J. Good (1963), Leo Goodman (1963), and Robin Plackett (1974), plus the availability of high-speed computers, led to an integrated theory and methodology for the analysis of contingency tables based on log-linear models, culminating in a series of books published in the 1970s. (Historical references can be found in various sources including Bishop, Fienberg, and Holland 1975, Carriquiry and Fienberg 1998, Fienberg 1980, and Haberman 1974.)

The next section outlines some of the basic results on likelihood estimation for log-linear models used to describe interactions in contingency tables, as the theory emerged by the early 1970s. I then briefly describe some of the major advances of the next three decades related to log-linear models. There is now an extensive literature on other classes of models and other methods of estimation, especially Bayesian, but I treat these only in passing, not because they are unimportant, but rather because they draw on similar foundations. Finally, I outline some important open research problems.

Stephen E. Fienberg is Maurice Falk University Professor of Statistics and Social Science, Department of Statistics and Center for Automated Learning and Discovery, Carnegie Mellon University, Pittsburgh, PA 15213 (E-mail: fienberg@stat.cmu.edu). This work was supported by National Science Foundation grant no. REC-9720374.

Many statisticians view the theory and methods of log-linear models for contingency tables as a special case of either exponential family theory or generalized linear models (GLMs) (Christensen 1996; McCullagh and Nelder 1989). It is true that computer programs for GLM often provide convenient and relatively efficient ways of implementing basic estimation and goodness-of-fit assessment. But adopting such a GLM approach leads the researcher to ignore the special features of log-linear models relating to interpretation in terms of cross-product ratios and their generalizations, crucial aspects of estimability and existence associated with patterns of zero cells, and the many innovative representations that flow naturally from the basic results linking sampling schemes. One very important development that I do not cover (due mainly to a lack of space) is the role of general estimating equations and marginal models for longitudinal data within the GLM framework (see, e.g., Diggle, Liang, and Zeger 1996).

2. SAMPLING MODELS AND BASIC LOG-LINEAR MODEL THEORY

Let $\mathbf{x}' = (x_1, x_2, \ldots, x_t)$ be a vector of observed counts for t cells, structured in the form of a cross-classification. Now let $\mathbf{m}' = (m_1, m_2, \ldots, m_t)$ be the vector of expected values that are assumed to be functions of unknown parameters $\boldsymbol{\theta}' = (\theta_1, \theta_2, \ldots, \theta_s)$, where $s < t$.

There are three standard sampling models for the observed counts in contingency tables. In the *Poisson model*, the $\{x_i\}$ are observations from independent Poisson random variables with means $\{m_i\}$, whereas in the *multinomial model*, the total count $N = \sum_{i=1}^{t} x_i$ is a random sample from an infinite population where the underlying cell probabilities are $\{m_i/N\}$. Finally, in the *product-multinomial model*, the cells are partitioned into sets, and each set has an independent multinomial structure, as in the multinomial model.

The following results hold under the Poisson and multinomial sampling schemes:

1. Corresponding to each parameter in $\boldsymbol{\theta}$ is a minimal sufficient statistic (MSS) that is expressible as a linear combination of the $\{x_i\}$. More formally, if $\mathcal{M}$ is used to denote the log-linear model specified by $\mathbf{m} = \mathbf{m}(\boldsymbol{\theta})$, then the MSSs are given by the projection of $\mathbf{x}$ onto $\mathcal{M}$, $P_{\mathcal{M}}\mathbf{x}$.

2. The maximum likelihood estimator (MLE), $\hat{\mathbf{m}}$, of $\mathbf{m}$, if it exists, is unique and satisfies the likelihood equations

$$P_{\mathcal{M}}\hat{\mathbf{m}} = P_{\mathcal{M}}\mathbf{x}. \tag{1}$$

Necessary and sufficient conditions for the existence of a solution to the likelihood equations, (1), are relatively complex (see, e.g., Haberman 1974). A sufficient condition is that all cell counts be positive (i.e., $\mathbf{x} > \mathbf{0}$), but MLEs for log-linear models exist in many sparse situations where a large fraction of the cells have zero counts.

For product-multinomial sampling situations, the basic multinomial constraints (i.e., that the counts must add up to the multinomial sample sizes) must be taken into account. Typically, some of the parameters in $\boldsymbol{\theta}$ that specify the log-linear model $\mathcal{M}$ [i.e., $\mathbf{m} = \mathbf{m}(\boldsymbol{\theta})$] are fixed by these constraints.

More formally, let $\mathcal{M}$ be a log-linear model for $\mathbf{m}$ under product-multinomial sampling that corresponds to a log-linear model $\mathcal{M}$ under Poisson sampling such that the multinomial constraints "fix" a subset of the parameters, $\boldsymbol{\theta}$, used to specify $\mathcal{M}$. Then the following result holds:

3. The MLE of $\mathbf{m}$ under product-multinomial sampling for the model $\mathcal{M}$ is the same as the MLE of $\mathbf{m}$ under Poisson sampling for the model $\mathcal{M}$.

The final basic result relates to assessing the fit of log-linear models:

4. Let ϕ be a real-valued parameter in the interval $-\infty < \phi < \infty$. If $\hat{\mathbf{m}}$ is the MLE of $\mathbf{m}$ under a log-linear model, and if the model is correct, then for each value of ϕ, the goodness-of-fit statistic,

$$K(\mathbf{x}, \hat{\mathbf{m}}) = \frac{2}{\phi(\phi + 1)} \sum_{i=1}^{t} x_i \left[\left(\frac{x_i}{\hat{m}_i} \right)^{\phi} - 1 \right], \qquad (2)$$

has an asymptotic chi-squared distribution with $t - s$ degrees of freedom as sample sizes tend to infinity, where s is the total number of independent constraints implied by the log-linear model and the multinomial sampling constraints (if any). If the model is not correct, then the distribution is stochastically larger than χ^2_{t-s}.

The usual Pearson chi-squared and likelihood-ratio chi-squared statistics are special cases of the family of *power-divergence statistics* defined by $K(\mathbf{x}, \hat{\mathbf{m}})$ in (2). The Pearson statistic chi-squared statistics corresponds to $\phi = 1$, and the statistic G^2 corresponds to the limit as $\phi \to 0$. (For further details on the properties of the general family of power divergence statistics, see Read and Cressie 1988.)

In the late 1970s, several authors attempted to address the problem of large sparse asymptotics; for example, for a sequence of multinomially structured tables in increasing size, where the sample size n and the number of cells t or the number of parameters s go to infinity in some fixed ratio. Results of Haberman (1977) and Koehler and Larntz (1980) provide some guidance to statistical practice and suggest that the usual advice that expected cell counts should be ≥ 5 is far too conservative and wasteful of information in large sparse tables.

Bartlett's (1935) no-second-order interaction model for the expected values in a $2 \times 2 \times 2$ table, with entries m_{ijk}, is based on equating the values of the cross-product ratio, α, in each layer of the table; that is,

$$\frac{m_{111}m_{221}}{m_{121}m_{211}} = \frac{m_{112}m_{222}}{m_{122}m_{212}}. \tag{3}$$

Expression (3) can be represented in log-linear form as

$$\log m_{ijk} = u + u_{1(i)} + u_{2(j)} + u_{3(k)} + u_{12(ij)} + u_{13(ik)} + u_{23(jk)}, \tag{4}$$

with suitable linear side constraints on the sets of u terms to achieve identifiability.

All of the parameters in (4) can be written as functions of cross-product ratios (see Bishop et al. 1975). Applying the basic results for the basic sampling schemes applied to the $2 \times 2 \times 2$ table, we have that the MSSs are the two-dimensional marginal totals, $\{x_{ij+}\}$, $\{x_{i+k}\}$, and $\{x_{+jk}\}$ (except for linearly redundant statistics included for purposes of symmetry), where a "+" indicates summation over the corresponding subscript. Further, the MLEs of the $\{m_{ijk}\}$ under model (4) must satisfy the likelihood equations

$$\hat{m}_{ij+} = x_{ij+}, \qquad i, j = 1, 2,$$

$$\hat{m}_{i+k} = x_{i+k}, \qquad i, k = 1, 2,$$

and

$$\hat{m}_{+jk} = x_{+jk}, \qquad j, k = 1, 2, \tag{5}$$

usually solved by some form of iterative procedure. For the example actually considered by Bartlett, the third set of equations in (5) corresponds to the binomial sampling constraints.

The class of log-linear models just described for the three-way table generalizes in a direct fashion to $k \geq 4$ dimensions. As long as the models retain a hierarchical structure (e.g., setting $u_{12(ij)} = 0$ for all i, j implies setting $u_{123(ijk)} = 0$ for all i, j, k), the MSSs are sets of marginal totals of the full table. Further, all independence or conditional independence relationships are representable as log-linear models, and these models have estimated expected values that can be computed directly. A somewhat larger class of log-linear models with this direct, or *decomposable*, representation is described later. All log-linear models that are not decomposable require an iterative solution of likelihood equations.

In a multiway contingency table, the model that results from setting exactly one two-factor term (and all of its higher-order relatives) equal to 0 is called a *partial association* model. For example, in four dimensions, if $u_{12(ij)} = 0$ for all i, j, then the MSSs are $\{x_{i+kl}\}$ and $\{x_{+jkl}\}$, and the resulting partial association model corresponds to the conditional independence of variables 1 and 2 given 3 and 4. The corresponding MLEs for the expected cell frequencies are

$$\hat{m}_{ijkl} = \frac{x_{i+kl}x_{+jkl}}{x_{++kl}} \qquad \forall \, i, j, k, l. \tag{6}$$

Bishop et al. (1975) and Whitaker (1990) provided more details on partial association models and their uses.

3. MAJOR SUBSEQUENT DEVELOPMENTS

3.1   The Graphical Subfamily of Log-Linear Models

A major innovation in log-linear model methods over the past 20 years has been the development of methods associated with a subfamily of log-linear models known as *graphical log-linear models*. Darroch, Lauritzen, and Speed (1980) first described these models, and the monographs by Lauritzen (1996) and Whitaker (1990) made accessible most of the subsequent results in the literature.

This approach uses the vertices of a graph to represent variables and the edges among them to represent relationships. Conditional independence relationships correspond to the absence of edges in such an undirected graph. Models defined solely in terms of such relationships are said to be *graphical*. For categorical random variables, all graphical models are log-linear. The subfamily of graphical log-linear models includes the class of decomposable models, but not all nondecomposable models are graphical. Various authors have used graphical log-linear models to simplify approaches to model search, and they are intimately related to an extensive literature on *collapsibility* and estimability of parameters via marginal tables.

3.2 p_1 Models for Social Networks

Holland and Leinhardt (1981) introduced a log-linear model for representing relationships among individuals in a social network. Their model has a graphical representation, but one that is different from that of the previous section, in that it links individuals instead of variables. Fienberg, Meyer, and Wasserman (1985) showed how to explicitly handle social network data and the Holland–Leinhart model and its extensions in contingency table form using basic log-linear-model tools. Wasserman and Pattison (1996) provided related logistic representations.

3.3   Latent Trait and Rasch Models

In psychological tests or attitude studies, one often is interested in quantifying the value of an unobservable *latent trait*, such as mathematical ability or manual dexterity, on a sample of individuals. Although latent traits are not directly measurable, one can attempt to assess indirectly a person's value for the latent trait from his or her responses to a set of well-chosen items on a test. The simplest model for doing so was introduced by Rasch (1960). Given responses for n individuals on k binary random variables, let $\mathbf{X}$ denote the $n \times k$ matrix of responses for n individuals on k binary variables, and let $\boldsymbol{\alpha}$ and $\boldsymbol{\theta}$ denote the vectors of item and individual parameters. Then the simple dichotomous Rasch model states that

$$\log[P\left(X_{ij} = 1|\theta_i, \alpha_j\right)/P\left(X_{ij} = 0|\theta_i, \alpha_j\right)] = \theta_i - \alpha_j. \tag{7}$$

This is a logit model for the log odds for $X_{ij} = 1$ versus $X_{ij} = 0$. We can recast the observed data x_{ij} in the form of a $n \times 2^k$ array, with exactly one observation for each level of the first variable.

In the 1980s, Duncan (1983) and Tjur (1982) recognized an important relationship between the Rasch model and log-linear models for the corresponding collapsed 2^k contingency table. Darroch (1986) and Fienberg and Meyer (1983) represented these models in terms of the log-linear models of quasi-symmetry, but ignored the moment constraints described by Cressie and Holland (1983). More recently, Agresti (1993a, 1993b) and others have carried these ideas further for other categorical data problems.

3.4 Multiple-Recapture Models for Population Estimation

If the members of a population are sampled k different times, the resulting re capture history data can be displayed in the form of a 2^k table with one missing cell, corresponding to those never sampled. Such an array is amenable to log-linear model analysis, the results of which can be used to project a value for the missing cell (as in Fienberg 1972). Major applications of capture–recapture methodology include estimating the undercount in the U.S. decennial census, where $k = 2$ (see, e.g., the articles in the special 1993 section of *JASA*), and the prevalence of various epidemiological conditions, where typically $k \geq 3$.

The use of standard log-linear models in this context presumes that capture probabilities are constant across the population. Agresti (1994) and Darroch, Fienberg, Glonek, and Junker (1993) used a variation of the Rasch model to allow for special multiplicative forms of heterogeneity. Fienberg, Johnson, and Junker (1999) integrated this form of heterogeneity into the log-linear framework and explicitly incorporated the moment constraints in a Bayesian implementation.

3.5 Association Models for Ordinal Variables

Log-linear models as described in this article ignore any structure linking the categories of variables, yet biostatistical problems often involve variables with ordered categories; for example, differing dosage levels for a drug or the severity of symptoms or side effects. Goodman (1979) provided a framework for extending standard log-linear models via multiplicative interaction terms of the form

$$u_{12(ij)} = u^*_{1(i)} u^*_{2(j)} \tag{8}$$

to represent a two-factor u-term. This extended class of models, known as *association models*, has close parallels with correspondence analysis models and both classes have been developed and extended by Clogg, Gilula, Goodman, and Haberman, among others. (For a detailed description of these and other methods for ordinal variables, see Agresti 1990 and Clogg and Shidadeh 1994.)

3.6 Gröbner Bases and Exact Distributions

Haberman (1974) actually gave the conditional distribution for a table under a log-linear model given the marginals which are the MSSs under the model. But actually calculating that conditional distribution is quite complex and most attempts to work with it have focused solely on the calculation of specific quantiles such as p values (see, e.g., Agresti 1992). Diaconis and Sturmfels (1998) provided an elegant solution to the computational problem of computing such "exact" distributions for multiway contingency tables, using the group theory structure of Gröbner bases and a Markov chain Monte Carlo algorithm. Applications of this technology for disclosure limitation can be found in the recent 1998 special issue of the *Journal of Official Statistics*, but realistic implementation for high-dimensional tables is still an open issue.

4. SOME CHALLENGING OPEN PROBLEMS

Although the basic theory of log-linear models and methods for the analysis of contingency tables was in place over 20 years ago, and there have been major advances in various related topics over the ensuing years, some problems have eluded satisfactory solution. First and foremost among these are diagnostics for model fit and graphical representations for model search. Typical GLM diagnostics are geared largely to the noncategorical data response situation and most of the other methods suggested to date are ad hoc at best. Similarly, although graphical model tools have helped to simplify model search, we have only limited graphical representations to link to model search methodologies.

Graphical log-linear models gave new impetus to the developments of log-linear model theory in the 1980s and 1990s, and there were related graphical representations for social network models linking individuals. But these two graphical representations remain unconnected. Elsewhere in multivariate analysis, researchers have exploited the duality between representations in spaces for individuals and for variables. Perhaps these ideas of duality of representations might allow us to link the two types of graphical structures into a new mathematical framework.

The problem of assessing bound for the entries of contingency tables given a set of marginals has a long statistical history going back to work done more than 50 years ago independently by Bonferroni, Fréchet, and Hoeffding on bounds for cumulative bivariate distribution functions given their univariate marginals (see Fienberg 1999 for a review of related literature). For the more general problem of a k-way contingency table given a set of possibly overlapping marginal totals, there are tantalizing links to the literature on log-linear models described in this article, including to the recent work on exact distributions and Gröbner bases described earlier. Implementation of bounds for large sparse tables is a special challenge.

More generally, as computer power and storage grows, researchers are attempting to work with larger and larger collections of categorical variables. We need new methods of model selection that scale up to situations where the dimensionality k of the table may exceed 100, and we need to revisit the asymptotics that are relevant for such situations.

REFERENCES

Agresti, A. (1990), *Categorical Data Analysis*, New York: Wiley.

——— (1992), "A Survey of Exact Inference for Contingency Tables" (with discussion), *Statistical Science*, 7, 131–177.

——— (1993a), "Computing Conditional Maximum Likelihood Estimates for Generalized Rasch Models Using Simple Loglinear Models with Diagonal Parameters, *Scandinavian Journal of Statistics*, 20, 63–72.

——— (1993b), "Distribution-free Fitting of Logit Models with Random Effects for Repeated Categorical Responses," *Statistics in Medicine*, 12, 1969–1987.

——— (1994), "Simple Capture-Recapture Models Permitting Unequal Catchability and Variable Sampling Effort," *Biometrics*, 50, 494–500.

Bartlett, M. S. (1935), "Contingency Table Interactions," *Journal of the Royal Statistical Society Supplement*, 2, 248–252.

Birch, M. W. (1963), "Maximum Likelihood in Three-Way Contingency Tables," *Journal of the Royal Statistical Society*, Ser. B, 25, 229–233.

Bishop, Y. M. M., Fienberg, S. E., and Holland, P. W. (1975), *Discrete Multivariate Analysis: Theory and Practice*, Cambridge, MA: MIT Press.

Carriquiry, A., and Fienberg, S. E. (1998), "Log-linear Models," in *Encyclopedia of Biostatistics*, Vol. 3, eds. P. Armitage and T. Colton, New York: Wiley, pp. 2333–2349.

Clogg, C. C., and Shidadeh, E. S. (1994), *Statistical Models for Ordinal Variables*, Thousand Oaks, CA: Sage.

Christensen, R. (1996), *Plane Answers to Complex Questions: The Theory of Linear Models* (2nd ed.), New York: Springer-Verlag.

Cressie, N. E., and Holland, P. W. (1983), "Characterizing the Manifest Probabilities of Latent Trait Models," *Psychometrika*, 48, 129–141.

Darroch, J. N. (1962), "Interaction in Multi-Factor Contingency Tables," *Journal of the Royal Statistical Society*, Ser. B, 24, 251–263.

—— (1986), "Quasi-Symmetry," in *Encyclopedia of Statistical Sciences*, Vol. 7, eds. S. Kotz and N. L. Johnson, New York: Wiley, pp. 469–473.

Darroch, J. N., Fienberg, S. E., Glonek, G., and Junker, B. (1993), "A Three-Sample Multiple-Recapture Approach to Census Population Estimation With Heterogeneous Catchability," *Journal of the American Statistical Association*, 88, 1137–1148.

Darroch, J. N., Lauritzen, S. L., and Speed, T. P. (1980), "Markov Fields and Log-Linear Interaction Models for Contingency Tables," *The Annals of Statistics*, 8, 522–539.

Diaconis, P., and Sturmfels, B. (1998), "Algebraic Algorithms for Sampling From Conditional Distributions," *The Annals of Statistics*, 26, 363–397.

Diggle, P. J., Liang, K-Y., and Zeger, S. L. (1996), *Analysis of Longitudinal Data*, New York: Oxford University Press.

Duncan, O. D. (1983), "Rasch Measurement: Further Examples and Discussion," in *Survey Measurement of Subjective Phenomena*, Vol. 2, eds. C. F. Turner and E. Martin, New York: Russell Sage, pp. 367–403.

Fienberg, S. E. (1972), "The Multiple Recapture Census for Closed Populations and Incomplete 2^k Contingency Tables," *Biometrika*, 59, 591–603.

—— (1980), *The Analysis of Cross-Classified Categorical Data* (2nd ed.), Cambridge, MA: MIT Press.

—— (1999), "Fréchet and Bonferroni Bounds for Multi-Way Tables of Counts With Applications to Disclosure Limitation," in *Statistical Data Protection: Proceedings of the Conference*, Luxembourg: Eurostat, 115–129.

Fienberg, S. E., Johnson, M. A., and Junker, B. (1999), "Classical Multi-Level and Bayesian Approaches to Population Size Estimation Using Data From Multiple Lists," *Journal of the Royal Statistical Society*, Ser. A, 162, 383–406.

Fienberg, S. E., and Meyer, M. M. (1983), "Loglinear Models and Categorical Data Analysis With Psychometric and Econometric Applications," *Journal of Econometrics*, 22, 191–214.

Fienberg, S. E., Meyer, M. M., and Wasserman, S. S. (1985), "Statistical Analysis of Multiple Sociometric Relations," *Journal of the American Statistical Association*, 80, 51–67.

Good, I. J. (1963), "Maximum Entropy for Hypothesis Formulation, Especially for Multidimensional Contingency Tables," *Annals of Mathematical Statistics*, 34, 911–934.

Goodman, L. A. (1963), "On Methods for Comparing Contingency Tables," *Journal of the Royal Statistical Society*, Ser. A, 126, 94–108.

—— (1979), "Simple Models for the Analysis of Association in Cross-Classifications Having Ordered Categories," *Journal of the American Statistical Association*, 74, 537–552.

Haberman, S. J. (1974), *The Analysis of Frequency Data*, Chicago: University of Chicago Press.

—— (1977), "Log-Linear Models and Frequency Tables With Small Expected Cell Counts," *The Annals of Statistics*, 5, 1148–1169.

Holland, P. W., and Leinhardt, S. (1981), "An Exponential Family of Probability Distributions for Directed Graphs" (with discussion), *Journal of the American Statistical Association*, 76, 33–65.

Koehler, K. J., and Larntz, K. (1980), "An Empirical Investigation of Goodness-of-Fit Statistics for Sparse Multinomials," *Journal of the American Statistical Association*, 75, 336–344.

Lauritzen, S. L. (1996), *Graphical Models*, New York: Oxford University Press.

McCullagh, P., and Nelder, J. A. (1989), *Generalized Linear Models* (2nd ed.), London: Chapman and Hall.

Plackett, R. L. (1974), *The Analysis of Categorical Data*, London: Charles Griffin.

Rasch, G. (1960), *Probabilistic Models for Some Intelligence and Attainment Tests*, The Danish Institute of Educational Research, expanded ed. (1980), Chicago: The University of Chicago Press.

Read, T. R. C., and Cressie, N. A. C. (1988), *Goodness-of-Fit Statistics for Discrete Multivariate Data*,

New York: Springer-Verlag.

Roy, S. N., and Kastenbaum, M. A. (1956), "On the Hypothesis of No Interaction in a Multi-way Contingency Table," *Annals of Mathematical Statistics*, 27, 749–757.

Tjur, T. (1982), "A Connection Between Rasch's Item Analysis Model and a Multiplicative Poisson Model," *Scandinavian Journal of Statistics*, 9, 23–30.

Wasserman, S., and Pattison, P. (1996), "Logit Models and Logistic Regressions for Social Networks: I. An Introduction to Markov Graphs and p^*," *Psychometrika*, 61, 401–425.

Whitaker, J. (1990), *Graphical Models in Applied Multivariate Statistics*, New York: Wiley.

Causal Inference in the Social Sciences

Michael E. Sobel

The subject of causation is controversial. For several hundred years, since Hume, who essentially equated causation with constant conjunction ("same cause, same effect"), until recently, theories that view causation as a form of empirical regularity (predictability) have dominated philosophical discussions of the causal relation. Many writers who subscribe to this view have also argued that not all regular sequences evidence causation. Thus Hume's account is regarded as incomplete (nor would this account be completed by including Hume's other criteria, temporal priority and contiguity), and it is important to understand just what separates causal from noncausal sequences. One idea that has been put forth (it can even be found in Hume, though he appears not to have realized its implications) is that whereas both types of sequences exhibit de facto regularity, only causal sequences are counterfactually regular. That is, in both cases, one can state that when the object or event X is present $(X = 1)$, it is followed by the effect Y $(Y = 1)$, but in the case of a causal sequence, one can also state that if X is not present, then Y also is not present.

These notions raise epistemological problems. First, the relationships that scientists study rarely hold in every instance. Thus it seems reasonable to entertain notions of causation that allow us to speak of causation in individual instances and/or that feature so-called "probabilistic causation." Second, if a causal relationship must sustain a counterfactually conditional statement, then additional problems arise because one can observe the actual states of X and Y, but not the state of Y that would occur were X in a state other than its actual state. Thus it is possible that $Y = 1$ in those instances where $X = 1$, but this might be true even if $X = 0$. Consider the case of weight loss among motivated persons who take diet pills $(X = 1)$. Intuition suggests that by controlling who receives (does not receive) diet pills, the possibility of encountering this situation is reduced. In other words, experiments might be useful for inferring causation.

In the social sciences, where many of the questions asked do not appear amenable to experimentation, approaches to inferring causation have been proposed that differed from those that emerged in the statistical literature on experimental design. Prior to the development of simultaneous equation models by econometricians and the related work by Simon (1954) on spurious correlation (which suggested the possibility

Michael E. Sobel is Professor, Department of Sociology, Columbia University, New York, NY 10027 (E-mail: mes105@columbia.edu). The author thanks Sander Greenland, Donald Rubin, and an anonymous reviewer for helpful comments.

of distinguishing between causal and noncausal associations), many quantitative social scientists eschewed causal language. Bernert (1983) described the situation in American sociology. Subsequently, sociologists (e.g., Duncan 1966) discovered path analysis and disseminated it to other social sciences, promoting its use for drawing causal inferences and, more generally, promoting the use of causal language. Later, in an important generalization, Jöreskog (1977) embedded factor analysis from psychology and simultaneous equation models from economics into the covariance structure model, seemingly allowing causal inference with latent causes and effects. (These models are now commonly called structural equation models or causal models.) More recently, a literature on graphical models (e.g., Whittaker 1990) has emerged. Use of these models for drawing causal inferences (e.g., Pearl 1998; Spirtes, Glymour, and Scheines 1993) is another outgrowth of the foregoing approach and the literature on path analysis. However, some authors (e.g., Cox and Wermuth 1993) have deliberately avoided using causal language in conjunction with such models. Other extensions of Simon's work are the methods that economists have proposed for assessing so-called "spurious causation" in time series (Geweke 1984; Granger 1969).

Although the earlier developers of these methods tended, in tandem with the prevailing currents in philosophy in their day, to equate causality with prediction, subsequent writers, in tune with later currents, have tended toward the view that causal relationships sustain counterfactual conditionals. Thus in the literature on simultaneous equation models, exogeneous variables (errors are defined so as to be uncorrelated with these exogenous variables) came to be regarded as potentially manipulable inputs, and model parameters were interpreted as indicative of what would happen under interventions (see, e.g., Goldberger 1964, p. 375). Similarly, sociologists, psychologists, and others speak of parameters in structural equation models as effects, interpreting these as indicative of what would happen to a response if the independent variables were manipulated. (For references to this literature, see Sobel 1990.) Similar remarks apply to some treatments of Granger causality. Although it is recognized that such interpretations are extramathematical and not always warranted, it is widely believed that these interpretations are licensed when the variables in the model are ordered properly and the "causal theory" underlying the model is "valid," irrespective of study design.

It seems evident that if causation is equated with predictability, then causal inference can be approached using nonexperimental studies and statistical methods for studying association (as above). In retrospect, it also seems evident that under a counterfactual account of causation, which involves comparing an observed response with a response that would have occurred under conditions that were not observed, an analysis of the causal relationship needs to begin with the use of something like the notation invented by Neyman (1923) to discuss potential yields in an agricultural experiment. This notation allows for unambiguous definitions of effects, enabling evaluation of methods for estimating these.

To illustrate, suppose that the response (Y) is to be compared under treatment and no treatment, and there is no interference among units; that is, the response of a unit does not depend on the treatment received by other units (Cox 1958). Then either y_{ic}, the response of unit i in the absence of treatment, or y_{it}, the response of unit i under treatment, can be observed. The unit (causal) effect is defined as a function of the two responses; for example, $y_{it} - y_{ic}$. Although this cannot be observed, Neyman showed (under a model of assignment equivalent to randomization) that the difference between treatment and control group means is an unbiased estimator of the average of the unit effects—the average treatment effect (ATE). Notice that here study design does enter the picture.

Although the potential outcomes notation was subsequently used in the statistical literature on experiments, Rubin (1974, 1977, 1978, 1980, 1991) applied it to observational studies to define effects that accorded with a counterfactual account of causation and to state sufficient conditions (ignorability and strong ignorability) for estimating these effects consistently (Rosenbaum and Rubin 1983a). Building on this work, Rubin and coworkers clarified a number of issues important to social and behavioral scientists, including Lord's paradox (Holland and Rubin 1983), causal inference in case-control studies (Holland and Rubin 1988) and causal inference in recursive structural equation models (Holland 1988).

Increasingly, social statisticians are finding that use of the foregoing framework above results in greater clarity, enabling precise definitions of causal estimands of interest and evaluation of methods traditionally used to make causal inferences. An excellent example of this is the work on instrumental variables by Angrist and Imbens (1995) and Angrist, Imbens, and Rubin (1996). (For related material, see also Robins and Greenland 1992.) The idea behind the instrumental variables estimator, which goes back at least to Wright (1928), is (in econometric parlance) to estimate the effect of a variable correlated with the error term in a regression (i.e., the variable is not "exogenous") by using another variable that is correlated with the response, but that does not directly affect the response. Angrist et al. (1996) studied this estimator in a randomized study with noncompliance, showing, under assumptions that they clearly spelled out, that this estimates a parameter they called the local average treatment effect (LATE). LATE is the average treatment effect among compliers (those who both comply with their actual assignment and who would comply with the assignment not assigned). Although the subpopulation of compliers is not observable, and other effects may often be of greater interest (e.g., the average effect among the treated), it nevertheless is of great value to know what the instrumental variables estimator is actually estimating, and what assumptions are implicit when this estimator is taken to estimate other parameters; for example, the average treatment effect among the treated. One of the important assumptions examined in this work is the exclusion restriction, namely that treatment assignment (as versus treatment received) does not affect the response. (This restriction appears also in the econometric literature and can also be formulated in the literature on structural equation models in terms of direct effects.) Another important assumption, used in conjunction with the exclusion restriction, is the monotonicity assumption; under this assumption, the latent subpopulation of subjects who would not take treatment when treatment is assigned and who would take treatment when it is not assigned is nonexistent. Monotonicity is realized in many policy and prevention studies where only subjects assigned to treatment are allowed to enroll in the intervention program.

Subsequently, Imbens and Rubin (1997) extended this work, applying likelihood and Bayesian procedures to estimate LATE in both the case where the exclusion restriction holds and the case where it does not; they referred to LATE as CACE ("complier average causal effect"). Little and Yau (1998) estimated LATE by maximum likelihood in a randomized study of unemployed workers, where the treatment consists of several training sessions and the response is the reduction in depression; in this study, almost half of the subjects assigned to treatment did not actually take the treatment. Whereas Little and Yau assumed that the exclusion restriction holds, Jo (1999), arguing that there should be a direct effect of treatment assignment on the response among noncompliers, extended this work to that case.

In observational studies, the assumption that treatment assignment is random, given observed covariates (strong ignorability), allows consistent estimation of effects. Matching or stratifying on these covariates is desirable, if possible, facilitating estimation with minimal additional assumptions. If this is not possible, Rosenbaum and Rubin (1983a) showed that it suffices to match or stratify on the propensity score. (In a study with treatment or no treatment, this is the probability of treatment, given the covariates.) Alternatively, model-based adjustments may be made. However, rarely is the state of knowledge in the social sciences adequate for a researcher to feel confident that he or she has measured all of the relevant covariates. The possibility of hidden bias (relevant unobserved covariates) is great. Thus it is important to develop methods for assessing the sensitivity of an analysis to this possibility, as in Rosenbaum and Rubin (1983b). (For further material, see Rosenbaum 1995, who has done extensive work on this topic.) Economists, who sometimes refer to strong ignorability as selection on observables (Heckman and Hotz 1989), have also pointed to the importance of unobserved variables (selection on unobservables), proposing various methods for using longitudinal data, in conjunction with modeling assumptions, to estimate treatment effects that correct for this type of selection (Heckman and Robb 1985, 1988). (For a more extensive review of this literature, see Winship and Morgan 1999.)

Another useful approach develops upper and lower bounds on the values of effects of interest (assuming that the response is bounded). Although these effects are not identified in observational studies, components of the effects are identified. Assumptions about the upper and lower bounds on the nonidentified components are then used in conjunction with the identified components to obtain the bounds. When the nonidentified components take on the maximum and the minimum values allowable, the bounds can be quite wide. Nonetheless, these can be narrowed considerably under quite reasonable assumptions compatible with the state of existing knowledge. Robins (1989) first developed these bounds. Manski (1995) also used these bounds, and also developed bounds for several other contexts of special relevance to social scientists.

Although the interplay between statistics and social sciences has proven very useful, many issues need further work. In Holland's (1988) work on path analysis, subjects are randomized to a group encouraged to study and a group not encouraged to study. The amount of time studied and the score on a subsequent test are recorded. Assuming that the "direct effect" of encouragement on test scores is 0 for all subjects (an exclusion restriction) and that the effect of time studied on the response is identical for all subjects (the assumption of constant effect), the value of this effect is the instrumental variable estimand. It would be useful to extend this work to psychological prevention studies, where the treatment targets several mediators, which in turn affect the response, and researchers inappropriately use structural equation models to estimate the effects of the mediators on the response. It would also be useful to do this without the strong assumption of constant effect typically made in conjunction with linear models. New study designs for estimating the mediated effects could then be developed. Second, in psychology and sociology, researchers often use latent variables derived from factor analyses as causes, covariates, and effects. Little attention has been paid to the issues that arise in attempting to make appropriate causal inferences in such studies (Sobel 1994, 1997). Finally, social scientists are often concerned with interdependence among units. One form occurs when the response of a subject under a given condition depends on the conditions to which other subjects are assigned; the assumption of no interference between units, discussed by Cox (1958) or, in Rubin's terms, SUTVA or stability, does not hold. Violations of stability certainly have been alluded to by social scientists (Garfinkel, Manski, and Michalopolous 1992; Sobel 1995), and some work on this subject has been done in epidemiology (Halloran and Struchiner 1995), but much remains to be done in both experimental and nonexperimental contexts.

In most disciplines within the social sciences, empirical work has not yet been informed by the aforementioned developments. As this changes, a dramatic transformation should be seen in the way social research is conducted and the purposes that this research serves. To take but one example, much of quantitative political science and sociology may be characterized as a highly stylized search for new causes of effects. Researchers typically begin with one or more outcomes and a list of causes identified by previous workers. Potentially new causes are then listed; if these account ("explain") for additional variability in the response, then the new causes are held to affect the outcome, and the significant coefficients in the new model are endowed with a causal interpretation. The process is repeated by subsequent workers, resulting in further "progress." When researchers realize that they are merely adding more and more variables to a predictive conditioning set, one wonders what will take the place of the thousands of purported (causal) effects that currently fill the journals.

REFERENCES

Angrist, J. D., and Imbens, G. W. (1995), "Two-Stage Least Squares Estimation of Average Causal Effects in Models With Variable Treatment Intensity," *Journal of the American Statistical Association*, 90, 431–442.

Angrist, J. D., Imbens, G. W., and Rubin, D. B. (1996), "Identification of Causal Effects Using Instrumental Variables" (with discussion), *Journal of the American Statistical Association*, 91, 444–472.

Bernert, C. (1983), "The Career of Causal Analysis in American Sociology," *British Journal of Sociology*, 34, 230–254.

Cox, D. R. (1958), *The Planning of Experiments*, New York: Wiley.

Cox, D. R., and Wermuth, N. (1993), "Linear Dependencies Represented by Chain Graphs" (with discussion), *Statistical Science*, 9, 204–218, 247–283.

Duncan, O. D. (1966), "Path Analysis: Sociological Examples," *American Journal of Sociology*, 72, 1–16.

Garfinkel, I., Manski, C., and Michalopolous, C. (1992), "Micro-Experiments and Macro Effects," in *Evaluating Welfare and Training Programs*, eds. C. Manski and I. Garfinkel, Cambridge, MA: Harvard University Press, pp. 253–273.

Geweke, J. (1984), "Inference and Causality in Economic Time Series Models," in *Handbook of Econometrics*, Vol. 2, eds. Z. Griliches and M. D. Intriligator, Amsterdam: North-Holland, pp. 1101–1144.

Goldberger, A. S. (1964), *Econometric Theory*, New York: Wiley.

Granger, C. W. (1969), "Investigating Causal Relations by Econometric Models and Cross-Spectral Methods," *Econometrica*, 37, 424–438.

Halloran, M. E., and Struchiner, C. J. (1995), "Causal Inference in Infectious Diseases," *Epidemiology*, 6, 142–151.

Heckman, J. J., and Hotz, V. J. (1989), "Choosing Among Alternative Nonexperimental Methods for Estimating the Impact of Social Programs: The Case of Manpower Training" (with discussion), *Journal of the American Statistical Association*, 84, 862–880.

Heckman, J. J., and Robb, R. (1985), "Alternative Methods for Evaluating the Impact of Interventions," in *Longitudinal Analysis of Labor Market Data*, eds. J. J. Heckman and B. Singer, Cambridge, U.K.: Cambridge University Press, pp. 156–245.

——— (1988), "The Value of Longitudinal Data for Solving the Problem of Selection Bias in Evaluating the Impact of Treatment on Outcomes," in *Panel Surveys*, eds. G. Duncan and G. Kalton, New York: Wiley, pp. 512–538.

Holland, P. W. (1988), "Causal Inference, Path Analysis, and Recursive Structural Equation Models" (with discussion), in *Sociological Methodology, 1988*, ed. C. C. Clogg, Washington, DC: American Sociological Association, pp. 449–493.

Holland, P. W., and Rubin, D. B. (1983), "On Lord's Paradox," in *Principals of Modern Psychological Measurement*, eds. H. Wainer and S. Messnick, Hillsdale, NJ: Lawrence Erlbaum, pp. 3–35.

——— (1988), "Causal Inference in Retrospective Studies," *Evaluation Review*, 12, 203–231.

Imbens, G. W., and Rubin, D. B. (1997), "Bayesian Inference for Causal Effects in Randomized Experiments With Non-Compliance," *The Annals of Statistics*, 25, 305–327.

Jo, B. (1999), "Estimation of Intervention Effects With Noncompliance: Model Misspecification Sensitivity Analysis," unpublished manuscript.

Jöreskog, K. G. (1977), "Structural Equation Models in the Social Sciences: Specification, Estimation and Testing," in *Applications of Statistics*, ed. P. R. Krishnaiah, Amsterdam: North-Holland, pp. 265–287.

Little, R. J., and Yau, L. H. Y. (1998), "Statistical Techniques for Analyzing Data From Prevention Trials: Treatment of No-Shows Using Rubin's Causal Model," *Psychological Methods*, 3, 147–159.

Manski, C. F. (1995), *Identification Problems in the Social Sciences*, Cambridge, MA: Harvard University Press.

Neyman, J. (1923), "On the Application of Probability Theory to Agricultural Experiments. Essay on Principles. Section 9," translated into English and edited by D. M. Dabrowska and T. P. Speed (1990), *Statistical Science*, 5, 463–472.

Pearl, J. (1998), "Graphs, Causality, and Structural Equation Models," *Sociological Methods and Research*, 27, 226–284.

Robins, J. M. (1989), "The Analysis of Randomized and Nonrandomized AIDS Treatment Trials Using a New Approach to Causal Inference in Longitudinal Studies," in *Health Services Research Methodology: A Focus on AIDS*, eds. L. Sechrest, H. Freedman, and A. Mulley, Rockville, MD: U.S. Department of Health and Human Services, pp. 113–159.

Robins, J. M., and Greenland, S. (1992), "Identifiability and Exchangeability for Direct and Indirect Effects," *Epidemiology*, 3, 143–155.

Rosenbaum, P. R. (1995), *Observational Studies*, New York: Springer-Verlag.

Rosenbaum, P. R., and Rubin, D. B. (1983a), "The Central Role of the Propensity Score in Observational Studies for Causal Effects," *Biometrika*, 70, 41–55.

―――― (1983b), "Assessing Sensitivity to an Unobserved Binary Covariate in an Observational Study With Binary Outcome," *Journal of the Royal Statistical Society*, Ser. B, 212–218.

Rubin, D. B. (1974), "Estimating Causal Effects of Treatments in Randomized and Nonrandomized Studies," *Journal of Educational Psychology*, 66, 688–701.

―――― (1977), "Assignment to Treatment Groups on the Basis of a Covariate," *Journal of Educational Statistics*, 2, 1–26.

―――― (1978), "Bayesian Inference for Causal Effects: The Role of Randomization," *The Annals of Statistics*, 6, 34–58.

―――― (1980), Comment on "Randomization Analysis of Experimental Data: The Fisher Randomization Test," by D. Basu, *Journal of the American Statistical Association*, 75, 591–593.

―――― (1991), "Practical Implications of Modes of Statistical Inference for Causal Effects and the Critical Role of the Assignment Mechanism," *Biometrics*, 47, 1213–1234.

Simon, H. A. (1954), "Spurious Correlation: A Causal Interpretation," *Journal of the American Statistical Association*, 49, 467–492.

Sobel, M. E. (1990), "Effect Analysis and Causation in Linear Structural Equation Models," *Psychometrika*, 55, 495–515.

―――― (1994), "Causal Inference in Latent Variable Models," in *Latent Variables Analysis: Applications for Developmental Research*, eds. A. von Eye and C. C. Clogg, Thousand Oaks, CA: Sage, pp. 3–35.

―――― (1995), "Causal Inference in the Social and Behavioral Sciences," in *Handbook of Statistical Modeling for the Social and Behavioral Sciences*, eds. G. Arminger, C. C. Clogg, and M. E. Sobel, New York: Plenum, pp. 1–38.

―――― (1997), "Measurement, Causation, and Local Independence," in *Latent Variable Modeling and Applications to Causality*, ed. M. Berkane, New York: Springer-Verlag, pp. 11–28.

Spirtes, P., Glymour, C., and Scheines, R. (1993), *Causation, Prediction and Search*, New York: Springer-Verlag.

Whittaker, J. (1990), *Graphical Models in Applied Multivariate Statistics*, New York: Wiley.

Winship, C., and Morgan, S. L. (1999), "The Estimation of Causal Effects From Observational Data," *Annual Review of Sociology*, 25, 659–706.

Wright, S. (1928), "Appendix," in *The Tariff on Animal and Vegetable Oil*, by P. G. Wright, New York: MacMillan.

Political Methodology: A Welcoming Discipline

Nathaniel L. Beck

1. INTRODUCTION

Although empirical political science can be dated as far back as Aristotle, and some of the earliest work in statistics was on political methodology (Petty 1690), the self-definition of political science as a science goes back only a century or so (with a convenient date being the founding of a "scientific" department at Columbia University in 1880). Political methodology is a young subfield within political science. The coming of age of that subfield is best evidenced by the arrival of the journal *Political Methodology*, now named *Political Analysis*. By that date, the subfield of political methodology is about 25 years old.

Although serious quantitative political analyses were conducted in the 19th and early 20th centuries (Gow 1985), the increased use of quantitative analysis went hand-in-hand with the post-World War II "behavioral revolution" in political science (Dahl 1961). The behavioralists had begun using more and more sophisticated methods, though few, if any, would have labelled themselves as "methodologists." The most sophisticated behavioralists acquired their methodological training from other disciplines (first sociology, later economics) and usually did not teach courses in methodology to pass on their sophistication to their graduate students. But starting some time in the late 1960s, we did begin to develop a small cadre of people who did at least a portion of their scholarship as methodologists.

Chris Achen, in his final editorial for *Political Methodology*, summed up the situation:

> When [*Political Methodology*] began in the mid-1970s, methodology was more often an avocation than a vocation. No political science journal welcomed methodological articles, and many journals rejected them out of hand. Certainly no Political Methodology Society existed to give shape and organization to the needs of political methodologists. In the face of these difficulties, John Sullivan and George Marcus created *Political Methodology*.. . .By [the 1980s], the field had come of age, and the rapid development of the last few years were possible (Achen 1985).

Nathaniel L. Beck is Professor, Department of Political Science, University of California, San Diego, La Jolla, CA 92093 (E-mail: beck@ucsd.edu) and Editor of *Political Analysis*. The author thanks Henry Brady and Gary King for helpful comments.

In this essay, I look at, from my own perspective, a few major features of this burgeoning of political methodology and how they might play out in the early part of the new millennium. In particular, I look at the relationship of political methodology to other social science disciplines and to statistics. I then look at the changing nature of data collection.

2. POLITICAL METHODOLOGY AND OTHER SOCIAL SCIENCES

Political science as a discipline is substantively, not methodologically, defined. Political scientists use a variety of methods to attack questions related to political institutions and behavior. Although the methodological issues are defined by our political questions, we freely use whatever methodological solutions are available. Thus political methodology has freely drawn on insights from econometrics, psychometrics, sociology, and statistics. One clear trend, however, is our increased reliance on our own methodological expertise.

Quantitative political analysis in the 1950s and early 1960s was largely the analysis of contingency tables. By the 1960s, path analysis, imported from sociology, was becoming the state of the art. By the 1970s, multiple regression had become the dominant tool, as it remains, and political methodology turned more and more to econometrics for its basic tools (and to econometricians for basic training). (Overviews of the political methodology subfield have been given in Achen 1983 and Bartels and Brady 1993.)

Whereas political methodologists have always been a diverse lot, the modal political methodologist looked a lot like an applied econometrician (and had received advanced training by taking the econometrics sequence at top departments). The ease of use of modern software has brought a decreasing lag between the introduction of a new method in econometrics and its use in political science. Thus political science articles now routinely have citations to state-of-the-art econometrics papers in such areas as *time series analysis*, *analysis of choice* (and other limited dependent-variable methods), and *analysis of duration data*.

Although it might be tempting to belittle political methodology in the 1970s and 1980s as a "borrowing" discipline, a fairer assessment would be that it was a "welcoming" discipline. Given its substantive political orientation, political methodology has always been willing to import methods, as long as those methods contributed to our understanding of political phenomena. Starting in the 1980s, political science departments even began to hire econometricians (and a few statisticians) as full-time department members. But these econometricians were judged on how they solved political science problems, not on the prettiness of their econometrics.

As political methodologists have become better trained, we are now creating as well as borrowing solutions. The best example of this is the ecological inference problem. One of the oldest political methodology problems, ecological inference involves using data observed only on geographic groups to make inferences about individual-level behavior. Political scientists have always had plentiful ecological data and have long searched for methods of using such data to assess questions of interest.

The use of ecological data declined with the publication of Robinson's (1950) classic article showing that correlating ecological data did not yield good estimates of the underlying individual correlations. But the Voting Rights Act, which demanded analyses of turnout and vote by race, brought a resurgence of interest in ecological inference. Unfortunately, the leading technique for analysis, Goodman's (1953) regression, was known to produce incorrect individual level-estimates. Thus by the 1980s, the search was on for a good way to do ecological inference.

Although scholars from many disciplines made contributions, a generally workable "solution" was recently proposed by the political scientist Gary King (1998). This solution proceeds by joining the random parameter models with the specific constraints imposed on those parameters by the nature of ecological data. King's method seems to yield good estimates of individual-level parameters and also provides diagnostics for when the method will fail, and it is now widely used by political scientists investigating electoral data. It is fitting that political methodologists solved the oldest problem in political methodology.

My own work has also moved in this direction. Econometricians and biometricians have shown how to analyze longitudinal data; that is, (a small number of) repeated observations on a (large) sample of individuals. But political scientists are often concerned with analyzing time series–cross-sectional data on a small number of countries over a long time span. These data may show temporal and spatial dependence. But the units are fixed and not sampled, and so many of the issues on which econometricians and biometricians focused are irrelevant; conversely, the political data have unique features—namely, some form of spatial autocorrelation. To make matters more complicated, many analyses in international relations use a binary dependent variable (e.g., war or no war). Rather than simply import the inappropriate but known methods from econometrics and biometrics, a methodology for estimating the types of models encountered in studies of comparative politics and international relations was developed (Beck and Katz 1995; Beck, Katz, and Tucker 1998). Political science datasets from comparative politics and international relations are different than datasets from epidemiology, so it makes sense that they should be analyzed using different methods, and the appropriate methods for the political datasets should be studied by political methodologists.

Political methodologists have done more than come up with appropriate techniques for analyzing political data. They have also been actively involved in looking at the underlying unity of political research. Much of political research, particularly in comparative politics, is inherently qualitative. Early work in political methodology more or less ignored those doing qualitative work, leaving qualitative analysts ignorant about issues of inference. But a recent and welcome trend is for political methodologists to show that the underlying logic of research does not depend on whether it is qualitative or quantitative and that both are subject to the same methodological issues, particularly the issue of sampling and case selection (Geddes 1990; King, Keohane, and Verba 1994). Much of political science is qualitative, but it is still science, and it is good that political methodologists are taking seriously the methodological problems faced by qualitative analysts. One consequence of this is the recent explosive growth in first-rate scientific studies of comparative politics, be they quantitative or qualitative.

3. POLITICAL METHODOLOGY AND STATISTICS

Whereas political methodology is creating methods to solve political problems, it still remains a welcoming discipline, drawing heavily on allied disciplines, primarily econometrics. There is a heartening tendency for political methodology to also draw heavily on modern statistical methods. The uneasy relationship between political methodology and statistics is not accidental. To oversimply, statisticians work hard to get the data to speak, whereas political scientists (and econometricians) are more interested in testing theory (and hence more interested in whether a model parameter is large or small, rather than in the exact relationship between a dependent variable and an independent variable). Much of statistics is data driven; political methodology is theory driven.

Modern statistics has influenced political methodology in a few ways. Perhaps the most positive recent innovation has been the acceptance of Bayesian ideas without the theological wrangling that often accompanies those ideas. Thus King's work on ecological analysis was at least partly a function of his use of Bayesian random parameter models in common use in statistics (Rubin 1981).

The random parameter model will also hopefully solve some of the thornier problems in comparative politics. Students of comparative politics have typically assumed that all units are incomparable, and hence did case studies, or have made the opposite assumption and assumed that all units followed the same underlying process, and hence used simple regression models. But, as Western (1998) showed, the random parameter (or hierarchical) model provides a very nice compromise between these two extreme positions. Countries may differ, but the parameters pertaining to a given country all reflect draws from a common distribution.

Although both of these methods are Bayesian, both use relatively uninformative priors and have proven quite acceptable to the typical classicist political methodologist. It appears that the debate between classicists and Bayesians may be much more severe when it comes to first principles than it is for actual applications. Whereas political methodology still follows classical rather than Bayesian principles, we have become more at ease in thinking about "posterior" distributions, even if we tend not to call them that. We have also become more at ease with using Bayesian ideas such as simulation, whether from a "classical" (Herron 2000; King, Tomz, and Wittenberg 2000) or a Bayesian (Jackman 2000) perspective.

Political methodologists have been loathe to abandon the simplicity of the linear model; again, this is due at least as much to the nature of our theoretical interests as to our unwillingness to master new and complicated techniques. But even if we are not interested in every nuance of the relationship between y and x, the assumption of linearity still may be much too strong. Beck and Jackman (1998) tried to convince political scientists that they could improve on their work by starting with Hastie and Tibshirani's (1990) generalized additive model rather than the more ubiquitous generalized linear model. Similarly, Beck, King, and Zeng (2000) examined neural networks as a substitute for standard linear models in the study of international conflict, where, it was conjectured, the correct model is one of massive interaction rather than linear additivity; the forecasting properties of the neural network model bore out this conjecture. Although political methodologists are turning to modern statistics only slowly, the trend seems clear.

4. DATA

As important as analysis is, no field is better than its data. Political methodology has been at the forefront of data collection and data dissemination. The Interuniversity Consortium for Political and Social Research was founded in 1962 at the University of Michigan. Its mission was to archive and disseminate datasets that could be used by all researchers. Long before the world wide web made data dissemination easy, the Consortium housed an enormous archive of data that was made available to all political scientists. The youngest graduate student had access to the same data as the most veteran researcher.

Since the 1950s, the single most-used data source in political science has been the National Election Study's survey conducted for each Presidential and Congressional election (often with a panel component). These data were collected according to the highest standards, with diverse political methodologists having input into the design of the study. By having a common source of data, political methodologists could attack research questions without worrying about whether different findings might be a result of different survey methods or, worse, of inferior survey methods.

Although this high-quality and publicly available dataset allowed for advances in the discipline, it also imposed a certain uniformity. Political methodologists have quickly taken advantage of modern advances in computer-assisted telephone (CAT) interviewing to answer questions that could never have been answered before.

The CAT technology has greatly expanded the ability to do survey experiments that join the rigor of the experiment with the relevance of the field study (Sniderman et al. 1991). Thus Johnston, Blais, Brady, and Crete (1992), for example, were able to convincingly demonstrate via survey experiment that the Liberals in Canada had a rhetorical advantage early in the 1988 Parliamentary election campaign, but that the Conservatives effectively countered this over the course of the campaign. This type of analysis has dramatically enriched our understanding of the electoral process.

5. CONCLUSION

The past quarter-century has seen political methodology come into its own. Political science data shares some features with other types of data, and so it is appropriate that political methodologists learn from other methodological disciplines. But our data also has some unique properties, so it is critical that political methodologists analyze and solve the problems that characterize political datasets.

Since Achen wrote his editorial in 1985, the subfield of political methodology has flourished. A small summer meeting in 1984 of about a dozen scholars at the University of Michigan has become an annual event attended by more than 100 faculty and graduate students. Political Methodology has become one of the five largest organized sections in the American Political Science Association; this section publishes the quarterly journal *Political Analysis*. Scholars can now publish methodological articles not only in *Political Analysis*, but also in all of the leading disciplinary journals. Departments now advertise for methodologists, and all major departments have faculty whose principal graduate teaching is devoted to methodology. A good, substantively trained graduate student will know much more methodology than the most sophisticated methodologists of 1980.

The challenge comes in remaining open to new methods, particularly related to the intensive computer analysis of datasets. The cutting-edge political methodologists have been open to such innovations, but it remains to be seen if they will be successful in importing such methods into general research.

One such challenge in the next millennium is a sensible treatment of missing data. Political methodologists have ignored that problem until very recently, but some recent breakthroughs (King, Honacker, Joseph, and Scheve 1998; Little and Rubin 1987) make it possible that all political scientists can improve their treatment of missing data. These two works show an interesting relationship between statisticians and political methodologists. The statistical work, though mathematically persuasive, had literally no impact on political science practice. But the political science work, building on the earlier statistical work, clearly showed political scientists how inefficient their practices were, and how the practice of ignoring missing data led to incorrect inferences in important substantive arenas. Equally important, the work by the political methodologists promises to provide a practical solution that can be used by almost any political scientist with missing data. Political methodology is happy to import methods (and scholars) from other disciplines, but those ideas must be shown to solve important problems in political science.

REFERENCES

Achen, C. (1983), "Towards Theories of Data," in *Political Science: The State of the Discipline*, ed. A. Finifter, Washington, DC: American Political Science Association, pp. 69–93.

——— (1985), "Editorial," *Political Methodology*, 11.

Bartels, L., and Brady, H. (1993), "The State of Quantitative Political Methodology," in *Political Science: The State of the Discipline II*, ed. A Finifter, Washington, DC: American Political Science Association, pp. 121–159.

Beck, N., and Jackman, S. (1998), "Beyond Linearity by Default: Generalized Additive Models," *American Journal of Political Science*, 42, 596–627.

Beck, N., and Katz, J. N. (1995), "What to do (and not to do) With Times Series Cross-Section Data," *American Political Science Review*, 89, 634–647.

Beck, N., Katz, J. N., and Tucker, R. (1998), "Taking Time Seriously: Time Series–Cross-Section Analysis With a Binary Dependent Variable," *American Journal of Political Science*, 42, 1260–1288.

Beck, N., King, G. M., and Zeng, L. (2000), "Improving Quantitative Studies of International Conflict: A Conjecture," *American Political Science Review*, 94, 21–35.

Dahl, R. (1961), "The Behavioral Approach in Political Science: Epitaph for a Monument to a Successful Protest," *American Political Science Review*, 55, 763–772.

Geddes, B. (1990), "How the Cases You Choose Affect the Answers You Get: Selection Bias in Comparative Politics," *Political Analysis*, 2, 131–150.

Goodman, L. (1953), "Ecological Regressions and the Behavior of Individuals," *American Sociological Review*, 18, 663–666.

Gow, D. J. (1985), "Quantification and Statistics in the Early Years of American Political Science, 1880–1922," *Political Methodology*, 11, 1–18.

Hastie, T. J., and Tibshirani, R. J. (1990), *Generalized Additive Models*, New York: Chapman and Hall.

Herron, M. C. (2000), "Post-Estimation Uncertainty in Limited Dependent Variable Models," *Political Analysis*, 8, 83–98.

Jackman, S. (2000), "Estimation and Inference via Bayesian Simulation: An Introduction to Markov Chain Monte Carlo," *American Journal of Political Science*, 44, 375–404.

Johnston, R., Blais, A., Brady, H. E., and Crete, J. (1992), *Letting the People Decide: Dynamics of a Canadian Election*, Montreal: McGill-Queen's University Press.

King, G. (1998), *A Solution to the Ecological Inference Problem*, Princeton, NJ: Princeton University Press.

King, G., Honaker, J., Joseph, A., and Scheve, K. (1998), "Analyzing Incomplete Political Science Data: An Alternative Algorithm for Multiple Imputation," unpublished paper presented at the Annual Meetings of the American Political Science Association, Boston.

King, G., Keohane, R. O., and Verba, S. (1994), *Designing Social Inquiry*, Princeton, NJ: Princeton University Press.

King, G., Tomz, M., and Wittenberg, J. (2000), "Making the Most of Statistical Analyses: Improving Interpretation and Presentation," *American Journal of Political Science*, 44, 341–355.

Little, R. J. A., and Rubin, D. B. (1987), *Statistical Analysis With Missing Data*, New York: Wiley.

Petty, W. (1690), *Political Arithmetick*, London: Robert Clavel and Hen. Mortlock.

Robinson, W. S. (1950), "Ecological Correlation and the Behavior of Individuals," *American Sociological Review*, 15, 351–357.

Rubin, D. R. (1981), "Estimation in Parallel Randomized Experiments," *Journal of Educational Statistics*, 6, 377–400.

Sniderman, P. M., Brody, R. A., and Tetlock, P. E. (1991), *Reasoning and Choice: Explorations in Political Psychology*, New York: Cambridge University Press.

Western, B. (1998), "Causal Heterogeneity in Comparative Research: A Bayesian Hierarchical Modelling Approach," *American Journal of Political Science*, 42, 1233–1259.

Statistics in Sociology, 1950–2000

Adrian E. Raftery

1. INTRODUCTION

Sociology is the scientific study of modern industrial society. Example questions include: What determines how well people succeed in life, occupationally and otherwise? What factors affect variations in crime rates between different countries, cities, and neighborhoods? What are the causes of the increasing U.S. divorce rate? What are the main factors driving fertility decline in developing countries? Why have social revolutions been successful in some countries but not in others?

The roots of sociology go back to the mid-19th century and to seminal work by Auguste Comte, Karl Marx, Max Weber, and Emile Durkheim on the kind of society newly emerging from the industrial revolution. Sociology has used quantitative methods and data from the beginning, but before World War II the data tended to be fragmentary and the statistical methods simple and descriptive.

Since then, the available data have grown in complexity, and statistical methods have been developed to deal with them, with the sociologists themselves often leading the way (Clogg 1992). The trend has been toward increasingly rigorous formulation of hypotheses, larger and more detailed datasets, more complex statistical models to match the data, and a higher level of statistical analysis in the major sociological journals.

Statistical methods have had a successful half-century in sociology, contributing to a greatly improved standard of scientific rigor in the discipline. Sociology has made use of a wide variety of statistical methods and models, but I focus here on the ones developed by sociologists, motivated by sociological problems, or first published in sociological journals. I distinguish three postwar generations of statistical methods in sociology, each defined by the kind of data it addresses. The first generation of methods, starting after World War II, deals with cross-tabulations of counts from surveys

Adrian E. Raftery is Professor of Statistics and Sociology, Department of Statistics, University of Washington, Seattle, WA 98195 (E-mail: raftery@stat.washington.edu). The author is very grateful to Andrew Abbott, Mark Becker, Timothy Biblarz, Kenneth Bollen, Leo Goodman, Mark Handcock, Robert Hauser, Charles Hirschman, Daphne Kuo, Paul LePore, David Madigan, Peter Marsden, Stephen Matthews, Martina Morris, Michael Seltzer, Burton Singer, Michael Sobel, Elizabeth Thompson, Rob Warren, Stanley Wasserman, and Kazuo Yamaguchi for extremely helpful comments. The fact that so many colleagues helped testifies to the vibrancy of the sociological methodology community whose work is the core of this discussion. The views expressed remain resolutely personal.

and censuses by a small number of discrete variables such as sex, age group, and occupational category; social mobility tables provide a canonical example. Schuessler (1980) gave a survey that largely reflects this first-generation work.

The second generation, starting in the early 1960s, deals with unit-level data from surveys that include many variables. This generation was galvanized by Blau and Duncan's (1967) highly influential book *The American Occupational Structure,* and by the establishment of *Sociological Methodology* in 1969 and *Sociological Methods and Research* in 1972 as publication outlets. These developments marked the coming of age of research on quantitative methodology in sociology. The third generation of methods, starting in the late 1980s, deals with data that are not usually thought of as cross-tabulations or data matrices, either because the data take different forms, such as texts or narratives, or because dependence is a crucial aspect. These generations do not have clear starting points and all remain active today; like real generations, they overlap.

Today, much sociological research is based on the reanalysis of large high-quality survey sample datasets, usually collected with public funds and publicly available to researchers, with typical sample sizes in the range of 5,000–20,000. This has opened the way to easy replication of results and has helped produce standards of scientific rigor in sociology comparable to those in many of the natural sciences. Social statistics is expanding rapidly as a research area, and several major institutions have recently launched initiatives in this area.

2. THE FIRST GENERATION: CROSS-TABULATIONS

2.1 Categorical Data Analysis

Initially, much of the data that quantitative sociologists had to work with came in the form of cross-classified tables, and so it is not surprising that this is perhaps the area of statistics to which sociology has contributed the most. A canonical example has been the analysis of social mobility tables, two-way tables of father's against respondent's occupational category; typically the number of categories used is between 5 and 17.

At first, the focus was on measures of association, or mobility indices as they were called in the social mobility context (Glass 1954; Rogoff 1953), but these indices failed to do the job of separating structural mobility from exchange (or circulation) mobility. It was Birch (1963) who proposed the log-linear model for the observed counts $\{x_{ij}\}$, given by

$$\log(E[x_{ij}]) = u + u_{1(i)} + u_{2(j)} + u_{12(ij)}, \tag{1}$$

where i indexes rows and j indexes columns, $u_{1(i)}$ and $u_{2(j)}$ are the main effects for the rows and columns, and $u_{12(ij)}$ is the interaction term, measuring departures from

Table 1. Observed Counts From the Largest U.S. Social Mobility Study and Expected Values
From a Goodman Association Model With 4 Degrees of Freedom, Sample Size 19,912

Father's occupation	Son's occupation									
	Upper nonmanual		Lower nonmanual		Upper manual		Lower manual		Farm	
	Obs.	Exp.	Obs.	Exp.	Obs.	Exp.	Obs.	Exp.	Obs.	Exp.
Upper nonmanual	1,414	1,414	521	534	302	278	643	652	40	42
Lower nonmanual	724	716	524	524	254	272	703	698	48	43
Upper manual	798	790	648	662	856	856	1,676	1,666	108	112
Lower manual	756	794	914	835	771	813	3,325	3,325	237	236
Farm	409	386	357	409	441	405	1,611	1,617	1,832	1,832

NOTE: Data from Hout (1983).

independence. The difficulty with (1) for social mobility and similar tables is that the number of parameters is too large for inference and interpretation; for example, in the U.S. datasets 17 categories were used, so the interaction term involves $16^2 = 256$ parameters.

A successful general approach to modeling the interaction term parsimoniously is the association model of Duncan (1979) and Goodman (1979),

$$u_{12(ij)} = \sum_{k=1}^{K} \gamma_k \alpha_i^{(k)} \beta_j^{(k)} + \phi_i \delta(i, j),$$ (2)

where $\delta(i, j) = 1$ if $i = j$ and 0 otherwise. In (2), $\alpha_i^{(k)}$ is the score for the ith row on the kth scoring dimension, and $\beta_j^{(k)}$ is the corresponding score for the jth column; these can be either specified in advance or estimated from the data. The last term allows a different strength of association on the diagonal. [Model (2) is unidentified as written; various identifying constraints are possible.] In most applications to date, $K = 1$. Goodman (1979) initially derived this model as a way of describing association in terms of local odds ratios. He (Goodman 1985) later showed that this model is closely related to canonical correlations and to correspondence analysis (Benzécri 1976), and provided an inferential framework for these methodologies. Table 1 shows the actual counts for a reduced version of the most extensive U.S. social mobility study and the fitted values from an association model; the model accounts for 99.6% of the association in the table, and its success is evident.

Hout (1984) extended the range of application of these models by modeling the scores and diagonal terms in (2) as sums or products of covariates, such as characteristics of the occupational categories in question; this is an extension of Birch's (1965) linear-by-linear interaction model. This has led to important discoveries, including Hout's (1988) finding that social mobility is on the increase in the United States. Biblarz and Raftery (1993) adapted Hout's models to higher-dimensional tables to study social mobility in nonintact families, finding that occupational resemblance is weaker there than in intact families. From sociology, these ideas have diffused to other disciplines, such as epidemiology (Becker 1989).

An appealing alternative formulation of the basic ideas underlying (1) and (2) is in terms of marginal *distributions* rather than the main effects in (1). The resulting marginal models specify a model for the marginal distributions and a model for the odds ratios, and this implies a model for the joint distribution that is not log-linear (Becker 1994; Becker and Yang 1998; Lang and Agresti 1994).

An alternative approach that answers different questions is the latent class model (Goodman 1974; Lazarsfeld 1950). This represents the distribution of counts as a finite mixture of distributions in each of which the different variables are independent. An interesting recent application to criminology was described by Roeder, Lynch, and Nagin (1999).

2.2   Hypothesis Testing and Model Selection

Sociologists often have sample sizes in the thousands, and so they came up early and hard against the problem that standard p values can indicate rejection of null hypotheses in large samples, even when the null model seems reasonable theoretically and inspection of the data fails to reveal any striking discrepancies with it. The problem is compounded by the fact that there are often many models rather than just the two envisaged by significance tests, and by the need to use stepwise or other multiple-comparison methods for model selection (e.g., Goodman 1971). By the early 1980s, some sociologists were dealing with this problem by ignoring the results of p value-based tests when they seemed counterintuitive and by basing model selection instead on theoretical considerations and informal assessment of discrepancies between model and data (e.g., Fienberg and Mason 1979; Grusky and Hauser 1984; Hout 1983, 1984).

Then it was pointed out that this problem could be alleviated by instead basing model selection on Bayes factors (Raftery 1986), and that this can be simply approximated for log-linear models by preferring a model if the Bayes information criterion (BIC) = deviance − (degrees of freedom) $\log(n)$ is smaller (Schwarz 1978). For nested hypotheses, this can be viewed as defining a significance level for a test that decreases automatically with sample size. Since then, this approach has been used in many sociological applications of log-linear models. Kass and Wasserman (1995) showed that the approximation is quite accurate if the Bayesian prior used for the model parameters is a unit information prior, and Raftery (1995) indicated how the methodology can be extended to a range of other models. Weakliem (1999) criticized the use of BIC on the grounds that the unit information prior to which it corresponds may be too diffuse in practice. This points toward using Bayes factors based on priors that reflect the actual information available; this is easy to do for log-linear and other generalized linear models (Raftery 1996).

3. THE SECOND GENERATION: UNIT-LEVEL SURVEY DATA

The second generation of statistical models responded to the availability of unit-level survey data in the form of large data matrices of independent cases. The methods that have proved successful for answering questions about such data have mainly been based on the linear regression model and its extensions to path models, structural equation models, generalized linear models, and event history models. For questions about the *distribution* of variables rather than their predicted value, however, nonparametric methods have proven useful (Handcock and Morris 1998; Morris, Bernhardt, and Handcock 1994).

3.1   Measuring Occupational Status

Occupational status is an important concept in sociology, and developing a useful continuous measure of it was a signal achievement of the field. Initially, the status of an occupation was equated with its perceived prestige, as measured in surveys. However, surveys could measure the prestige of only a small number of the 800 or so occupations identified in the Census. To fill in the missing prestige scores, Duncan (1961) regressed the prestige scores for the occupations for which they were available on measures of the average education and average income of incumbents of the occupation. He found that the predictions were very good $(R^2 = .91)$, and that the two predictors were about equally weighted. Based on this, he created a predicted prestige score for all occupations, which became known as the Duncan socioeconomic index (SEI). The SEI later turned out to be a better predictor of various social outcomes than the prestige scores themselves. Duncan's initial work has been updated several times (Hauser and Warren 1997).

In much social research, particularly in economics, current income is used as a predictor of social outcomes, but there are good reasons to prefer occupational status. It has proven to be a good predictor of many social outcomes. Jobs and occupations can be measured accurately, in contrast to income or wealth, whose measurement is plagued by problems of refusal, recall, and reliability. Also, occupational status is more stable over time than income, both within careers and between generations. This suggests that occupational status may actually be a better indicator of long-term or permanent income than current income itself. The status of occupations tends to be fairly constant both in time and across countries (Treiman 1977).

3.2   The Many Uses of Structural Equation Models

Figure 1 shows the basic path model of occupational attainment at the heart of the work of Blau and Duncan (1967) (see Duncan 1966). Wright (1921) introduced

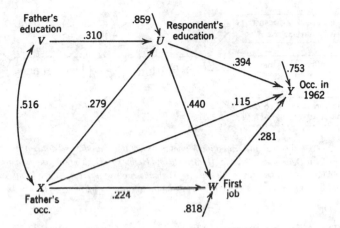

Figure 1. A Famous Path Model: The process of stratification, U.S. 1962 (Blau and Duncan 1967). The numbers on the arrows from one variable to another are regression coefficients, .516 is the correlation between V and X, and the numbers on the arrows with no sources are residual standard deviations. All the variables have been centered and scaled.

path analysis, and Blalock (1964) gave it a causal interpretation in a social science context. (See Freedman 1987 and Sobel 1998 for critique and discussion, and Abbott 1998, Pearl 1998, and Sobel 2000 for histories of causality in social science.)

Often, variables of interest in a causal model are not observed directly, but other variables are observed that can be viewed as measurements of the variables, or "constructs" of interest, such as prejudice, alienation, conservatism, self-esteem, discrimination, motivation, or ability. Jöreskog (1973) dealt with this by maximum likelihood estimation of a structural equation model with latent variables; this is sometimes called a LISREL model, from the name of Jöreskog's software. Figure 2 shows a typical model of this kind; the goal of the analysis is testing and estimating the strength of the relationship between the unobserved latent variables represented by the thick arrow. Diagrams such as Figures 1 and 2 have proven useful to sociologists for specifying theories and hypotheses and for building causal models.

The LISREL framework has been extended and used ingeniously for purposes beyond those for which it was originally intended. Muthén (1983) extended it to categorical variables, and later (Muthén 1997) showed how it can be used to represent longitudinal data, growth curve models, and multilevel data. Kuo and Hauser (1996) used data on siblings to control for unobserved family effects on socioeconomic outcomes, and cast the resulting random effects model in a LISREL framework. Warren, LePore, and Mare (2001) considered the relationship between the number of hours that high school students work and their grades; a common assumption might be that working many hours tends to depress grades. They found that although number of hours and grades do indeed tend to covary (negatively), the causal direction is the opposite: Low grades leads to many hours worked, rather than the other way round.

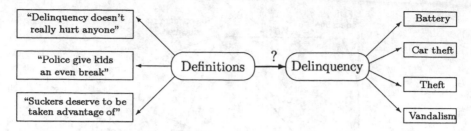

Figure 2. Part of a Structural Equation Model to Assess the Hypothesis That Learned Definitions of Delinquency Cause Delinquent Behavior (Matsueda and Heimer 1987). The key goal is testing and estimating the relationship represented by the thick arrow. The constructs of interest, "Definitions" and "Delinquency", are not measured directly. The variables inside the rectangles are measured.

The advent of graphical Markov models (Spiegelhalter, Dawid, Lauritzen, and Cowell 1993), specified by conditional independencies rather than by regression-like relationships, is important for the analysis of multivariate dependencies, although they can seem less interpretable to sociologists. The relationship between these and structural equation models has begun to be understood (Spirtes et al. 1998). Also, the LISREL model seems ideally suited to Markov chain Monte Carlo (MCMC) methods (Gilks, Richardson, and Spiegelhalter 1996), and this is likely to permit useful extensions of the framework (Arminger 1998; Raftery 1991; Scheines, Hoijtink, and Boomsma 1999).

3.3   Event History Analysis

Unit-level survey data often include or allow the reconstruction of life histories. These include the times of crucial events such as marriages, divorces, births, commitals to and releases from prison, job changes, and going on or off welfare.

The analysis of factors influencing the time to a single event such as death was revolutionized by the introduction of the Cox (1972) proportional hazards model. Tuma and Hannan (1984) generalized this approach to allow for repeated events, for multiple types of events, such as marriages and divorces, and for events consisting of movement between different types of states, such as different job categories.

Uses of the Cox model in medicine have tended to treat the baseline hazard nonparametrically, but in social science it has sometimes been found useful to model it parametrically. For example, Yamaguchi (1992) analyzed permanent employment in Japan where the surviving fraction (those who never change jobs) and its determinants are of key interest; he found that covariates were associated both with the timing of job change and with the surviving fraction.

Social science event history data are often recorded in discrete time (e.g., by year), either because events tend to happen at particular times of year (e.g., graduating) or because of measurement constraints. As a result, discrete-time event history models have been popular (Allison 1982; Xie 1994), and in some ways these are easier to handle than their continuous-time analogs. Ways of dealing with multilevel event history data, smoothly time-varying covariates, and other complications have been introduced in this context (e.g., Fahrmeir and Knorr-Held 1997; Raftery, Lewis, and Aghajanian 1995).

One problem with social science event history data is that dropping out can be related to the event of interest. For example, people may tend to leave a study shortly before a divorce, which will play havoc with estimation of divorce rates. The problem seems almost insoluble at first sight, but Hill (1997) produced an elegant solution using the shared unmeasured risk factor (SURF) model of Hill, Axinn, and Thornton (1993). The basic trick is to observe that although one does not know which of the people who dropped out actually got divorced soon afterward, one can estimate which ones were most at risk of divorcing.

4. THE THIRD GENERATION: NEW DATA, NEW CHALLENGES, NEW METHODS

4.1   Social Networks and Spatial Data

Social networks consist of sets of pairwise connections, such as friendships between adolescents, sexual relationships between adults, and political alliances and patterns of marriage between social groups. The analysis of data about such networks has a long history (Wasserman and Faust 1994). Frank and Strauss (1986) developed formal statistical models for such networks related to the Markov random field models used in Bayesian image analysis and derived using the Hammsersley–Clifford theorem (Besag 1974). This has led to the promising "p^*" class of models for social networks (Wasserman and Pattison 1996).

Methods for the analysis of social networks have focused mostly on small datasets with complete data. In practical applications, however, such as the effect of sexual network patterns on the spread of sexually transmitted diseases (Morris 1997), the datasets tend to be large and very incomplete, and current methods are somewhat at a loss. This is the stage that pedigree analysis in statistical genetics was at some years ago, but the use of likelihood and MCMC methods have led to major progress since then (Thompson 1998). Social networks are more complex than pedigrees in one way, because pedigrees tend to have a tree structure whereas social networks often have cycles, but progress does seem possible.

Most social data are spatial, but this fact has been largely ignored in sociological research. A major exception is Massey and Denton's (1993) study of residential segregation by race, reviving a much older sociological tradition of spatial analysis in American society (e.g., Duncan and Duncan 1957). More recently, the field of research on fertility and contraception in Asia (several major projects focused on China, Thailand, and Nepal) has been making fruitful use of satellite image and Geographic Information System (GIS) data (e.g., Entwisle, Rindfuss, Walsh, Evans, and Curran 1997). More extensive use of spatial statistics in sociology seems likely.

4.2   Textual Data

In its rawest form, a great deal of sociological data is textual; for example, interviews, answers to open-ended questions in surveys, enthnographic accounts. How to analyze such data and draw inference from it is a largely open question. Efforts at formal analysis have focused on standard content analysis, consisting mainly of counting words in the text in different ways. It seems likely that using the context in which words and clauses appear would yield better results. Promising recent efforts to do just this include Carley's (1993) map analysis, Franzosi's (1994) set-theoretic approach, and Roberts's (1997) generic semantic grammar, but the surface has only been scratched. The human mind is very good at analyzing individual texts, but computers are not (at least as yet); in this way the analysis of textual data may be like other problems such as image analysis and speech recognition. A similar challenge is faced on a massive scale by information retrieval for the Web (Jones and Willett 1997), where most search engines are based on simple content analysis methods. The more contextual methods being developed in sociology might be useful in this area as well.

Singer, Ryff, Carr, and Magee (1998) have made an intriguing use of textual data analysis, blending quantitative and qualitative approaches. They took a standard unit-level dataset with more than 250 variables per person and converted them into written "biographies." They then examined the biographies for common features and thinned them to more generic descriptions.

4.3   Narrative and Sequence Analysis

Life histories are typically analyzed by reducing them to variables and doing regression and multivariate analysis, or by event history analysis. Abbott and Hrycak (1990) argued that these standard approaches obscure vital aspects of a life history (such as a professional career) that emerge when it is considered as a whole. They proposed viewing life histories of this kind as analogous to DNA or protein sequences, using optimal alignment methods adapted from molecular biology (Sankoff and Kruskal 1983), followed by cluster analysis, to detect patterns common to groups of careers. Stovel, Savage, and Bearman (1996) used these methods to describe changes in career systems at Lloyds Banks over the past century.

Subsequently, Dijkstra and Taris (1995) extended the ideas to include independent variables, and Abbott and Barman (1997) applied the Gibbs sampling sequence detection method of Lawrence et al. (1993), originally also developed for microbiology; this seems to work very well. The approach is interesting, and there are many open statistical questions.

4.4 Simulation Models

Another way to represent a social process in more detail is via a macrosimulation or microsimulation model. Such models are often deterministic and quite complicated, representing systems by different compartments that interact, and each compartment by a set of differential or difference equations. They have been used to, for example, explore the implications of different theories about how domestic politics and war interact (Hanneman, Collins, and Mordt 1995), the social dynamics of collective action (Kim and Bearman 1997), and the role of sexual networks in the spread of HIV (Morris 1997 and references therein).

A difficulty with such models is that ways of estimating the many parameters involved, of assessing the fit of the model, and of comparing competing models are not well established; all of this tends to be done by informal trial and error. Methods being developed to put inference for such models on a solid statistical footing in other disciplines may prove helpful in sociology as well (Guttorp and Walden 1987; Poole and Raftery 2000; Raftery, Givens, and Zeh 1995).

4.5 Macrosociology

Macrosociology deals with large entities, such as states and their interactions. As a result, the number of cases tends to be small, and the use of standard statistical methods such as regression is difficult. This was pointed out trenchantly by Ragin (1987) in an influential book. His own proposed alternative, qualitative comparative analysis, seems unsatisfactory, because it does not allow for variability of any kind and so is sensitive to small changes in the data and in the way the method is applied (Lieberson 1994).

One solution to the problem is to obtain an at least moderately large sample size, as Bollen and Appold (1993) were able to do, for example. Often, however, this is not possible, so this is not a general solution. Another approach is to use standard regression-type models, but to do Bayesian estimation with strong prior information if available, which it often is from the practice, common in this area, of analyzing specific cases in great detail (Western and Jackman 1994). Bayes factors may also help, as they tend to be less stringent than standard significance tests in small samples and allow a calibrated assessment of evidence rather than forcing the rejection or acceptance of a hypothesis (Kass and Raftery 1995). They also provide a way of accounting for model uncertainty, which can be quite large in this context (Western 1996).

5. DISCUSSION

Statistical methodology has had a successful half-century in sociology, leading the way in providing models for cross classifications and developing well-adapted methods for unit-level datasets. This has contributed to the greatly improved level of scientific rigor in sociology today.

New kinds of data and new challenges abound, and the area is ripe for statistical research. Several major institutions are launching initiatives in the area. The University of Washington has just established a new Center for Statistics and the Social Sciences, UCLA's new Statistics Department grew out of social statistics, and there are other initiatives at the University of Michigan, Columbia University, UC Santa Barbara, and the universities in North Carolina's Research Triangle. Harvard's new Center for Basic Research in the Social Sciences also emphasizes social statistics. They all join the most successful effort of this kind to date, the Social Statistics Department at the University of Southampton.

REFERENCES

Abbott, A. (1998), "The Causal Devolution," *Sociological Methods and Research*, 27, 148–181.

Abbott, A., and Barman, E. (1997), "Sequence Comparison via Alignment and Gibbs Sampling: A Formal Analysis of the Emergence of the Modern Sociological Article," *Sociological Methodology*, 27, 47–88.

Abbott, A., and Hrycak, A. (1990), "Measuring Sequence Resemblance," *American Journal of Sociology*, 96, 144–185.

Allison, P. (1982), "Discrete-Time Methods for the Analysis of Event Histories," *Sociological Methodology*, 13, 61–98.

Arminger, G. (1998), "A Bayesian Approach to Nonlinear Latent Variable Models Using the Gibbs Sampler and the Metropolis–Hastings Algorithm," *Psychometrika*, 63, 271–300.

Becker, M. P. (1989), "Using Association Models to Analyze Agreement Data: Two Examples," *Statistics in Medicine*, 8, 1199–1207.

——— (1994), "Analysis of Cross-Classifications of Counts Using Models for Marginal Distributions: An Application to Trends in Attitudes on Legalized Abortion," *Sociological Methodology*, 24, 229–265.

Becker, M. P., and Yang, I. (1998), "Latent Class Marginal Models for Cross-Classifications of Counts," *Sociological Methodology*, 28, 293–326.

Benzécri, J-P. (1976), *L'Analyse des Données* (2nd ed.), Paris: Dunod.

Besag, J. E. (1974), "Spatial Interaction and the Statistical Analysis of Lattice Systems" (with discussion), *Journal of the Royal Statistical Society*, Ser. B, 36, 192–236.

Biblarz, T. J., and Raftery, A. E. (1993), "The Effects of Family Disruption on Social Mobility," *American Sociological Review*, 58, 97–109.

Birch, M. W. (1963), "Maximum Likelihood in Three-Way Tables," *Journal of the Royal Statistical Society*, Ser. B, 25, 220–233.

——— (1965), "The Detection of Partial Association, II: The General Case," *Journal of the Royal Statistical Society*, Ser. B, 27, 111–124.

Blalock, H. M. (1964), *Causal Inference in Non-Experimental Research*, New York: Harcourt, Brace.

Blau, P. M., and Duncan, O. D. (1967), *The American Occupational Structure*, New York: Free Press.

Bollen, K. A., and Appold, S. J. (1993), "National Industrial-Structure and the Global System," *American Sociological Review*, 58, 283–301.

Carley, K. M. (1993), "Coding Choices for Textual Analysis: A Comparison of Content Analysis and Map Analysis," *Sociological Methodology*, 23, 75–126.

Clogg, C. C. (1992), "The Impact of Sociological Methodology on Statistical Methodology" (with discussion), *Statistical Science*, 7, 183–207.

Cox, D. R. (1972), "Regression Models and Life Tables" (with discussion), *Journal of the Royal Statistical Society*, Ser. B, 34, 187–220.

Dijkstra, W., and Taris, T. (1995), "Measuring the Agreement Between Sequences," *Sociological Methods and Research*, 24, 214–231.

Duncan, O. D. (1961), "A Socioeconomic Index for All Occupations," in *Occupations and Social Status*, ed. A. J. Reiss, New York: Free Press, pp. 109–138.

——— (1966), "Path Analysis," *American Journal of Sociology*, 72, 1–16.

——— (1979), "How Destination Depends on Origin in the Occupational Mobility Table," *American Journal of Sociology*, 84, 793–803.

Duncan, O. D., and Duncan, B. (1957), *The Negro Population of Chicago*, Chicago: University of Chicago Press.

Entwisle, B., Rindfuss, R. R., Walsh, S. J., Evans, T. P., and Curran, S. R. (1997), "Geographic Information Systems, Spatial Network Analysis, and Contraceptive Choice," *Demography*, 34, 171–187.

Fahrmeir, L., and Knorr-Held, L. (1997), "Dynamic Discrete-Time Duration Models: Estimation via Markov Chain Monte Carlo," *Sociological Methodology*, 27, 417–452.

Fienberg, S. E., and Mason, W. M. (1979), "Identification and Estimation of Age-Period-Cohort Effects in the Analysis of Discrete Archical Data," *Sociological Methodology*, 10, 1–67.

Frank, O., and Strauss, D. (1986), "Markov Graphs," *Journal of the American Statistical Association*, 81, 832–842.

Franzosi, R. (1994), "From Words to Numbers: A Set Theory Framework for the Collection, Organization and Analysis of Narrative Data," *Sociological Methodology*, 24, 105–136.

Freedman, D. A. (1987), "As Others See Us" (with discussion), *Journal of Educational Statistics*, 12, 101–223.

Gilks, W. R., Richardson, S., and Spiegelhalter, D. J. (1996), *Markov Chain Monte Carlo in Practice*, London: Chapman and Hall.

Glass, D. V. (1954), *Social Mobility in Britain*, Glencoe, IL: Free Press.

Goodman, L. A. (1971), "The Analysis of Multidimensional Contingency Tables: Stepwise Procedures and Direct Estimation Methods for Building Models for Multiple Classifications," *Technometrics*, 13, 33–61.

——— (1974), "The Analysis of Systems of Qualitative Variables When Some of the Variables are Unobservable," *American Journal of Sociology*, 79, 1179–1259.

——— (1979), "Simple Models for the Analysis of Association in Cross-Classifications Having Ordered Categories," *Journal of the American Statistical Association*, 74, 537–352.

——— (1985), "The Analysis of Cross-Classified Data Having Ordered and/or Unordered Categories," *The Annals of Statistics*, 13, 10–69.

Grusky, D. B., and Hauser, R. M. (1984), "Comparative Social Mobility Revisited: Models of Convergence and Divergence in Sixteen Countries," *American Sociological Review*, 49, 19–38.

Guttorp, P., and Walden, A. T. (1987), "On the Evaluation of Geophysical Models," *Geophysical Journal of the Royal Astronomical Society*, 91, 201–210.

Handcock, M. S., and Morris, M. (1998), "Relative Distribution Methods," *Sociological Methodology*, 28, 53–98.

Hanneman, R. A., Collins, R., and Mordt, G. (1995), "Discovering Theory Dynamics by Computer Simulation: Experiments on State Legitimacy and Imperialist Capitalism," *Sociological Methodology*, 25, 1–46.

Hauser, R. M., and Warren, J. R. (1997), "Socioeconomic Indexes for Occupations: A Review, Update and Critique," *Sociological Methodology*, 27, 177–298.

Hill, D. H. (1997), "Adjusting for Attrition in Event-History Analysis," *Sociological Methodology*, 27, 393–416.

Hill, D. H., Axinn, W. G., and Thornton, A. (1993), "Competing Hazards With Shared Unmeasured Risk Factors," *Sociological Methodology*, 23, 245–277.

Hout, M. (1983), *Mobility Tables*, Beverly Hills, CA: Sage.

——— (1984), "Status, Autonomy and Training in Occupational Mobility," *American Journal of Sociology*, 89, 1379–1409.

——— (1988), "Expanding Universalism, Less Structural Mobility: The American Occupational Structural in the 1980s," *American Journal of Sociology*, 93, 1358–1400.

Jones, K. S., and Willett, P. (1997), *Readings in Information Retrieval*, San Francisco: Morgan Kaufman.

Jöreskog, K. G. (1973), "A General Method for Estimating a Linear Structural Equation System," in *Structural Equation Models in the Social Sciences*, ed. A. S. Goldberger and O. D. Duncan, New York: Seminar, pp. 85–112.

Kass, R. E., and Raftery, A. E. (1995), "Bayes Factors," *Journal of the American Statistical Association*, 90, 773–795.

Kass, R. E., and Wasserman, L. (1995), "A Reference Bayesian Test for Nested Hypotheses and Its Relationship to the Schwarz Criterion," *Journal of the American Statistical Association*, 90, 928–934.

Kim, H., and Bearman, P. S. (1997), "The Structure and Dynamics of Movement Participation," *American Sociological Review*, 62, 70–93.

Kuo, H. H. D., and Hauser, R. M. (1996), "Gender, Family Configuration, and the Effect of Family Background on Educational Attainment," *Social Biology*, 43, 98–131.

Lang, J. B., and Agresti, A. (1994), "Simultaneously Modeling Joint and Marginal Distributions of Multivariate Categorical Responses," *Journal of the American Statistical Association*, 89, 625–632.

Lawrence, C. E., Altschul, S. F., Boguski, M. S., Liu, J. S., Neuwald, A. F., and Wooton, J. C. (1993), "Detecting Subtle Sequence Signals," *Science*, 262, 208–214.

Lazarsfeld, P. F. (1950), "The Logical and Mathematical Foundation of Latent Structure Analysis," in *Studies in Social Psychology in World War II. Vol. 4: Measurement and Prediction*, eds. E. A. Schulman, P. F. Lazarsfeld, S. A. Starr, and J. A. Clausen, Princeton, NJ: Princeton University Press, pp. 362–412.

Lieberson, S. L. (1994), "More on the Uneasy Case for Using Mill-Type Methods in Small-N Comparative Studies," *Social Forces*, 72, 1225–1237.

Massey, D. S., and Denton, N. A. (1993), *American Apartheid: Segregation and the Making of the Underclass*, Cambridge, MA: Harvard University Press.

Matsueda, R. L., and Heimer, K. (1987), "Race, Family Structure, and Delinquency: A Test of Differential Association and Social Control Theories," *American Sociological Review*, 52, 826–840.

Morris, M. (1997), "Sexual Networks and HIV," *AIDS*, 11, S209–S216.

Morris, M., Bernhardt, A. D., and Handcock, M. S. (1994), "Economic Inequality: New Methods for New Trends," *American Sociological Review*, 59, 205–219.

Muthén, B. (1983), "Latent Variable Structure Equation Modeling With Categorical Data," *Journal of Econometrics*, 22, 43–65.

———— (1997), "Latent Variable Modeling of Longitudinal and Multilevel Data," *Sociological Methodology*, 27, 453–480.

Pearl, J. (1998), "Graphs, Causality, and Structural Equation Models," *Sociological Methods and Research*, 27, 226–284.

Poole, D., and Raftery, A. E. (2000), "Inference from Deterministic Simulation Models: The Bayesian Melding Approach," *Journal of the American Statistical Association*, 95, 1244–1255.

Raftery, A. E. (1986), "Choosing Models for Cross-Classifications," *American Sociological Review*, 51, 145–146.

———— (1991), "Bayesian Model Selection and Gibbs Sampling in Covariance Structure Models," Working Paper 92-4, University of Washington, Center for Studies in Demography and Ecology.

———— (1995), "Bayesian Model Selection in Social Research" (with discussion), *Sociological Methodology*, 25, 111–193.

———— (1996), "Approximate Bayes Factors and Accounting for Model Uncertainty in Generalized Linear Models," *Biometrika*, 83, 251–266.

Raftery, A. E., Givens, G. H., and Zeh, J. E. (1995), "Inference From a Deterministic Population Dynamics Model for Bowhead Whales" (with discussion), *Journal of the American Statistical Association*, 90, 402–430.

Raftery, A. E., Lewis, S. M., and Aghajanian, A. (1995), "Demand or Ideation? Evidence From the Iranian Marital Fertility Decline," *Demography*, 32, 159–182.

Ragin, C. (1987), *The Comparative Method: Moving Beyond Qualitative and Quantitative Strategies*, Berkeley, CA: University of California Press.

Roberts, C. W. (1997), "A Generic Semantic Grammar for Quantitative Text Analysis: Applications to East and West Berline News Content From 1979," *Sociological Methodology*, 27, 89–130.

Roeder, K., Lynch, K. G., and Nagin, D. S. (1999), "Modeling Uncertainty in Latent Class Membership: A Case Study in Criminology," *Journal of the American Statistical Association*, 94, 766–776.

Rogoff, N. (1953), *Recent Trends in Occupational Mobility*, Glencoe, IL: Free Press.

Sankoff, D., and Kruskal, J. B. (1983), *Time Warps, String Edits, and Macromolecules*, Reading, MA: Addison-Wesley.

Scheines, R., Hoijtink, H., and Boomsma, A. (1999), "Bayesian Estimation and Testing of Structural Equation Models," *Psychometrika*, 64, 37–52.

Schuessler, K. F. (1980), "Quantitative Methodology in Sociology: The Last 25 Years," *American Behavioral Scientist*, 23, 835–860.

Schwarz, G. (1978), "Estimating the Dimension of a Model," *The Annals of Statistics*, 6, 461–464.

Singer, B., Ryff, C. D., Carr, D., and Magee, W. J. (1998), "Linking Life Histories and Mental Health: A Person-Centered Strategy," *Sociological Methodology*, 28, 1–52.

Sobel, M. E. (1998), "Causal Inference in Statistical Models of the Process of Socioeconomic Achievement: A Case Study," *Sociological Methods and Research*, 27, 318–348.

———— (2000), "Causal Inference in the Social Sciences," *Journal of the American Statistical Association*, 95, 647–651.

Spiegelhalter, D., Dawid, P., Lauritzen, S., and Cowell, R. (1993), "Bayesian Analysis in Expert Systems," *Statistical Science*, 8, 219–282.

Spirtes, P., Richardson, T. S., Meek, C., Scheires, R., and Glymour, C. (1998), "Using Path Diagrams as a Structural Equation Modeling Tool," *Sociological Methods and Research*, 27, 182–225.

Stovel, K., Savage, M., and Bearman, P. (1996), "Ascription into Achievement: Models of Career Systems at Lloyds Bank, 1890–1970," *American Journal of Sociology*, 102, 358–399.

Thompson, E. A. (1998), "Inferring Gene Ancestry: Estimating Gene Descent," *International Statistical Review*, 66, 29–40.

Treiman, D. J. (1977), *Occupational Prestige in Comparative Perspective*, New York: Academic Press.

Tuma, N. B., and Hannan, M. T. (1984), *Social Dynamics: Models and Methods*, Orlando, Fla.: Academic Press.

Warren, J. R., LePore, P. C., and Mare, R. D. (2001), "Employment During High School: Consequences for Students' Grades in Academic Courses," *American Educational Research Journal*, to appear.

Wasserman, S., and Faust, K. (1994), *Social Network Analysis: Methods and Applications*, Cambridge, U.K.: Cambridge University Press.

Wasserman, S., and Pattison, P. (1996), "Logit Models and Logistic Regressions for Social Networks. 1. An Introduction to Markov Graphs and p," *Psychometrika*, 61, 401–425.

Weakliem, D. L. (1999), "A Critique of the Bayesian Information Criterion for Model Selection" (with discussion), *Sociological Methods and Research*, 27, 359–443.

Western, B. (1996), "Vague Theory and Model Uncertainty in Macrosociology," *Sociological Methodology*, 26, 165–192.

Western, B., and Jackman, S. (1994), "Bayesian Inference for Comparative Research," *American Political Science Review*, 88, 412–423.

Wright, S. (1921), "Correlation and Causation," *Journal of Agricultural Research*, 20, 557–585.

Xie, Y. (1994), "Log-Multiplicative Models for Discrete-Time, Discrete-Covariate Event History Data," *Sociological Methodology*, 24, 301–340.

Yamaguchi, K. (1992), "Accelerated Failure-Time Regression Models With a Regression Model of Surviving Fraction: An Application to the Analysis of 'Permanent Employment' in Japan," *Journal of the American Statistical Association*, 87, 284–292.

Psychometrics

Michael W. Browne

1. INTRODUCTION

To progress, a scientific discipline must develop methodology for obtaining measurements of relevant constructs and to extract meaning from the measurements it does have. This is not a straightforward matter in psychology. Typically, constructs of interest are not clearly defined and cannot be measured directly. In addition, the measurements that are available are subject to substantial measurement error. Consequently, the measurement process often consists of repeated attempts to measure the same construct in different ways. When the relationship between several constructs is under investigation, each of the constructs is measured repeatedly, resulting in a substantial number of measurements. Thus the statistical methodology developed for the analysis of psychological measurements is typically multivariate.

Because constructs are not clearly defined, the investigator is often not sure exactly what is being measured. This has led to the concept of a latent or hidden variable that is not measured directly. Inferences about the latent variable are deduced from interrelationships between manifest, or observed, variables.

In a broad sense, psychometrics may be regarded as the discipline concerned with the quantification and analysis of human differences. This involves both the construction of procedures for measuring psychological constructs and the analysis of data consisting of the measurements made. In this sense, both the construction of a psychological attitude scale and the analysis of the data resulting from its application may be regarded as part of psychometrics. In a narrower sense, psychometrics is often regarded as the development of mathematical or statistical methodology for the analysis of measurement data in psychology. This methodology is primarily multivariate, and latent variables feature strongly. It is this aspect of psychometrics that I consider here.

Although many of the techniques that currently constitute psychometrics date far earlier, psychometrics emerged as a formal discipline with the formation of the Psychometric Society in 1935, with L. L. Thurstone as its first president. Thurstone

Michael W. Browne is Professor in the Department of Psychology and in the Department of Statistics, The Ohio State University, Columbus, OH 43210 (E-mail: browne.4@osu.edu).
The author thanks the associate editor, Mark Becker, who, in collaboration with three anonymous referees, provided constructive suggestions for improvement. He is also grateful to Michael Walker and to Robert MacCallum for helpful comments that have positively influenced this vignette.

had played a leadership role at the University of Chicago in providing methodology for analyzing measurements of psychological constructs, and students of his featured strongly among the founding members of the society (Horst and Stalnaker 1986). From the beginning, the Psychometric Society has had an international membership, and the proportion of members outside the United States has grown steadily over the years.

The Psychometric Society produced a journal, *Psychometrika*, whose stated aim was the "development of psychology as a quantitative rational science." Subsequently, *Psychometrika* has been joined by other journals with some overlap of subject matter: *Educational and Psychological Measurement, Journal of Educational Measurement, Journal of Educational and Behavioral Statistics* (formerly *Journal of Educational Statistics*), *Applied Psychological Measurement, Journal of Classification, Psychological Methods* (formerly the Quantitative Methods in Psychology section of *Psychological Bulletin*) and *Journal of Mathematical Psychology*. In addition, two journals with similar content to *Psychometrika* have been founded overseas: *British Journal of Mathematical and Statistical Psychology* (formerly *British Journal of Statistical Psychology*) and *Behaviormetrika*, a Japanese journal published in the English language.

It is conventional to classify psychometrics into three major areas: mental test theory, factor analysis and associated methods, and multidimensional scaling. I consider each of these separately, and examine some significant work.

2. MENTAL TEST THEORY

Mental test theory is concerned with methodology developed for the analysis of mental tests. Two general approaches currently in use for the analysis of tests consisting of dichotomous items are classical test theory and item response theory. The foundations of classical test theory were laid by Charles Spearman early in the 20th century, and the subject was developed extensively thereafter by a number of authors. A unified account was first presented by Gulliksen (1950). Some years later, an advanced treatment was given by Lord and Novick (1968).

The model of classical test theory is of the form

$$X = T + E,$$

where the manifest variate X represents scores on a test, T is a latent variate representing unobservable true scores, and E represents errors of measurement. This is essentially a random-effects one-way analysis of variance model, where T is a random treatment effect. A central concept of mental test theory is the reliability of a test or ratio of true score variance, σ_T^2, to the variance of the test, σ_X^2. Because true scores are unobservable, additional assumptions are necessary for estimating the reliability. An example of the type of assumption made to make the estimation of reliability possible is that of parallel tests; namely, tests with the same true score and with independently distributed errors with the same variance. These tests have the same reliability, and the correlation between them provides an estimate of this reliability. Methods are also available for estimating the reliability of a sum of parallel measurements.

The properties of a test will depend on properties of the items. Classical item analysis makes use of properties of items such as item difficulty, item variance, and the item–test correlation. These item properties are dependent on the population. Consequently, properties of the test change from one population to another.

An advantage of item response theory is that item properties do not depend on the population under consideration. Consider the normal ogive item characteristic curve treated extensively by Lord in a series of papers (starting with Lord 1952). The probability of an examinee correctly answering the gth item in a test is

$$P_g(\theta) = \int_{-\infty}^{a_g(\theta - b_g)} \frac{1}{\sqrt{2\pi}} e^{-(t^2/2)} dt, \qquad (1)$$

where θ represents the ability of the examinee, a_g is a parameter representing the discriminating power of item g, and b_g represents the item's difficulty. The item parameters, a_g and b_g, are invariant over different populations of examinees. An overview has been given by Lord and Novick (1968).

There is a relationship between the item characteristic curve of item response theory and the stimulus response curve of bioassay. Equation (1) is familiar as the equation of probit analysis where θ represents the amount of a dose, $P_g(\theta)$ represents the probability of survival, and the parameters a_g and b_g represent the intercept and slope of the linear regression of the probit transformation on the dose. The fundamental difference between the two applications of (1) is that in probit analysis the value of θ is known, whereas in item response theory θ represents a latent variable. Because of this difference, the methodologies of item response theory and bioassay differ. In the early stages of the development of item response theory (Lawley 1943), θ was approximated by the total score of a test, and there were strong similarities to bioassay. As pointed out by Finney (1971, sec. 3.6), bioassay and item response theory had their origins in psychophysics.

Birnbaum (1968) suggested using the logistic cumulative distribution function for the item characteristic curve as a more tractable alternative to the normal ogive. He provided an algorithm for simultaneously estimating the person parameters, $\theta_i, i = 1, \ldots, N$, and the item parameters, a_g and $b_g, g = 1, \ldots, p$, given scores on each of p items obtained from a sample of N persons. In this approach the number of parameters increases as N increases, leading to computational difficulties in very large samples and invalidating the asymptotic theory associated with maximum likelihood estimates. A more tractable logistic item characteristic curve with a single difficulty parameter was justified by Rasch (1960) and has received considerable attention since then.

Bock and Aitkin (1981) made use of the normal ogive model and avoided the estimation of person parameters by regarding them as unobserved latent variables with a specified distribution. Marginal maximum likelihood estimates were obtained using an EM algorithm.

Modern kernel smoothing methods for nonparametric function estimation are now being used to advantage in item response theory. Ramsay (1991) has developed a nonparametric approach that requires prior estimates of examinee ability but avoids restrictive assumptions on the item characteristic curve. It provides helpful additional information in the form of option characteristic curves for each option of a multiple choice item.

The availability of small, powerful, and reasonably priced computers is currently spurring the computerized administration of adaptive tests. Overall testing time is reduced by only administering items that are suited to an examinee's ability level and avoiding items that are either too difficult or too easy. Methodology (Lord 1980, chap. 10) based on item response theory makes the use of this computerized approach possible. Because adaptive tests require substantial pools of items, Embretson (1999) is developing models, based on cognitive psychological theory, for the computerized generation of items.

A comprehensive coverage of modern developments in item response theory may be found in the volume edited by van der Linden and Hambleton (1996).

3. FACTOR ANALYSIS AND ASSOCIATED MODELS

The originator of factor analysis is generally recognized as Charles Spearman, who developed a factor analysis model with a single common factor representing general intelligence during the first quarter of the 20th century. Over time, it became apparent that a single common factor was not sufficient to account for interrelationships between variables. Several persons advocated the generalization of factor analysis to multiple common factors. This multiple factor analysis model is a multivariate multiple regression model with manifest dependent variates and latent explanatory variates. The additional assumption made to enable estimation of the regression weights on unobservable latent variates, or factor loadings, is that the regression residual variates are independently distributed.

Prominent among the proponents of the multiple factor analysis model was L. L. Thurstone (1935), who suggested the centroid method for estimating factor loadings. The maximum likelihood estimation of factor loadings was first investigated by Lawley (1940). Effective computational procedures became available only about 30 years later when nested algorithms involving eigenvalues and eigenvectors and imposing inequality constraints on unique variance estimates were discovered independently by Jöreskog (1967) and by Jennrich and Robinson (1969).

The generalization of factor analysis to several factors introduced a rotational indeterminacy, and it became necessary to choose a particular solution from an infinite class of equivalent solutions. Thurstone (1935) suggested the principle of simple structure for choosing an interpretable rotation of the factor matrix and listed five defining properties. Graphical methods for carrying out rotations by hand were developed but were very time consuming. With the advent of electronic computers, methods of rotation that optimized a "simplicity" function were developed. Effective computerized methods for orthogonal rotation, defining factors to be uncorrelated, were first developed. The most successful simplicity function for orthogonal rotation was the Varimax criterion proposed by Kaiser (1958) and used to this day. But Thurstone's original hand rotation methods were predominantly for oblique rotation, where common factors are allowed to be correlated. The problem of developing an effective computerized method for oblique rotation resisted solution for some years but was finally solved by Jennrich and Sampson (1966). Because of the iterative nature of the rotation process, it seemed impossible, at first, to obtain standard errors for rotated factor loadings. Solutions were, however, provided by Jennrich and his coauthors in a series of papers culminating with the one by Jennrich and Clarkson (1980).

One manner in which difficulties associated with the rotation process can be avoided is by carrying out a confirmatory factor analysis. This approach bypasses the rotational indeterminacy of the factor analysis model by prespecifying the values of some factor loadings. It then becomes possible to generalize the factor analysis model by imposing a structural equation model on the common factors. The main initial contributions in this area were made by Jöreskog (1973) with the LISREL model. Subsequent developments have been concerned with adapting the model to discrete manifest variables, investigating properties of the estimators, and developing alternative fitting procedures.

Factor analysis has also been extended to the analysis of three-way matrices, starting with the work of Tucker (1966) and followed by substantial subsequent developments. Another innovation by Tucker (1958) related to factor analysis is the latent curve approach for studying interindividual variation in change over time. This topic is currently being pursued further using structural equation modeling approaches.

It has been known for many years that asymptotic properties of the method of maximum likelihood applied to covariance structures are nonrobust to some violations of normality assumptions. In two technical reports by Anderson and Amemiya that appeared in 1985, surprising results were presented where maximum likelihood applied to factor analysis yields methodology that is robust to nonnormality under certain circumstances. (Details and further references may be found in Amemiya and Anderson 1990.) The Anderson and Amemiya technical reports immediately generated a spurt of research on this topic, much of it strongly influenced by a comprehensive mathematical framework for robustness in the analysis of covariance structures provided by Shapiro (1987). A balanced overview and a synthesis of different approaches was given by Kano (1993).

At the present time the most challenging problems are generalizations of structural equation modeling that involve nonlinear functions of latent variables. These models involve an additional problem, not present in classical linear factor analysis, in that incorrect assumptions about the distribution of latent variables invalidate the covariance structure and result in inconsistent estimators. An example is a polynomial factor analysis model proposed by Kenny and Judd (1984) that has received substantial attention. Difficulties associated with an inappropriate choice of distribution for the latent variables in the Kenny–Judd model have now been solved by Wall and Amemiya (2000).

4. MULTIDIMENSIONAL SCALING

Multidimensional scaling was developed for the investigation of distances between psychological attributes. A method for obtaining coordinates of points from a matrix of interpoint distances was given by Young and Householder (1938) and adapted by Torgerson (1958) to the practical situation where distances are measured subject to error. Shepard (1962) introduced the concept of an arbitrary monotonic relationship between observed dissimilarities and true distances, and Kruskal (1964) formalized this approach and provided an algorithm to perform nonmetric multidimensional scaling. Guttman (1968) proposed a rank order approach to deal with the monotonic relationship between dissimilarities and true distances. His multidimensional scaling methods were used to investigate his radex theory. Guttman (1954) proposed it as a competitor to factor analysis. Carroll and Chang (1970) provided a solution to the simultaneous scaling of a number of dissimilarity matrices, thereby enabling the investigation of interpersonal differences in perceived dissimilarities.

These multidimensional scaling methods were primarily model approximation procedures that did not make distributional assumptions. They were applied in the same way to all dissimilarity measures without taking into account how they were derived from data. In a series of articles, Takane and coauthors developed specific models with a multidimensional scaling component for different data collection methods. These models involved distributional assumptions and the method of maximum likelihood was used to obtain parameter estimates and associated goodness-of-fit tests. Takane and Sergent (1983) provided an example of this approach and gave further references. Modern developments in multidimensional scaling were surveyed in the recent book by Borg and Groenen (1997).

5. CONCLUDING COMMENTS

During the course of the 20th century, psychometrics has developed into a sophisticated, mathematically oriented discipline aimed at providing methodology for handling the particularities of psychological measurement. To do this it has become strongly computer based and algorithm development has been a component of much of the research carried out. Much methodology that arose out of psychometrics has been of use elsewhere. In particular, factor analysis, structural equation modeling with latent variables and multidimensional scaling are being widely used in a variety of disciplines and are included in most modern statistical computer packages. Conversely, standard statistical methodology is routinely used by research psychologists. In particular, psychology has a rich tradition of experimental design, and texts on the topic, such as that of Winer (1971), have had a substantial influence on psychological research.

REFERENCES

Amemiya, Y., and Anderson, T. W. (1990), "Asymptotic Chi-Squared Tests for a Large Family of Factor Analysis Models," *The Annals of Statistics*, 18, 1453–1463.

Birnbaum, A. (1968), "Some Latent Trait Models and Their Use in Inferring an Examinee's Ability," in *Statistical Theories of Mental Test Scores*, eds. F. M. Lord and M. R. Novick, Reading, MA: Addison-Wesley, pp. 397–479.

Bock, R. D., and Aitkin, M. (1981), "Marginal Maximum Likelihood Estimation of Item Parameters: Application of an EM Algorithm," *Psychometrika*, 46, 443–445.

Borg, I., and Groenen, P. (1997), *Modern Multidimensional Scaling: Theory and Applications*, New York: Springer-Verlag.

Carroll, J. D., and Chang, J. J. (1970), "Analysis of Individual Differences in Multidimensional Scaling via an *N*-Way Decomposition of 'Eckart–Young' Decomposition," *Psychometrika*, 35, 283–320.

Embretson, S. (1999), "Generating Items During Testing: Psychometric Issues and Models," *Psychometrika*, 64, 407–433.

Finney, D. J. (1971), *Probit Analysis* (3rd ed.), Cambridge, U.K.: Cambridge University Press.

Gulliksen, H. (1950), *Theory of Mental Tests*, New York: Wiley.

Guttman, L. (1954), "A New Approach to Factor Analysis: The Radex," in *Mathematical Thinking in the Behavioral Sciences*, ed. P. Lazarsfeld, New York: Free Press, pp. 258–348.

――― (1968), "A General Nonmetric Technique for Finding the Smallest Coordinate Space for a Configuration of Points," *Psychometrika*, 33, 469–506.

Horst, P., and Stalnaker, J. (1986), "Present at the Birth," *Psychometrika*, 51, 3–6.

Jennrich, R. I., and Clarkson, D. B. (1980), "A Feasible Method for Standard Errors of Estimate in Maximum Likelihood Factor Analysis," *Psychometrika*, 45, 237–247.

Jennrich, R. I., and Robinson, S. M. (1969), "A Newton–Raphson Algorithm for Maximum Likelihood Factor Analysis," *Psychometrika*, 34, 111–123.

Jennrich, R. I., and Sampson, P. F. (1966), "Rotation for Simple Loadings," *Psychometrika*, 31, 313–323.

Jöreskog, K. G. (1967), "Some Contributions to Maximum Likelihood Factor Analysis," *Psychometrika*, 32, 443–458.

――― (1973), "A General Method for Estimating a Linear Structural Equation System," in *Structural Equation Models in the Social Sciences*, eds. A. S. Goldberger and O. D. Duncan, New York: Academic Press, pp. 85–112.

Kaiser, H. F. (1958), "The Varimax Criterion for Analytic Rotation in Factor Analysis," *Psychometrika*, 23, 187–200.

Kano, Y. (1993), "Asymptotic Properties of Statistical Inference Based on Fisher-Consistent Estimators in the Analysis of Covariance Structures," in *Statistical Modelling and Latent Variables*, eds. K. Haagen, C. J. Bartholomew, and M. Deistler, New York: Elsevier.

Kenny, D. A., and Judd, C. M. (1984), "Estimating the Nonlinear and Interactive Effects of Latent Variables," *Psychological Bulletin*, 96, 201–210.

Kruskal, J. B. (1964), "Nonmetric Multidimensional Scaling: A Numerical Method," *Psychometrika*, 29, 115–129.

Lawley, D. N. (1940), "The Estimation of Factor Loadings by the Method of Maximum Likelihood," *Proceedings of the Royal Society of Edinburgh*, Ser. A, 60, 64–82.

―――― (1943), "On Problems Connected With Item Selection and Test Construction," *Proceedings of the Royal Society of Edinburgh*, Ser. A, 61, 273–287.

Lord, F. M. (1952), *A Theory of Test Scores*, Psychometric Monograph No. 7, Psychometric Society.

―――― (1980), *Applications of Item Response Theory to Practical Testing Problems*, Hillsdale, NJ: Erlbaum.

Lord, F. M., and Novick (1968), *Statistical Theories of Mental Test Scores*, Reading, MA: Addison-Wesley.

Ramsay, J. O. (1991), "Kernel Smoothing Approaches to Nonparametric Item Characteristic Curve Estimation," *Psychometrika*, 56, 611–630.

Rasch, G. (1960), *Probabilistic Models for Some Intelligence and Attainment Tests*, Copenhagen: Danish Institute for Educational Research.

Shapiro, A. (1987), "Robustness Properties of the MDF Analysis of Moment Structures," *South African Statistical Journal*, 21, 39–62.

Shepard, R. N. (1962), "The Analysis of Proximities: Multidimensional Scaling With an Unknown Distance Function, I & II," *Psychometrika*, 27, 125–140, 219–246.

Takane, Y., and Sergent, J. (1983), "Multidimensional Scaling Models for Reaction Times and Same-Different Judgements," *Psychometrika*, 48, 393–423.

Thurstone, L. L. (1935), *The Vectors of Mind*, Chicago: University of Chicago Press.

Torgerson, W. S. (1958), *Theory and Methods of Scaling*, New York: Wiley.

Tucker, L. R. (1958), "Determination of Parameters of a Functional Relation by Factor Analysis," *Psychometrika*, 23, 19–23.

―――― (1966), "Some Mathematical Notes on Three-Mode Factor Analysis," *Psychometrika*, 31, 279–311.

van der Linden, W. J., and Hambleton, R. K. (Eds.) (1996), *Handbook of Modern Item Response Theory*, New York: Springer-Verlag.

Wall, M. M., and Amemiya, Y. (2000), "Estimation for Polynomial Structural Equation Models," *Journal of the American Statistical Association*, 95, 929–940.

Winer, B. J. (1971), *Statistical Principles in Experimental Design*, New York: McGraw-Hill.

Young, G., and Householder, A. S. (1938), "Discussion of a Set of Points in Terms of Their Mutual Distances," *Psychometrika*, 3, 19–22.

Empirical Methods and the Law

Theodore Eisenberg

> For the rational study of the law the black letter man may be the man of the present, but
> the man of the future is the man of statistics and the master of economics.
>
> Oliver Wendell Holmes,
> The Path of the Law (1897)

One can divide empirical analysis of legal issues into three major branches: (1) the use of scientific empirical analysis by litigants to attempt to prevail in individual cases, (2) the use of *social* scientific empirical analysis in individual cases, and (3) the use of empirical methods to describe the legal system's operation. The first two uses present difficulties that reflect a fundamental limitation on using statistical methods in law: the difference between establishing statistical association and establishing actual causation in an individual case filtered through our adversary legal system. The third use encounters no such obstacle and can aid understanding of how the legal system operates and inform policymakers. Accurate description of the legal system's operation can in turn influence the outcome of specific cases. More important, accurate description of the legal system can supply the information necessary for sound policymaking; for example, a substantial body of evidence suggests that our civil justice system performs quite well.

1. "HARD" SCIENTIFIC EMPIRICAL ANALYSIS

Scientific statistical analysis of evidentiary issues in individual cases is common in criminal and civil cases. The most important class of scientific evidence relates to forensic identification. In this field, DNA testing has fostered a revolution. DNA testing has been subjected to levels of rigorous analysis not previously applied to forensic identification reasoning. The debates generated by DNA testing may reverberate back onto more traditional forensic identification methods. Pure statistical questions also arise in a variety of other contexts.

1.1 DNA Testing

DNA matching is now common in criminal and civil cases. Many articles and books on the subject are available (e.g., Aitken 1995; Roberston and Vignaux 1995; Schum 1994). Likelihood ratios measure the strength of DNA evidence supporting

Theodore Eisenberg is Henry Allen Mark Professor of Law, Cornell Law School, Ithaca, NY 14853 (E-mail: eisnberg@law.mail.cornell.edu).

the hypothesis that a suspect is the source of a sample, as against the alternative hypothesis that someone other than the suspect is the source (e.g., Mellen and Royall 1997). Proper execution of DNA analysis has triggered substantial debate (e.g., Aitken 1995, pp. 92–106; Evett 1991) and has led to two National Research Council (NRC) reports (NRC 1992, 1996) and to critiques of the NRC reports (e.g., Donnelly and Friedman 1999; Lempert 1997).

In addition to baseline discussions about DNA evidence, increasing use of DNA databases in England, Wales, and the United States has generated questions about the NRC's approved methods of analyzing and reporting data when a suspect has been identified through a DNA database search and not from independent related evidence. The NRC (1996) recommended that when a suspect is found by a search of a DNA database, the random-match probability "should be multiplied by...the number of persons in the database" (p. 40). Donnelly and Friedman (1999) suggested that the NRC approach is incorrect (p. 944). In their view, the fact that a DNA match is "found only after a search does not diminish the value of the evidence;" the NRC (1996) requires "a testifying expert to drastically understate the value of the evidence" (p. 933). They base their differences with the NRC on the tendency of statisticians "to export to the legal context methods that were developed to assist scientific inquiries and that appear more suitable in that context than in adjudication" (p. 934).

1.2 Implications of the DNA Testing Debates

Notwithstanding the debates, DNA inquiries tend to be regarded as purely scientific. But DNA testing can be viewed as an instance of a larger class of forensic identification techniques. For example, the question whether two fingerprints, handwritings, gun barrels, or hair samples match can be approached through reasoning applied in the DNA context. Of particular importance is quantifying the probability of erroneous matches. But this has not been done. Instead, many forensic scientists are content to assert that no two of these objects can be alike (Saks 1998, p. 1082). Except for DNA-type matching, forensic identification sciences "have not taken the trouble to collect data on populations of forensically relevant objects so that the probability of erroneous matches can be calculated" (pp. 1087–1088). It has been suggested that the data-based and probabilistic approach of DNA typing will become the norm of all forensic identification sciences (Saks and Koehler 1991). Seeds of this theme with respect to human hair comparisons have been given by Fienberg (1989, pp. 64–67).

1.3 Scientific Analysis in Other Contexts

Reasonably pure statistical inquiries arise in many contexts. Illustrations of scientific inquiries include regression estimates of damages in price-fixing cases (Finkelstein and Levenbach 1983), estimates of the quantity of drugs handled using predictive distributions (Aitken, Bring, Leonard, and Papasouliotis 1997), and other sampling issues, including those relating to tax allocations of sales to different governmental authorities, sampling for public confusion in trademark cases, sampling to measure royalties by the American Society of Composers, Authors & Publishers (ASCAP), and census undercount issues (Finkelstein and Levin 1980, pp. 269–283). The data available in legal cases often are not ideal. Bayesian techniques for dealing with missing data in trial testimony have been proposed (Kadane and Terrin 1997).

2. "SOFT" SOCIAL SCIENTIFIC EMPIRICAL ANALYSIS

Often the empirical question in a case is not of the form typified by whether DNA matches are the proper way to estimate damages. Rather, it is whether human beings have behaved in a particular manner that cannot be proven by reference to physical evidence. In analyzing human behavior in legal disputes, statisticians need to be aware of their own limitations and the special nature of the adversary process (Fienberg 1997).

Much of the law relating to the use of social scientific statistics in individual cases has developed in discrimination cases. Some of the crudest forms of discrimination can be identified through simple description, without the need for statistical analysis. For example, the absence over a period of years of any blacks on juries in counties with substantial black populations establishes the existence of discrimination in juror selection. (Norris v. State of Alabama 1935). But modern discrimination cases are rarely so simple.

2.1 The Leading Modern Empirical Case: McCleskey v. Kemp

The most visible modern case relating to use of social scientific statistical proof in specific cases is a death penalty case, McCleskey v. Kemp (1987). McCleskey, a Georgia death row inmate, relied on one of the most comprehensive law-related social science studies ever performed to claim that race had played an improper role in his case. The study, led by Dr. David Baldus, examined more than 2,000 murders that occurred in Georgia and considered 230 variables that could have explained the data on nonracial grounds. The study concluded that defendants charged with killing white victims were 4.3 times more likely to receive the death penalty than defendants charged with killing blacks and that black defendants were 1.1 times more likely to receive the death penalty than other defendants. (Baldus, Woodworth, and Pulaski 1990). Thus a black defendant who killed a white victim had the greatest likelihood of receiving a death sentence. Given the many other studies finding a relation between race of capital case decision making (U.S. General Accounting Office 1990), McCleskey's case presented the Supreme Court with a plausible vehicle in which to rely on empirical social science methodology to influence judicial decision making in a vitally important area.

Despite the evidence, the McCleskey case foundered on the difference between a statistical tendency and the legal system's demand for proof of causation in each case. The legal system claims to deal in certainty, or at least in direct causation. Social scientists and statisticians often rely on statistical association without substantial concern about whether the association corresponds with actual causation in a particular observation.

The Supreme Court relied on this difference to discount McCleskey's statistical showing. In a 5–4 decision, in an opinion authored by Justice Powell, the Court held that general statistical evidence showing that a particular state's capital punishment scheme operated in a discriminatory manner does not establish a constitutional violation. The Court found that the empirical study established "at most. . .a discrepancy that appears to correlate with race" (481 U.S. at 291 n. 7). The majority opinion stated that "even a sophisticated multiple-regression analysis such as the Baldus study can demonstrate only a *risk* that the factor of race entered into some capital sentencing decisions and a necessarily lesser risk that race entered into any particular sentencing decision."

Because the Supreme Court had sometimes found in jury selection cases that statistics standing alone were sufficient proof of discriminatory intent, Justice Powell's opinion had to distinguish these cases. He explained that the prior jury selection case had involved less complicated decisions and fewer decision makers. The many actors and factors in a capital sentencing case increased the likelihood that other factors were responsible for racial effects, and thus rendered the jury selection precedents inapplicable. Moreover, this complexity would increase the rebuttal burden intolerably if the state were required to explain statewide statistics. Because the statistical showing did not constitute a pattern as stark as in early jury discrimination cases, the Court dismissed the statewide statistics as insufficient to support an inference that the decision makers in the McCleskey case acted with a discriminatory purpose.

The ruling's effect on later race-based claims in capital cases has been devastating. No race-based challenge to a capital sentence has been sustained in the 13 years since the McCleskey case. In a poignant change of view, Justice Powell reported that he wished that he could change his deciding vote in the McCleskey case (Von Drehle 1994). In the post-McCleskey era, it has been suggested that analyzing individual prosecutorial offices' decision making, rather than statewide data, could overcome the Court's concerns in McCleskey (Blume, Eisenberg, and Johnson 1998).

2.2 Voting Rights and Employment Cases

The legal standard applicable to race-based constitutional challenges rendered the empirical showing in the McCleskey case vulnerable. In two other classes of discrimination cases, voting discrimination and employment discrimination, litigants use sophisticated statistical methods, including regression, with somewhat greater success. One form of voting discrimination, racial dilution claims, usually involves establishing that black voters are less able than white voters to elect the candidate of their choice or to otherwise influence the political process. The standard method of proof in such cases, referred to by the Supreme Court in Thornburg v. Gingles (1986) and widely relied on by lower courts, is ecological regression. Despite its acceptance in court, this technique remains controversial within the statistical community (Freedman, Ostland, Roberts, and Klein 1999; King 1999).

In employment discrimination cases brought under federal statutes such as Title VII of the Civil Rights Act of 1964, the plaintiff need only establish a statistical disparity to put the burden on the defendant of justifying its hiring practices. In employment discrimination cases, the Supreme Court has articulated a sensible attitude about the use of statistics and regression analysis. Under Bazemore v. Friday (1986), an employment discrimination plaintiff need not find information about all possibly relevant variables in a statistical model. The Court stated the following:

> While the omission of variables from a regression analysis may render the analysis less probative than it otherwise might be, it can hardly be said, absent some other infirmity, that an analysis which accounts for the major factors "must be considered unacceptable as evidence of discrimination." . . . Normally, failure to include variables will affect the analysis' probativeness, not its admissibility.
>
> Importantly, it is clear that a regression analysis that includes less than "all measurable variables" may serve to prove a plaintiff's case. A plaintiff in a Title VII suit need not prove discrimination with scientific certainty; rather, his or her burden is to prove discrimination by a preponderance of the evidence. . . . [478 U.S. at 400-01 (footnote & citations omitted)].

Cases that seek to establish violations of Title VII using statistics are called disparate impact cases. However promising the disparate impact standard as articulated in Bazemore sounds, it is not so easy to prevail in such cases in practice. The Supreme Court has not approved a showing of disparate impact in an employment case in many years, although plaintiffs have prevailed in cases in lower courts.

2.3 Empirical Methods in Nondiscrimination Cases

Although discrimination cases receive most of the legal system's attention, social scientific statistical analysis plays an important role in other litigation contexts. For example, with the growing use of trial consultants, defendants who believe that they cannot obtain a fair trial in the plaintiff's venue of choice, or who merely wish to delay proceedings, usually try to demonstrate to the court the existence of local bias. Statistical analysis using logistic regression suggests that forum in litigation does matter (Clermont and Eisenberg 1995, 1998). Attempts to demonstrate bias can be done through statistical analysis of public opinion polls showing the existence of bias through analysis of simple tables and hypothesis tests using chi-squared distributions. Such polls are paid for and conducted by the defendant and can be highly suspect. Published opinions suggest that successful motions based on such polls are difficult to obtain (First Heights Bank, FSB v. Gutierrez 1993; Rutledge v. Arizona Board of Regents 1985; State v. Rice 1993). But much of the activity in this field occurs in preliminary motions that are unlikely to be published and may not be discussed on appeal.

The dearth of thoroughly reported proceedings enhances the opportunities for gamesmanship. Leading litigation consulting firms have been known to change the order and phrasing of questions to elicit the paying client's desired response. These results are presented to the court without noting the prior inconsistent practice (Poe, d/b/a Lake Guide Service v. PPG Industries, Inc. et al. 1994).

3. EMPIRICAL ANALYSIS OF THE LEGAL SYSTEM

Legal, economic, sociological, psychological, and other scholars regularly use social science methodology to describe and analyze the legal system. One can confidently forecast increasing use of law-related empirical and statistical analysis. Scholars unconstrained by the need to satisfy legal standards applicable to individual cases have broad and inexpensive resources available. Growth in online databases, their increasingly facile availability over the internet, and the availability of inexpensive and sophisticated commercial statistical programs promise to improve statistical training at the undergraduate level and to facilitate empirical legal research at all levels.

The most important contribution of empirical legal research will not be statistical analyses that help determine individual cases. Rather, by providing an accurate portrayal of how the legal system operates, empirical legal analysis can influence not only individual cases, but also larger policy questions. Much room for progress exists, because misperceptions about the legal system are common.

3.1 A Nonlitigious Citizenry

Recent studies reveal much about how the system of civil justice works, often with surprising results. Contrary to popular and professional belief, our legal system is not dominated by a highly litigious citizenry. Surveys of the general public and detailed analysis of hospital records show that fewer than 10% of malpractice and product liability victims initiate legal action (Danzon 1985; Hensler et al. 1991; Report of the Harvard Medical Practice Study 1990).

3.2 Surprisingly Sober and Predictable Juries

Studies also show that the law functions in a sober and predictable manner in important and controversial areas. Medical malpractice studies using contingency tables and regression models find that the quality of medical care is an important determinant of a defendant's medical malpractice liability and that care quality and injury severity are extremely important in determining expected settlements (Farber and White 1991, 1994; Vidmar 1995). Time trend analysis of product liability cases shows no pro-plaintiff trend during a period when the product liability system was said to be out of control in favor of plaintiffs (Eisenberg 1999; Eisenberg and Henderson 1992). Regression analysis of the relation between punitive damages and compensatory damages shows that punitive awards are strongly tied to the level of compensatory awards and that not only are jurors not pulling numbers out of the air, but they also are behaving about the same as judges do in punitive damages cases (Eisenberg, LaFountain, Ostrom, Rottman, and Wells 2000; Eisenberg, Goerdt, Ostrom, Rottman, and Wells 1997). Interview evidence and controlled experiments suggest that jurors are far from being out-of-control Robin Hoods seeking to steal from the rich and give to the poor (Hans and Lofquist 1992; Hans and Vidmar 1986; MacCoun 1996).

3.3 Deeper Understanding of Capital Cases

Statistical, psychological, and empirical legal work is yielding deeper understanding of behavior in capital cases. Widespread beliefs about states' propensity to impose the death penalty are shaped by states with large death row populations, such as California, Florida, and Texas. Yet none of these states imposes a high number of death penalties per murder (Blume and Eisenberg 1999, p. 500). Nevada and Oklahoma impose the death penalty at more than twice the rate per murder than any of these larger states. And, despite the belief of many that the judiciary is largely politicized, Cox proportional hazard models of which defendants obtain relief from death row show no effects based on a state's method of selecting judges.

Empirical studies at the case level also have been revealing. Using Bayesian hierarchical models, Stasny, Kadane, and Fritsch (1998) reported support for the belief that jurors who would be excluded from death penalty juries behave differently from other jurors in non–capital offense trials. Surveys and experiments yield similar results (Cowan, Thompson, and Ellsworth 1984; Fitzgerald and Ellsworth 1984; Haney, Hurtado, and Vega 1994). Eisenberg and Wells (1993) used simple hypothesis tests to show that jurors' misperceptions about how long a murderer will actually serve in prison can lead them to impose the death penalty when they otherwise would not, a result partly relied on by the Supreme Court in Simmons v. South Carolina (1994). Eisenberg, Garvey, and Wells (2000) analyzed survey data based on interviews with actual capital case jurors to construct ordered probit models that explain the influence of race, religion, and attitude toward the death penalty on jurors' first votes in capital cases.

3.4   Interpreting Case Outcomes

A persistent issue in interpreting even sound legal system data is that the vast majority of cases settle and that most settlements are not publicly available. Observation often is limited to cases that reach trial or appeal. Yet such cases need not be random samples of all cases. Drawing inferences about the system as a whole can be difficult when one tends to observe a small and biased subset of cases (Priest and Klein 1984). Despite this selection effect, substantial progress can be made in interpreting the mass of case outcomes. Iteratively reweighted least squares analysis using delta method linearization shows a strong correlation, by subject area, between plaintiff win rates at trial and plaintiff win rates in pretrial stages of adjudication (Eisenberg 1991). A similar correlation exists between adjudicated and settled cases (Eisenberg 1994, pp. 292–293). Thus observers of case outcomes in a subject area at one procedural stage can make informed assessments of how that category fares at other stages. Sensitivity to the selection effect does not necessarily preclude the drawing of reasonable conclusions from observing win rates in adjudicated cases (Clermont and Eisenberg 1998).

A further selection effect arises because only a small fraction of the vast number of potential claims result in the filing of a lawsuit. Yet the process through which cases are selected for litigation cannot be ignored because it does not yield a random selection of claims. Probit, bivariate probit, and competing risk models that account for case selection confirms that plaintiff win rates and trial rates can be viewed as partly a consequence of the selection of disputes for filing (Eisenberg and Farber 1997, 1999).

4. DESIRABLE STATISTICAL CONTRIBUTIONS

Both case-specific and system-wide statistical analysis of legal issues would benefit from advances in statistical methodology. Because many law-related datasets have sparse cells, there is a need for advances in inferences involving small samples. Legal data are often categorical, and models often suffer from problems of endogeneity. Advances in the ease of use of tools for factor analysis of categorical data and for solving systems of equations involving ordered and categorical response variables are also needed. Data often cluster on more than one variable; readily accessible methods for estimating hierarchical models with random coefficients are needed. These methodological improvements need to be accompanied by greater determination on the part of government authorities to collect systematic data about the legal system.

REFERENCES

Aitken, C. G. G. (1995), *Statistics and the Evaluation of Evidence for Forensic Scientists*, Chichester, U.K.: Wiley.

Aitken, C. G. G., Bring, J., Leonard, T., and Papasouliotis, O. (1997), "Estimation of Quantities of Drugs Handled and the Burden of Proof," *Journal of the Royal Statistical Society*, Ser. A, 160, 333–350.

Baldus, D. C., Woodworth, G. G., and Pulaski, C. A. (1990), *Equal Justice and the Death Penalty*, Boston: Northeastern University Press.

Bazemore v. Friday, 478 U.S. 385 (1986).

Blume, J. H., and Eisenberg, T. (1999), "Judicial Politics, Death Penalty Appeals, and Case Selection: An Empirical Study," *University of Southern California Law Review*, 72, 465–503.

Blume, J. H., Eisenberg, T., and Johnson, S. L. (1998), "Post-McCleskey Racial Discrimination Claims in Capital Cases," *Cornell Law Review*, 83, 1771–1810.

Clermont, K. M., and Eisenberg, T. (1998), "Do Case Outcomes Really Reveal Anything About the Legal System? Win Rates and Removal Jurisdiction," *Cornell Law Review*, 83, 581–607.

—— (1995), "Exorcising the Evil of Forum Shopping," *Cornell Law Review*, 80, 1507–1535.

Cowan, C. L., Thompson, W. C., and Ellsworth, P. C. (1984), "The Effects of Death Qualification on Jurors' Predisposition to Convict and on the Quality of Deliberation," *Law and Human Behavior*, 8, 53–80.

Danzon, P. M. (1985), *Medical Malpractice: Theory, Evidence and Public Policy*, Cambridge, MA: Harvard University Press.

Donnelly, P., and Friedman, R. D. (1999), "DNA Database Searches and the Legal Consumption of Scientific Evidence," *Michigan Law Review*, 97, 931–984.

Eisenberg, T. (1991), "The Relationship Between Plaintiff Success Rates Before Trial and at Trial," *Journal of the Royal Statistical Society*, Ser. A, 154, 111–116.

—— (1994), "Negotiation, Lawyering, and Adjudication: Kritzer on Brokers and Deals," *Law & Social Inquiry*, 19, 275–299.

—— (1999), "Judicial Decisionmaking in Federal Products Liability Cases, 1978–1997," *DePaul Law Review*, 49, 323–333.

Eisenberg, T., and Farber, H. S. (1997), "The Litigious Plaintiff Hypothesis: Case Selection and Resolution," *RAND Journal of Economics*, 28, S92–S112.

—— (1999), "The Government as Litigant: Further Tests of the Case Selection Model," National Bureau of Economic Research Working Paper 7296, Cambridge, MA.

Eisenberg, T., Garvey, S. P., and Wells, M. T. (2000), "The Effects of Juror Race, Religion, and Attitude Towards the Death Penalty on Capital Case Sentencing Outcomes," unpublished manuscript.

Eisenberg, T., Goerdt, J., Ostrom, B., Rottman, D., and Wells, M. T. (1997), "The Predictability of Punitive Damages," *Journal of Legal Studies*, 26, 623–661.

Eisenberg, T., and Henderson, J. A. (1992), "Inside the Quiet Revolution in Products Liability," *UCLA Law Review*, 39, 731–810.

Eisenberg, T., LaFountain, N., Ostrom, B., Rottman, D., and Wells, M. T. (2000), "Judges, Juries, and Punitive Damages: An Empirical Study," unpublished manuscript.

Eisenberg, T., and Wells, M. T. (1993), "Deadly Confusion: Juror Instructions in Capital Cases," *Cornell Law Review*, 79, 1–17.

Evett, I. W. (1991), "Interpretation: A Personal Odyssey," in *The Use of Statistics in Forensic Science*, eds. C. G. G. Aitken and D. A. Stoney, New York: E. Horwood.

Farber, H. S., and White, M. J. (1991), "Medical Malpractice: An Empirical Examination of the Litigation Process," *Rand Journal of Economics*, 22, 199–217.

—— (1994), "A Comparison of Formal and Informal Dispute Resolution in Medical Malpractice," *Journal of Legal Studies*, 23, 777–806.

Fienberg, S. E. (Ed.) (1989), *The Evolving Role of Statistical Assessments in the Courts*, New York: Springer-Verlag.

────── (1997), "Ethics and the Expert Witness: Statistics on Trial," *Journal of the Royal Statistical Society*, Ser. A, 160, 321–331.

Finkelstein, M. O., and Levenbach, H. (1983), "Regression Estimates of Damages in Price-Fixing Cases," *Law and Contemporary Problems*, 46, 145–169.

Finkelstein, M. O., and Levin, B. (1980), *Statistics for Lawyers*, New York: Springer-Verlag.

First Heights Bank, FSB v. Gutierrez, 852 S.W.2d 596 (Texas 1993).

Fitzgerald, R., and Ellsworth, P. C. (1984), "Due Process vs. Crime Control: Death Qualification and Juror Attitudes," *Law and Human Behavior*, 8, 31–52.

Freedman, D. A., Ostland, M., Roberts, M. R., and Klein, S. P. (1999), Rejoinder, *Journal of the American Statistical Association*, 94, 355–357.

Haney, C., Hurtado, A., and Vega, L. (1994), "Modern Death Qualification: New Data on Its Biasing Effects," *Law and Human Behavior*, 18, 619–633.

Hans, V. P., and Lofquist, W. S. (1992), "Jurors' Judgements of Business Liability in Tort Cases: Implications for the Litigation Explosion Debate," *Law and Society Review*, 26, 85–115.

Hans, V., and Vidmar, N. (1986), *Judging the Jury*, New York: Plenum Press.

Hensler, D. R., Marquis, M. S., Abrahamse, A. F., Berry, S. H., Ebener, P., Lewis, E. G., Lind, E. A., MacCoun, R. J., Manning, W. G., Rogowski, J. A., and Vaiana, M. E. (1991), *Compensation for Accidental Injuries in the United States*, Santa Monica, CA: RAND Institute for Civil Justice.

Kadane, J. B., and Terrin, N. (1997), "Missing Data in the Forensic Context," *Journal of the Royal Statistical Society*, Ser. A, 160, 351–357.

King, G. (1999), Comment on "Rejoinder", by Freedman, Ostland, Roberts, and Klein, *Journal of the American Statistical Association*, 94, 352–355.

Lempert, R. (1997), "After the DNA Wars: Skirmishing with NRC II," *Jurimetrics Journal*, 37, 439–462.

MacCoun, R. J. (1996), "Differential Treatment of Corporate Defendants by Juries: An Examination of the 'Deep-Pockets' Hypothesis," *Law and Society Review*, 30, 121–161.

McCleskey v. Kemp, 481 U.S. 279 (1987).

Mellen, B. G., and Royall, R. M. (1997), "Measuring the Strength of Deoxyribonucleic Acid Evidence, and Probabilities of Strong Implicating Evidence," *Journal of the Royal Statistical Society*, Ser. A, 160, 305–320.

National Research Council (1992), *DNA Technology in Forensic Science*, Washington, DC: National Research Council, Committee on DNA Technology in Forensic Science.

National Research Council (1996), *The Evaluation of Forensic DNA Evidence*, Washington, DC: National Research Council, Committee on DNA Forensic Evidence.

Norris v. State of Alabama, 294 U.S. 587 (1935).

Poe, d/b/a Lake Guide Service v. PPG Industries, Inc. et al., No. 10-11774, Cameron Parish, LA. Nov. 14–16, 1994 (hearing on motion to change venue).

Priest, G. L., and Klein, B. (1984), "The Selection of Disputes for Litigation," *Journal of Legal Studies*, 13, 1–56.

Report of the Harvard Medical Practice Study (1990), *Patients, Doctors, and Lawyers: Medical Injury, Malpractice Litigation, and Patient Compensation in New York*. Cambridge, MA: President and Fellows of Harvard College.

Robertson, B., and Vignaux, G. A. (1995), *Interpreting Evidence: Evaluating Forensic Science in the Courtroom*. Chichester, U.K.: Wiley.

Rutledge v. Arizona Board of Regents, 147 Ariz. 534, 711 P.2d 1207 (Ariz. App. 1985).

Saks, M. J. (1998), "Merlin and Solomon: Lessons From the Law's Normative Encounters With Forensic Identification Science," *Hastings Law Journal*, 49, 1069–1141.

Saks, M. J., and Koehler, J. J. (1991), "What DNA 'Fingerprinting' Can Teach the Law About the Rest of Forensic Science," *Cardozo Law Review*, 13, 361–372.

Schum, D. A. (1994), *Evidential Foundations of Probabilistic Reasoning*, New York: Wiley.

Simmons v. South Carolina, 512 U.S. 154 (1994).

Stasny, E. A., Kadane, J. B., and Fritsch, K. S. (1998), "On the Fairness of Death-Penalty Jurors: A Comparison of Bayesian Models With Different Levels of Hierarchy and Various Missing-Data Mechanisms," *Journal of the American Statistical Association*, 93, 464–477.

State v. Rice, 120 Wash.2d 549, 844 P.2d 416 (Wash. 1993) (en banc).

Thornburg v. Gingles, 487 U.S. 30 (1986).

U.S. General Accounting Office (1990), *Death Penalty Sentencing: Research Indicates Pattern of Racial Disparities*, Report to the Senate and House Committees on the Judiciary.

Vidmar, N. (1995), *Medical Malpractice and the American Jury: Confronting the Myths About Jury Incompetence, Deep Pockets, and Outrageous Damage Awards*, Ann Arbor, MI: University of Michigan Press.

Von Drehle, D. (1994), "Retired Justice Changes Stand on Death Penalty," *Washington Post*, June 10, 1994, at A1.

Demography: Past, Present, and Future

Yu Xie

In a classic statement, Hauser and Duncan (1959) defined demography as "the study of the size, territorial distribution, and composition of population, changes therein, and the components of such changes" (p. 2). It was fortunate that Hauser and Duncan explicitly included "composition of population" and "changes therein" in their definition, for their inclusion has broadened demography to encompass two types of demography: formal demography and population studies. Formal demography, whose origin can be traced to John Graunt in 1662, is concerned with fertility, mortality, age structure, and spatial distribution of human populations. Population studies is concerned with population compositions and changes from substantive viewpoints anchored in another discipline, be it sociological, economic, biological, or anthropological; its origin can be traced to Thomas Malthus in 1798. By definition, population studies is interdisciplinary, bordering between formal demography and a substantive discipline that is often, but not necessarily, a social science.

Defined in this way, demography provides the empirical foundation on which other social sciences are built. It is hard to imagine that a social science can advance steadily without first knowing the basic information about the human population that it studies. As a field of inquiry, demography enjoyed a rapid growth in the 20th century. For example, the membership of the Population Association of America (PAA), the primary association for demographers in the U.S. founded in 1931, grew from fewer than 500 in 1956 to more than 3,000 in 1999. This growth is remarkable given the virtual absence of demography departments at American universities (with a few exceptions, such as the University of California Berkeley). To recognize contributions made by demographers, one only needs to be reminded of factual information about contemporary societies. Much of what we know as "statistical facts" about American society, for instance, has been provided or studied by demographers. Examples include socioeconomic inequalities by race (Farley 1984) and gender (Bianchi and Spain 1986), residential segregation by race (Duncan 1957; Massey and Denton 1993), intergenerational social mobility (Blau and Duncan 1967; Featherman and Hauser 1978), increasing trends of divorce (Sweet and Bumpass 1987) and cohabitation (Bumpass, Sweet, and Cherlin 1991), consequences of single parenthood for children (McLanahan and Sandefur 1994), rising income inequality (Danziger and Gottschalk

Yu Xie is Frederick G. L. Huetwell Professor of Sociology, Department of Sociology, University of Michigan, Ann Arbor, MI 48109 (E-mail: yuxie@umich.edu). The author thanks Jennifer Barber, Mark Becker, Aimee Dechter, Tom Fricke, David Harris, Robert Hauser, Robert Mare, and Pam Smock for their comments on an earlier draft of this article.

1995), and increasing economic returns for college education (Mare 1995).

Besides providing factual information, demography has also been fundamental in forecasting future states of human societies. Although demographic forecasting is subject to uncertainty, as any type of forecasting, demographers are able to predict future population sizes by age with a high degree of confidence, utilizing information pertaining to past fertility, regularity in age patterns of mortality, and likely future levels of mortality. A notable example of demographic forecasting is the work by Lee and Tuljapurkar (1994, 1997), who demonstrated how demographic forces (i.e., projected improvements in longevity) dramatically impact future demands on social security.

Formal demography and population studies not only take on different subject matters, but also rely on different methodological approaches. Characteristically, formal demography is built on mathematics and thus is closely tied to mathematical demography. It has a rich arsenal of powerful research tools, such as life tables and stable population theory, the latter of which is usually accredited to Alfred J. Lokta in 1922. Note that mathematical models in formal demography sometimes incorporate stochastic processes. The refinement and formalization of mathematical demography and its successful application to human populations can be found in works by Coale (1972), Keyfitz (1985), Preston and Campbell (1993), Rogers (1975), and Sheps and Menken (1973). In its applications, mathematical demography typically presumes access to population data and handles heterogeneity through disaggregation (i.e., dividing a population into subpopulations).

Methods used in population studies are eclectic, borrowing heavily from substantive social science disciplines. Given the widespread use of survey data and the predominant role of statistical inference in all social science disciplines since the 1960s, it should come as no surprise that the characteristic method in population studies is statistical. [It should be noted that qualitative methods are also used in demography (Kertzer and Fricke 1997).] The types of statistical methods used by demographers vary a great deal and change quickly, ranging from path analysis and structural equations (Duncan 1975) and log-linear models (Goodman 1984) to econometric models (Heckman 1979; Willis and Rosen 1979) and event history models (Yamaguchi 1991). Substantive research in population studies usually involves statistical analysis of sample data (as opposed to population data) in a multivariate framework; sometimes, researchers develop statistical models to test hypotheses derived from an individual-level behavioral model. For this reason, population studies is closely tied to statistical demography.

Although it is useful to draw the distinction between formal demography and population studies, the boundary between the two is neither fixed nor impermeable. Indeed, this boundary presents many exciting topics for research. For example, it is possible, and indeed desirable, to estimate demographic rates from sample data with statistical models before feeding them as input for mathematical analysis in formal demography (e.g., Clogg and Eliason 1988; Hoem 1987; Land, Guralnik, and Blazer 1994; Xie 1990; Xie and Pimentel 1992). There are good reasons for using statistical tools in combination with mathematical models. First, the advancement of demography has brought with it more, richer, and better data in the form of sample data; the use of sample data requires statistical tools; and treating sample data as exactly known quantities runs the danger of being contaminated by sampling errors. Second, statistical models are better suited for examining group differences through the use of covariates, for the method of disaggregation presumes a full interaction model and may lead to inaccurate estimation due to small group sizes. Third, because observed data may sometimes be irregular or simply missing, statistical models can help smooth or impute data (e.g., Little and Rubin 1987). Conversely, techniques in formal demography (e.g., indirect estimation, intercensal cohort comparisons, model life tables) can also be utilized to improve statistical analysis. For example, mathematical relationships for demographic components are often used to correct faulty data or provide missing data before they are analyzed with statistical techniques.

Let me now turn to the methodological implications of Hauser and Duncan's two important phrases "composition of population" and "changes therein" in their definition of demography. The first phrase refers to population heterogeneity; the second, to dynamic processes. Population heterogeneity or variability is fundamental to all social science. As articulated by Mayr (1982), one of the most important and long-lasting influences of Darwin's evolution theory is "population thinking," as distinguished from "typological thinking" prevalent in physical science. Briefly stated, typological thinking attempts to discover the "essence" or "truth" in society and human nature. Once well understood, a scientific concept or law is always generalizable to all settings at all times. If the observed phenomena appear in disarray, then averages are taken and deviations are disregarded. In the history of statistics, typological thinking is associated with the "average man" in Quetelet's social physics. In contrast, population thinking makes *variations* the very subject matter to be studied. It was Francis Galton, Darwin's cousin, who introduced population thinking into the field of statistics and in so doing discovered correlation and regression (see Xie 1988 and references therein). Demography should pride itself for approaching social phenomena through population thinking by focusing on variations by group and individual characteristics. For example, demographers have long been interested in how life chances differ by gender, age, race, region, and family origin, as well as historical and cultural context.

In studying social changes, demographers have advocated the perspectives of cohort (Ryder 1965) and life course (Elder 1985). Whereas the latter is concerned with the occurrences of significant events and transitions over individuals' lives and the dependence of these events and transitions on earlier experiences and societal forces, the former relates individuals' different experiences at the micro level to social changes at the macro level. That is, demographers are especially interested in both age-graded intracohort changes and intercohort changes resulting from cohort replacement. In addition, demographers also accept the notion that certain temporal effects are period based, affecting all individuals regardless of age. Due to the linear dependency of the three measures, however, the conceptual model of age, period, and cohort is intrinsically underidentified, and its statistical implementation requires constraining assumptions based on substantive considerations (Mason and Fienberg 1985).

Many of the advances in demography over the past few decades cannot be separated from the field's close relationship to statistics. Indeed, it is the integration of statistical methods into demographic research that has enabled demographers to study population heterogeneity and change with sample data. A prime example where statistical methodology has significantly impacted demography is event history analysis (also called survival models or hazard models). As argued by Tuma and Hannan (1984), event history analysis is ideally suited for studying social dynamics and social processes, because it deals with the timing of transitions or event occurrences. Although event history analysis was invented by demographers in the form of life tables, the life table approach in mathematical demography assumes population data and cannot easily incorporate population heterogeneity except by disaggregation. Cox's (1972) influential work bridged the gap between life tables and regression analysis by focusing on estimation of the effects of covariates using sample data. The influence of Cox (1972) in demography goes beyond his innovative estimation method, for it represents a shift in methodological orientation toward statistical demography. With this new orientation, event history analysis is seen as a statistical operationalization of life tables for sample data, with attention paid primarily to group differences or the effects of covariates.

Studying population heterogeneity by observed group or individual characteristics with event history analysis has proven fruitful; see, for example, Thornton, Axinn, and Teachman's (1995) study of the effects of education on the likelihood of cohabitation and marriage and Wu's (1996) work on the determinants of premarital childbearing. In addition, demographers have a long-standing concern with unobserved heterogeneity in event history analysis and life tables (e.g., Sheps and Menken 1973). Unlike in the case of linear models, unobserved heterogeneity uncorrelated with covariates in event history data could bias estimated hazards through altering the composition of the exposed population at risk (Vaupel and Yashin 1985). For example, demographers have long been puzzled as to whether the well-documented black/white mortality crossover is due to biological selection, in which old blacks are more fit than old whites (Manton, Poss, and Wing 1979), or due to age misreporting (Preston, Elo, Rosenwaike, and Hill 1996). There are two broad approaches to handling unobserved heterogeneity: parametric mixture models (see Powers and Xie 2000) and Heckman and Singer's (1984) nonparametric method that is analogous to latent class models for contingency tables. To improve identification, demographers have also capitalized on pooled information from clustered data structures, such as siblings or twins, assuming unobserved heterogeneity at the cluster level instead of at the individual level (Guo and Rodriguez 1992; Yashin and Iachine 1997).

As stated earlier, population studies is interdisciplinary, with sociological and economic demography taking center stage. In both sociology and economics, there has been a methodological tradition of structural equation models simultaneously representing multiple processes with strong theoretical priors. Event history analysis has been incorporated into this framework, with the hazard rate treated as a limited dependent variable. In this framework, demographers have gone beyond assuming unobserved heterogeneity orthogonal to covariates. Instead, they consider unobserved heterogeneity that is selective with respect to observed variables (Lillard, Brien, and Waite 1995). But identification of such models is difficult and requires pooled information from repeated events and strong parametric assumptions.

Where will demography go from here? Although predicting the future inherently carries some risks, past experience provides some basis for making cautionary conjectures about demography's future. First, demography, especially in the United States, has been a very successful interdisciplinary enterprise and will continue to be one. Not only will it retain sociologists and economists who already strongly identify themselves with demography, but it will also attract the interest of scholars in other social and biological sciences. In particular, incorporation of biological approaches in demographic research looks promising (for a recent example, see Yashin and Iachine 1997). Second, the interdisciplinary nature of demography makes it an ideal locus for innovating, testing, and popularizing new statistical methods. This means closer ties between statistical science and demography, as statistical demography continues to affirm its prominent role in demography. Third, in the burgeoning "information age," the public will demand more easily accessible information about the society in which they live, as exemplified by the strong public interest in the implementation of the 2000 U.S. Census. Demography can play an important role in providing that information. The internet is and will continue to be a powerful tool for bringing demographic information to the general public and policymakers alike.

One substantive area where demography is rapidly making progress is in research on aging. As the elderly comprise an increasingly large portion of the population in many societies, including the United States, demographers have become increasingly interested in aging (e.g., Wachter and Finch 1997). Economists and sociologists are already busy studying social and economic aspects of aging, such as retirement, health, residence, economic security, and family support. The demography of aging has also benefited from contributions from scholars in biology, psychology, public health, gerontology, and geriatrics. This is an area where the aforementioned conjectures about the future of demography will likely be manifested, as we will see in the demography of aging more interdisciplinary work, newer and better applications of statistical methodology, and provision of more public and scientific information.

REFERENCES

Bianchi, S. M., and Spain, D. (1986), *American Women in Transition*, New York: Russell Sage Foundation.

Blau, P. M., and Duncan, O. D. (1967), *The American Occupational Structure*, New York: Wiley.

Bumpass, L. L., Sweet, J. A., and Cherlin, A. (1991), "The Role of Cohabitation in Declining Rates of Marriage," *Journal of Marriage and the Family*, 53, 913–927.

Clogg, C. C., and Eliason, S. R. (1988), "A Flexible Procedure for Adjusting Rates and Proportions, Including Statistical Methods for Group Comparisons," *American Sociological Review*, 53, 267–283.

Coale, A. J. (1972), *The Growth and Structure of Human Populations: A Mathematical Investigation*. Princeton, NJ: Princeton University Press.

Cox, D. R. (1972), "Regression Models and Life-Tables," *Journal of the Royal Statistical Society*, Ser. B, 34, 187–220.

Danziger, S., and Gottschalk, P. (1995), *America Unequal*, New York: Russell Sage.

Duncan, O. D. (1957), *The Negro Population of Chicago: A Study of Residential Succession*, Chicago: University of Chicago Press.

―――― (1975), *Introduction to Structural Equation Models*, New York: Academic Press.

Elder, G. H., Jr. (1985), "Perspectives on the Life Course," in *Life Course Dynamics: Trajectories and Transitions, 1968–1980*, ed. G. H. Elder, Jr., Ithaca, NY: Cornell University Press, pp. 23–49.

Farley, R. (1984), *Blacks and Whites: Narrowing the Gap?*, Cambridge, MA: Harvard University Press.

Featherman, D. L., and Hauser, R. M. (1978), *Opportunity and Change*, New York: Academic Press.

Goodman, L. A. (1984), *The Analysis of Cross-Classified Data Having Ordered Categories*, Cambridge, MA: Harvard University Press.

Guo, G., and Rodríguez, G. (1992), "Estimating a Multivariate Proportional Hazards Model for Clustered Data Using the EM Algorithm, With an Application to Child Survival in Guatemala," *Journal of the American Statistical Association*, 87, 969–976.

Hauser, P. M., and Duncan, O. D. (Eds.) (1959), *The Study of Population: An Inventory and Appraisal*, Chicago: University of Chicago Press.

Heckman, J. J. (1979), "Sample Selection Bias as a Specification Error," *Econometrica*, 47, 153–161.

Heckman, J. J., and Singer, B. (1984), "A Method for Minimizing the Impact of Distributional Assumptions in Econometric Models for Duration Data," *Econometrica*, 52, 271–320.

Hoem, J. M. (1987), "Statistical Analysis of a Multiplicative Model and Its Application to the Standardization of Vital Rates: A Review," *International Statistical Review*, 55, 119–152.

Kertzer, D. I., and Fricke, T. (Eds.) (1997), *Anthropological Demography: Toward a New Synthesis*, Chicago: University of Chicago Press.

Keyfitz, N. (1985), *Applied Mathematical Demography* (2nd ed.), New York: Springer-Verlag.

Land, K. C., Guralnik, J. M., and Blazer, D. G. (1994), "Estimating Increment-Decrement Life Tables With Multiple Covariates From Panel Data: The Case of Active Life Expectancy," *Demography*, 31, 297–319.

Lee, R. D., and Tuljapurkar, S. (1994), "Stochastic Population Forecasts for the United States: Beyond High, Medium, and Low," *Journal of the American Statistical Association*, 89, 1175–1189.

―――― (1997), "Death and Taxes: Longer Life, Consumption, and Social Security," *Demography*, 34, 67–81.

Lillard, L. A., Brien, M. J., and Waite, L. J. (1995), "Premarital Cohabitation and Subsequent Marital Dissolution: A Matter of Self-Selection?," *Demography*, 32, 437–457.

Little, R. J. A., and Rubin, D. B. (1987), *Statistical Analysis With Missing Data*, New York: Wiley.

Manton, K. G., Poss, S. S., and Wing, S. (1979), "The Black/White Mortality Crossover: Investigation From the Perspective of the Components of Aging," *The Gerontologist*, 19, 291–300.

Mare, R. D. (1995), "Changes in Educational Attainment and School Enrollment," in *State of the Union: America in the 1990s*, Vol. II, ed. R. Farley, New York: Russell Sage.

Mason, W. M., and Fienberg, S. E. (1985), *Cohort Analysis in Social Research: Beyond the Identification Problem*, New York: Springer-Verlag.

Massey, D. S., and Denton, N. A. (1993), *American Apartheid: Segregation and the Making of the Underclass*, Cambridge, MA: Harvard University Press.

Mayr, E. (1982), *The Growth of Biological Thought: Diversity, Evolution, and Inheritance*, Cambridge, MA: Belknap Press.

McLanahan, S. S., and Sandefur, G. (1994), *Growing Up With a Single Parent: What Hurts, What Helps*, Cambridge, MA: Harvard University Press.

Powers, D. A., and Xie, Y. (2000), *Statistical Methods for Categorical Data Analysis*, New York: Academic Press.

Preston, S. H., and Campbell, C. (1993), "Differential Fertility and the Distribution of Traits: The Case of IQ," *American Journal of Sociology*, 98, 997–1019.

Preston, S. H., Elo, I. T., Rosenwaike, I., and Hill, M. (1996), "African-American Mortality at Older Ages: Results of a Matching Study," *Demography*, 33, 193–209.

Rogers, A. (1975), *Introduction to Multiregional Mathematical Demography*, New York: Wiley.

Ryder, N. B. (1965), "The Cohort as a Concept in the Study of Social Change," *American Sociological Review*, 30, 843–861.

Sheps, M. C., and Menken, J. A. (1973), *Mathematical Models of Conception and Birth*, Chicago: University of Chicago Press.

Sweet, J. A., and Bumpass, L. L. (1987), *American Families and Households*, New York: Russell Sage.

Thornton, A., Axinn, W., and Teachman, J. D. (1995), "The Influence of School Environment and Accumulation on Cohabitation and Marriage in Early Adulthood," *American Sociological Review*, 60, 762–774.

Tuma, N. B., and Hannan, M. T. (1984), *Social Dynamics: Models and Methods*, San Francisco: Academic Press.

Vaupel, J. W., and Yashin, A. I. (1985), "Heterogeneity's Ruses: Some Surprising Effects of Selection on Population Dynamics," *The American Statistician*, 39, 176–185.

Wachter, K. W., and Finch, C. E. (Eds.) (1997), *Between Zeus and the Salmon*, Washington, DC: National Academy Press.

Willis, R. J., and Rosen, S. (1979), "Education and Self-Selection," *Journal of Political Economy*, 87(5), S7–S36.

Wu, L. (1996), "Effects of Family Instability, Income, and Income Instability on the Risk of a Premarital Birth," *American Sociological Review*, 61, 386–406.

Xie, Y. (1988), "Franz Boas and Statistics," *Annals of Scholarship*, 5, 269–296.

────── (1990), "What is Natural Fertility? The Remodelling of a Concept," *Population Index*, 56, 656–663.

Xie, Y., and Pimentel, E. E. (1992), "Age Patterns of Marital Fertility: Revising the Coale–Trussell Method," *Journal of the American Statistical Association*, 87, 977–984.

Yamaguchi, K. (1991), *Event History Analysis*, London: Sage.

Yashin, A. I., and Iachine, I. I. (1997), "How Frailty Models can be Used for Evaluating Longevity Limits: Taking Advantage of an Interdisciplinary Approach," *Demography*, 34, 31–48.

Chapter 3

Statistics in the Physical Sciences and Engineering

Diane Lambert

No doubt much of the progress in statistics in the 1900s can be traced back to statisticians who grappled with solving real problems, many of which have roots in the physical sciences and engineering. For example, George Box developed response surface designs working with chemical engineers, John Tukey developed exploratory data analysis working with telecommunications engineers, and Abraham Wald developed sequential testing working with military engineers. These statisticians had a strong sense of what was important in the area of application, as well as what statistics could provide. The beginning of the 2000s is a good time to reflect on some of the current problems in the physical sciences and engineering, and how they might lead to new advances in statistics—or, at the least, what statisticians can contribute to solving these problems. My hope is that this set of vignettes will convey some sense of the excitement over the opportunities for statistics and statisticians that those of us who work with physical scientists and engineers feel.

The vignettes are loosely grouped by field application, starting with earth sciences and then moving on to telecommunications, quality control, drug screening, and advanced manufacturing. Superficially, these areas have little in common, but they do share some deep similarities. Most of these areas, for example, face new opportunities and challenges because of our increasing ability to collect tremendous amounts of data. More and more often, the unit of sample size in the physical sciences and engineering is not the number of observations, but rather the number of gigabytes of space needed to store the data. Despite the tremendous advances in raw computational power, processing so much data can still be painful, and visualizing, exploring, and modeling the data can be much worse. As Doug Nychka points out, though, one advantage to working with physical scientists and engineers is that many of them have years of experience designing systems for collecting, processing, analyzing, and modeling massive sets of data, and statisticians can learn from their experience. Moreover, as Bert Gunter and Dan Holder point out, many of the advances are encouraging

Diane Lambert is Head, Statistics Research Department, Bell Labs, Lucent Technologies, Murray Hill, NJ 07974 (E-mail: dl@bell-labs.com).

statisticians to collaborate not only with subject matter specialists, but also with computer scientists.

Another theme in several of the vignettes is that statistical models alone are likely to be insufficient. What is needed instead are models that incorporate both scientific understanding and randomness. David Vere-Jones, for example, writes that progress in earthquake prediction requires advances in understanding the geophysics that produce earthquakes and progress in building statistical models that respect the geophysics and are appropriate for highly clustered, self-similar data. Cleveland and Sun make the same point in the context of internet data: Models are much more likely to be successful and to play a role in networking if they take into account the protocols used to pass data around the internet. On a slightly different note, Rissanen and Yu show that the information theory that underlies signal compression and channel coding (used in wireless communications, for example) both draws from and adds to statistics, especially statistical model selection.

Even areas of engineering in which statistics has a long history, like quality control and reliability, offer new challenges to statistics. As Jerry Lawless points out, product degradation in dynamic environments leads to models like multivariate point processes for which inference is barely developed. He also points out that even the basic steps of defining and measuring quality are problematic in some contexts, such as software. Zach Stoumbos and his co-authors point out that the gap between the methods in statistical quality control that statisticians think are needed and what actually are needed is large and increasing. Finally, Vijay Nair, Mark Hansen, and Jan Shi make clear that the spatial and temporal statistical models needed in advanced manufacturing are perhaps more like those needed in geophysics than the statistical models commonly viewed as relevant to process monitoring.

This set of vignettes shows the enormous opportunities for statisticians to contribute to advances in the physical sciences and engineering while working at the forefront of statistical methodology. There is a growing sense in some areas of the sciences and engineering that branches of computer science, such as data mining, are more appropriate than statistics for understanding massive, complex data. But this is not so much a threat as an opportunity to learn new ways of working with data and building models. A continuing, and perhaps more serious threat is the view that statistical theory is harder, loftier, and a more worthy intellectual pursuit than is working on serious statistical applications. This easily slips into the view that statisticians involved in applications need to keep up with theory, but statisticians involved in building theory do not need to be involved in applications. These vignettes, like the others before it on the life sciences and the business and social sciences in preceding issues of *JASA*, show that grappling with methodology to solve real problems in real applications is as worthy and challenging an intellectual pursuit as can be and the best way to ensure that theory is worthwhile.

Challenges in Understanding the Atmosphere

Doug Nychka

Understanding climate change, forecasting the weather for next week, or predicting turbulence for an aircraft flight— each of these activities combines a knowledge of the atmosphere with data analysis and statistical science. The practical benefits of these activities to our society and economy are balanced by the fascinating interplay between the models that describe geophysical processes and need for statistics to quantify uncertainty and to assimilate observational data.

To provide an overview of statistics in the atmospheric sciences, I have organized some examples around different scales, from large-scale problems connected with our climate to the localized phenomenon of clear air turbulence. This scope will showcase the role of statistics, progressing from a support role in the development and assessment of numerical climate models, through a partnership in forecasting weather, and to a central role in predicting small-scale phenomena such as clear air turbulence. (As a companion to this short article, the reader is referred to Berliner, Nychka, and Hoar 1999 for a larger set of examples with more detail and an extensive bibliography.) Before giving these examples, however, I outline some of the physical principles that underlie atmospheric science.

1. PHYSICAL EQUATIONS FOR THE ATMOSPHERE

The motion and evolution of the atmosphere can be determined from a system of several nonlinear partial differential equations. These equations, derived from classical results in thermodynamics and fluid mechanics on a rotating sphere, are well accepted as the fundamental description of atmospheric flow (see Salby 1996 for an introduction) and are referred to as the primitive equations. So if a complete physical description is known, then why is statistics needed? The short answer is that these equations are simply too complicated to solve in an exact manner for any practical modeling of the atmosphere. A more deliberate answer is related to scales. The primitive equations afford links between large-scale and small-scale motions. Even if one wanted to focus on large-scale motions, say in a climate study, it would be necessary to solve the equations at a much finer scale. This is because energy from smaller scales feeds back into the larger scale motion.

Another property of the primitive equations is a sensitive dependence on initial

Doug Nychka is Section Head, Geophysical Statistics Project, National Center for Atmospheric Research, Boulder, CO 80307.

conditions. This feature comes under the rubric of chaos (see Berliner 1992) and was noted by E. Lorenz in the 1960s. Suppose that one attempted to solve the primitive equations using two slightly different initial conditions. The nonlinear components in the primitive equations would tend to amplify small differences at $t = 0$ into significant departures as t increased. Eventually the two solutions would appear to be independent of one another. Of course this has tremendous practical significance for forecasting weather. One never knows the state of the atmosphere perfectly, and although this discrepancy may be small, differences between the forecast and the true state will diverge exponentially as a function of the forecast time. This property also has implications for variability among climate experiments, because a climate situation also depends on initial conditions of the atmosphere and ocean.

2. CLIMATE CHANGE

2.1 General Circulation Models

A focus for climate studies is the use of a general circulation model (GCM). The development and evaluation of these models provides a rich area for statistical research. A GCM is an adaptation of the primitive equations to a spatial grid and discrete steps in time. The initials GCM are sometimes mistaken for "global climate model," which is also an apt description. Long-term changes in the earth's atmosphere require modeling the large-scale, global motions. It is important for the statistical community to realize that such models represent the confluence of many scientists' efforts, running to many millions of lines of code and requiring high-performance computing to run at useful levels of resolution. One example, the Community Climate Model (Kiehl et al. 1998), is typically run on a spatial grid of approximately 8,000 points at 18 vertical levels with a time step of 20 minutes. A typical climate experiment might involve stepping the model every 20 minutes for hundreds of years, producing large and, depending on your definition, massive datasets. Because of the necessity of dealing with model output, atmospheric scientists are very data savvy, and statisticians can benefit from their experience in manipulating large datasets. The statistics used to summarize model output are often rudimentary, however, and one challenge for statistical research is to provide a more comprehensive analysis of GCM output beyond means and variances. For example, the societal impacts of climate change might be tied to the occurrence of extreme events, not shifts in mean level. At a more fundamental level, a GCM is a dynamical system. Delineating the attractor, finding regimes where the system is nearly linear, and reducing the dimension of the state vector are important theoretical questions but are made difficult by the sheer size of the model.

2.2   Subgrid Scale Processes and Parameterizations

Despite its large state vector, a typical GCM such as the Community Climate Model has grid boxes that are 300 km × 300 km (2.8° × 2.8°) at the equator. Features smaller than these horizontal dimensions cannot be resolved by the model and are referred to as subgrid scale processes. Although some features of the atmosphere are of large scale, there are crucial components that are smaller than this grid size; for example, the strong convection events associated with thunderstorms. In the tropics these storms generate strong updrafts and are a mechanism for vertical transport in the atmosphere. Convection also produces clouds that in turn influence the amount of radiation reaching the surface and the amount trapped by the atmosphere.

Clearly, any model of the atmosphere's general circulation must account for localized strong convection, and any model for radiation must include the influence of clouds. But both of these are subgrid scale processes not directly represented by the model state vector. These processes are accounted for through a technique known as *parameterization*. (This is different from a statistician's use of this term.) A parameterization is a function that expresses a spatial average (or higher moments) of the subgrid scale process in terms of variables that are part of the model's state vector. The parameterization is often grounded on physical arguments but can often involve simplifying approximations and empirically derived relationships. In many cases GCMs are sensitive to how subgrid scale processes are parameterized, and in particular climate results hinge on how clouds reflect and absorb radiation.

Statistics has a clear role to play in improving parameterizations. Building on conventional approaches, one can use observational data to estimate an empirical regression relationship between subgrid scale processes and the grid values. One successful example of these ideas is using neural network models to parameterize the fraction of clouds in a grid box (Bailey, Berliner, Collins, and Kiehl 1999).

The fact that a statistical approach, and most other schemes, does not give perfect predictions of subgrid scale features raises a deeper conceptual issue concerning geophysical parameterizations. The subgrid scale processes may be modeled more realistically with the inclusion of a random component. At some level, the subgrid scale effects and their resulting feedback to larger scales are not predictable, and adding a random component is an efficient way to model this uncertainty. GCMs are deterministic, and so adding a stochastic component would demand a different perspective. One potential benefit is that the increased variability in the model due to stochastic terms may provide better simulations for tail behavior and extremes of climate variables.

2.3   Assessing Climate Change

The ingredients for studying climate change involve both observational data and climate model output. (Current climate models combine a GCM with an ocean

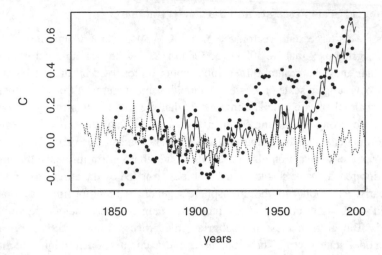

*Figure 1. A Comparison of Observation and Climate Model Results for Annual, Global Average Temper-
ature Anomalies. Deviations in annual global temperature relative to the observed temperatures over the
period 1861–1900. Points are observed average temperatures. Dashed and solid lines are experimental
results from the Climate System Model (CSM), National Center for Atmospheric Research. The dashed line
represents 1870 constant levels of greenhouse gas aerosols; the solid line represents increasing greenhouse
gases and aerosols.*

circulation model, a land surface model, and also a model for sea ice.) A summary of
an experiment using the National Center for Atmospheric Research Climate System
Model (Boville and Gent 1998) is given in Figure 1. A "fingerprint" of climate change
(e.g., a spatial pattern of warming temperatures) is estimated from different runs of the
climate model, and the observed data are then regressed on one or more fingerprints.
The strength of the regression
relationship is used to draw inferences for climate change. Many methodological issues are involved in carrying out a standard fingerprint analysis; an important recent
extension is to consider the analysis within a Bayesian framework (Berliner, Levine,
and Shea 1999). The observational record for temperatures is a rich area for spatial
statistics and statistical modeling. Just as the GCMs have model error, the observational record, especially in the 19th and early 20th centuries, is both irregular and
may include systematic biases. The compilation of monthly or annual means on a
regular spatial grid with reliable measures of uncertainty is a difficult problem, but is
important in any attempt to compare the climate of this century with climate model
simulations.

3. WEATHER FORECASTING

The need for accurate forecasts of the weather is undisputed. In this short overview,
I focus on the statistical foundations of weather forecasts rather than on operational

aspects of implementation or evaluation. Let x_t denote the state of the atmosphere at time t. A caricature of a forecast system is a complicated nonlinear function that evolves the current state forward one-time step, $x_{t+1} = g(x_t)$. Without any other information, one would use this relation to make a forecast at time $t + 1$, plugging in our best estimate of the current state. Now suppose that atmospheric data, y_t, are available at time t and follow the model $y_t = h(x_t) + e_t$. Here h is a function based on the measurement process and e_t is an error term. For most weather forecasts, the primary observations are vertical profiles of the atmosphere obtained from weather balloons (rawinsondes), but other information can come from satellite instruments, Doppler radar, or other surface measurements. Making a forecast has two steps. The first step *assimilates* the data to give an updated and hopefully a better estimate of the current state. In Bayes language, with a prior for x_t and a model for y_t given x_t, one finds a posterior distribution of x_t. The next step is to propagate the state forward, using g to get a forecast. Formally this is just a change of variables: Given the posterior distribution for x_t, find the distribution of $g(x_t)$.

The atmospheric sciences have a mature literature on this basic problem and solutions when distributions are Gaussian and g and h are approximately linear. Even with these assumptions, the calculations are formidable: For operational forecasts, x has a dimension on the order of 10^6 and y on the order of 10^5. Gaussian prior distributions on x depend on covariance matrices that have complex, state-dependent structure. Nonetheless, these covariance models must have a computationally efficient representation to make the assimilation with data possible. Statistics can benefit from the computational tools developed to handle large assimilation problems and can make contributions to extending the distributional models to non-Gaussian data, such as precipitation measurements.

The posterior distribution for $g(x_t)$, essentially a probabilistic forecast, usually does not have a closed form. Current practice is to represent the distribution by a sample of forecasts termed an *ensemble*. If the ensemble is constructed correctly, then the variability among ensemble members should reflect the actual uncertainty in the forecast. For a (Bayesian) statistician, the simplest way to construct the ensemble is to draw a random sample from the posterior of x_t and then propagate each member of the sample forward by g. There are also more dynamically based strategies that also involve the form of g. In any case, the key issue is whether measures of uncertainty for the forecasts stand up to a frequentist criterion.

An important component of the forecast problem is that today's posterior becomes tomorrow's prior. Thus ensemble of forecasts should be constructed so that they can function as a prior distribution for x_{t+1} during the next forecast period. In general, repetition of the assimilation/forecast cycle is a fertile ground for studying adaptive estimation, and poses theoretical questions concerning the convergence and the limiting properties of approximate Bayes procedures.

At a deeper level, the assimilation and forecast problems in meteorology raise issues of dimension reduction. The folklore in atmospheric dynamics is that the high-dimensional state vector used to describe the atmosphere at a grid of points actually lives in much a lower-dimensional manifold. Finding representations of this attractor would help create ensembles and guide the choice of ensemble size.

4. PREDICTING CLEAR AIR TURBULENCE

Turbulence in the atmosphere and ocean is an area with still many open questions. In most cases turbulence can be readily identified, but a simple and rigorous definition of turbulent flow is still elusive. High-altitude, clear air turbulence (CAT) is a serious problem for aviation. The challenge is to forecast CAT accurately so that aircraft can be diverted or at least crews can be warned of turbulent regions. CAT is a small-scale phenomenon often linked to vertical wind shear at the boundaries of the jet stream. But its exact causes are still uncertain. To make forecasts of CAT, atmospheric scientists have proposed a number of indices derived from the output of numerical weather prediction models; for example, Ellrod's number is a product of the vertical wind shear and a measure of the horizontal shearing and stretching of the wind field. This index and others have been found to have some correlation with observed instances of CAT but also have limitations because of a lack of a physical theory and because they are calculated from numerical model output run at coarser grids than the spatial scale of CAT.

A statistical approach to this problem is to use the indices derived on physical grounds and improved by empirical studies to build a forecast model. The basic form is a discriminant model where the indices serve as potential covariates and the training data for model estimation is a heterogeneous mix of pilot reports and accelerometer instruments on selected aircraft. Figure 2 is an example of a 3-hour period of data

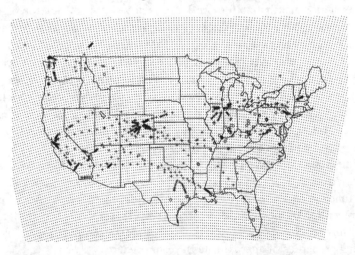

Figure 2. Pilot Reports of Turbulence (○) and Accelerometer Readings of No Turbulence (+) for December 12, 1997 above 20,000 ft. The regular grid is the resolution of the Rapid Update Cycle-60 numerical model used to construct indicators of clear air turbulence.

indicating locations of observations and the forecast grid used by the numerical model. The details of an approach that uses flexible discriminant analysis (linear discriminant analysis combined with multivariate adaptive regression splines) have been described by Tebaldi, Nychka, Brown, and Sharman (1999). This study indicates that a statistical model achieves levels of detection that approach acceptable thresholds for its use as an operational forecast.

Although we have focused on a particular study involving CAT, many small-scale features of the atmosphere are difficult or expensive to resolve with traditional numerical models. Model selection and validation, dimension reduction, departures from Gaussian distributions, and nonparametric function estimation are all relevant tools for building forecast models and are also areas of active statistical research. Note that unlike more generic forecasting or classification problems, problems in the atmospheric sciences often have a physical underpinning that guides selection of covariates and model building. Similar to weather forecasting, these problems also have a sequential aspect and provide the opportunity to use adaptive methods that improve over time.

5. CLOSING REMARKS

Popular views of statistics in the geophysical sciences often focus on spatial methods (e.g., Kriging) and time series. Of course, these remain standard methods for data analysis, and their extension to nonstationary and non-Gaussian processes poses new research problems. Recently, Bayesian hierarchical models, coupled with Markov chain Monte Carlo for sampling posteriors, have gained prominence as important modeling tools (Wikle, Milliff, Nychka, and Berliner, unpublished manuscript, 2000). One strength of hierarchical models is the ease with which they incorporate the physical constraints of geophysical processes and their clarity of interpretation. I hope that this vignette balances some conventional views of geostatistics with an emphasis on the emerging areas in atmospheric science.

Due to limited space, I have focused solely on the atmosphere. But a complete understanding of the earth's environment must include chemical and biological processes along with study of the sun. Overlaid on this natural framework is the influence on physical systems by human activities. These areas challenge statisticians with the need to work closely with substantive numerical models and large, complicated observational datasets. I hope that it is clear that no single discipline alone can approach these problems: The easy stuff has been done! Progress in understanding and forecasting the earth's systems requires collaborative effort among teams of scientists, including statisticians.

REFERENCES

Bailey, B. A., Berliner, L. M., Collins, W., and Kiehl, J. T. (1999), "Neural Networks: Cloud Parameterizations," in *Studies in the Atmospheric Sciences*, eds. L. M. Berliner, D. W. Nychka, and T. Hoar, New York: Springer-Verlag.

Berliner, L. M. (1992), "Probability, Statistics, and Chaos," *Statistical Science*, 7, 69–122.

Berliner, L. M., Levine, R., and Shea, D. (1999), "Bayesian Detection of Climate Change," unpublished manuscript.

Berliner, L. M., Nychka, D. W., and Hoar, T. (1999), *Studies in the Atmospheric Sciences*, New York: Springer-Verlag.

Boville, B. A., and Gent, P. (1998), "The NCAR Climate System Model, Version One," *Journal of Climate*, 11, 1115–1130.

Kiehl, J. T., Hack, J. J., Bonan, G. B., Boville, B. A., Williamson, D. L., and Rasch, P. J. (1998), "The National Center for Atmospheric Research Community Climate Model: CCM3," *Journal of Climate*, 11, 1131–1149.

Salby, M. L. (1996), *Fundamentals of Atmospheric Physics*, San Diego: Academic Press.

Tebaldi, C., Nychka, D., Brown, B., and Sharman, R. (1999), "Predicting Clear-Air Turbulence," in *Studies in the Atmospheric Sciences*, eds. L. M. Berliner, D. W. Nychka, and T. Hoar, New York: Springer-Verlag.

Seismology—A Statistical Vignette

David Vere-Jones

1. INTRODUCTION

Geophysics, and seismology in particular, has a somewhat uneasy relationship with statistics. On one side, geophysicists have always been passionate collectors, processors, and interpreters of observational data. From Halley in the 17th century to Harold Jeffreys in the 20th century, its leading practitioners have also pioneered important developments in statistics—in graphical and numerical methods, in the treatment of errors, in time series analysis, and in many more specialized topics. On the other side, geophysicists, in common with other physical scientists, tend to cling to a view of the universe as governed by deterministic differential equations. Probability models tend to be relegated to the role of describing observational errors, even where, as in describing the occurrence times of earthquakes, the sources of uncertainty lie considerably deeper. The upshot is that the general level of statistical usage among geophysicists, and among seismologists in particular, is very uneven, from contributions of fundamental importance to disappointing misunderstandings.

The role of seismology within geophysics is greater than its rather special subject matter might suggest. This is chiefly because for many years measurement of the reflections and refractions of earthquake waves passing through the earth has provided an important tool for probing the earth's inner structure. At the same time, the subject is kept in the public eye through its applications to engineering, building codes, insurance, and earthquake prediction. In all of these applications, and at the heart of the subject itself, statistical problems abound, few of them easy and some challenging the limits of current statistical methodology. In this vignette I attempt to indicate the nature of these problems, first by tracing a brief history of seismology, and then by selecting a few special issues of current interest. Bolt (1988) and Bullen (1963) have provided useful general introductions to the subject.

2. A BRIEF HISTORY OF SEISMOLOGY, WITH A STATISTICAL BIAS

2.1 First Stages: 1890–1920

Seismology has little statistical history before the development of the Milne–Shaw seismograph in the 1890s. Somewhat earlier the theory of wave propagation

David Vere-Jones is Emeritus Professor, School of Mathematical and Computing Sciences, Victoria University, Wellington, New Zealand.

in an elastic medium had been worked out, a qualitative intensity scale for assessing ground motion had been developed, and there had been attempts to compile lists of large historical earthquakes. But the Milne–Shaw instrument was the first that was compact and accurate enough to allow objective measurements of earthquake wave motion in many different places to be made and compared. Early instruments were deployed not only in Europe and the United States, but also in Japan, New Zealand, and China.

With the first reliable instruments came the first reliable data and the beginnings of seismology as a quantitative scientific discipline. As early as 1894, the Japanese seismologist Omori, studying the aftershocks of large Japanese earthquakes, formulated the first empirical law of seismology, *Omori's law*: the frequency $\lambda(t)$ of aftershocks at time t after the main shock decays hyperbolically. This is most commonly quoted in the generalized form

$$\lambda(t) = \frac{A}{c} \left(1 + \frac{t}{c} \right)^{-(1+\delta)}.$$

Remarkably, to this day, Omori's law remains without any clear physical explanation. It can be modeled to a first approximation as a nonhomogeneous Poisson process (Jeffreys 1938), and it is used in Ogata's epidemic-type aftershock sequence (ETAS) model (Ogata 1988), where every event is supposed to trigger its own aftershocks ("offspring"), but the models remain at a descriptive level.

The next major stimulus was the 1906 San Francisco earthquake. Challenging U.S. technical "know-how" at one of its major centers, the earthquake drew forth a massive technical report (Lawson 1908) that documented the extent and character of damage and of displacements along and off the San Andreas fault, and thereby laid the foundations for the field of engineering seismology. In a sequel to this report, H. F. Reid (1911) set out his *elastic rebound theory* of earthquakes: On either side of a major fault, large-scale forces operate to cause relative motion between the two sides of the fault. Friction opposes the motion. As time passes, the rock material deforms elastically and strain (deformation), and hence stress (elastic force), accumulate, until ultimately the strength of the fault is exceeded and the two sides slip in an earthquake; then the process starts again. Detailed measurements from surveys before and after the 1906 earthquake strongly supported this hypothesis, which both explained the origin of the earthquake waves and gave some basis for regarding earthquakes as a recurrent process.

In broad terms, Reid's hypothesis has dominated thinking about earthquake mechanisms ever since its formulation. It lies behind recent stochastic models for earthquake occurrence, such as renewal or semi-Markov processes with log-normal interevent times, or the stress-release model, in which the conditional intensity at time t has the form

$$\lambda(t) = \exp[a + bX(t)],$$

where $X(t) = X(0) + \rho(t - \gamma \sum_{t_i < t} X_i)$ is a measure of the current stress level in the region, the X_i being the stresses released in previous events. Once again, however, the models remain at a broadly descriptive level.

2.2   The Classical Period: 1920–1950

The decades following the 1906 San Francisco earthquake were marked by steady improvements in instrumentation and data collection. Networks of stations were established, and information began to be collected at both global and local levels. The principal theme was not the study of the earthquakes themselves, however, but rather the information that they provided about the earth's interior. As early as 1909, Mohorovicic had observed waves apparently reflected from an internal boundary some kilometers below the earth's surface, and evidence for other boundaries, including that of a central core, accumulated.

The disentanglement of such data is a classical inversion problem, the basic unknown being the velocity structure inside the earth. Many seismologists contributed to these issues, but the most profound contributions were those of Harold Jeffreys, later assisted by K. E. Bullen. The notable feature of Jeffreys's work was its careful attention to statistical procedures. Like that of Laplace and Gauss before him, Jeffreys's fundamental work on probability and statistical inference (Jeffreys 1939) was underlain by the better part of a decade of experience in the reduction of physical data, earthquake travel-time data and the establishment of improved procedures for epicenter and hypocenter (three-dimensional) location. The work culminated in 1940 with the publication of the Jeffreys Bullen global travel-time tables (Jeffreys and Bullen 1940), which are still used for calculating travel times based on the assumption of a spherically symmetric globe.

Another important step taken during this period was Richter's (1935) development of an earthquake magnitude scale. Based on the logarithm of the maximum amplitude recorded on a standard instrument, and adjusted to a standard distance from the source, this was the first objective measure of the size of an earthquake. Despite its limitations, such a measure remains an almost indispensible tool for the quantitative analysis of earthquake catalog data.

Hard on the heels of the magnitude scale came the second empirical law of seismology, the *Gutenberg–Richter frequency-magnitude law*. In statistical terms, this law asserts that magnitudes follow an exponential distribution. If the magnitudes are related back to physical variables such as the seismic energy release, then this translates to a power law (Pareto's law) distribution for the physical variable. In particular, it suggests that the tails of the energy distribution are of the form

$$\text{pr}(\text{energy} > E) \propto E^{-\alpha},$$

where the exponent α is in the range .4–.8.

This law also remains without any universally accepted explanation, although the problem here is less an absence than a proliferation of models. It puts earthquakes squarely into the realm of phase–change-like phenomena, associated with features such as power law distributions of size, long-range spatial and temporal correlations, and self-similarity. But here there are many possible models, and the Gutenberg–Richter (G–R) law by itself is not enough to distinguish between them. Nevertheless, it and Omori's law provide constraints that any successful model of the earthquake fracture process must satisfy.

2.3 Time Series Analysis and Explosions: 1945–1970

Time series and geophysics have grown up together, each contributing to the development of the other, and seismology has been an integral part of this process. Early time-series work in seismology related to the largely fruitless study of hidden periodicities in earthquake occurrence. In the period following the Second World War, however, a number of practical problems pushed time-series methods into the center of seismological research. The most important of these (not in the least because it led to substantial increases in funding for seismology) was the problem of detecting underground nuclear explosions and distinguishing them from earthquakes. The analysis of data from seismic arrays (i.e., instruments set up in a grid or other structured pattern) required the solution of further problems. The same period saw the growing use of explosion seismology (recording of waves from deliberate explosions) to investigate subsurface structures for oil exploration and other purposes, and of spectral methods to analyze the response of buildings and other structures to earthquake waves. As seismic networks became more highly automated, questions arose concerning the automatic triggering of unmanned equipment and the effective analysis and storage of data from such equipment. All of these issues required the solution of difficult, often highly technical problems in time series analysis and engaged the attention of leading experts in both fields. New ideas arose, such as maximum entropy methods, and the links between the disciplines remain very close.

2.4 Plate Tectonics and Earthquake Prediction: 1970–Present

Plate tectonics is one of the scientific success stories of the second half of the 20th century. For seismology, it provided a unifying principle that helped explain many incompletely resolved issues. It gave meaning to the highly irregular distribution of seismically active zones around the world, and indicated the nature of the "large-scale forces" required by Reid's elastic rebound theory—plate motions, impelled by convection processes in the earth's mantle. Collision and subduction zones, rigid plates and fractured plate-boundaries, mid-ocean ridges, and heat flow and gravity anomalies were concepts illuminated and coordinated by plate tectonics.

This somewhat euphoric period also saw the first steps in what was to prove a salutory reminder that the earth does not yield its secrets cheaply. In the lull before the Chinese cultural revolution, intriguing rumors of eccentric animal behavior and anomolous physical measurements before large earthquakes emanated from behind the "bamboo curtain." These culminated in 1976 with the claimed prediction of the Haicheng earthquake, leading to the evacuation of residents from their homes and the consequent saving of many lives. Observers from the international seismological community visited and confirmed much of the story. American and Japanese scientists, conscious of the superiority of their technical equipment, were spurred to emulate the Chinese, again assisted by offerings of additional funds. Only 2 years afterward, however, the Chinese program suffered a severe reversal with the devastating 1978 Tangshan earthquake. No formal predictions were claimed, and massive losses of both life and property were incurred. Such has been the progress of earthquake prediction ever since; each claimed success has been matched by an embarrassing failure to predict or a false alarm. The unpredicted earthquakes in Northridge, California and Kobe, Japan, each in the heart of earthquake research territory, did little to help matters. Funding started to dry up, the credibility of scientists working on prediction was threatened, and serious doubts were entertained as to whether earthquake prediction was a feasible or even a desirable accomplishment.

I am more optimistic over these matters than the last paragraph might suggest (Vere-Jones 1995). The viewpoint is slowly gaining acceptance that predictions must be couched in terms of probabilities of occurrence. Many embarrassments might have been avoided had this viewpoint prevailed sooner. Currently, there seem to be two main stumbling blocks. The first is in the physics, in the lack of an adequate theory of earthquake genesis and growth. The second is the lack of statistical models for the highly clustered, self-similar types of data from earthquake patterns. In both areas there is considerable room for improvement and some indication that despite current pessimism, the problems are starting to yield. Features such as local activation, foreshocks and precursory swarms, accelerated moment release, and precursory quiescence do provide some degree of enhancement of background probabilities, and suggest that the accumulation of stress before a large event may be detectable. However, the factors are not yet large enough, and the models are not well enough established, for them to be useful in direct practical applications. Improvements in data quality and the range of characteristics studied can only lead to improvements in this situation.

In the meantime, seismology offers statisticians the opportunity to collaborate in an extremely diverse range of problems. Let me conclude by quoting a few examples of recent or current work which happen to have caught my interest (but are not claimed to be representative).

3. SOME RECENT EXAMPLES

Dating of Events Along the New Zealand Alpine Fault.   Although this fault marks a major plate boundary, it has been a seismically quiet zone ever since Europeans arrived about 2 centuries ago. A central issue was to determine whether large earthquakes had occurred along the fault, and if so, when. A recent workshop brought together scientists who had been tackling this issue from different points of view: carbon dating from peat residues, tree ring data, data from lichens growing on the underside of fallen rocks. As the workshop progressed, the uncertainties in one technique were resolved by information provided by another. By the time it finished, a clear answer had emerged—the last major earthquake had occurred in 1717. Before then, two or three further dates were established, with less certainty, at intervals of from 100 to 300 years.

A Time Series Problem: Identification of Preseismic and Coseismic Changes in Water Well Levels.   Water level changes have long been touted as an earthquake precursor. Kitagawa and Matsumoto (1996) finally married data of sufficient quality to noise-reduction techniques (involving nonlinear filtering) of sufficient sensitivity to isolate coseismic and some small preseismic signals.

Inversion Problems: Mapping of Slip at Depth and Gravity Anomalies.   The 1989 Loma Prieta earthquake was widely felt and caused moderate damage; it did not, however, break the surface. Using a combination of statistical and geophysical arguments, Arnodottir, Segall, and Matthews (1992) were able to reconstruct the area on the fault plane that slipped, from measurements taken on the surface across and alongside the fault. In somewhat related work, Bayesian smoothing methods developed by Akaike and coworkers have been used to tackle a wide range of geophysical inversion problems; one example is their use by Murata (1992) to map Bouguer density anomalies.

Forecasting of Large Aftershocks.   Matsu'ura (1986) studied the forecasting of large aftershocks using Ogata's techniques, based on the ETAS model, for detecting precursory relative quiescence. This was one of the few techniques to perform credibly in the Kobe earthquake, in which it gave real-time warnings of major aftershocks.

Fundamental Theory: Mode-Switching in Complex Earthquake Models.   A key theoretical issue is the tussle between rebound-type arguments, suggesting "characteristic earthquakes" at regular intervals, and the G–R law, suggesting extreme irregularity. In a series of recent articles (e.g., Dahmen, Ertas, and Ben-Zion 1998), Ben-Zion and colleagues in the United States have demonstrated the existence of complex systems, imitating features of tectonically driven fault structures, and capable of existing in two possible modes. The first mode produces more or less stationary sequences of events following a standard G–R law. The second mode exhibits near periodic behavior of the elastic-rebound type, with regular occurrence of large "characteristic" events outside the G–R range. The process can flip from one mode to the other at apparently random instants of time. Does the geological evidence support the existence of such behavior in real fault systems?

Statistical Seismology and an S-PLUS-Based Software Environment.   Recent work of our own group has focussed on reviewing the applicability of catalog-based prediction methods to New Zealand data. Working jointly with a group headed by Professor Ma Li from the Chinese Seismological Bureau, we have developed an S-PLUS–based software environment (SSLIB; see Harte 1999) for subsetting and displaying catalog data and for fitting, simulating, predicting, and evaluating a range of conditional-intensity models.

REFERENCES

Arnadottir, T., Segall, P., and Matthews, M. (1992), "Resolving the Discrepancy Between Geodetic and Seismic Fault Models for the 1989 Loma Prieta, California, Earthquake," *Bulletin of the Seismologic Society of America*, 82, 2248–2255.

Bolt, B. A. (1988), *Earthquakes* (2nd ed.), San Francisco: W. H. Freeman.

Bullen, K. E. (1963), *An Introduction to the Theory of Seismology* (3rd ed.), Cambridge, U.K.: Cambridge University Press.

Dahmen, K., Ertas, D., and Ben-Zion, Y. (1998), "Gutenberg Richter and Characteristic Earthquake Behavior in Simple Mean-Field Models of Heterogeneous Faults," *Physical Reviews, E*, 58, 1494–1501.

Harte, D. (1999), "Documentation for the Statistical Seismology Library," Research Report 98-10, School of Mathematics and Computer Science, Victoria University, Wellington.

Jeffreys, H. (1938), "Aftershocks and Periodicity in Earthquakes," *Gerlands Beitrage zur Geophysik*, 53, 111–139.

——— (1939), *The Theory of Probability*, Oxford, U.K.: Oxford University Press.

Jeffreys, H., and Bullen, K. E. (1940), *Seismological Tables*, British Association, Gray-Milne Trust.

Kitagawa, G., and Matsumoto, N. (1996), "Detection of Coseismic Changes of Underground Water Level," *Journal of American Statistical Association*, 91, 521–528.

Lawson, A. C. (Chairman) (1908), *The California Earthquake of April 18th 1906. Report of the State Earthquake Investigation Commission*, Vols. 1 and 2, Washington, Carnegie Institute of Washington.

Matsu'ura, R. S. (1986), "Precursory Quiescence and Recovery of Aftershock Activities Before Some Large Aftershocks," *Bulletin of the Earthquake Research Institute, University of Tokyo*, 61, 1–65.

Murata, Y. (1992), "Estimation of Optimum Surface Density Distribution Only From Gravitational Data: An Objective Bayesian Approach," *J. Geophys. Res.*, 12097–12109.

Ogata, Y. (1988), "Statistical Models for Earthquake Occurrences and Residual Analysis for Point Processes," *Journal of American Statistical Association*, 83, 9–27.

Omori, F. (1894), "On After-Shocks of Earthquakes," *Journal of the College of Sciences*, Tokyo Imperial University, 7, 111–200.

Reid, H. F. (1911), "The Elastic Rebound Theory of Earthquakes," *Bulletin of the Department of Geology, University of California*, 6, 413–444.

Richter, F. (1935), "An Instrumental Earthquake Scale," *Bulletin of the Seismologic Society of America*, 25, 1–32.

Vere-Jones, D. (1995), "Forecasting Earthquakes and Earthquake Risk," *International Journal of Forecasting*, 11, 503–538.

Vere-Jones, D., and Smith, E. G. C. (1981), "Statistics in Seismology," *Communications in Statistics, Part A—Theory and Methods*, A10, 1559–1585.

Internet Traffic Data

William S. Cleveland and Don X. Sun

Internet engineering and management depend on an understanding of the characteristics of network traffic. Statistical models are needed that can generate traffic that mimics closely the observed behavior on live Internet wires. Models can be used on their own for some tasks and combined with network simulators for others. But the challenge of model development is immense. Internet traffic data are ferocious. Their statistical properties are complex, databases are very large, Internet network topology is vast, and the engineering mechanism is intricate and introduces feedback into the traffic. Packet header collection and organization of the headers into connection flows yields data rich in information about traffic characteristics and serves as an excellent framework for modeling. Many existing statistical tools and models—especially those for time series, point processes, and marked point processes—can be used to describe and model the statistical characteristics, taking into account the structure of the Internet, but new tools and models are needed.

1. THE INTERNET

Internet traffic data are exciting because they measure an intricate, fast-growing network connecting up the world and transforming culture, politics, and business. A deep understanding of Internet traffic can contribute substantially to network performance monitoring, equipment planning, quality of service, security, and the engineering of Internet communications technology. Two ingredients are required for this understanding: frameworks for traffic measurement that produce data bearing on the Internet issues, and statistical models for these data. Measurement has received expert, effective attention. A good start has been made in modeling, but much more can be done. This first section contains a description of some of the fundamentals of Internet communication, serving as a basis for the discussions of measurement and modeling in later sections.

1.1 Packets and TCP/IP

Each Internet communication consists of a transfer of information from one computer to another; examples are the downloading of a Web page and the sending

William S. Cleveland and Don X. Sun are Members of Technical Staff, Statistics and Data Mining Research, Bell Labs, Murray Hill, NJ 07901 (E-mail: wsc@bell-labs.com and dxsun@bell-labs.com). The authors thank Mark Hansen, Jim Landwehr, and Diane Lambert, whose comments were very helpful for formulating and executing this article.

214

of an e-mail message. When a file is transferred, it is not sent across the Internet as a continuous block of bits. Rather, the file is broken up into pieces called *packets*, and each packet is sent individually. Many different *protocols* collectively carry out the transfer. An Internet protocol is simply a set of rules for communication between computers. The two core protocols are the TCP (Transmission Control Protocol) and IP (Internet Protocol).

TCP runs on both computers. It breaks up a file to be transferred into packets, sends them out from the source headed for the destination, receives them at the destination, and reassembles them into their proper order. A typical packet size is 1,460 bytes. So if the abstract of this article were sent by e-mail as a text file, about 1,000 bytes, it would fit into a single packet. If the entire article were sent as a postscript file, about 250 kilobytes, this would take 170 packets.

TCP does the transfer by establishing a *connection* between the computers. The connection is not a physical path; rather, it is simply TCP software executing on the computers in a coordinated fashion, with each aware that it is working with the other. The connection continues until both sides agree that it is over, or until one side fails to hear from the other for a specified amount of time. Packets go back and forth between the two computers. For a packet, the *source* computer is the one that sends it out, and the *destination* computer is the one that receives it.

IP is in charge of routing TCP packets across the Internet. Each computer has an IP address, a unique 32-bit number. Often, the number is displayed by dividing the sequence of bits into four 8-bit fields, writing each field as a decimal number between 0 and 255, and then displaying the four numbers separated by dots; an example is 135.104.13.160. The number of possible IP addresses is 2^{32}, about 4 billion, but to show how fast the Internet is growing, plans have been made to switch to 128-bit addresses.

An IP header is added to each packet. The header includes, among other things, the source IP address, the destination IP address, and the packet size. *Routers* are Internet devices that get the packet to its destination. The packet moves from one router to the next, each reading the destination address in the header and looking in a table to find the router to which it should forward the packet. The source, the series of routers, and the destination form a path across the Internet. The packets of a single file transfer do not have to follow the same path; for example, if a path in use at the beginning of the transfer becomes unavailable, then tables are updated by communication among the routers, and a new path is used.

A TCP header is also added to each packet. The header contains information to control the connection and to reassemble packets. TCP also creates packets that contain only headers, but no file information. Control is their only purpose, and they are generated by both computers. For example, a connection is initiated by one computer transmitting a *synchronization*, or SYN, packet to the other. The control information in headers plays a role in a TCP feedback mechanism that is fundamental to the operation of the Internet and has a major effect on traffic. TCP sends packets at a conservative rate at the onset of a connection and then progressively increases the rate. If a router receives packets faster than it can forward them, it places the overflow in a buffer; if the buffer overflows, packets are dropped. If packets are dropped, TCP decreases the transmission rate, retransmits the dropped packets, and then progressively increases the rate again.

Each TCP connection is set up at the request of an Internet application. For example, the application HTTP (Hypertext Transfer Protocol) transfers a World Wide Web page from a *server* computer to a *client* computer. SMTP (Simple Mail Transfer Protocol) sends e-mail. Telnet enables logging on to remote computers, transferring input on the client to the remote computer, and transferring responses back. FTP (File Transfer Protocol) transfers files between local and remote computers. A single invocation of an application can result in many TCP connections. For example, a Web page request using the most common version of HTTP results in a connection to transfer the linked file, and a connection for the transfer of each embedded image file in the linked file.

1.2   Network Topology and End-to-End Connections

The Internet is a network of networks. A simplified but useful view of it involves end-points, end-networks, and a core. The end-networks are the networks of companies, universities, and other organizations. The end-points are individual home users, typically with a single connected computer, although little end-networks are forming in homes as well. The core is a collection of ISPs (Internet Service Providers) such as AOL-Time-Warner, AT&T, BBN, and MCI. Each end-point or end-network connects to an ISP, and the ISPs have interconnection points.

Suppose that a home user downloads a Web page from a large university computer. An *end-to-end* connection is established between client and server. The first packet, the SYN, starts out at the client and goes over a wire to the ISP. Today this wire might be a phone line or a television cable, and the transmission rate can range from about 28.8 kilobits/sec to 2 megabits/sec. These transmission rates refer to the speed at which the two devices at the two ends of the wire can send and receive bits; once the bits get on the wire, they travel at the speed of light. Then the packet moves through the ISP's network, perhaps transfers to another ISP, but eventually hits the ISP providing service to the university. Today ISP backbone wires, which are fiber, have transmission rates of up to 10 gigabits/sec. Then the packet enters the university network and travels through the university's wires and network devices until it is routed to the university computer with the Web page. Transmission speeds on the university network might vary over the different links from 10 megabits/sec to 1 gigabit/sec. The travel time of the packet depends on the distance of the university from the home, the speed of the link on the path, and the congestion encountered along the way, but times ranging from tens of milliseconds to the low hundreds are common. When the server receives the SYN, it sends a control packet back to the client, acknowledging the SYN; the client then sends a control packet back to the server, acknowledging the acknowledgment, and the real job begins, transferring information.

2. FLOW MEASUREMENT

One effective framework for traffic measurement is TCP/IP packet header collection, and organization of the headers into TCP connection *flows*. The framework has been in place throughout much of the short history of the Internet, and important fundamental work has arisen from it (Caceres, Danzig, Jamin, and Mitzel 1991; Claffy, Braun, and Polyzos 1995; Mogul 1992; Paxson 1997a). This section discusses the framework and its uses.

2.1 Connection Flows

Consider a wire carrying Internet traffic. An example is the wire that connects a Bell Labs Research network of about 3,000 machines to the rest of the Internet. We collect headers on this wire. There are two directions: packets coming into the Bell Labs network from computers outside, and packets traveling outside from computers inside. Packets pass by one by one, and packets from different TCP connections are superposed in the sense that at any one time there can be many TCP connections in progress, so we might see a packet from one connection, then a packet from a different connection, and then a packet from the first. We read and copy the TCP/IP headers of each packet and add a timestamp, the arrival time of the packet. Then an algorithm is used to disentangle the packets and form individual TCP connection flows. The headers for each connection flow are stored together in the database in order of arrival time.

2.2 End-to-End Characteristics

TCP connection flows provide a large amount of information. Each flow is an end-to-end connection traversing the Internet. For the Bell Labs wire, one end is a computer inside Bell Labs, and the other end is a computer outside. The TCP/IP headers contain the IP addresses of the two computers, so we know their location in the vast Internet topology. Thus flows can be used to study network-wide characteristics. Here is one example involving HTTP. For this Bell Labs wire, there are no incoming Web page requests, just outgoing; traffic to and from Bell Labs Web servers passes over other wires. On the collection wire, the client computers for HTTP (the ones that request pages) are inside, and the server computers (the ones that supply pages) are outside. For each HTTP flow, we can compute its duration, the time of the last packet minus the time of the first packet, and the size of the transferred file. Then we can divide size by time to get throughput in bytes/sec. This throughput variable is one measure of the quality of the Web transfers that are requested by users in the Bell Labs network. Bigger is better. We can study this quality metric and, using the destination IP address, determine how it varies topologically across the Internet, or adding the timestamp information, how it varies with time of day and topology. Of course, we can do the same for other applications as well.

2.3   Aggregate Traffic on the Wire

A TCP connection flow database also provides information about the traffic on the wire. The TCP/IP headers have the size of each packet in bytes, so together with the timestamps, we have the aggregated packet process: the arrival times and sizes of all packets. Studying aggregates is important, because the devices at each end must handle packets, in time order, and the performance of the devices depends on the packet interarrival times and the packet sizes. Forming the aggregate of all packets from the flows takes us back to the packet information in its original state: packets in time order. But storage by connection flow is still important, because we often study subaggregate traffic: time-ordered packets from a subset of the flows. For example, each flow results from an application such as HTTP, FTP, SMTP, or Telnet requesting a connection and transfer of information; it is important to study aggregate traffic by application because the packet processes for different applications are different. We can also study derived processes formed from any subaggregate. A common process is byte counts, in which time is divided up into intervals of equal length, and the number of bytes of packets arriving in each interval is computed.

3. MODELING

Internet traffic data are ferocious. Their statistical properties are complex, and databases are very large. The protocols are complex and introduce feedback into the traffic system. Added to this is the vastness of the Internet network topology. This challenges analysis and modeling. Most Internet traffic data can be thought of as time data: a point process, a marked point process, or a time series. The start times of TCP connection flows for HTTP on an Internet wire are a point process. If we add to each of these start times the file size downloaded from the server to the client, the result is a marked point process. Byte counts of aggregate traffic summed over equally spaced intervals are a time series. Modeling Internet traffic data will require new approaches, new tools, and new models for time data. In this section we look back at the current body of literature on Internet traffic studies to formulate issues for this new work.

Other such discussions may be found in interesting articles by Willinger and Paxson (1997) and Floyd and Paxson (1999).

3.1   HTTP Start Times

We begin with an example of traffic data and their analysis, to serve as a backdrop for the discussion (Cleveland, Lin, and Sun 2000a). The data are HTTP start times on the Bell Labs wire described earlier. When a Bell Labs user clicks on a link, the linked file is downloaded from the server outside to the client inside; the client's HTTP sees the names of embedded image files and requests that they be downloaded. For the most prevalent version of HTTP, each file is transferred by its own TCP connection, so a single click can open many TCP connection flows. The SYN packet sent from the client that begins the connection travels through the Bell Labs network and arrives on the measurement wire. As part of the overall header collection, we read and copy the TCP/IP headers on the SYN, and record the arrival time. The measured HTTP start time on the wire is a bit later than the actual start time on the client, because it takes time for the packet to travel from the client to the wire, but there is typically only a small delay inside the Bell Labs network compared to the travel time outside.

The HTTP start times are a superposition point process. Each user generates an HTTP start time point process on the wire, and the aggregate start time process is the superposition of the user point processes. The data used in the study cover the period November 18, 1998–July 10, 1999. There are 23,008,664 measured start times during this period. We organized the data into 15-minute blocks. Not every block during the measurement period appears due to monitor downtime or low usage. We eliminated blocks for which the monitoring was not operational more than 5% of the full 15 minutes, blocks with fewer than 50 flows, and connections from certain hosts that developed problems. The result was 10,704 blocks of start times. We assume that the point process of start times is stationary within a block.

Figure 1(b), an *interarrival plot* (Cleveland et al. 2000a), displays the 2,515 HTTP start times for one block, 7:45 A.M. to 8:00 A.M. on December 11, 1998. Let s_k for $k = 1$ to 2,515 be the start times, and let $t_k = s_k - s_{k-1}$ for $k = 2$ to 2,515 be the interarrival times. On the plot, $l_k = \log_2(t_k)$ is graphed against s_k, where $\log_2$ is the log base 2. The log on the vertical scale is vital, because interarrivals can vary over 16 powers of 2, a factor of about 64,000, and small intervals are as important as large ones; the vertical scale provides the requisite resolution to see this variation. The horizontal scale, however, conveys arrivals and interarrivals on the original scale. The connection rate over this block in connections per second, or c/s, is 2,515/900 c/s = 2.8 c/s. Figure 1(a) is an interarrival plot for the block 12:15 P.M. to 12:30 P.M. on the same day. The connection rate, 20.1 c/s, is greater than that for (b), because more users are browsing the Web from 12:15 P.M. to 12:30 P.M. than from 7:45 A.M. to 8:00 A.M.; in other words, the amount of superposition is greater in the later period.

Both (a) and (b) show discreteness on the vertical scale, with interarrivals piling

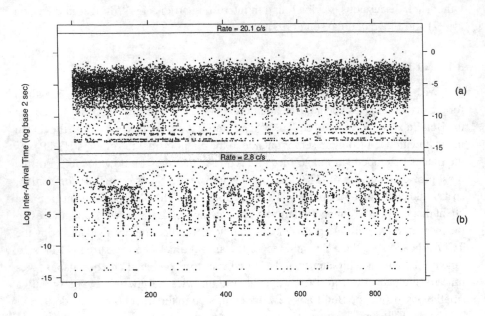

Figure 1. HTTP Start Times. Log base 2 interarrival time is graphed against start time for two 15-minute blocks of HTTP start times. The connection rate in the (a) is greater than in (b) because there is more Web usage. The higher rate reduces the amount of autocorrelation of the log interarrival time sequence and makes the Weibull marginal distribution of the interarrival times closer to an exponential.

up at a few values. This is a network effect, a small delay; each accumulation point is a time equal to the time it takes to process a packet in the network. For example, suppose that two SYN packets are back to back, which happens a small fraction of the time. They arrive on the wire; each is timestamped and then read by the PC. The interarrival time is the time it takes to read the first packet, which is the packet size times the wire speed; at the time of collection, the speed was 10 megabits/sec.

The different levels of superposition in Figure 1(a) and 1(b) dramatically change the statistical properties of the interarrival times. In 1(b), the data form distinct, narrow vertical bands for which the l_k range from about $-8 \log_2$ sec to about $-2 \log_2$ sec. The bands are bursts of connections caused by single clicks of individual users. The number of connections in such bursts can be large because the number of embedded files has a distribution with a long upper tail (Barford and Crovella 1998; Feldmann 1998). In 1(a), the bursty behavior has disappeared. Because the connection rate is much higher, the SYNs of more users intermingle, and the behavior of individual users is broken up. There is another difference between the two panels. The variance of the l_k is less for the higher-rate start times than for the lower. For both cases, the l_k sequence exhibits *long-range persistence*: a slowly decaying positive autocorrelation function. In (a), because of the increased superposition, the autocorrelation is much reduced compared to (b). The marginal empirical distribution of the t_k for both (a) and (b) is well approximated by a Weibull (except for the discreteness), with a shape parameter λ less than 1, the shape for an exponential, but the value of λ for (b) is smaller than for (a).

The analysis of the HTTP start times produced the following conclusions (Cleveland et al. 2000a). The marginal distribution of the t_k is approximately Weibull with λ less than 1, and as the connection rate increases, λ tends toward 1. The change in the distribution over the intervals is large; the estimate of λ ranges from about .4 to .9. The autocorrelation of the l_k is described by an exceedingly simple model with two parameters: white noise plus a long-range persistent time series. The two parameters are the variances of the two series. As the connection rate increases, the variance of the persistent series tends toward 0, and the l_k tend toward independence. The change in the autocorrelation function over the intervals is large; the fraction of the variance of l_k accounted for by the persistent series ranges from about .5 to .1.

3.2 Superposition

Aggregated Internet traffic is a superposition of traffic sources. It is vital to exploit superposition to uncover the characteristics of Internet traffic. In so doing, we exploit the fundamental structure of the traffic. We can operate mathematically, using the theory of superposition of point processes, marked point processes, and time series. We can operate empirically, studying the data and how it changes as the number of sources changes.

The notion of how we define a source for analysis purposes needs more thought and trial with data. For example, for the Bell Labs HTTP start times, we can take sources to be users. However, the Bell Labs network is actually a network of subnetworks. So we could take each source to be the start times of one subnetwork. There is another method of approaching superposition that avoids explicit identification of sources. This is *rate superposition*; traffic rates are used as a measure of the number of traffic sources (Cao, Cleveland, Lin, and Sun 2000; Cleveland et al. 2000a). This method was used in the analysis and modeling of the HTTP start times. A low base rate was selected, r_0, close to the minimum observed rate. Let k be a positive integer. The start times over a period of 15 minutes in an interval with rate kr_0 were taken to be the superposition of k independent start-time point processes, each with the statistical properties of the point process in an interval with the base rate. Theoretical results were derived based on the superposition theory of point processes. The results were compared to the empirical results, the analyses of the 10,704 intervals of start times; agreement was excellent. The conclusions about the HTTP start times just presented resulted from this combination of theory and empiricism.

3.3   Long-Range Persistence and Long Upper Tails

Long-range persistence is pervasive in Internet traffic data. But the pervasiveness had to be discovered (Leland, Taqqu, Willinger, and Wilson 1994; Paxson and Floyd 1995; Willinger, Taqqu, Leland, and Wilson 1995). Traffic models for voice traffic developed over the years to serve the telephone network did not apply as might have been hoped, because voice traffic does not give rise to the same traffic characteristics as Internet data traffic, which is burstier. As we have seen, the HTTP start times reveal persistence.

The reason for the increased burstiness in some cases is a combination of superposition and distributions with very long tails (Crovella and Bestavros 1996; Willinger, Taqqu, Sherman, and Wilson 1997). Suppose that we look at HTTP byte counts over equal-length intervals on an Internet wire carrying Web traffic. The counts for each interval result from the superposition of TCP connection flows, each transferring a file. So the byte counts in an interval are the sum of byte counts in the interval for a number of ongoing connections. The distribution of Web file sizes has an extremely long upper tail; in fact, the density falls off for large x like $x^{-\beta}$, where $1 < \beta < 2$, which means the size distribution has an infinite variance (Crovella and Bestavros 1996). So interspersed among the common-sized files will be immense files; their transfers raise the byte count level over a very long period, creating positive autocorrelation extending over a long range of lags.

The discovery of long-range persistence was an important piece of basic science for the analysis of network traffic, not just because it served to establish a fundamental characteristic of Internet data, but also because it served notice that the common practice of simply assuming a model for network traffic was defective and that looking at data was important. It was also an important discovery for the engineering of the Internet; the burstiness has required communication algorithms in some cases different from those that would be needed for much less bursty data (Erramilli, Narayan, and Willinger 1996) although not in all cases (Ryu and Elwalid 1996).

After the discovery of long-range persistence, a number of methods for characterizing the dependence were used that were appropriate to self-similar processes and that came from the creative work of Mandelbrot (1968). Consider a stationary time series y_t for integer t. Form a new time series,

$$y_v^{(m)} = m^{-H} \sum_{k=1}^{m} y_{vm+k},$$

for integer v and positive integer m. Then y_t is self-similar with parameter H if for all m, the finite-sample distributions of the $y_v^{(m)}$ are the same as those of y_t. Self-similar processes with $.5 < H < 1$ are persistent processes, and the autocorrelation at lag k decays like $k^{2(H-1)}$.

But self-similarity in the strict sense is exceedingly restrictive, and Internet data are not self-similar (Feldman, Gilbert, and Willinger 1998; Floyd and Paxson 1999;

Ribeiro, Riedi, Crouse, and Baraniuk 1999). At very small time scales, hundreds of milliseconds and less, the behavior of Internet protocols can be dominant, and at time scales of tens of minutes and more, diurnal variation can be dominant. In between are the time scales of the persistent processes. Thus there are various components, and some are persistent. Multifractal wavelet methods have become a widely used tool to study the components (Feldman et al. 1998; Ribeiro et al. 1999; Riedi and Vehel 1997). Another approach is to build time domain models and include long-range persistent components as is done in other disciplines (Haslett and Raftery 1989; Hosking 1981). This latter approach was used in the analysis of the Bell Labs HTTP start times.

3.4 Nonstationarity

Nonstationarity is as pervasive in Internet traffic data as long-range persistence (Cao et al. 2000; Cleveland et al. 2000a). However, nonstationarity has received much less attention. More attention is surely needed, because the nonstationarity can be major. For example, for the Bell Labs HTTP start times, the autocorrelation function and univariate distributions changed substantially. The reason for the nonstationarity is quite simple—changing amounts of superposition. As the number of traffic sources increases and decreases, statistical properties can change. Byte counts on a wire are sums of byte counts of the individual sources, and the finite-sample distributions of sums of random variables can change with the number of terms in the sums. HTTP start times on a wire are a superposition of the individual traffic sources, and the finite-sample distributions of the interarrival times of superposed point processes can change with the number of superposed processes.

3.5 Accounting for Structure and Linking Up With Network Simulators

Willinger and Paxson (1997) argue effectively against black box modeling that ignores Internet structure. At its most elemental, Internet traffic on a wire consists of packets arriving through time, a marked point process. The point process is the packet arrival times, and the mark at each point is multivariate: all of the information in the TCP/IP headers. And because each packet traverses the Internet end-to-end, we must add the vast Internet topology to the structure. But just how much of the structure that we want to invoke in a statistical model will depend on the goal and on the practicality.

One thing does seem clear, however: If we want to study an aspect of the network that requires a model to account for TCP feedback at an individual connection level, then statistical models alone are unlikely to be able to account for packets properly. But linking up statistical models with network simulators could produce a highly effective hybrid (Joo, Ribeiro, Feldman, Gilbert, and Willinger 1999). Network simulators such as NS (McCanne and Floyd 1998) are a big achievement of Internet analysis. To run the simulator, a network with routers and router algorithms is specified. TCP is simulated over this network, and packets are produced. There are two places where statistics can help. One is statistical models of source traffic, which serve as input to the simulation. The other is statistical models for traffic aspects such as packet interarrival times and byte counts, built to reflect the characteristics of data on live Internet wires; these models can be used to validate the simulator output.

4. VERY LARGE DATABASES

Packet header collection can result in a very large database, or VLDB, because packets come across Internet wires continuously. Even if the total throughput on a wire is small, the data grow and grow, so the database eventually gets large if collection continues. Our Bell Labs wire is just a trickle compared with high throughput wires of Internet ISPs. But after 1 year of collection, our databases of 328 million TCP connection flows with 6,866 million TCP/IP packet-header pairs took up about 350 gigabytes. An Internet wire with 100 times the throughput of the Bell Labs wire would reach the same size in about 3.6 days.

4.1   S-Net: A Low-Cost, Distributed Data Analysis Computing Environment

The success of analyzing Internet traffic data depends heavily on an ability to analyze the traffic VLDB in great detail. We need to explore the raw data in its full complexity; relying only on summaries is inadequate. We need to study packet-level processes taking many variables into account; studying only byte counts in equally spaced intervals is inadequate. Success in detailed intensive analysis depends on the analyst's computing environment.

To cope with the VLDB created by the Bell Labs traffic measurements, we developed S-Net (Cleveland et al. 2000b), a traffic collection and analysis system that begins with packet collection on a network link and ends with data analysis on a cluster of Linux PCs running S, a language and system for organizing, visualizing, and analyzing data (Chambers 1998). Packet capture uses the program tcpdump (Jacobson, Leres, and McCanne 1998) running on a PC with Berkeley Unix, an altered kernel to enhance performance, time-stamping based on global positioning system (GPS) clock discipline, and attention to packet drops. The compressed header files are moved to the cluster of Linux PCs, which are linked by fast switches. Each PC has one, two, or four processors, and they all have large amounts of disk space. An algorithm then organizes the header information by TCP connection flow, and the flows are processed to create flow objects in S. Analysis is carried out in S.

Flows and S flow objects are computed in parallel on all of the PC processors and are stored on the disks of all machines. S is run on high-end PCs with large amounts of memory. Each analyst has a low-end PC that stores that user's S directories. The analyst logs onto a high-end machine from the home machine to run S, mounting the home S directories as well as the directories across the cluster housing the S objects. In other words, each data analysis session is distributed across the cluster.

S-Net has worked quite well. Because the PCs and switches can be inexpensive and Linux is free, the cluster has a low overall cost. The cluster architecture scales readily; in our case, PCs and disks have been added and replaced incrementally as our database has grown. The S flow objects vary according to the specific analysis tasks; each is designed to enhance computational performance and to make the S commands that carry out the analysis as simple as possible. S is well suited to the task of analyzing Internet traffic data; its elegant design, which won it the ACM Software System Award for 1999, allows very rapid development of new tools.

But surely we can do better than S-Net. We need a whole new architecture for software for data analysis in networked environments that takes into account, from the ground up, the distributed nature of the environment. One effort, the Omega Project (www.omegahat.org), is underway; if it succeeds, then data analysis of kinds, including Internet traffic modeling, will benefit.

4.2 Visualization Tools: Multipage, Multipanel Displays

As for most databases, visualization tools are vital for analyzing a VLDB. Analytic visualization tools support model development. A vital aspect is screen real estate. Because Internet databases are large and the structure is complex, we must accept the notion that displays need to cover tens and perhaps hundreds of pages with many panels on each page. Data visualization is often limited to a display of a set of data that can be placed all at once in our visual field; so, it can be shocking at first to contemplate looking at so many pages. But using the structure of trellis display, it is easy to generate many pages (Becker, Cleveland, and Shyu 1996). Using a document viewer, it is possible to learn a great deal about Internet traffic data from these *multipage, multipanel* (MPMP) displays. Figure 1 is a trellis display with one page and two panels; it shows the HTTP start times for 2 of 10,704 blocks. It was immensely informative for modeling the start times to see such displays for hundreds of blocks. And the only practical medium to communicate these MPMP displays is our medium of study, the Internet.

4.3 Synchronized Measurement

We can raise the measurement bar even higher. Suppose that traffic is measured on two or more wires that have some traffic in common, and that time is measured accurately, perhaps by using a feed from a GPS satellite. Such synchronized measurements can reveal much about the movement of traffic across the Internet (Paxson 1997b). But there has been very little synchronized collection, because it greatly increases the already substantial burden of measurement, database management, and data analysis.

5. CONCLUSIONS

Packet header collection with timestamps and TCP connection flow formation provide an effective framework for measuring Internet traffic. But the resulting data are ferocious. They are nonstationary and long-range persistent, and distributions can have immensely long upper tails. There are many header variables, such as the source and destination IP addresses, and many variables can be derived from the header variables, such as the throughput. The structure of these variables is complex. Added to this is the vastness of the Internet topology and the intricacy and feedback of Internet protocols.

Challenges for producing tools and models for meeting the ferocity abound; here is a short list:

- statistical tools and models for point processes, marked point processes, and time series that account for nonstationarity, persistence, and distributions with long upper tails
- frameworks for incorporating the structure of the Internet into traffic models and analyses
- theoretical and empirical exploitations of the superposition of Internet traffic
- methods for measuring and characterizing vast, complex network topologies
- integration of statistical models with network simulators
- synchronized network measurement, and tools and models for comprehending the results
- methods for viewing MPMP data displays
- low-cost, distributed computing environments for the analysis of very large databases.

REFERENCES

Barford, P., and Crovella, M. (1998), "Generating Representative Web Workloads for Network and Server Performance Evaluation," in *Proceedings of ACM SIGMETRICS '98*, pp. 151–160.

Becker, R. A., Cleveland, W. S., and Shyu, M. J. (1996), "The Design and Control of Trellis Display," *Journal of Computational and Statistical Graphics*, 5, 123–155.

Caceres, R., Danzig, P., Jamin, S., and Mitzel, D. (1991), "Characteristics of Wide-Area TCP/IP Conversations," in *Proceedings of ACM SIGCOMM '91*, pp. 101–112.

Cao, J., Cleveland, W. S., Lin, D., and Sun, D. X. (2000), "On the Nonstationarity of Internet Traffic," technical report, Bell Labs, Murray Hill, NJ.

Chambers, J. M. (1998), *Programming With Data*, New York: Springer.

Claffy, K., Braun, H.-W., and Polyzos, G. (1995), "A Parameterizable Methodology for Internet Traffic Flow Profiling," *IEEE Journal on Selected Areas in Communications*, 13, 1481–1494.

Cleveland, W. S., Lin, D., and Sun, D. X. (2000a), "Network Simulation: Modeling the Nonstationarity and Long-Range Dependence of Client TCP Connection Start Times Under HTTP," in *Proceedings of ACM SIGMETRICS '00*, pp. 166–177.

―――― (2000b), "S-Net: An Internet Traffic Collection and Analysis System," technical report, Bell Labs, Murray Hill, NJ.

Crovella, M. E., and Bestavros, A. (1996), "Self-Similarity in World Wide Web Traffic: Evidence and Possible Causes," in *Proceedings of ACM SIGMETRICS '96*, pp. 160–169.

Erramilli, A. O., O Narayan, and W. Willinger (1996), "Experimental Queueing Analysis with Long-Range Dependent Packet Traffic," *IEEE/ACM Transactions on Networking*, 4, 209–223.

Feldman, A., Gilbert, A. A., and Willinger, W. (1998), "Data Networks as Cascades: Explaining the Multifractal Nature of Internet WAN Traffic," in *Proceedings of ACM SIGCOMM '98*, pp. 42–55.

Feldmann, A. (1998), "Characteristics of TCP Connection Arrivals," technical report, AT&T Labs Research.

Floyd, S., and Paxson, V. (1999), "Why We Don't Know How to Simulate the Internet," technical report, LBL Network Research Group.

Haslett, J., and Raftery, A. (1989), "Space-Time Modeling with Long Memory Dependence: Assessing Ireland's Wind Power Resource" (with discussion), *Journal of the Royal Statistical Society*, Ser. C, 38, 1–50.

Hosking, J. R. M. (1981), "Fractional Differencing," *Biometrika*, 68, 165–176.

Jacobson, V., C. Leres, and S. McCanne (1998), tcpdump-3.4. http://ftp.ee.lbl.gov/nrg.html.

Joo, Y., Ribeiro, V., Feldman, A., Gilbert, A. C., and Willinger, W. (1999), "On the Impact of Variability on the Buffer Dynamics in IP Networks," in *Proceedings of the 37th Annual Allerton Conference on Communication, Control, and Computing*, To appear.

Leland, W., Taqqu, M., Willinger, W., and Wilson, D. (1994), "On the Self-Similar Nature of Ethernet Traffic," *IEEE/ACM Transactions on Networking*, 2, 1–15.

Mandelbrot, B. B. (1968), "Noah, Joseph, and Operational Hydrology," *Water Sciences Research*, 4, 909–918.

McCanne, S., and Floyd, S. (1998), UCB/LBNL Network Simulator-ns (Version 2), http://www-mash.cs.berkeley.edu/ns/.

Mogul, J. (1992), "Observing TCP Dynamics in Real Networks," in *Proceedings of ACM SIGCOMM '92*, pp. 305–317.

Paxson, V. (1997a), "Automated Packet Trace Analysis of TCP Implementations," in *Proceedings of ACM SIGCOMM '97*, pp. 167–179.

―――― (1997b), "End-to-End Internet Packet Dynamics," in *Proceedings of ACM SIGCOMM '97*, pp. 139–152.

Paxson, V., and S. Floyd (1995), "Wide-Area Traffic: The Failure of Poisson Modeling," *IEEE/ACM Transactions on Networking*, 3, 226–244.

Ribeiro, V. J., Riedi, R. H., Crouse, M. S., and Baraniuk, R. G. (1999), "Simulation of Non-Gaussian Long-Range-Dependent Traffic Using Wavelets," in *Proceedings of ACM SIGMETRICS '99*, pp. 1–12.

Riedi, R., and Vehel, J. L. (1997), "Multifractal Properties of TCP Traffic: A Numerical Study," technical report, DSP Group, Rice University.

Ryu, B. K., and A. Elwalid (1996), "The Importance of Long-Range Dependence of VBR Traffic in ATM Traffic Engineering: Myths and Realities," in *Proceedings of ACM SIGCOMM '96*, pp. 3–14.

Willinger, W., and Paxson, V. (1997), discussion of "Heavy-Tailed Modeling and Teletraffic Data" by S. I. Resnick, *The Annals of Statistics*, 25, 1805–1869.

Willinger, W., Taqqu, M. S., Leland, W. E., and Wilson, D. V. (1995), "Self-Similarity in High-Speed Packet Traffic: Analysis and Modeling of Ethernet Traffic Measurements," *Statistical Science*, 10, 67–85.

Willinger, W., Taqqu, M. S., Sherman, R., and Wilson, D. V. (1997), "Self-Similarity Through High-Variability: Statistical Analysis of Ethernet LAN Traffic at the Source Level," *IEEE/ACM Transactions on Networking*, 5, 71–86.

Coding and Compression: A Happy Union of Theory and Practice

Jorma Rissanen and Bin Yu

1. INTRODUCTION

The mathematical theory behind coding and compression began a little more than 50 years ago with the publication of Claude Shannon's (1948) "A Mathematical Theory of Communication" in the *Bell Systems Technical Journal*. This article laid the foundation for what is now known as information theory in a mathematical framework that is *probabilistic* (see, e.g., Cover and Thomas 1991; Verdú 1998); that is, Shannon modeled the signal or message process by a random process and a communication channel by a random transition matrix that may distort the message. In the five decades that followed, information theory provided fundamental limits for communication in general and coding and compression in particular. These limits, predicted by information theory under probabilistic models, are now being approached in real products such as computer modems. Because these limits or fundamental communication quantities, such as entropy and channel capacity, vary from signal process to signal process or from channel to channel, they must be estimated for each communication setup. In this sense, information theory is intrinsically *statistical*. Moreover, the algorithmic theory of information has inspired an extension of Shannon's ideas that provides a formal measure of information of the kind long sought in statistical inference and modeling. This measure has led to the minimum description length (MDL) principle for modeling in general and model selection in particular (Barron, Rissanen, and Yu 1998; Hansen and Yu 1998; Rissanen 1978, 1989).

A coding or compression algorithm is used when one surfs the web, listens to a CD, uses a cellular phone, or works on a computer. In particular, when a music file is downloaded through the internet, a losslessly compressed file (often having a much smaller size) is transmitted instead of the original file. Lossless compression works because the music signal is statistically redundant, and this redundancy can be removed through statistical prediction. For digital signals, integer prediction can be easily done based on the past signals that are available to both the sender and receiver,

Jorma Rissanen is Fellow, IBM Research, San Jose, CA (E-mail: rissanen@almaden.ibm.com). Bin Yu is Member of Technical Staff, Bell Laboratories, Lucent Technologies, and Associate Professor of Statistics, University of California, Berkeley, CA (E-mail: binyu@stat.berkeley.edu). Her research is supported in part by National Science Foundation grant DMS-9803063. The authors are grateful to Diane Lambert for her very helpful comments.

and so we need to transmit only the residuals from the prediction. These residuals can be coded at a much lower rate than the original signal (see, e.g., Edler, Huang, Schuller, and Yu 2000).

2. ENTROPY AND LOSSLESS CODING

Shannon considered messages or signals to be concatenations of symbols from a set $A = \{a_1, \ldots, a_m\}$, called an *alphabet*. For example, the alphabet A for an English message contains the Roman letters and grammatic separation symbols. For an 8-bit digital music signal, A contains the integers from 0 to 255. A lossless *code* is an invertible function $C: A \to B^*$, the set of binary strings or *codewords*, and can be represented by nodes in a binary tree as in Figure 1(a). It can be extended to sequences $x^n = x_1, \ldots, x_n$, also written as x,

$$C: A^* \to B^*,$$

by *concatenation*: $C(xx_{n+1}) = C(x)C(x_{n+1})$. To make the extended code uniquely decodable without the use of separating commas, we impose the restriction that *no codeword is a prefix of another*. Each codeword node in the tree is then a leaf or an end-node; Figure 1(b) gives an example.

This *prefix* requirement implies the important Kraft inequality (see Cover and Thomas 1991)

$$\sum_i 2^{-n_i} \leq 1, \tag{1}$$

where $n_i = |C(a_i)|$ denotes the length of the codeword $C(a_i)$ in units of *bits* for binary digits, a term suggested by John W. Tukey. The Kraft inequality holds even for countable alphabets. Because of this inequality, the codeword lengths of a prefix code define a (sub) probability distribution with $Q(a_i) \sim 2^{-n_i}$. Even the converse is true, in the sense that for any set of integers $n_1, \ldots, n_m$ satisfying the Kraft inequality, and in particular for $n_i = -\lceil \log Q(a_i) \rceil$ obtained from any distribution, there exists a prefix code with the codeword lengths defined by the integers. This (sub) probability Q should be viewed as a means for designing a prefix code; it is not necessarily the message-generating distribution. When the data or message is assumed to be an independently and identically distributed (iid) sequence with distribution P, an important practical coding problem is to design a prefix code C with a minimum expected code length,

$$L^*(P) = \min_C \sum_i P(a_i)|C(a_i)|.$$

The optimal lengths can be found by Huffman's algorithm (see Cover and Thomas 1991), but far more important is the following remarkable property, proved readily

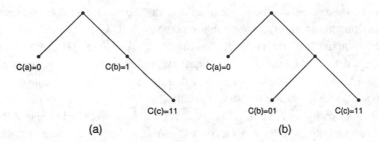

Figure 1.   Examples of (a) Nonprefix and (b) Prefix Codes for Alphabet $A = \{a, b, c\}$.

with Jensen's inequality. For any prefix code, the expected code length $L(Q)$ satisfies

$$L(Q) \geq -\sum_i P(a_i) \log_2 P(a_i) \equiv H(P), \tag{2}$$

with equality holding if and only if $Q = P$ or $|C(a_i)| = -\log_2 P(a_i)$ for all a_i, taking $0 \log_2 0 = 0$. The lower bound $H(P)$ is the *entropy*. Because the integers $\lceil -\log P(a_i) \rceil$ defining a prefix code exceed the *ideal* code length $-\log P(a_i)$ by at most one bit, the optimal code satisfies the inequalities $H(P) \leq L^*(P) \leq H(P) + 1$. In terms of the n-tuples or the product alphabet, the joint distribution has entropy $nH(P)$, within one bit of $L^*(P^n)$. Thus $H(P)$ gives the lossless compression limit (per symbol).

Concatenation codes are not well suited for small alphabets or for data modeled by a nonindependent random process even when the data-generating distribution P is known. But a different kind of code, *arithmetic code*, is well suited for coding binary alphabets and all types of random processes (see Cover and Thomas 1991). The original version was introduced by Rissanen (1976); a more practical version was given by Rissanen and Mohiuddin (1989).

For example, consider a binary alphabet A and let $\{P_n(x^n)\}$ denote a collection of nonzero probability functions that define a random process on A, written simply as $P(x^n)$. When the strings x^n are sorted alphabetically, the cumulative probability $C(x^n) = \sum_{y^n < x^n} P(y^n)$ is a strictly increasing function and can be taken as the code of x^n. A binary tree with $s0$ as the left descendant node of s gives the recursion

$$C(x^n 0) = C(x^n), \tag{3}$$

$$C(x^n 1) = C(x^n) + P(x^n 0), \tag{4}$$

and

$$P(x^n i) = P(x^n) P(i|x^n), \qquad i = 0, 1. \tag{5}$$

A minor drawback of this code is that it is one-to-one only for strings that end in 1. The main drawback is that even if we begin with conditional probabilities $P(i|x^n)$, each written to a finite precision, the multiplication in the third equation keeps on increasing the precision, and soon no register can hold the product. To overcome this, the exact multiplication in (5) is replaced by following $\bar{P}(x^n i) = \lfloor \bar{P}(x^n)\bar{P}(i|x^n) \rfloor_q, i = 0, 1,$ where $\lfloor z \rfloor_q$ denotes the truncation of a fractional binary number z to q digits after the first 1, as in $0 \ldots 01x \ldots x$. It is then clear that the addition in the update of the code string can also be done in a register of width $q + 1$, except for a possible overflow, which is dealt with through a device called "bit stuffing." We see in (5) that the code length is essentially the number of leading 0's in $\bar{P}(x^n 0)$ plus the size of the register q needed to write down the probability itself. As a result, the mean per symbol code length exceeds the entropy by no more than 2^{-q-1}.

When the data-generating distribution P is unknown, an adaptive arithmetic code can be designed based on an estimator of P, provided that the estimator itself is coded and transmitted. This leads naturally to the topic of the next section.

3. UNIVERSAL CODING AND THE MINIMUM DESCRIPTION LENGTH PRINCIPLE

As we have seen, Shannon's theory of lossless coding requires knowledge of the probability distribution generating the data. The theory of *universal* coding applies when a family of distributions is selected. Consequently, statistical modeling is indispensable in universal coding. A two-stage universal code based on a family of distributions comprises the coding of an estimated distribution (the overhead) and the coding of data using the estimated distribution. A predictive universal coding scheme requires a predictor based on the past and the coding of the residuals, but no overhead is needed.

The performance of the distribution estimator or the predictor is evaluated from the coding point of view. The first question is whether any universal codes exist such that the per symbol code length, either in the mean or almost surely, approaches the entropy no matter which distribution in the family generates the data. As a functional of the process, entropy is similar to the ergodic mean. Thus the existence of a universal code requires much less than the estimability of the joint distribution of the process. Two main types of universal codes are in use for data compression. The first, due to Ziv and Lempel (LZ) (1977, 1978), is fast and efficient for data arising from written natural languages and is used extensively for file compression (e.g., *compress* in the unix system). LZ is based on an efficient parsing algorithm that builds an increasing tree of phrases. So far, no explicit probabilistic model has been found for the LZ algorithm. The second type of universal code is based on an explicit probability model. One good example is Rissanen's context algorithm, which builds on the finite state machine defined by context trees. It is slower than LZ, but provides better compression for image data. Weinberger, Seroussi, and Sapiro's (1998) LOCO algorithm is based on the context modeling idea and has been selected as the new lossless (or nearly lossless) image compression standard JPEG-LS-2000. The mean per symbol length of the LZ code approaches the entropy at the rate of $O(1/\log n)$, n being the length of the string, no matter which process in a large class of stationary ergodic processes generates the data. Rissanen's context code for Markov processes converges at rate $.5k\log n/n$, where k denotes the number of free parameters. The faster rate of Rissanen's algorithm under the context model and the fact that it works for image data suggest that the context model is adequate for images, but not for written natural languages. On the other hand, the superior empirical performance of LZ for written natural languages contrasts with its slow rate under stationary and ergodic assumptions. This raises the question of the appropriateness of such assumptions or even the whole analysis framework. Perhaps instead of a fixed alphabet, an expanding alphabet must be modeled to reflect what LZ actually does. This calls for a new class of statistical analyses based on array asymptotic results.

The theory of universal codes suggests that amount of information in a finite set of data, given a class of models, can be measured by the shortest universal code length, called the *stochastic complexity*, for that class (see Barron et al. 1998). Recent results show that the stochastic complexity is the minimum in expectation for the worst-case model. The best universal code defines a *universal model* that decomposes the data into noninformative "noise" that cannot be compressed with the given models and the signal defined by an information-bearing optimal model. This idea generalizes the usual decomposition of the sufficient statistics. Given several classes of models for the signal, the *minimum description length* (MDL) principle searches for the class with the smallest stochastic complexity, which implies the most efficient removal of incompressible noise.

Various universal codes have been found to approximate the stochastic complexity to first order ($.5k\log n/n$) for iid Markov time series and other regular parametric models of dimension k (see Barron et al. 1998). Three such codes are the two-stage, predictive, and mixture codes. In a two-stage code, the estimated distribution is the plug-in maximum likelihood estimator (MLE) distribution. In a predictive code, the predictive distribution uses the MLE plug-in estimator based on the past. A mixture code is based on the marginal distribution of a model class with respect to a regular prior. There is an important connection between universal coding and model selection. Using MDL with a two-stage code gives the conventional Bayesian information criterion (BIC) model selection criterion, MDL with a predictive gives the accumulated prediction error criterion, and MDL with a mixture code generalizes the Bayes factor to more than two classes. Thus MDL provides a way to compare frequentist and Bayesian model selection criteria (see Hansen and Yu 1998).

How MDL can be used to select among complex models such as Bayesian belief networks remains an open question. Both information theory results (e.g., universal coding theorems) and statistical evaluation analyses are anticipated.

4. RATE-DISTORTION FUNCTION AND LOSSY COMPRESSION

Entropy gives the limiting rate for lossless compression, but often the practical bandwidth of a communication channel or the available disk space for storage demands a rate lower than the entropy. A compromise is to sacrifice fidelity in the original message or data in exchange for a lower rate. The mathematical study of this trade-off is the Shannon's rate-distortion theory (see Cover and Thomas 1991), and the practice of coding with loss of information is called lossy compression. For example, an image retrieved from a remote website is compressed via the lossy compression standard JPEG. (A more complete review of this area has been given in Berger and Gibson 1998.)

The rate-distortion function is a sort of entropy function when distortion is allowed between the original data and the coded data. Given a distortion distance d and a threshold D, the rate-distortion function is defined in the iid case as

$$R(D) := \min_{\hat{X}} I(X, \hat{X}),$$

where

$$I(X, \hat{X}) = \int \int f(x, \hat{x}) \log f(x, \hat{x}) / [f(x) f(\hat{x})] \, dx \, d\hat{x}$$

defines the *mutual information*. Here X is the random variable generating the source messages, and $\hat{X}$ is another coded message variable that is within distance D from X on average. It is a remarkable fact that $R(D)$ is the minimum achievable rate at distortion D. Calculating $R(D)$ even in the iid case can be a daunting task, and finding codes that achieve $R(D)$ is beyond reach except for special cases, such as Bernoulli and Gaussian. And, of course, speech and image signals are never well modeled by such processes. Prediction based on time series models and lossy coding of residuals, which are much closer to iid, are the basic ingredients behind speech coders such as ADPCM, which codes our telephone conversations.

In statistics, data can be transformed to greatly expand the applicability of conventional models. For decades, transform coding has also been the major approach to lossy image compression. The current JPEG standard uses discrete cosine transform (DCT) in a blockwise manner; the next generation, JPEG2000, will be wavelet-based. In the transform domain, local iid statistical modeling or context-based prediction and arithmetic coding of the scalar-quantized coefficients has given rise to the state-of-the art wavelet image coders such as those of LoPresto, Ramchandran, and Orchard (1997) and Yoo, Ortega, and Yu (1999).

Statistical modeling will also continue to play an essential role in lossy compression. Better predictions based on more sophisticated models will further decrease the entropy of the residuals to be coded and hence reduce the coding rate. This has spurred attempts to develop an expansion of the information-theoretic framework to encompass both estimation and coding limits.

5. CHANNEL CODING AND CHANNEL CAPACITY

Channel coding is about designing codes and efficient decoding algorithms to combat channel noise. Its goal is to spread out the codewords so that they can be distinguished with high probability when the noisy signal is received. The theory of algebraic coding relies on the same combinatorics as does experimental design. For a channel code, the recovery of codewords from the received signal amounts to hypothesis testing. Given a received signal, maximum likelihood corresponds to hard decoding, which uses the most likely code word, and Bayesian methods correspond to soft decoding, which gives a probability distribution over all possible codewords.

Channel coding was considered in parallel with source coding in Shannon's 1948 work. He modeled a (single-user) communication channel by a conditional probability distribution $f(y|x)$ on the output message y given the input message x. He defined *channel capacity* C to characterize an (iid) probabilistic channel as $C = \max_{w(x)} I(X, Y)$, where the maximization is over input distributions w, which the designer can select, and the channel output Y is determined by $w(x)f(y|x)$.

Shannon's channel coding theorem showed, contrary to the popular belief at the time, that if and only if the average transmission rate is below C, iid messages can be transmitted with error probability decreasing to 0 by increasing the size of the message block. This theorem gives the operational meaning of C as the upper bound at which a channel can carry reliable communication. Shannon (1948) calculated the capacity of the additive Gaussian white noise channel with a power constraint. This capacity is being approached by sophisticated coding methods in applications such as space and satellite communications, audio/video transmission, and CDs and CD-ROMs.

Models for wireless communication channels, such as those in cellular phones, must be time-varying and must accommodate multiple users (Biglieri, Proakis, and Shamai 1998; Lapidoth and Narayan 1998). Their decoding is essentially statistical inference with time-varying time series models. Joint source-channel coding gives better data throughput, but complicates decoding even further. With limited bandwidth to share, any decoding mileage is worthwhile. This demands efficient statistical inference procedures for special time-varying time series models, and fast algorithms in applications such as cellular phones. It is safe to say that wireless communication promises to be another good example of the happy union of theory and practice.

REFERENCES

Barron, A. R., Rissanen, J., and Yu, B. (1998), "The MDL Principle in Modeling and Coding," *IEEE Transactions on Information Theory*, 44, 2743–2760.

Berger, T., and Gibson, J. D. (1998), "Lossy Source Coding," *IEEE Transactions on Information Theory*, 44, 2693–2723.

Biglieri, E., Proakis, T., and Shamai, S. (1998), "Fading Channels: Information Theoretic and Communication Aspects," *IEEE Transactions on Information Theory*, 44, 2619–2692.

Cover, T. M., and Thomas, J. A. (1991), *Elements of Information Theory*, New York: Wiley.

Edler, B., Huang, D., Schuller, G., and Yu, B. (2000), "Perceptual Coding Using Adaptive Pre and Post-filters and Lossless Compression," submitted to *IEEE Transactions in Signal Processing*.

Hansen, M., and Yu, B. (1998), "Model Selection and Minimum Description Length Principle," submitted to *Journal of the American Statistical Association*.

Lapidoth, A., and Narayan, P. (1998), "Reliable Communication Under Channel Uncertainty," *IEEE Transactions on Information Theory*, 44, 2148–2177.

LoPresto, S., Ramchandran, K., and Orchard, M. (1997), "Image Coding Based on Mixture Modeling of Wavelet Coefficients and a Fast Estimation-Quantization Framework," in *Proceedings of the Data Compression Conference*, Snowbird, Utah.

Rissanen, J. (1976), "Generalized Kraft Inequality and Arithmetic Coding," *IBM Journal of Research and Development*, 20, 198–203.

—— (1978), "Modeling by Shortest Data Description," *Automatica*, 14, 465–471.

—— (1989), *Stochastic Complexity and Statistical Inquiry*, Singapore: World Scientific.

Rissanen, J., and Mohiuddin, K. (1989), "A Multiplication-Free Multialphabet Arithmetic Code," *IEEE Transactions on Communications*, 37, 93–98.

Shannon, C. E. (1948), "A Mathematical Theory of Communication," *Bell Systems Technical Journal*, 27, 379–423, 623–656.

Verdú, S. (1998), "Fifty Years of Information Theory," *IEEE Transactions on Information Theory*, 44, 2057–2078.

Weinberger, M. J., Seroussi, G., and Sapiro, G. (1998), "The LOCO-I Lossless Image Compression Algorithm: Principles and Standardization Into JPEG-LS," Technical Report HPL-98-193, Hewlett-Packard Laboratories.

Yoo, Y., Ortega, A., and Yu, B. (1999), "Image Subband Coding Using Context-Based Classification and Adaptive Quantization," *IEEE Transactions on Image Processing*, 8, 1702–1715.

Ziv, J., and Lempel, A. (1977), "A Universal Algorithm for Sequential Data Compression," *IEEE Transactions in Information Theory*, 33, 337–343.

—— (1978), "Compression of Individual Sequences via Variable-Rate Coding," 33, 530–536.

Statistics in Reliability

Jerry Lawless

1. INTRODUCTION

The term *reliability* refers to the proper functioning of equipment and systems and thus encompasses hardware, software, human, and environmental factors. Important aspects range from the development and improvement of products or systems to maintenance and performance measurement in the field. This vignette deals with statistics in reliability, so it emphasizes areas where data collection and analysis play a major role. Such topics as systems design and qualitative tools such as fault trees or failure mode and effects analysis are not considered, although they are important components of reliability engineering (e.g., Hoyland and Rausand 1994).

Historically, both the statistical and engineering development of reliability has focused on adverse events, or *failures*, which have undesirable consequences and are to be prevented or avoided. Consequences may range from rather trivial, as is usually the case when, for example, a light bulb fails, to grave, as in the case of the Challenger space shuttle disaster. These examples represent two extremes: events of small impact that occur frequently in a large population or system and very rare events that have severe consequences. Only for the former are substantial data on failures available.

As the ability to prevent failures increases, fewer failures are observed. In this case (and more generally), it is important to measure other things besides physical failure—in particular, variation in the *condition* of units or systems over time. "Condition" refers to factors associated with the system's performance or likelihood of failure; for example, the amount of physical or functional degradation in a battery or integrated circuit. Failure is often defined in terms of degradation reaching a certain level (e.g., Meeker and Escobar 1998, chap. 13).

Narrow technical definitions of reliability refer to the probability a system functions "satisfactorily" over some time period, but various aspects are involved. For manufactured products, reliability and durability are important dimensions of quality. Reliability is closely connected to risk and safety in settings where failures may have dire consequences. Economic considerations are often close at hand: in service systems, the failure to link a customer with a server can negatively impact customer loyalty; in manufacturing systems, stoppages decrease productivity.

Jerry Lawless is Professor of Statistics and holds the GM-NSERC Industrial Research Chair, Department of Statistics and Actuarial Science, University of Waterloo, Canada.

In this vignette I take a broad view of reliability within the context of system performance. Statistical science provides tools for understanding, improving, and maintaining reliability. In the next two sections I review past contributions, then consider current trends and challenges. The view is necessarily brief and personal, and space permits only a few references; many are to books or review articles from which further information can be obtained.

2. HISTORICAL PERSPECTIVE

Many statistical contributions to reliability have revolved around the analysis of failure or life-time data. Early work (e.g., Ansell and Phillips 1989; Lawless 1983) emphasized parametric methods based on distributions such as the exponential and the Weibull. Formal hypothesis tests and estimation were developed for life testing and reliability demonstration, and statistical procedures were embedded in reliability standards. Regression methodology provided tools for explanatory analysis of failures, accelerated testing, reliability improvement experiments, and prediction. Nonparametric and semiparametric methods of lifetime data analysis were developed. The seminal work of Cox (1972) introduced the proportional hazards regression model and also drew attention to time-varying covariates. This is tremendously significant for reliability, because the conditions under which systems operate usually vary over time. Lifetime data methodology is now included in all of the major statistical software packages and is featured in many books (see, e.g., Crowder, Kimber, Smith, and Sweeting 1991; Kalbfleisch and Prentice 1980; Lawless 1982; Meeker and Escobar 1998; Nelson 1982, 1990).

More or less in parallel with lifetime data analysis, point process methods for repeated failures in repairable systems were developed (Ascher and Feingold 1984; Cox and Lewis 1966). Unfortunately, methods based on Poisson and renewal processes are not featured explicitly in most software packages, although existing software can be coerced for analysis in some cases. Recent developments include robust methods for analyzing rates of occurrence and mean functions (Lawless and Nadeau 1995).

Another important area concerns degradation and, more generally, system condition. There is a long history of work on phenomena such as wear and material deterioration, and there is increased emphasis on degradation as physical processes become better understood. Knowledge of degradation processes allows better prediction and prevention of failures, especially when failure is (close to) a deterministic function of measured degradation. Meeker and Escobar (1998, chaps. 13 and 21) described some methodology and applications. A more general concept is the condition of a unit or system, mentioned in Section 1. Condition measures are used for maintenance scheduling; for example, analysis of metal fragments in oil is used with diesel engines, and ultrasonic scans and eddy current testing are used to detect material flaws or deterioration in railway track and aircraft bodies (Olin and Meeker 1996).

Most statistical work to date has addressed the often complex measurement and flaw detection processes, rather than the linking of condition measurements to reliability (but see Makis and Jardine 1992; Whitmore, Crowder, and Lawless 1998).

A final consideration is the environment in which a unit or system operates. Regression analysis for lifetimes or repeated failures allows factors such as stress, intensity of use, or climate to be considered, and there is extensive development of specialized areas (e.g., Nelson 1990). Recently, attention has been directed at time-varying factors or dynamic environments (e.g., Singpurwalla 1995).

In addition to modeling and analysis, the prominent role of study design must be noted. A well-crafted strategy for experimentation or data collection is essential and can lead to product or system improvement without much formal analysis. Finally, an important area that is implicit but not discussed directly here is system structure; that is, the relationship of components or modules within a system. Work linking component and system reliability has been useful in system design, reliability prediction, and maintenance. (Barlow and Proschan 1975 is a fundamental reference; for a more recent exposition, see Hoyland and Rausand 1994.)

I now turn to some recent trends and challenges. I begin by indicating how developments in stochastic modeling and event history analysis over the past 30 years have provided a unified framework for the various aspects of reliability described earlier.

3. TRENDS AND CHALLENGES

Consider a process or system with the following ingredients, operating on some time scale $t \geq 0$: (a) J different types of events that may occur; (b) a vector $\mathbf{Y}(t)$ that describes the condition of the system, including factors such as degradation; and (c) a vector $\mathbf{x}(t)$ describing fixed or external time-varying factors (e.g., environmental conditions, stress, load, demand) that affect the system. For simplicity, suppose that system reliability or performance is measured in terms of the numbers of events of various types that occur, and for $j = 1, \ldots, J$, define $N_j(s, t)$ as the number of events of type j that occur over the time interval $[s, t]$. More generally, however, we may associate costs or other values with events and use them to quantify reliability.

The field of multivariate counting processes provides models for event occurrence (e.g., Andersen, Borgan, Gill, and Keiding 1993). Let $H(t)$ denote relevant aspects of the "history" of events in a system prior to time t, as well as the values of $\mathbf{Y}(t)$ and $\mathbf{x}(t)$. Then for $j = 1, \ldots, J$,

$$\lambda_j(t|H(t)) = \lim_{\Delta t \to 0} \frac{\Pr\{N_j(t, t + \Delta t) = 1|H(t)\}}{\Delta t} \tag{1}$$

are called event intensity functions. If no condition measures are present, then, under mild assumptions, the intensities (1) specify the process of event occurrences, given a covariate history $\{\mathbf{x}(t), t \geq 0\}$. If condition measures such as degradation are of interest, then it is necessary to consider the joint process for event occurrences and $\mathbf{Y}(t), t \geq 0$. If events are deterministic functions of condition, then only $\mathbf{Y}(t), t \geq 0$ must be modeled (Meeker and Escobar 1998, chap. 13). The general situation is more difficult to handle and is a subject of current research (e.g., Singpurwalla 1995; Whitmore et al. 1998).

Note that (1) covers single failure times, repeated failures of the same type, or multiple events of different types. By a suitable specification of intensities, we can allow dependence on prior events in any part of the system and on fixed or time-varying environmental factors. There is an explicit form for the probability distribution of any sequence of events (Andersen et al. 1993, sec. 6.1), so that if complete data are available, then maximum likelihood or Bayesian methods can be readily developed. For certain models, lifetime data or generalized linear model analysis software may be coerced for estimation (e.g., Berman and Turner 1992; Therneau 1997).

I now consider some challenges. Promoting existing methodology is, of course, important, but because failure time analysis and controlled life testing are well understood, I focus mainly on methodological needs involving field data and complex systems or processes.

Physical Models for Degradation and Failure. In many areas, the physics of degradation and failure are sufficiently understood so that statistical modeling and experimentation can aid reliability improvement (e.g., Meeker and Escobar 1998, chap. 13; Nelson 1990, chap. 11). Statistical scientists have not been highly involved in this work, but many opportunities to contribute exist. Major challenges include the development of methodology for kinetic models with covariates and random effects.

Reliability in Dynamic and Heterogeneous Environments. Systems and products are affected by the conditions under which they operate, and so data on the environment as well as on system condition or failures must be obtained. It is a statistical challenge to obtain high-quality data relevant to reliability improvement, and to develop methodology for utilizing these data. To plan warranty coverage for manufactured products, for example, we need information about variations in usage across the population of customers. Prior events may be relevant; in software development, the analysis of faults detected may take into account code modifications made in response to faults discovered earlier (e.g., Dalal and McIntosh 1994). The framework in (1) provides a general basis for modeling. In testing software, for example, we might consider an intensity function for the detection of new faults of the form

$$\lambda(t|H(t)) = \nu(t) \left[g(t) + \int_0^t h(t, u) \, dC(u) \right], \tag{2}$$

where $\nu(t)$ represents the intensity of testing at time t, $C(t)$ is the cumulative amount of code modification up to time t, and $g(t)$ and $h(t, u)$ are parametrically specified functions.

Condition-Based and Usage-Based Monitoring and Maintenance. As condition measurement technology evolves, opportunities arise for statistical development. Although the framework in (1) allows one to study the effect of condition measures on failure, there is little methodology for prediction or decision because of the need to also model the condition measure process (e.g., Makis and Jardine 1992). In many settings, a better understanding of the underlying physical processes is needed. A related issue is the recognition of environmental factors in scheduling maintenance. One approach concerns time scales, the idea being to select a scale in which failure time is highly predictable. Operational scales that incorporate stress or usage measures (e.g., Duchesne and Lawless 2000) are of particular interest.

Field Data. As the ability to capture field data has increased, opportunities for management and system improvement arise. Warranty or field return data on products such as appliances and vehicles are essential for cost prediction and are an important source of information for reliability improvement (Lawless 1998). Manufacturing, service, and telecommunications systems generate vast amounts of data that may be used for reliability improvement, maintenance, or system management. But relevant data are often missing or mismeasured, and there is no guarantee that factors affecting reliability are even being monitored. Sometimes service system or network reliability is assessed in terms of system performance measures (e.g., the fraction of successfully completed transactions) or some service time distribution (e.g., Becker, Clark, and Lambert 1998). Challenges in this area include the design of good data collection processes and the provision of methodology dealing with large volumes of data, selective observation, missing data, and errors of measurement.

Complex Systems and Networks. The stochastic modeling of systems and networks has received considerable study (e.g., Barlow and Proschan 1975; references in Becker et al. 1998), but the literature on data collection and analysis is sparse. Recent work on networks and graphical models provides insight and some methodology, but it is an important challenge to develop data analytic tools for studying the reliability of complex systems and processes. Software reliability is an important special area.

REFERENCES

Andersen, P. K., Borgan, O., Gill, R. D., and Keiding, N. (1993), *Statistical Models Based on Counting Processes*, New York: Springer.

Ansell, J. I., and Phillips, M. J. (1989), "Practical Problems in the Statistical Analysis of Reliability Data" (with discussion), *Applied Statistics*, 38, 205–247.

Ascher, H., and Feingold, H. (1984), *Repairable Systems Reliability*, New York: Marcel Dekker.

Barlow, R. E., and Proschan, F. (1975), *Statistical Theory of Reliability and Life Testing*, New York: Holt, Rinehart and Winston.

Becker, R. A., Clark, L. A., and Lambert, D. (1998), "Events Defined by Duration and Severity, With an Application to Network Reliability" (with discussion), *Technometrics*, 40, 177–194.

Berman, M., and Turner, T. R. (1992), "Approximating Point Process Likelihoods With GLIM," *Applied Statistics*, 41, 31–38.

Cox, D. R. (1972), "Regression Models and Life Tables" (with discussion), *Journal of the Royal Statistical Society*, Ser. B, 34, 187–220.

Cox, D. R., and Lewis, P. A. W. (1966), *The Statistical Analysis of Series of Events*, London: Methuen.

Crowder, M. C., Kimber, A. C., Smith, R. L., and Sweeting, T. J. (1991), *Statistical Analysis of Reliability Data*, London: Chapman and Hall.

Dalal, S. R., and McIntosh, A. A. (1994), "When to Stop Testing for Large Software Systems With Changing Code," *IEEE Transactions on Software Engineering*, 20, 318–323.

Duchesne, T., and Lawless, J. (2000), "Alternative Time Scales and Failure Time Models," *Lifetime Data Analysis*, 6, 157–179.

Hoyland, A., and Rausand, M. (1994), *System Reliability Theory*, New York: Wiley.

Kalbfleisch, J. D., and Prentice, R. L. (1980), *The Statistical Analysis of Failure Time Data*, New York: Wiley.

Lawless, J. F. (1982), *Statistical Models and Methods For Lifetime Data*, New York: Wiley.

——— (1983), "Statistical Methods in Reliability" (with discussion), *Technometrics*, 25, 305–335.

——— (1998), "Statistical Analysis of Product Warranty Data," *International Statistical Review*, 66, 41–60.

Lawless, J. F., and Nadeau, C. (1995), "Some Simple Robust Methods for the Analysis of Recurrent Events," *Technometrics*, 37, 158–168.

Makis, V., and Jardine, A. K. S. (1992), "Computation of Optimal Policies in Replacement Models," *Institute of Mathematics and its Applications Journal of Mathematics Applied in Business and Industry*, 3, 169–175.

Meeker, W. Q., and Escobar, L. A. (1998), *Statistical Methods for Reliability Data*, New York: Wiley.

Nelson, W. (1982), *Applied Life Data Analysis*, New York: Wiley.

——— (1990), *Accelerated Testing*, New York: Wiley.

Olin, B. D., and Meeker, W. Q. (1996), "Applications of Statistical Methods to Nondestructive Evaluation" (with discussion), *Technometrics*, 38, 95–130.

Singpurwalla, N. D. (1995), "Survival in Dynamic Environments," *Statistical Science*, 10, 86–103.

Therneau, T. M. (1997), "Extending the Cox Model," in *Proceedings of the First Seattle Symposium on Biostatistics*, eds. D. Y. Lin and T. R. Fleming, New York: Springer, pp. 51–84.

Whitmore, G. A., Crowder, M. J., and Lawless, J. F. (1998), "Failure Inference from a Marker Process Based on a Bivariate Wiener Model," *Lifetime Data Analysis*, 4, 229–251.

The State of Statistical Process Control as We Proceed into the 21st Century

Zachary G. Stoumbos, Marion R. Reynolds, Jr.,
Thomas P. Ryan, and William H. Woodall

1. INTRODUCTION

Statistical process control (SPC) refers to some statistical methods used extensively to monitor and improve the quality and productivity of manufacturing processes and service operations. SPC primarily involves the implementation of *control charts*, which are used to detect any change in a process that may affect the quality of the output. Control charts are among the most important and widely used tools in statistics. Their applications have now moved far beyond manufacturing into engineering, environmental science, biology, genetics, epidemiology, medicine, finance, and even law enforcement and athletics (see Lai 1995; Montgomery 1997; Ryan 2000). C. R. Rao (1989) stated: "It is not surprising that a recent book on modern inventions lists statistical quality control as one of the technological inventions of the past century. Indeed, there has rarely been a technological invention like statistical quality control, which is so wide in its application yet so simple in theory, which is so effective in its results yet so easy to adopt and which yields so high a return yet needs so low an investment."

The first control charts were developed by Walter A. Shewhart in the 1920s (see Shewhart 1931). These simple Shewhart charts have dominated applications to date. Much research has been done on control charts over the last 50 years, but the diffusion of this research to applications has been very slow. As Crowder, Hawkins, Reynolds, and Yashchin (1997) noted, "There are few areas of statistical application with a wider gap between methodological development and application than is seen in SPC."

Zachary G. Stoumbos is Associate Professor, Department of Management Science and Information Systems and Rutgers Center for Operations Research, Rutgers, The State University of New Jersey, Newark, NJ 07102 (E-mail: stoumbos@andromeda.rutgers.edu). Marion R. Reynolds, Jr. is Professor, Departments of Statistics and Forestry, Virginia Polytechnic Institute and State University, Blacksburg, VA 24061. Thomas P. Ryan is Visiting Professor, Department of Statistics, University of Michigan, Ann Arbor, MI 48109. William H. Woodall is Professor, Department of Statistics, Virginia Polytechnic Institute and State University, Blacksburg, VA 24061. The authors wish to express appreciation to the following individuals for their comments on the manuscript: J. Stuart Hunter, Princeton University; C. R. Rao, Penn State University; Chris P. Tsokos, University of South Florida; and Emmanuel Yashchin, IBM Research. Zachary G. Stoumbos' work was supported in part by 1999 and 2000 Rutgers Faculty of Management Research Fellowships. William H. Woodall's work was supported in part by National Science Foundation grant DMI 9908013.

An examination of what is used in practice and what appears in the SPC literature shows that there are actually two gaps. There is one gap between applications and applied research in journals such as the *Journal of Quality Technology* and *Technometrics* and another gap between this applied research and the research in some of the more theoretical statistics journals. The existence of these gaps is disturbing, because it means that most practitioners have received little of the potential benefit from the technical advances made in SPC over the last half-century. Here we discuss in detail the current state of SPC and give our views on some important future research topics. These topics have good potential to narrow the gaps between applications and applied and theoretical SPC research.

2. THE PROCESS-MONITORING PROBLEM

The process-monitoring problem can be described in general terms as follows. Let X represent a quality variable of interest, and suppose that $f_\theta(x)$, the distribution function of X, is indexed by θ, a vector of one or more parameters. A stable process that is operating with $\theta = \theta_0$ is said to be in *statistical control*. The value of θ_0 may or may not correspond to an ideal (or target) value.

"Murphy's law" explains the purpose of process monitoring; over time, something will inevitably change and possibly cause deterioration in process quality. Something that affects process quality is assumed to be reflected by a change in θ from the value θ_0, so the basic goal of process monitoring is to detect changes in θ that can occur at unknown times. Many types of changes in θ could occur, such as brief self-correcting changes or shifts and drifts that persist over long periods if undetected.

Control charts for monitoring θ are based on taking samples from the process and observing the values of X. A *control statistic*, say Y, is computed after each sample and plotted in time order on a control chart. *Control limits* are constructed such that a value of Y is very unlikely to fall outside of them when $\theta = \theta_0$. A value of Y that falls outside the control limits is taken as a signal that a change in θ has occurred, and that some appropriate action is required.

At the start of the process, the values of some or all of the components of θ_0 may be unknown, and thus a preliminary phase of collecting process data, estimating parameters, and testing for process stability may be required. Process monitoring can begin after θ_0 is estimated in this preliminary phase.

The crisis and subsequent quality revolution in U.S. industry in the 1980s triggered an increasing emphasis on actively working to improve quality (Deming 1986). Thus, in addition to detecting undesirable changes in θ, control charts also should be used to identify improvements in the process. For example, if X is normally distributed and $\theta = (\mu, \sigma)$, then improving quality might correspond to making process adjustments that will reduce σ (Reynolds and Stoumbos 2000a).

3. TRADITIONAL SHEWHART CONTROL CHARTS

The first control charts proposed by W. A. Shewhart in the 1920s remain in widespread use today. The Shewhart charts were designed to make it relatively easy for process personnel without statistical training to set up, apply, and interpret the charts using only a pencil and paper for calculations. Although it is not often explicitly stated, these charts are based on the assumption that $f_\theta(x)$ is one of a few standard distributions (normal for continuous data, and binomial or Poisson for discrete data), and that successive observations of X are independent. The control statistic Y_k computed after sample k is a function of the data in sample k only. Ease of computation is emphasized; so, for example, the sample range is typically used as the measure of dispersion. The design of Shewhart charts is traditionally based on simple heuristics, such as using samples of four or five observations at suitable sampling intervals, say every hour, and using "three-sigma" limits set three standard deviations away from the in-control mean of Y_k.

Shewhart charts have functioned as simple graphical tools in a wide variety of applications. It is not surprising, however, that such simple charts are usually far from optimal (see Secs. 4 and 5) and may even be inappropriate. For example, with three-sigma limits, the false-alarm rate is not adjusted to suit the specific conditions of different applications, and anomalies arise such as lack of lower control limits for nonnegative statistics. The use of the sample range has continued long after computational ease ceased to be a primary concern.

Modifying a Shewhart chart can alleviate some of the aforementioned problems, but a greater disadvantage is that these charts are inefficient for detecting all but relatively large changes in θ. The increasing emphasis today on high-quality products increases the importance of detecting small changes in θ.

4. MORE EFFICIENT CONTROL PROCEDURES

Efficient detection of small and moderate shifts in θ requires that the control statistic in some way accumulate information across past samples. *Runs rules*, which are based on patterns of points in a Shewhart chart, help to improve the ability of Shewhart charts to detect small shifts in θ (Champ and Woodall 1987), but using these rules is not the best method of detecting small shifts in θ.

A much better method of accumulating information across samples uses a control statistic that is an *exponentially weighted moving average* (EWMA) of current and past sample statistics. In particular, if I_k is the individual statistic for sample k, then the EWMA control statistic computed after sample k is

$$E_k = (1 - \lambda)^k E_0 + \sum_{i=1}^{k} (1 - \lambda)^{k-1} \lambda I_i = (1 - \lambda) E_{k-1} + \lambda I_k,$$

where E_0 is the starting value and $\lambda > 0$ is the smoothing parameter that determines the weight given to current data relative to past data. A signal is given if E_k falls outside of control limits.

The *cumulative sum* (CUSUM) chart is another highly efficient control chart that accumulates information over current and past samples. The CUSUM statistic for detecting a shift from θ_0 to a specified alternative θ_1 can be written as

$$C_k = \max\{0, C_{k-1}\} + \ln(f_{\theta_1}(x_k)/f_{\theta_0}(x_k)),$$

where C_0 is the starting value. A signal is given if C_k exceeds a control limit. Two-sided CUSUM charts are usually constructed by running two one-sided CUSUM charts simultaneously.

For a given false-alarm rate, both the EWMA and CUSUM charts are much better than a Shewhart chart for detecting small sustained shifts in θ. The CUSUM chart is optimal for detecting a shift from θ_0 to a specified θ_1 in that it minimizes the worst mean signal delay for a large class of signal rules with appropriately constrained false-alarm rates (Lorden 1971; Moustakides 1986; Ritov 1990). A good broad exposition on CUSUM charts from a more applied perspective was given by Hawkins and Olwell (1998), and discussion of some recent developments on CUSUM charts has been provided by Reynolds and Stoumbos (1999, 2000b). For the problem of monitoring the process mean μ, several studies have shown that the EWMA and CUSUM charts generally have similar detection efficiencies over a range of shifts in μ (see, e.g., Lucas and Saccucci 1990). The EWMA and CUSUM charts date from the 1950s (Page 1954 for the CUSUM and Roberts 1959 for the EWMA), but usage of these efficient charts in applications was very infrequent for many years. Their usage is steadily increasing, although still relatively low.

A number of *generalized* CUSUM schemes proposed over the last three decades allow for θ_1 to be any unknown value in a given interval. These schemes are based on a generalized likelihood ratio (GLR) or on integrating the likelihood ratio with respect to a probability distribution of θ (Basseville and Nikiforov 1993; Lai 1995; references therein). The latter are quasi-Bayesian schemes akin to the procedures discussed in Section 5. Most generalized CUSUM schemes proposed to date are of mainly theoretical interest, because they cannot be easily expressed by computationally convenient recursive forms (Lai 1995).

5. BAYESIAN PROCEDURES AND ECONOMIC MODELS

Bayesian procedures appear to be naturally suited for process monitoring. The application of Bayesian procedures requires recognition that the form of the a priori information about θ is not simply a prior distribution, as would be the case in traditional estimation. In process monitoring, it is assumed that θ will eventually change from the value θ_0 while monitoring is conducted. Thus the prior distribution must reflect when a change in θ will occur, as well as the type of change that will occur.

Bayesian methods for process monitoring have been available since the works of Girshick and Rubin (1952), Shiryaev (1963), and Roberts (1966), who placed a geometric prior distribution with parameter p on the unknown time T of the change in θ (*changepoint*), and independently derived what is commonly termed the *Shiryaev–Roberts* (S-R) procedure. The S-R control statistic for detecting a shift from θ_0 to a specified alternative θ_1 can be expressed in terms of the log-likelihood ratio as

$$R_k = \ln(p + e^{R_{k-1}}) - \ln(1 - p) + \ln(f_{\theta_1}(x_k)/f_{\theta_0}(x_k)),$$

where R_0 is the starting value. A signal is generated if R_k exceeds a control limit, which in the original work was determined to minimize a cost-based loss function.

Pollak (1985) proved the S-R signal rule to be asymptotically Bayes risk efficient as $p \to 0$. Basseville and Nikiforov (1993) and Lai (1995) provided good discussions of the S-R control chart. Bayesian procedures for process monitoring appear to be unknown to most applied statisticians and industrial engineers and thus are rarely used in SPC applications.

There is a relatively large volume of applied SPC literature on economic models (starting with Duncan 1956) related to Bayesian approaches. These economic models aim to find the optimum control chart design (sample size, sampling interval, and control limits) to minimize long-term expected costs. The control statistics used in these models are from standard control charts, such as Shewhart or CUSUM charts, and thus are not based on a posterior distribution. These models use prior distributions for the time and size of change in θ and a loss function that accounts for the costs of sampling, false alarms, and operating out of control. Thus these models have the key elements that would be used in a Bayesian approach to the problem, except that the control statistics are not based on a posterior distribution.

Economic models appear to provide a natural approach for process engineers to use in control chart design and application, because decisions are put into terms that managers understand (dollars), and the problem is framed in terms of designing traditional control charts. But, like many purely Bayesian methods, these models are rarely used by SPC practitioners. There is disagreement among SPC researchers about the general usefulness of the economic modeling approach. Some researchers (e.g., Woodall 1986) criticize these models and feel that future SPC research efforts would be more fruitful in other areas. Other researchers are actively working on these models and feel that they provide the best approach to control chart design for many applications (e.g., Keats, Del Castillo, von Collani, and Saniga 1997).

6. MORE EFFICIENT SAMPLING

The standard approach to sampling for a control chart is to use a *fixed sampling rate* (FSR) in which samples of fixed size are obtained using a fixed-length sampling interval. In recent years, *variable sampling rate* (VSR) control charts have been developed. VSR charts allow the sampling rate to vary as a function of the process data. When the data exhibit no evidence of a change in θ, a low sampling rate is used, but as soon as there is evidence of a possible change in θ, a high sampling rate is used. If the evidence of a change in θ is strong enough, a VSR chart signals in the same way as a traditional FSR chart. Using a high sampling rate when there is evidence of a change in θ results in much faster detection of most shifts in θ, compared to an FSR chart with the same average in-control sampling rate.

There are several ways to allow the sampling rate to vary as a function of the process data. One way is to allow the sampling interval to vary (Reynolds, Amin, and Arnold 1990; Reynolds and Stoumbos 2000a; references therein). Another way is to allow the sample size to vary. A particularly efficient way to allow the sample size to vary is to apply a *sequential probability ratio test* for testing θ_0 versus θ_1 at each sampling point (Reynolds and Stoumbos 1998; Stoumbos and Reynolds 1997, 2000a). Tagaras (1998) presented a review of VSR charts.

The great majority of VSR control charts in the literature, including those mentioned herein, have been developed for discrete-time models. Recently, several VSR control charts have been developed for the continuous-time problem of monitoring the drift coefficient of a Brownian motion process (Assaf, Pollak, and Ritov 1992; Assaf and Ritov 1989; Srivastava and Wu 1994). Assaf et al. (1992) and Srivastava and Wu (1994) noted that these continuous-time VSR control charts are quite complicated to implement in practice and were considered mainly from a theoretical viewpoint, using diffusion theory.

The disadvantage of VSR charts is, of course, the administrative inconvenience of the varying sampling rate. However, the ability to make better use of sampling resources by selectively allocating them to the time periods in which they will be most effective provides a powerful method for significantly increasing the efficiency of process monitoring.

7. MULTIVARIATE CONTROL CHARTS

In many SPC applications, the quality of the process will be characterized by multiple correlated quality variables, and in this situation both the quality characteristic **X** and parameter θ will be vectors. In multivariate SPC applications, the most common approach to process monitoring is to apply separate univariate control charts for each variable, ignoring the issue of their joint performance.

One approach to constructing a multivariate control chart is based on forming a single control statistic from the multivariate data in each sample. This control statistic would usually be a quadratic form involving summary statistics for each variable, and would be plotted on a Shewhart-type chart (Hotelling 1947). The resulting control chart has the disadvantage of all Shewhart-type charts; it is inefficient for detecting small and moderate-sized sustained shifts in θ.

A much better approach is to compute an EWMA or CUSUM statistic for each variable, and then use a quadratic form to combine these separate univariate statistics into a single control statistic to be plotted on a control chart as usual (Lowry, Woodall, Champ, and Rigdon 1992; Mason, Champ, Tracy, Wierda, and Young 1997; references therein).

8. AUTOCORRELATION

A basic assumption usually made in constructing and evaluating control charts is that the process data are independent. But autocorrelation is present in many applications, particularly in cases in which data are closely spaced in time. Relatively low levels of autocorrelation can have a significant impact on the statistical properties of standard control charts designed under the assumption of independence. For example, estimates of θ_0 can be severely biased, resulting in a much higher false-alarm rate than expected. It is not uncommon in applications for standard control charts to be applied to autocorrelated data. When these control charts do not work properly, ad hoc adjustments are made to the charts to try to compensate for the autocorrelation. This clearly is not the best approach to use for the problem of autocorrelation.

In recent years, two basic approaches to dealing with autocorrelation have been studied in the applied SPC literature. Under both approaches, an underlying time series model is assumed. The first approach uses the original data in a standard control chart, but adjusts the control limits to account for the autocorrelation. The second approach advocates plotting the residuals from the time series model on a standard control chart (Faltin, Mastrangelo, Runger, and Ryan 1997; Lu and Reynolds 1999; references therein).

Both of these approaches tend to make the problem of process monitoring appear simpler than it actually is. If a time series model really captures the in-control behavior of the process, then the parameters of this time series model become elements of θ. Thus the complexity of the process-monitoring problem is increased due to the increase in the number of parameters to estimate and monitor.

Several interesting extensions of CUSUM, GLR, and nonlikelihood control chart schemes for autocorrelated data have appeared in the engineering literature over the last three decades. Basseville and Nikiforov (1993) gave a comprehensive overview of these algorithms in the context of univariate as well as multivariate process-monitoring applications.

9. STATISTICAL PROCESS CONTROL AND AUTOMATIC PROCESS CONTROL

The basic philosophy of SPC for improving quality is to detect process changes, so that the cause(s) of the changes can be investigated. Another approach to improving quality, sometimes called *automatic process control* (APC), has been developed in the engineering literature. APC can be used in situations in which there is autocorrelation in the data and a mechanism is available to adjust the process when it appears to be deviating from the desired state. The approach used in APC is to forecast the next observation, and then use the adjustment mechanism to adjust the process so that the observation will be closer to the desired state. In APC the process is assumed to be wandering in some sense, and the adjustment mechanism compensates for this wandering. Thus the basic philosophy of APC is to compensate for undesirable process changes, rather than to detect and remove them as in SPC.

The determination of the best adjustment to make in APC requires a model for process behavior. The optimal adjustment chosen for this model may not work well in the presence of a process change that alters the model. The need to detect changes in the underlying model suggests combining SPC monitoring with APC adjustment to exploit their individual strengths. There has been recent work on combining SPC and APC, but more work is needed (Box and Luceño 1997; Tsung, Shi, and Wu 1999; references therein).

10. DIAGNOSTICS

Monitoring, in either the case of a time series model or the case of multiple variables, involves monitoring multiple parameters. When a control chart signals in this situation, the parameter(s) that have changed may be difficult to determine. In addition, with small parameter changes, pinpointing the time of the change may also be difficult. Thus an important issue with multiple parameters is the ability to diagnose the type and time of the parameter change that has occurred.

In the case of multivariate data, some work has been done on diagnostics, but little has been done in the case of time series models. The literature on estimating changepoints in a sequence of observations is substantial (Basseville and Nikiforov 1993 and references therein). But the problem of estimating the changepoint in process monitoring is different from the problem for a fixed sequence of observations because in process monitoring the estimation is done only after a signal by the control chart (Nishina 1992; Nishina and Peng-Hsiung 1996; references therein). The problem of diagnostics is one area where the use of Bayesian models, nonlinear filtering theory, and stochastic calculus seems quite natural and may prove very helpful (Lai 1995; Stoumbos 1999; Yashchin 1997).

11. PARAMETER ESTIMATION AND NONPARAMETRIC PROCEDURES

Control chart performance is very sensitive to errors in estimating θ_0. For example, the false-alarm rate may be much higher or lower than expected unless the dataset used in the initial phase of estimating θ_0 is quite large. The situation is even worse in more complex situations when multiple parameters must be estimated (Adams and

Tseng 1998; Lu and Reynolds 1999). The effect of errors in estimating θ_0 in complex models awaits additional study, and methods for compensating for these effects remain to be developed.

Traditional Shewhart-type charts are usually based on the assumption that if $f_\theta(x)$ is continuous, then it will be normal. Almost all work on multivariate control charts is based on the assumption that $f_\theta(x)$ is multivariate normal. In some cases, the central limit theorem can be used to justify approximate normality when monitoring means, but in numerous cases normality is an untenable assumption and one is unwilling to use another parametric model. A number of nonparametric methods are available for use in these cases. A more prominent role is expected for nonparametric methods. As data availability increases, nonparametric methods seem especially useful in multivariate applications where most methods proposed thus far rely on normality. Nonparametric multivariate control charts have been studied only very recently and much more research is needed (e.g., Liu 1995; Stoumbos and Jones 2000; Stoumbos, Jones, Woodall, and Reynolds 2000).

12. FUTURE DIRECTIONS

Advances in automated manufacturing systems coupled with advances in sensing and automatic inspection technology will continue to increase the volume of data available for drawing inferences about many processes. In some applications, this will change the inference problem from one dependent on scarce data to one based on plentiful data. However, the emphasis on higher quality will often require measuring more variables, which in some cases may be expensive and/or time-consuming. Thus we foresee no reduction in the need for efficient procedures for process monitoring.

In the future, we expect problems to be more diverse, with specialized monitoring methods required. Multiple quality variables, along with possible autocorrelation in these variables, will require more complex models with a large increase in the number of parameters to monitor.

The increasing complexity of problems encountered provides an opportunity to narrow the gaps between applications and applied and theoretical SPC research. The Shewhart charts that have dominated industrial applications over the past 75 years were designed to be extremely simple, with a one-size-fits-all approach to their design and implementation. In the case of relatively simple problems, arguments that CUSUM or EWMA charts have much better statistical properties convinced only some industrial practitioners to move beyond using the familiar Shewhart charts. As problems become more complex, the need for more sophisticated monitoring procedures will become critical and more obvious to all. Theoretical and applied research that addresses this need can have a major impact on applications.

The following are some additional research areas that we feel have very good potential for impacting applications (for some other research topics, see Montgomery and Woodall 1997 and Woodall and Montgomery 1999):

- Basic and applied research is needed on methods for monitoring multiple parameters that arise in models for the cases of single or multiple process variables and/or autocorrelation.
- When multiple parameters are to be monitored, methods are needed for diagnosing both the changepoint (Nishina 1992; Nishina and Peng-Hsiung 1996) and the specific parameter or parameters that have changed (Reynolds and Stoumbos 2000a). The application of monitoring procedures to complex problems may require sophisticated efforts in model fitting and parameter estimation, and the effects of the fitting and estimation on the procedures await much more study. We foresee that VSR approaches can significantly improve the effectiveness of change-point estimation.
- The robustness of fitted process models and monitoring procedures to model misspecification needs further study (Stoumbos and Reynolds 2000b). Nonparametric procedures for multivariate problems are an open field with great potential.
- Additional basic and applied research is needed on procedures that integrate SPC and APC methodology.
- The scope of SPC methods should be expanded to include the study of all variation over time throughout the entire production process (tracking), which usually includes numerous stages of process steps and product measurement (Agrawal, Lawless, and Mackay 1999; Hawkins 1991; Lawless, Mackay, and Robinson 1999).
- A greater synthesis of the theoretical changepoint and applied SPC literatures is very desirable. Moreover, excellent opportunities exist for cross-fertilization of ideas from other areas of statistics and stochastic modeling, including epidemiology, outlier detection, and especially stochastic calculus and financial mathematics (Stoumbos 1999).
- The use of complicated models will place even greater emphasis on the development of software. In many cases, software will need to be customized for particular applications.

REFERENCES

Adams, B. M., and Tseng, I. T. (1998), "Robustness of Forecast-Based Monitoring Schemes," *Journal of Quality Technology*, 30, 328–339.

Agrawal, R., Lawless, J. F., and Mackay, R. J. (1999), "Analysis of Variation Transmission in Manufacturing Processes—Part II," *Journal of Quality Technology*, 31, 143–154.

Assaf, D., Pollak, M., and Ritov, Y. (1992), "A New Look at Warning and Action Lines of Surveillance Schemes," *Journal of the American Statistical Association*, 87, 889–895.

Assaf, D., and Ritov, Y. (1989), "Dynamic Sampling Procedures for Detecting a Change in the Drift of a Brownian Motion: A Non-Bayesian Model," *The Annals of Statistics*, 17, 793–800.

Basseville, M., and Nikiforov, I. V. (1993), *Detection of Abrupt Changes: Theory and Applications*, Engelwood Cliffs, NJ: Prentice-Hall.

Box, G. E. P., and Luceño, A. (1997), *Statistical Control by Monitoring and Feedback Adjustment*, New York: Wiley.

Champ, C. W., and Woodall, W. H. (1987), "Exact Results for Shewhart Control Charts With Supplementary Runs Rules," *Technometrics*, 29, 393–399.

Crowder, S. V., Hawkins, D. M., Reynolds, M. R., Jr., and Yashchin, E. (1997), "Process Control and Statistical Inference," *Journal of Quality Technology*, 29, 134–139.

Deming, W. E. (1986), *Out of the Crisis*, Cambridge, MA: Center for Advanced Engineering Studies, Massachusetts Institute of Technology.

Duncan, A. J. (1956), "The Economic Design of $\bar{X}$ Charts Used to Maintain Current Control of a Process," *Journal of the American Statistical Association*, 51, 228–242.

Faltin, F. W., Mastrangelo, C. M., Runger, G. C., and Ryan, T. P. (1997), "Considerations in the Monitoring of Autocorrelated and Independent Data," *Journal of Quality Technology*, 29, 131–133.

Girshick, M. A., and Rubin, H. (1952), "A Bayes Approach to a Quality Control Model," *The Annals of Mathematical Statistics*, 23, 114–125.

Hawkins, D. M. (1991), "Multivariate Quality Control Based on Regression-Adjusted Variables," *Technometrics*, 33, 61–75.

Hawkins, D. M., and Olwell, D. H. (1998), *Cumulative Sum Control Charts and Charting for Quality Improvement*, New York: Springer-Verlag.

Hotelling, H. (1947), "Multivariate Quality Control," in *Techniques of Statistical Analysis*, eds. C. Eisenhart, M. W. Hastay, and W. A. Wallis, New York: McGraw-Hill.

Keats, J. B., Del Castillo, E., von Collani, E., and Saniga, E. (1997), "Economic Modeling for Statistical Process Control," *Journal of Quality Technology*, 29, 144–147.

Lai, T. L. (1995), "Sequential Changepoint Detection in Quality Control and Dynamical Systems" (with discussion), *Journal of the Royal Statistical Society*, Ser. B, 57, 613–658.

Lawless, J. F., Mackay, R. J., and Robinson, J. A. (1999), "Analysis of Variation Transmission in Manufacturing Processes—Part I," *Journal of Quality Technology*, 31, 131–142.

Liu, R. Y. (1995), "Control Charts for Multivariate Processes," *Journal of the American Statistical Association*, 90, 1380–1387.

Lorden, G. (1971), "Procedures for Reacting to a Change in Distribution," *The Annals of Mathematical Statistics*, 42, 1897–1908.

Lowry, C. A., Woodall, W. H., Champ, C. W., and Rigdon, S. E. (1992), "A Multivariate Exponentially Weighted Moving Average Control Chart," *Technometrics*, 34, 46–53.

Lu, C. W., and Reynolds, M. R., Jr. (1999), "EWMA Control Charts for Monitoring the Mean of Autocorrelated Processes," *Journal of Quality Technology*, 31, 166–188.

Lucas, J. M., and Saccucci, M. S. (1990), "Exponentially Weighted Moving Average Control Schemes: Properties and Enhancements," *Technometrics*, 32, 1–12.

Mason, R. L., Champ, C. W., Tracy, N. D., Wierda, S. J., and Young, J. C. (1997), "Assessment of Multivariate Process Control Techniques," *Journal of Quality Technology*, 29, 140–143.

Montgomery, D. C. (1997), *Introduction to Statistical Quality Control* (3rd ed.), New York: Wiley.

Montgomery, D. C., and Woodall, W. H. (1997), "A Discussion of Statistically-Based Process Monitoring and Control," *Journal of Quality Technology*, 29, 121–162.

Moustakides, G. V. (1986), "Optimal Stopping Times for Detecting Changes in Distributions," *The Annals of Statistics*, 14, 1379–1387.

Nishina, K. (1992), "A Comparison of Control Charts From the Viewpoint of Change-Point Estimation," *Quality and Reliability Engineering International*, 8, 537–541.

Nishina, K., and Peng-Hsiung, W. (1996), "Performance of CUSUM Charts From the Viewpoint of Change-Point Estimation in the Presence of Autocorrelation," *Quality and Reliability Engineering International*, 12, 3–8.

Page, E. S. (1954), "Continuous Inspection Schemes," *Biometrika*, 41, 100–114.

Pollak, M. (1985), "Optimal Detection of a Change in Distribution," *The Annals of Statistics*, 13, 206–227.

Rao, C. R. (1989), *Statistics and Truth: Putting Chance to Work*, Fairland, MD: International Co-Operative Publishing House.

Reynolds, M. R., Jr., Amin, R. W., and Arnold, J. C. (1990), "CUSUM Charts With Variable Sampling Intervals," *Technometrics*, 32, 371–384.

Reynolds, M. R., Jr., and Stoumbos, Z. G. (1998), "The SPRT Chart for Monitoring a Proportion," *IIE Transactions on Quality and Reliability Engineering*, 30, 545–561.

—— (1999), "A CUSUM Chart for Monitoring a Proportion When Inspecting Continuously," *Journal of Quality Technology*, 31, 87–108.

—— (2000a), "Monitoring the Process Mean and Variance Using Individual Observations and Variable Sampling Intervals," *Journal of Quality Technology*, (to appear).

—— (2000b), "A General Approach to Modeling CUSUM Charts for a Proportion," *IIE Transactions on Quality and Reliability Engineering, Special Issue on Emerging Trends in Quality Engineering*, 32, 515–535.

Ritov, Y. (1990), "Decision Theoretic Optimality of the CUSUM Procedure," *The Annals of Statistics*, 18, 1464–1469.

Roberts, S. W. (1959), "Control Charts Based on Geometric Moving Averages," *Technometrics*, 1, 239–250.

—— (1966), "A Comparison of Some Control Chart Procedures," *Technometrics*, 8, 411–430.

Ryan, T. P. (2000), *Statistical Methods for Quality Improvement* (2nd ed.), New York: Wiley.

Shewhart, W. A. (1931), *Economic Control of Quality of Manufactured Product*, New York: Van Nostrand.

Shiryaev, A. N. (1963), "On Optimum Methods in Quickest Detection Problems," *Theory of Probability and Applications*, 8, 22–46.

Srivastava, M. S., and Wu, Y. (1994), "Dynamic Sampling Plan in the Shiryaev–Roberts Procedure for Detecting a Change in the Drift of Brownian Motion," *The Annals of Statistics*, 22, 805–823.

Stoumbos, Z. G. (1999), "The Detection and Estimation of the Change Point in a Discrete-Time Stochastic System," *Stochastic Analysis and Applications*, 17, 637–649.

Stoumbos, Z. G., and Jones, L. A. (2000), "On the Properties and Design of Individuals Control Charts Based on Simplicial Depth," *Nonlinear Studies*, (to appear).

Stoumbos, Z. G., Jones, L. A., Woodall, W. H., and Reynolds, M. R., Jr. (2000), "On Shewhart-Type Nonparametric Multivariate Control Charts Based on Data Depth," in *Frontiers in Statistical Quality Control 6*, eds. H.-J. Lenz and P.-Th. Wilrich, Heidelberg, Germany: Springer-Verlag, 208–228.

Stoumbos, Z. G., and Reynolds, M. R., Jr. (1997), "Control Charts Applying a Sequential Test at Fixed Sampling Intervals," *Journal of Quality Technology*, 29, 21–40.

—— (2000a), "The SPRT Control Chart for the Process Mean With Samples Starting at Fixed Times," *Nonlinear Analysis*, (to appear).

—— (2000b), "Robustness to Non-Normality and Autocorrelation of Individual Control Charts," *Journal of Statistical Computation and Simulation, Special Issue on Statistics: Modeling of Processes in Science and Industry*, 66, 145–187.

Tagaras, G. (1998), "A Survey of Recent Developments in the Design of Adaptive Control Charts," *Journal of Quality Technology*, 30, 212–231.

Tsung, F., Shi, J., and Wu, C. F. J. (1999), "Joint Monitoring of PID-Controlled Processes," *Journal of Quality Technology*, 31, 275–285.

Woodall, W. H. (1986), "Weaknesses of the Economic Design of Control Charts" (letter to the editor), *Technometrics*, 28, 408–410.

Woodall, W. H., and Montgomery, D. C. (1999), "Research Issues and Ideas in Statistical Process Control," *Journal of Quality Technology*, 31, 376–386.

Yashchin, E. (1997), "Change-Point Models in Industrial Applications," *Nonlinear Analysis*, 30, 3997–4006.

Statistics in Preclinical Pharmaceutical Research and Development

Bert Gunter and Dan Holder

1. INTRODUCTION

Although most statisticians and the public at large are familiar with the role of statistics in human clinical drug trials, advances in the basic science and technology of drug research and development (R&D) have created equally challenging and important opportunities in the preclinical arena. Preclinical pharmaceutical research encompasses all aspects of drug discovery and development, from basic research into how cells and organs work and how disease processes disrupt that work to the development, formulation, and manufacture of drugs. The activities that fall under this rubric include biological and biochemical research using in vitro ("test tube"-based) and in vivo (whole animal) experiments; genomics, the study of gene expression in cells, organisms, and populations to determine the molecular biology of disease; proteomics, the study of protein expression patterns to understand how normal and disease processes differ; design, synthesis, and selection of diverse chemical and natural product "libraries" of compounds to screen for desirable biological activity, often via high throughput, "industrialized" drug screening assays; analytical development for drug research and manufacturing; animal testing of drug candidates for efficacy and metabolism and to determine drug toxicity, teratogenicity (fetal and growth effects), and carcinogenicity; development and scale-up of chemical and fermentation drug manufacturing processes; and drug formulation and stability testing. This list is far from complete.

To put these activities into perspective, it can easily cost more than $1 billion and require 10 to 15 years of R&D to bring out a single new drug, of which only the last 2–3 involve the FDA-reviewed human trials with which statisticians and the public are most familiar. So preclinical activities occupy the bulk of the time and scientific effort. The statistics that support this work cover a broad range of statistical methods. Sample sizes can range from longitudinal case-control studies of 10 or fewer animals (although they may produce thousands of data points from continuous monitoring using sophisticated instruments and telemetry) to hundreds of thousands

Bert Gunter is Director, Scientific Staff, Merck Research Laboratories, Rahway, NJ 08540 (E-mail: bert_gunter@merck.com) and Dan Holder is Associate Director, Scientific Staff, Merck Research Laboratories, West Point, PA 19486 (E-mail: dan_holder@merck.com).

or millions of multivariate records in drug screening and structure searches. All areas of statistics find useful application, but recent opportunities for nonparametric experimental design, linear and nonlinear longitudinal data modeling, high-dimensional exploration and visualization, inference using exact permutation methods and bootstrapping, and pattern recognition, classification, and clustering of large databases are perhaps noteworthy.

Clearly, in a brief survey like this we can highlight only a couple of examples. We have chosen chemometrics and genomics because they provide good examples of the kind of interdisciplinary, data-rich, and nonstandard issues that are increasingly at the forefront of modern pharmaceutical research. But these examples are just the tip of a vast and fascinating iceberg.

2. CHEMOMETRICS

Roughly speaking, chemometrics is the statistics of (analytical) chemistry data, especially spectroscopy data. Physics and chemistry have developed an arsenal of ingenious tools to probe chemical composition and structure. (A nice internet resource for spectroscopy is www.anachem.umu.se/jumpstation.htm.) These techniques can produce (one- and two-dimensional) spectra of exquisite resolution, often with hundreds or thousands of individual peaks. Digitizing translates them to multivariate vectors of that dimensionality. Chemometrics arose because classical multivariate normal statistical methods were inadequate for such data and related matters of calibration and quality control.

One typical application will give the flavor of the issues. Suppose that one has, say, 200 unknown natural chemical extracts from various biological sources that are tested for antibiotic activity against 30 different pathogens. (Many antibiotics originate from naturally occurring sources; penicillin and erithromycin are familiar examples.) Suppose, also, that ultraviolet spectra with (up to) 500 peaks are measured for these 200 extracts, producing a 200×500 matrix of spectra, X, and a 200×30 matrix of potencies, Y. Can we determine from these data whether and how the extracts cluster into groups that may reveal common active chemical components?

Various kinds of clustering might be tried, but the high dimensionality with so few data points will likely render such approaches ineffective. Rather, it is usually necessary to first project the data onto a low-dimensional manifold that captures the "essential" structure, for which variants of principal components analysis (PCA) are widely used. But what if one or more spectral values for many or even all the extracts are missing because of, for example, detector noise or sample contamination? Simple case deletion will lose most of the data, but what kind of imputation is appropriate? And how does one deal with the isolated outliers far from the main cluster(s) that are frequently of greatest interest? Standard PCA is a least-squares technique, and high-dimensional outliers are often missed by such methods. Computational issues are also critical; one must avoid, for example, calculating the full singular 500×500 covariance or correlation matrix. Although methods have recently been developed to deal with some of these issues individually (Little and Rubin 1987; Rousseeuw and Van Driessen 1999), these methods often do not deal with them together, especially when sample sizes grow to the thousands or hundreds of thousands, which will increasingly be the case as lab automation technology continues to develop.

Moreover, PCA makes use only of the X data—the spectra. More desirable would be clustering based on the relationship of the spectra to clusters of similar potency among the pathogens, which is what would be expected given that different pathogens are susceptible to different biochemical pathways and, therefore, structures. Indeed, often the smaller components (i.e., corresponding to smaller eigenvalues)— representing structures that occur relatively rarely—have the greatest antibiotic effect on subgroups of the pathogens. These could be overlooked as noise in conventional PCA. So what is needed is a kind of *simultaneous* projection procedure in both X and Y in which variation in the X projection is associated with structural variation in Y. Such an approach exists; it is called partial least squares (PLS), and was first developed by Herman Wold and his collaborators (Helland 1988, 1990; Wold 1976), but there is some controversy regarding its statistical effectiveness (Frank and Friedman 1993; Garthwaite 1994; Butler and Denham, 2000). Nevertheless, it is widely used (Chemometrics Web Resource; Thomas and Haaland 1988, 1990).

One can contemplate doing better by taking advantage of modern computational power to extend these least squares ideas from variance to more general projection indices, as with projection pursuit density estimation (Friedman and Stuetzle 1981; Friedman, Stuetzle, and Schroeder 1984). Wavelet approaches also appear promising and are being explored as a preprocessing step for initial "signal" compression to directly reduce dimensionality (Vidakovic 1999). Clearly, much of this is ad hoc, and it would be desirable to have better theory to guide the efforts, although actual application must always remain the critical testing ground. In particular, there should be a role for Bayesian approaches, as Friedman and Frank have already indicated. Clearly, then, important challenges remain on this interdisciplinary border of statistics, analytical chemistry, and computer science. The technology driving the needs will only accelerate.

3. GENOMICS

Advances in molecular genetics have provided various helpful tools for discovering effective therapies for human diseases. To exemplify how molecular genetics is used in the drug discovery and development process, we focus on one application where substantial progress has been made.

3.1 Human Gene Hunting

Currently, many researchers are attempting to discover genetic causes for human diseases. A primary motivation for this research is the belief that this knowledge will help identify persons at high risk for these diseases and lead to the development of effective therapies. The hunt for human disease genes is a complex multidisciplinary challenge in which statistics plays an important role. An approach called positional cloning has met with some success. This approach is based on analyzing the variation in molecular markers between diseased and nondiseased individuals. Molecular marker loci are regions of DNA that vary among individuals and have a known location in the human genome. Historically, many of the markers have levels (called alleles) based on the number of consecutive repeats of a particular DNA pattern. The markers themselves do not usually affect disease susceptibility; however, because DNA is passed from parents to offspring in contiguous chunks, loci in close physical proximity to each other on the same parental chromosome (called "linked" loci) will tend to be passed on together, whereas loci far apart or on different parental chromosomes are passed on independently. Although individuals have two copies of each chromosome, through a process called recombination, it is a combination of the pair, rather than entirely one or the other, that is transmitted to offspring. Linkage, or the closeness of two loci, is usually measured by the recombination fraction, Θ, which may be defined as the probability that the alleles at the two loci passed on to the next generation came from different chromosomes of the parental pair. When loci are very close to each other, they are said to be linked and almost always are passed on together ($\Theta \approx 0$). When loci are not very close to each other or are on different chromosomes, the probability of being transmitted together is close to $\frac{1}{2}$. The basic strategy is to continually narrow a region suspected of harboring a disease locus by repeatedly examining sets of markers, finding the ones that appear to be linked to the putative disease locus, and then looking for a denser set of markers in the areas of interest and repeating the process. Once a region is sufficiently narrowed, geneticists identify and examine DNA structures (e.g., genes, promoter regions) in the interval and try to find one or more etiologic mutations.

Initial searches often use a case-control design because of its relatively lower cost. Researchers look for associations between marker alleles and disease status using a group of affected individuals and either a random or a closely matched control group. The degree of allelic association between a two-level (0, 1) marker allele, M, and a disease allele, D (affected, not affected), may be defined as $\delta = \Pr(M = 1 \text{ and } D = \text{affected}) - \Pr(M = 1)\Pr(D = \text{affected})$. It can be shown that in a simple idealized large randomly mating population, δ will decrease by a factor of $(1 - \Theta)$ at each generation (Hartl 1988). Allelic association is often taken as evidence of linkage, because in this idealized setting at time t, $\delta_t = \delta_0(1 - \Theta)^t$. But, Ewens and Spielman (1995) showed that allelic association does not necessarily provide evidence of close linkage. Lack of random mating throughout the entire population (stratification) or the recent mixture of different populations (admixture) may result in the disease allele and the associated marker both being more common in some strata (or some populations) than in others. These types of population heterogeneity may result in spurious allelic associations when a sample is taken across different populations and/or strata. The extent to which allelic association in the absence of linkage is problematic is a matter of some controversy (see, e.g., Risch and Teng 1998).

Family-based studies are alternatives to the case-control design. In these studies, genotypes and disease statuses are collected for multiple generations in a number of families, and the transmission of marker alleles from parents to their offspring is analyzed. When an appropriate genetic model can be posited, a likelihood approach can be used. This idea is due to Morton (1955), who built on the probability ratio test of Haldane and Smith (1947) and the sequential sampling of Wald (1947). The basic idea of this approach is that within the framework of the genetic model for each family, one may construct a likelihood that is a function of the marker allele frequencies, the disease frequency, and the recombination fraction, Θ. A full likelihood is constructed by multiplying (or adding in the log scale) together the likelihood for all of the individual families. Linkage is assessed by comparing the ratio of the likelihood function evaluated at arbitrary values of Θ to the likelihood function evaluated at $\Theta = \frac{1}{2}$. (Details and refinements for this approach have been given in Ott 1991, 1992 and Terwilliger and Ott 1994.) This approach is very powerful when a genetic model can be constructed that accurately describes the number of alleles influencing disease susceptibility, the probability of the disease manifesting itself given the presence of a disease allele (penetrance), and the allele frequencies. Although such genetic models can be reasonably specified for some diseases caused by a single gene with high penetrance (i.e., a large proportion of individuals with the genetic defect get the disease), most diseases are likely to be caused by multiple genes, making it difficult to specify an appropriate genetic model. Hence much of the current interest is in methods that do not require a detailed specification of the genetic model.

Penrose (1953) suggested analyzing the degree of allele sharing among affected siblings. When two individuals share the same allele at the same locus, that allele is said to be identical by state (IBS). If the individuals have inherited the same allele from the same common ancestor, then the allele is also identical by descent (IBD). Due to Mendelian inheritance, for a given locus, on average $\frac{1}{4}$ of sib pairs will share 2 alleles IBD, $\frac{1}{2}$ will share 1 allele IBD, and $\frac{1}{4}$ will share no alleles IBD. The idea behind affected siblings analysis is that excess IBD sharing of a marker allele by affected sibs is evidence that the marker is linked to the disease allele. When IBD status can be determined, excess allele sharing can be tested using either a means test or a goodness-of-fit test (Blackwelder and Elston 1985). In practice, it is often not possible to determine whether shared alleles are IBD or merely IBS (e.g., if both mother and father have the allele). Methods of estimating the number of alleles that are IBD have been worked out by Amos, Dawson, and Elston (1990) and implemented using a procedure described by Haseman and Elston (1972). Risch (1990) took a likelihood approach to this problem, and Holmans (1993) showed how the power of Risch's method can be improved by restricting maximization to the set of possible haplotype-sharing probabilities. Kruglyak and Lander (1995) showed how the results of multiple markers can be used simultaneously. Weeks and Lange (1988) constructed a test comparing arbitrary affected pedigree members (not necessarily siblings) and using IBS.

Another group of tests focuses more directly on parental transmission of markers. Spielman, McGinnis, and Ewens (1993) showed that the probability that a parent whose two chromosomes carry different values for the marker transmits a specific marker allele to an affected offspring is a function of both allelic association and the recombination rate. Accordingly, they devised a McNemar-type test, the transmission disequilibrium test (TDT), which can be used to test linkage in the presence of allelic association or allelic association in the presence of linkage for the transmission to one affected offspring. Martin, Kaplan, and Weir (1997) showed how additional affected sibs can be incorporated into the test. Later, similar methods were used to construct tests of these hypotheses for discordant pairs of siblings (Curtis 1997; Spielman and Ewens 1998) and general sibships containing at least one affected and at least one unaffected member (Horvath and Laird 1998). Tests of discordant sibships are especially useful for late-onset diseases in which parental genotypes are often not easy to obtain. More recently, Schaid (1999) showed that a likelihood ratio test statistic may be more robust than the TDT over a wide range of genotype relative risk models.

Although positional cloning has already scored some major successes (cystic fibrosis, insulin-dependent diabetes mellitus, Alzheimer's), the best may be yet to come. Single nucleotide polymorphisms (SNPs) will provide a much denser set of markers, and DNA microchips will allow researchers to assay thousands of markers at one time. These advances, along with the likelihood of multiple interacting genes for complex diseases, will provide many future statistical challenges. A number of excellent web sites provide information on these topics and give links to software; three of the best are the Whitehead Institute (http://www-genome.wi.mit.edu/), Rockefeller University (http://linkage.rockefeller.edu/), and, for a wide range of statistical genetics, the University of Wisconsin (http://www.stat.wisc.edu/biosci/linkage.html#linkage).

REFERENCES

Amos, C. I., Dawson, D. V., and Elston, R. C. (1990), "The Probabilistic Determination of Identity-by-Descent Sharing for Pairs of Relatives From Pedigrees," *American Journal of Human Genetics*, 47, 842–853.

Blackwelder, W. C., and Elston, R. C. (1985), "A Comparison of Sib-Pair Linkage Tests for Disease Susceptibility Loci," *Genetic Epidemiology*, 2, 85–98.

Butler, N. A., and Denham, M. C. (2000), "The Peculiar Shrinkage Properties of Partial Least Squares Regression," *Journal of the Royal Statistical Society B*, 62, 585–593.

Chemometrics Web Resource: www.acc.umu.se/~tnkjtg/chemometrics/index.html.

Curtis, D. (1997), "Use of Siblings as Controls in Case-Control Association Studies," *American Journal of Human Genetics*, 61, 319–333.

Ewens, W. J., and Spielman, R. S. (1995), "The Transmission/Disequilibrium Test: History, Subdivision, and Admixture," *American Journal of Human Genetics*, 57, 455–464.

Frank, I., and Friedman, J. (1993), "A Statistical View of Some Chemometrics Regression Tools" (with discussion), *Technometrics*, 31, 3–40.

Friedman, J. H., and Stuetzle, W. (1981), "Projection Pursuit Regression," *Journal of the American Statistical Association*, 76, 817–823.

Friedman, J. H., Stuetzle, W., and Schroeder, A. (1984), "Projection Pursuit Density Estimation," *Journal of the American Statistical Association*, 79, 599–608.

Garthwaite, P. (1994), "An Interpretation of Partial Least Squares," *Journal of the American Statistical Association*, 89, 122–127.

Haldane, J. B. S., and Smith, C. A. B. (1947), "A New Estimate of the Linkage Between the Gene for Colour-Blindness and Hemophilia in Man," *Annals of Eugenics*, 14, 10–31.

Hartl, D. L. (1988), *Principles of Population Genetics*, Sunderland: Sinaur Associates, pp. 108–109.

Haseman, J. K., and Elston, R. C. (1972), "The Investigation of Linkage Between a Quantitative Trait and a Marker Locus," *Behavior Genetics*, 2, 3–19.

Helland, I. (1988), "On the Structure of Partial Least Squares Regression," *Communications in Statistics, Part B: Simulation and Computation*, 17, 581–607.

——— (1990), "Partial Least Squares Regression and Statistical Models," *Scandinavian Journal of Statistics*, 17, 97–114.

Holmans, P. (1993), "Asymptotic Properties of Affected Sib-Pair Linkage Analysis," *American Journal of Human Genetics*, 52, 362–374.

Horvath, S., and Laird, N. M. (1998), "A Discordant-Sibship Test for Disequilibrium and Linkage: No Need for Parental Data," *American Journal of Human Genetics*, 63, 1886–1897.

Kruglyak, L., and Lander, E. S. (1995), "Complete Multipoint Sib-Pair Analysis of Qualitative and Quantitative Traits," *American Journal of Human Genetics*, 57, 439–454.

Little, R., and Rubin, D. (1987), "Statistical Analysis With Missing Data," New York: Wiley.

Martin, E. R., Kaplan, N. L., and Weir, B. S. (1997), "Test for Linkage and Association in Nuclear Families," *American Journal of Human Genetics*, 61, 429–448.

Morton, N. E. (1955), "Sequential Tests for the Detection of Linkage," *American Journal of Human Genetics*, 7, 277–318.

Ott, J. (1991), *Analysis of Human Genetic Linkage*, Baltimore: Johns Hopkins University Press.

——— (1992), "Strategies for Characterizing Highly Polymorphic Markers in Human Gene Mapping," *American Journal of Human Genetics*, 41, 283–290.

Penrose, L. S. (1953), "The General-Purpose Sib-Pair Linkage Test," *Annals of Eugenics*, 18, 120–124.

Risch, N. (1990), "Linkage Strategies for Genetically Complex Traits III. The Effects of Marker Polymorphism Markers in Human Gene Mapping," *American Journal of Human Genetics*, 16, 242–253.

Risch, N., and Teng, J. (1998), "The Relative Power of Family-Based and Case-Control Designs for Linkage Disequilibrium Studies of Complex Human Diseases. I. DNA Pooling," *Genome Research*, 8, 1224.

Rousseeuw, P., and Van Driessen, K. (1999), "A Fast Algorithm for the Minimum Covariance Determinant Estimator," *Technometrics*, 41, 212–223.

Schaid, D. J. (1999), "Likelihoods and TDT for Case-Parents Design," *Genetic Epidemiology*, 16, 250–260.

Spielman, R. S., and Ewens, W. J. (1998), "A Sibship Test of Linkage in the Presence of Association: The Sib Transmission/Disequilibrium Test," *American Journal of Human Genetics*, 62, 450–458.

Spielman, R. S., McGinnis, R. E., and Ewens, W. J. (1993), "Transmission Test for Linkage Disequilibrium: The Insulin Gene Region and Insulin-Dependent Diabetes Mellitus (IDDM)," *American Journal of Human Genetics*, 52, 506–516.

Terwilliger, J. D., and Ott, J. (1994), *Handbook of Human Genetic Linkage*, Baltimore: Johns Hopkins University Press.

Thomas, E., and Halland, D. (1988), "Partial Least Squares Regression for Spectral Analyses, 1: Relation to Other Quantitative Calibration Methods and the Extraction of Qualitative Information," *Analytical Chemistry*, 60, 1193–1202.

——— (1990), "Comparison of Multivariate Calibration Methods for Quantitative Spectral Analysis," *Analytical Chemistry*, 62, 1091–1099.

Vidakovic, B. (1999), *Statistical Modeling by Wavelets*. New York: Wiley.

Wald, A. (1947), *Sequential Analysis*, New York: Wiley.

Weeks, D. E., and Lange, K. (1988), "The Affected Pedigree Member Method of Linkage Analysis," *American Journal of Human Genetics*, 42, 315–326.

Wold, H. (1976), "Soft Modelling by Latent Variables: The Non-Linear Iterative Partial Least Squares (NIPALS) Approach," in *Perspectives in Probability and Statistics, in Honor of M. S. Bartlett*, New York: Academic Press, pp. 117–144.

Statistics in Advanced Manufacturing

Vijay Nair, Mark Hansen, and Jan Shi

Statistical concepts and methods have played a critical role in speeding the pace of industrial development over the last century. In return, industrial applications have provided statisticians with incredible opportunities for methodological research. The richness and variety of these applications have had a major influence on the development of statistics as a discipline; consider, for example, the extensive research in statistical process control (SPC) and changepoint detection, dating back to the pioneering work of Shewhart in the 1920s, and developments in automatic process control, design of experiments, sequential analysis, reliability, and so on. Recent efforts by manufacturers to adopt sustained quality and productivity improvement programs have generated a renewed interest in and appreciation for statistics in industry. In fact, fundamental statistical concepts such as understanding and managing variation form the backbone of popular quality management paradigms and practices.

Many of the traditional SPC concepts and techniques grew in response to the manufacturing environments prevalent several decades ago. Current advanced manufacturing and high-technology industries, however, operate under a much more complex and diverse set of conditions. These changes have important implications for research directions in industrial statistics, not only in terms of identifying new problems and developing new methods, but also in reevaluating the paradigms that inspired earlier approaches. In this vignette we use applications from automotive and semiconductor manufacturing to illustrate various issues and to discuss future research needs and directions. The discussion is limited to a few selected topics and is inevitably slanted toward our own experiences.

1. THE ENVIRONMENT OF ADVANCED MANUFACTURING

Pressures from the competitive marketplace are forcing manufacturers to continuously reduce product development cycle times. In parallel, the underlying technology of products and processes are becoming increasingly complex to keep up with consumer demands. Thus manufacturers frequently move from design to full-scale

Vijay Nair is Professor of Statistics and Professor of Industrial and Operations Engineering, University of Michigan, Ann Arbor. Mark Hansen is Member of Technical Staff at Bell Laboratories, Lucent Technologies. Jan Shi is Associate Professor of Industrial and Operations Engineering, University of Michigan, Ann Arbor.

production well before the technology and the fabrication processes are completely understood.

Consider the manufacture of semiconductor devices. The critical dimensions of integrated circuits (ICs, or chips) are very fine: newly developed ICs have features as small as .16 μm, and any company hoping to remain competitive will have plans to reduce these sizes to as little as .01 μm in the next 5–10 years. Given the scale of these devices, ICs are fabricated in a "clean room" through a process involving hundreds of separate steps and lasting several weeks. The various steps are rarely stable, as is commonly assumed in the SPC paradigm, and they frequently interact in unexpected ways. Such complexity and instability are typical in advanced manufacturing applications.

2. VOLUME AND COMPLEXITY OF THE DATA

Massive amounts of process and product quality data are now collected routinely, made possible by advances in computing and data acquisition technologies. Much of these data have special structures; images, functional data, marked point processes, and high-dimensional time series are all common.

In IC fabrication, several hundred chips are fabricated simultaneously on a *wafer,* and the wafers are themselves processed in groups called *lots.* Large manufacturing plants can start thousands of wafers each week. A wide range of measurements are made on each wafer, including particle data, in-line electrical measurements, and final probe test data even before the wafers are shipped to be packaged. The final probe test alone generates a vector of 15- to 20-dimensional measurements for each chip. In all, as much as 1.5 Gb of data per week are gathered in a typical fabrication. It is well known that problems in different manufacturing steps will leave telltale spatial signatures (which can vary within and across lots), so the wafer map data must be treated as inherently spatial objects.

In the automotive industry, reducing auto body "dimensional" variation is a major quality challenge. Auto body assembly involves several hundred parts and more than 100 assembly stations. With the implementation of in-line optical coordinate measurement machines (OCMM), tremendous amounts of dimensional data are now routinely collected. The OCMM measures 100–150 points on each major assembly with a 100% inspection rate. These data exhibit both spatial and temporal dependence. Both the volume and the complexity of the data dictate the need for fast and flexible methods of analysis, as well as appropriate environments for computing and visualization.

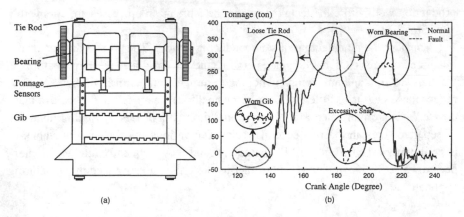

Figure 1. A Tonnage Signal Collected From a Sheet Metal Stamping Process. One complete forming cycle is displayed (b). The response is the total forming force measured by all the tonnage sensors mounted on the stamping press (a).

3. PROCESS MONITORING FOR DATA WITH COMPLEX STRUCTURE

Although there has been a tremendous amount of research in process monitoring, much of it has focused on new and more powerful tests for detecting changes in process means and variances. The real need in advanced manufacturing applications, however, is in dealing with processes where the observations have complex structure. Due to the lack of appropriate statistical methods and software tools for analyzing such data, practitioners typically force the problem into a more traditional framework, often resorting to simple overall summary measures that ignore the structure in the data. In so doing, valuable opportunities for process improvement are lost.

Figure 1 illustrates a tonnage signal collected from a sheet metal stamping process. The signal corresponds to a complete forming cycle and measures the total forming force from all of the tonnage sensors mounted on the stamping press. This is a typical example of the kind of *functional* data that are now being collected and used to monitor and diagnose problems in manufacturing processes. A traditional approach that treats the data as a vector of multidimensional observations and applies standard multivariate SPC techniques has been shown to be quite inadequate in this application (see Jin and Shi 1999, 2000). We return to this application later in the vignette.

Figure 2 demonstrates why spatial patterns are important in IC data. The rightmost wafer in this figure is a display of binary (pass/fail) probe test results collected at the end of the fabrication process. This map can be viewed as the superposition of the two wafer maps on the left: a cluster of defective chips representing a special or assignable cause, with a pattern that helps identify the responsible machine or area, and (essentially) random defects resulting from the overall cleanliness of the fabrication

Figure 2. A Graphical Model for Overall Wafer Yield. The defective chips in the middle wafer occur essentially at random, whereas those on the left are "clustered," reflecting a process problem. The final probe map is a superposition of these two processes.

line. A traditional SPC approach would simply summarize the observed test results with a single measure, wafer yield. Such an analysis clearly misses critical spatial information about yield loss. Hansen, Nair, and Friedman (1997) described methods for monitoring binary spatial processes to detect the presence of spatial clustering. Because the null hypothesis of complete spatial randomness is too simplistic, they use a Markov random field to characterize null situations with mild spatial clustering. This spatial process monitoring effectively complements the information from control charts that track only yield. However, as was shown by Hansen et al. (1999), this spatial monitoring tool is only the first step in fully exploiting the spatial character of these data.

4. BEYOND PROCESS MONITORING: THE REAL OPPORTUNITIES FOR STATISTICS

4.1 Intelligent Failure Diagnostics

The primary emphasis in SPC has been on monitoring and changepoint detection. The development of failure diagnostics and root cause determination have typically been considered to be the domain of experts with subject matter knowledge. In advanced manufacturing applications where the processes are not well understood, even the subject matter experts are increasingly relying on in-process and product quality data for diagnosing process problems. From a statistical viewpoint, there is more information in the data when a process goes out of control than when it is in control. This is where the real opportunities are and where we can make important contributions. Unfortunately, statisticians have been slow in recognizing these opportunities, and much of the exciting work is being done by nonstatisticians in the engineering community.

To provide a concrete example of the role that statistics can play in this area, consider again the IC application. The monitoring scheme for spatial data discussed earlier will trigger an alarm when there is significant spatial clustering as in Figure 2, but it will not provide any information about the *nature* of the clusters. As discussed by Hansen et al. (1999), spatial patterns provide tremendous information about potential process problems. For example, if the defects concentrate in the center of the wafer, then there is likely a problem in controlling the thickness of a chemical "resist" deposited on the surface of the wafer prior to lithography. Hansen et al. (1999) described statistical techniques for extracting *spatial signatures* of defect patterns and using them to immediately identify one or more likely root causes. These methods have been successfully deployed within Lucent Microelectronics. Since our original involvement in this area, many new techniques have appeared, and SEMATECH recently sponsored an entire conference on spatial statistics and pattern recognition in IC fabrication (see www.sematech.org/public/resources/stats/Symposium/1999/index.htm).

4.2 Combining Detection and Diagnosis

In high-volume manufacturing environments, there is a need to diagnose process problems as they occur in real time. The usual two-stage approach of detection followed by diagnosis, typically done off-line, is not adequate in this setting. Ideally, we should integrate statistical methods with underlying engineering knowledge about potential faults and failure diagnostics to develop a combined approach to detection and diagnosis. It is reasonable to view this as a classification problem in which the different classes represent different possible faults or states of the "system" (including the null state). An initial specification of these states can be obtained during the design and development stage, and the specifications can be constantly updated as on-line process and product quality data are collected and analyzed.

Jin and Shi (1999, 2000) provided a good example of this in the context of the tonnage signal data for stamping processes (see Figure 1). Their methodology segments the tonnage signal according to the different stages of the forming cycle and exploits information about the potential faults and how they will manifest themselves on the tonnage signal. For example, a flat peak is the result of a loose tie rod, whereas an oscillating peak indicates a worn bearing (see Figure 1). Jin and Shi (1999, 2000) described a wavelet-based statistical analysis for feature extraction and used these features to do process monitoring and fault diagnosis.

Ceglarek and Shi (1996) and Apley and Shi (1998) described another application from the automotive industry involving "fixture" failure diagnostics. Design and maintenance of the fixturing process is an important problem, as the "dimensional" variation in the auto body panels depends critically on the quality of the fixturing process. Ceglarek and Shi (1996) described how the engineering knowledge about the fixture geometry and tooling design layout can be used to develop a model that relates variations in the sensor measurements to different fixture faults. Apley and Shi (1998) and Dong (1999) showed that incorporating this information provides tremendous advantage over traditional SPC methods, not only in being able to detect the faults quickly, but also in being able to diagnose the problems in real time.

In ongoing research on IC applications, we are studying methods for developing a library of spatial templates that characterize different process problems and for classifying observed wafer maps with spatial clustering into one of the templates.

5. DATABASES, COMPUTING, AND VISUALIZATION

As mentioned earlier, advances in sensing and data acquisition technologies have made it possible to routinely collect massive amounts of data about manufacturing processes. Statisticians must be involved in all aspects of this data collection process, ensuring that the right kinds of data are being collected and stored, helping design appropriate measurement systems (choices of sensors, their location and number, etc.), assuring data quality, and so on. We have seen two trends in data acquisition and storage over the last few years.

First, the various streams of data currently pool into different databases. Process control engineers have one source for machine-level routing and maintenance information, whereas yield enhancement engineers pull postproduction tests results from another source. Recent years have brought a move to centralize data collection, management, and access, so that factory-wide information soon may be readily applied to process improvement efforts. Besides creating methods that make use of these new data sources (an incredible challenge in itself), statisticians have an important role to play in helping design and implement effective data warehousing solutions.

The second trend that we have observed relates to the *information content* being stored in factory databases. As both the complexity and volume of data increase, space and computing considerations dictate some form of reduction before storage in a database. At a practical level, this can represent a gain in efficiency, because the reduced or *compressed* data might be more readily amenable to SPC methods. (See, e.g., Jin and Shi 1999, 2000 for the use of statistical techniques for feature-based data compression in stamping processes.) In general, we expect that statisticians will be called on to design specialized compression techniques for storing only the most relevant information. This poses an incredible challenge in that unlike traditional notions of "sufficiency" for a parametric model, departures from standard operating conditions can be quite complex and often difficult to anticipate.

Finally, we comment on the role of statistical computing and graphics in advanced manufacturing. Naturally, the practical success of any industrial application depends largely on acceptance of new statistical methods by engineers in the factory. In our IC work, visualization was critical, as was the development of a convenient computing environment in which to express quantitative ideas about in-line and postproduction fabrication data. This led to the creation of a software platform, called S-wafers, that makes use of the object-based facilities of the S language. As a vehicle for technology transfer, S-wafers supports a range of tasks from generating automated reports (disseminated via the Lucent intranet in the form of HTML documents and Java applets) to interactive, specialized analyses on one or more lots. Engineers familiar with S can immediately augment their routine data analysis (based on a summary-like yield) with spatial information. We believe that our experience is not unique and that industrial statisticians who take the challenges of technology transfer seriously are regularly called on to make use of and to create new and novel tools for computing and visualization that can be transferred to the engineers responsible for manufacturing.

6. SUMMARY

We close by mentioning two other challenges that face industrial statisticians. First, a major consequence of the drive to reduce product development cycle times is that manufacturers are moving away from physical experimentation and testing to computer modeling and CAD/CAM tools. This presents a wide array of research opportunities for statisticians, ranging from model validation and verification to efficient design and analysis of very high-dimensional computer (or virtual) experiments.

Second, statisticians have too often worked in isolation and developed fragmented approaches that ignore important information and interactions present in sequential, multistage manufacturing processes. The complexities of advanced manufacturing environments dictate that we must work closely with engineers and take a systems approach to process improvement. Ceglarek, Shi, and Wu (1994), for example, have described a knowledge-based system for "dimensional" control that has been very effective in improving the quality of new auto body assembly processes. The methodology captures information about the multilevel assembly process and factory layout through a hierarchical structural model. Combining statistical analysis of the in-process data with information about the assembly architecture and sequence allows the root causes of process variability to be diagnosed quickly and efficiently.

As all of these examples suggest, statistics has an extremely important role to play in industry as we move into the data-rich 21st century. These are indeed exciting times for our profession, with a wide array of interesting research opportunities.

REFERENCES

Apley, D., and Shi, J. (1998), "Diagnosis of Multiple Fixture Faults in Panel Assembly," *ASME Transactions, Journal of Manufacturing Science and Engineering*, 120, 793–801.

Ceglarek, D., and Shi, J. (1996), "Fixture Failure Diagnostics for Autobody Assembly Using Pattern Recognition," *Journal of Engineering for Industry*, 118, 55–66.

Ceglarek, D., Shi, J., and Wu, S. M. (1994), "A Knowledge-Based Diagnostic Approach for the Launch of the Auto-Body Assembly Process," *ASME Transactions, Journal of Engineering for Industry*, 116, 491–499.

Dong, X. (1999), "Contributions to Statistical Process Control and Robust Parameter Design," unpublished doctoral dissertation, University of Michigan.

Hansen, M. H., Nair, V. N., and Friedman, D. J. (1997), "Monitoring Wafer Map Data in Integrated Circuit Fabrication for Spatially Clustered Defects," *Technometrics*, 39, 241–253.

——— (1999), "Process Improvement Through the Analysis of Spatially Clustered Defects on Wafer Maps," Technical Report No. 333, Department of Statistics, University of Michigan.

Jin, J., and Shi, J. (1999), "Feature-Preserving Data Compression of Stamping Tonnage Information Using Wavelets," *Technometrics*, 41, 327–339.

——— (2000), "Diagnostic Feature Extraction From a Stamping Tonnage Signal Based on Design of Experiments," *ASME Transactions, Journal of Manufacturing Science and Engineering*, 122, to appear.

Chapter 4

Theory and Methods of Statistics

George Casella

The vignette series concludes with 22 contributions in Theory and Methods. It is, of course, impossible to cover all of theory and methods with so few articles, but we hope that a snapshot of what was, and what may be, is achieved. This is the essence of "vignettes" which, according to the *American Heritage Dictionary,* are "literary sketches having the intimate charm and subtlety attributed to vignette portraits."

Through solicitation and announcement, we have collected the Theory and Methods vignettes presented here. The scope is broad, but by no means exhaustive. Exclusion of a topic does not bear on its importance, but rather on the inability to secure a vignette. Many topics were solicited, and a general call was put in *Amstat News.* Nonetheless, we could not get every topic covered (among other factors, time was very tight).

There is some overlap in the vignettes, as the authors, although aware of the other topics, were working independently, and there are places where information in one vignette complements that in another. Rather than edit out some of the overlap, I have tried to signpost these instances with cross-references allowing the reader the luxury of seeing two (or even three) views on a topic. Such diverse accounts can help to enhance our understanding.

As I hope you will agree, the resulting collection is nothing short of marvelous. The writers are all experts in their fields, and bring a perception and view that truly highlights each subject area. My goal in this introduction is not to summarize what is contained in the following pages, but rather to entice you to spend some time looking through the vignettes. At the very least, you will find some wonderful stories about the history and development of our subject. (For example, see the vignettes by McCulloch and Meng for different histories of the EM algorithm.) Some of the speculation may

George Casella is Arun Varma Commemorative Term Professor and Chair, Department of Statistics, University of Florida, Griffin-Floyd Hall, Gainesville, FL 32611 (Email: casella@stat.ufl.edu). This work was supported by National Science Foundation grant DMS-9971586.

even inspire you to try your hand, either in developing the theory or applying the methodology.

The question of in which order to present the vignettes was one that I thought hard about. First, I tried to put them in a subject-oriented order, to create some sort of smooth flow throughout. This turned out to be impossible, as the connections between topics are not linear. Moreover, any absolute ordering could carry a connotation of importance of the topics, a judgment that I don't feel qualified to make. (Indeed, such a judgment may be impossible to make.) So in the end I settled for an alphabetical ordering according to author name. This is not only objective, but also makes the various vignettes a bit easier to find.

Bayesian Analysis: A Look at Today and Thoughts of Tomorrow

James O. Berger

1. INTRODUCTION

Life was simple when I became a Bayesian in the 1970s; it was possible to track virtually all Bayesian activity. Preparing this paper on Bayesian statistics was humbling, as I realized that I have lately been aware of only about 10% of the ongoing activity in Bayesian analysis. One goal of this article is thus to provide an overview of, and access to, a significant portion of this current activity. Necessarily, the overview will be extremely brief; indeed, an entire area of Bayesian activity might only be mentioned in one sentence and with a single reference. Moreover, many areas of activity are ignored altogether, either due to ignorance on my part or because no single reference provides access to the literature.

A second goal is to highlight issues or controversies that may shape the way that Bayesian analysis develops. This material is somewhat self-indulgent and should not be taken too seriously; for instance, if I had been asked to write such an article 10 years ago, I would have missed the mark by not anticipating the extensive development of Markov chain Monte Carlo (MCMC) and its enormous impact on Bayesian statistics.

Section 2 provides a brief snapshot of the existing Bayesian activity and emphasizes its dramatic growth in the 1990s, both inside and outside statistics. I found myself simultaneously rejoicing and being disturbed at the level of Bayesian activity. As a Bayesian, I rejoiced to see the extensive utilization of the paradigm, especially among nonstatisticians. As a statistician, I worried that our profession may not be adapting fast enough to this dramatic change; we may be in danger of "losing" Bayesian analysis to other disciplines (as we have "lost" other areas of statistics). In this regard, it is astonishing that most statistics and biostatistics departments in the United States do not even regularly offer a single Bayesian statistics course.

Section 3 is organized by approaches to Bayesian analysis—in particular the objective, subjective, robust, frequentist-Bayes, and what I term quasi-Bayes approaches. This section contains most of my musings about the current and future state of Bayesian statistics. Section 4 briefly discusses the critical issues of computation and software.

James O. Berger is Arts and Sciences Professor of Statistics, Institute of Statistics and Decision Sciences, Duke University, Durham, NC 27708 (E-mail: berger@stat.duke.edu). Preparation was supported by National Science Foundation grant DMS-9802261. The author is grateful to George Casella, Dalene Stangl, and Michael Lavine for helpful suggestions.

2. BAYESIAN ACTIVITY

2.1 Numbers and Organizations

The dramatically increasing level of Bayesian activity can be seen in part through the raw numbers. Harry Martz (personal communication) studied the SciSearch database at Los Alamos National Laboratories to determine the increase in frequency of articles involving Bayesian analysis over the last 25 years. From 1974 to 1994, the trend was linear, with roughly a doubling of articles every 10 years. In the last 5 years, however, there has been a very dramatic upswing in both the number and the rate of increase of Bayesian articles.

This same phenomenon is also visible by looking at the number of books written on Bayesian analysis. During the first 200 years of Bayesian analysis (1769–1969), there were perhaps 15 books written on Bayesian statistics. Over the next 20 years (1970–1989), a guess as to the number of Bayesian books produced is 30. Over the last 10 years (1990–1999), roughly 60 Bayesian books have been written, not counting the many dozens of Bayesian conference proceedings and collections of papers. Bayesian books in particular subject areas are listed in Sections 2.2 and 2.3. A selection of general Bayesian books is given in Appendix A.

Another aspect of Bayesian activity is the diversity of existing organizations that are significantly Bayesian in nature, including the following (those with an active website): International Society of Bayesian Analysis (http://www.bayesian.org), ASA Section on Bayesian Statistical Science (http://www.stat.duke.edu/sbss/sbss.html), Decision Analysis Society of INFORMS (http://www.informs.org/society/da), and ASA Section on Risk Analysis (http://www.isds.duke.edu/riskanalysis/ras.html).

In addition to the activities and meetings of these societies, the following are long-standing series of prominent Bayesian meetings that are not organized explicitly by societies: Valencia Meetings on Bayesian Statistics (http://www.uv.es/~bernardo/valenciam.html), Conferences on Maximum Entropy and Bayesian Methods (http://omega.albany.edu:8008/maxent.html), CMU Workshops on Bayesian Case Studies (http://lib.stat.cmu.edu/bayesworkshop/), and RSS Conferences on Practical Bayesian Statistics. The average number of Bayesian meetings per year is now well over 10, with at least an equal number of meetings being held that have a strong Bayesian component.

2.2 Interdisciplinary Activities and Applications

Applications of Bayesian analysis in industry and government are rapidly increasing but hard to document, as they are often "in-house" developments. It is far easier to document the extensive Bayesian activity in other disciplines; indeed, in many fields of the sciences and engineering, there are now active groups of Bayesian researchers. Here we can do little more than list various fields that have seen a considerable amount of Bayesian activity, and present a few references to access the corresponding literature. Most of the listed references are books on Bayesian statistics in the given field, emphasizing that the activity in the field has reached the level wherein books are being written. Indeed, this was the criterion for listing an area, although fields in which there is a commensurate amount of activity, but no book, are also listed. (It would be hard to find an area of human investigation in which there does not exist some level of Bayesian work, so many fields of application are omitted.)

For **archaeology,** see Buck, Cavanaugh, and Litton (1996); **atmospheric sciences,** see Berliner, Royle, Wikle, and Milliff (1999); **economics and econometrics,** see Cyert and DeGroot (1987), Poirier (1995), Perlman and Blaug (1997), Kim, Shephard and Chib (1998), and Geweke (1999); **education,** see Johnson (1997); **epidemiology,** see Greenland (1998); **engineering,** see Godsill and Rayner (1998); **genetics,** see Iverson, Parmigiani, and Berry (1998), Dawid and Pueschel (1999), and Liu, Neuwald, and Lawrence (1999); **hydrology,** see Parent, Hubert, Bobée, and Miquel (1998); **law,** see DeGroot, Fienberg, and Kadane (1986) and Kadane and Schuan (1996); **measurement and assay,** see Brown (1993) and http://www.pnl.gov/bayesian/; **medicine,** see Berry and Stangl (1996) and Stangl and Berry (1998); **physical sciences,** see Bretthorst (1988), Jaynes (1999), and http://www.astro.cornell.edu/staff/loredo/bayes/; **quality management,** see Moreno and Rios-Insua (1999); **social sciences,** see Pollard (1986) and Johnson and Albert (1999).

2.3 Areas of Bayesian Statistics

Here, Bayesian activity is listed by statistical area. Again, the criterion for inclusion of an area is primarily the amount of Bayesian work being done in that area, as evidenced by books being written (or a corresponding level of papers).

For **biostatistics,** see Berry and Stangl (1996), Carlin and Louis (1996), and Kadane (1996); **causality,** see Spirtes, Glymour, and Scheines (1993) and Glymour and Cooper (1999); **classification, discrimination, neural nets,** and so on, see Neal (1996, 1999), Müller and Rios-Insua (1998), and the vignette by George; **contingency tables,** see the vignette by Fienberg; **decision analysis and decision theory,** see Smith (1988), Robert (1994), Clemen (1996), and the vignette by Brown; **design,** see Pilz (1991), Chaloner and Verdinelli (1995), and Müller (1999); **empirical Bayes,** see Carlin and Louis (1996) and the vignette by Carlin and Louis; **exchangeability and other foundations,** see Good (1983), Regazzini (1999), Kadane, Schervish, and Seidenfeld (1999), and the vignette by Robins and Wasserman; **finite-population sampling,** see Bolfarine and Zacks (1992) and Mukhopadhyay (1998); **generalized linear models,** see Dey, Ghosh, and Mallick (2000); **graphical models and Bayesian networks,** see Pearl (1988), Jensen (1996), Lauritzen (1996), Jordan (1998), and Cowell, Dawid, Lauritzen, and Spiegelhalter (1999); **hierarchical (multilevel) modeling,** see the vignette by Hobert; **image processing,** see Fitzgerald, Godsill, Kokaram, and Stark (1999); **information,** see Barron, Rissanen, and Yu (1998) and the vignette by Soofi; **missing data,** see Rubin (1987) and the vignette by Meng; **nonparametrics and function estimation,** see Dey, Müller, and Sinha (1998), Müller and Vidakovic (1999), and the vignette by Robins and Wasserman; **ordinal data,** see Johnson and Albert (1999); **predictive inference and model averaging,** see Aitchison and Dunsmore (1975), Leamer (1978), Geisser (1993), Draper (1995), Clyde (1999), and the BMA website under "software"; **reliability and survival analysis,** see Clarotti, Barlow, and Spizzichino (1993) and Sinha and Dey (1999); **sequential analysis,** see Carlin, Kadane, and Gelfand (1998) and Qian and Brown (1999); **signal processing,** see Ó Ruanaidh and Fitzgerald (1996) and Fitzgerald, Godsill, Kokaram, and Stark (1999); **spatial statistics,** see Wolpert and Ickstadt (1998) and Besag and Higdon (1999); **testing, model selection, and variable selection,** see Kass and Raftery (1995), O'Hagan (1995), Berger and Pericchi (1996), Berger (1998), Racugno (1998), Sellke, Bayarri, and Berger (1999), Thiesson, Meek, Chickering, and Heckerman (1999), and the vignette by George; **time series,** see Pole, West, and Harrison (1995), Kitagawa and Gersch (1996) and West and Harrison (1997).

3. APPROACHES TO BAYESIAN ANALYSIS

This section presents a rather personal view of the status and future of five approaches to Bayesian analysis, termed the objective, subjective, robust, frequentist-Bayes, and quasi-Bayes approaches. This is neither a complete list of the approaches to Bayesian analysis nor a broad discussion of the considered approaches. The section's main purpose is to emphasize the variety of different and viable Bayesian approaches to statistics, each of which can be of great value in certain situations and for certain users. We should be aware of the strengths and weaknesses of each approach, as all will be with us in the future and should be respected as part of the Bayesian paradigm.

3.1 Objective Bayesian Analysis

It is a common perception that Bayesian analysis is primarily a subjective theory. This is true neither historically nor in practice. The first Bayesians, Thomas Bayes (see Bayes 1783) and Laplace (see Laplace 1812), performed Bayesian analysis using a constant prior distribution for unknown parameters. Indeed, this approach to statistics, then called "inverse probability" (see Dale 1991) was very prominent for most of the 19th century and was highly influential in the early part of this century. Criticisms of the use of a constant prior distribution caused Jeffreys to introduce significant refinements of this theory (see Jeffreys 1961). Most of the applied Bayesian analyses I see today follow the Laplace–Jeffreys objective school of Bayesian analysis, possibly with additional modern refinements. (Of course, others may see subjective Bayesian applications more often, depending on the area in which they work.)

Many Bayesians object to the label "objective Bayes," claiming that it is misleading to say that any statistical analysis can be truly objective. Though agreeing with this at a philosophical level (Berger and Berry 1988), I feel that there are a host of practical and sociological reasons to use the label; statisticians must get over their aversion to calling good things by attractive names.

The most familiar element of the objective Bayesian school is the use of *noninformative* or *default* prior distributions. The most famous of these is the *Jeffreys prior* (see Jeffreys 1961). *Maximum entropy* priors are another well-known type of noninformative prior (although they often also reflect certain informative features of the system being analyzed). The more recent statistical literature emphasizes what are called *reference priors* (Bernardo 1979; Yang and Berger 1997), which prove remarkably successful from both Bayesian and non-Bayesian perspectives. Kass and Wasserman (1996) provided a recent review of methods for selecting noninformative priors.

A quite different area of the objective Bayesian school is that concerned with techniques for default model selection and hypothesis testing. Successful developments in this direction are much more recent (Berger and Pericchi 1996; Kass and Raftery 1995; O'Hagan 1995; Sellke, Bayarri, and Berger 1999). Indeed, there is still considerable ongoing discussion as to which default methods are to be preferred for these problems (see Racugno 1998).

The main concern with objective Bayesian procedures is that they often utilize improper prior distributions, and so do not automatically have desirable Bayesian properties, such as coherency. Also, a poor choice of improper priors can even lead to improper posteriors. Thus proposed objective Bayesian procedures are typically studied to ensure that such problems do not arise. Also, objective Bayesian procedures are often evaluated from non-Bayesian perspectives, and usually turn out to be stunningly effective from these perspectives.

3.2 Subjective Bayesian Analysis

Although comparatively new on the Bayesian scene, subjective Bayesian analysis is currently viewed by many Bayesians to be the "soul" of Bayesian statistics. Its philosophical appeal is undeniable, and few statisticians would argue against its use when the needed inputs (models and subjective prior distributions) can be fully and accurately specified. The difficulty in such specification (Kahneman, Slovic, and Tversky 1986) often limits application of the approach, but there has been a considerable research effort to further develop elicitation techniques for subjective Bayesian analysis (Lad, 1996; French and Smith 1997; *The Statistician,* 47, 1998).

In many problems, use of subjective prior information is clearly essential, and in others it is readily available; use of subjective Bayesian analysis for such problems can provide dramatic gains. Even when a complete subjective analysis is not feasible, judicious use of partly subjective and partly objective prior distributions is often attractive (Andrews, Berger, and Smith 1993).

3.3   Robust Bayesian Analysis

Robust Bayesian analysis recognizes the impossibility of complete subjective specification of the model and prior distribution; after all, complete specification would involve an infinite number of assessments, even in the simplest situations. The idea is thus to work with classes of models and classes of prior distributions, with the classes reflecting the uncertainty remaining after the (finite) elicitation efforts. (Classes could also reflect the differing judgments of various individuals involved in the decision process.)

The foundational arguments for robust Bayesian analysis are compelling (Kadane 1984; Walley 1991), and there is an extensive literature on the development of robust Bayesian methodology, including Berger (1985, 1994), Berger et al. (1996), and Rios-Insua (1990). Routine practical implementation of robust Bayesian analysis will require development of appropriate software, however.

Robust Bayesian analysis is also an attractive technology for actually implementing a general subjective Bayesian elicitation program. Resources (time and money) for subjective elicitation typically are very limited in practice, and need to be optimally utilized. Robust Bayesian analysis can, in principle, be used to direct the elicitation effort, by first assessing if the current information (elicitations and data) is sufficient for solving the problem and then, if not, determining which additional elicitations would be most valuable (Liseo, Petrella, and Salinetti 1996).

3.4   Frequentist-Bayes Analysis

It is hard to imagine that the current situation, with several competing foundations for statistics, will exist indefinitely. Assuming that a unified foundation is inevitable, what will it be? Today, an increasing number of statisticians envisage that this unified foundation will be a mix of Bayesian and frequentist ideas (with elements of the current likelihood theory thrown in; see the vignette by Reid). Here is my view of what this mixture will be.

First, the language of statistics will be Bayesian. Statistics is about measuring uncertainty, and over 50 years of efforts to prove otherwise have convincingly demonstrated that the only coherent language in which to discuss uncertainty is the Bayesian language. In addition, the Bayesian language is an order of magnitude easier to understand than the classical language (witness the p value controversy; Sellke et al. 1999), so that a switch to the Bayesian language should considerably increase the attractiveness of statistics. Note that, as discussed earlier, this is not about subjectivity or objectivity; the Bayesian language can be used for either subjective or objective statistical analysis.

On the other hand, from a methodological perspective, it is becoming clear that both Bayesian and frequentist methodology is going to be important. For parametric problems, Bayesian analysis seems to have a clear methodological edge, but frequentist concepts can be very useful, especially in determining good objective Bayesian procedures (see, e.g., the vignette by Reid).

In nonparametric analysis, it has long been known (Diaconis and Freedman 1986) that Bayesian procedures can behave poorly from a frequentist perspective. Although poor frequentist performance is not necessarily damning to a Bayesian, it typically should be viewed as a warning sign that something is amiss, especially when the prior distribution used contains more "hidden" information than elicited information (as is virtually always the case with nonparametric priors).

Furthermore, there are an increasing number of examples in which frequentist arguments yield satisfactory answers quite directly, whereas Bayesian analysis requires a formidable amount of extra work. (The simplest such example is MCMC itself, in which one evaluates an integral by a sample average, and not by a formal Bayesian estimate; see the vignette by Robins and Wasserman for other examples). In such cases, I believe that the frequentist answer can be accepted by Bayesians as an approximate Bayesian answer, although it is not clear in general how this can be formally verified.

This discussion of unification has been primarily from a Bayesian perspective. From a frequentist perspective, unification also seems inevitable. It has long been known that "optimal" unconditional frequentist procedures must be Bayesian (Berger 1985), and there is growing evidence that this must be so even from a conditional frequentist perspective (Berger, Boukai, and Wang 1997).

Note that I am *not* arguing for an eclectic attitude toward statistics here; indeed, I think the general refusal in our field to strive for a unified perspective has been the single biggest impediment to its advancement. I am simply saying that any unification that will be achieved will almost necessarily have frequentist components to it.

3.5   Quasi-Bayesian Analysis

There is another type of Bayesian analysis that one increasingly sees being performed, and that can be unsettling to "pure" Bayesians and many non-Bayesians. In this type of analysis, priors are chosen in various ad hoc fashions, including choosing vague proper priors, choosing priors to "span" the range of the likelihood, and choosing priors with tuning parameters that are adjusted until the answer "looks nice." I call such analyses *quasi-Bayes* because, although they utilize Bayesian machinery, they do not carry the guarantees of good performance that come with either true subjective analysis or (well-studied) objective Bayesian analysis. It is useful to briefly discuss the possible problems with each of these quasi-Bayes procedures.

Using vague proper priors will work well when the vague proper prior is a good approximation to a good objective prior, but this can fail to be the case. For instance, in normal hierarchical models with a "higher-level" variance V, it is quite common to use the vague proper prior density $\pi(V) \propto V^{-(\varepsilon+1)} \exp(-\varepsilon'/V)$, with ε and ε' small. However, as $\varepsilon \to 0$, it is typically the case in these models that the posterior distribution for V will pile up its mass near 0, so that the answer can be ridiculous if ε is too small. An objective Bayesian who incorrectly used the related prior $\pi(V) \propto V^{-1}$ would typically become aware of the problem, because the posterior would not converge (as it will with the vague proper prior). The common perception that using a vague proper prior is safer than using improper priors, or conveys some type of guarantee of good performance, is simply wrong.

The second common quasi-Bayes procedure is to choose priors that span the range of the likelihood function. For instance, one might choose a uniform prior over a range that includes most of the "mass" of the likelihood function but that does not extend too far (thus hopefully avoiding the problem of using a "too vague" proper prior). Another version of this procedure is to use conjugate priors, with parameters chosen so that the prior is considerably more spread out than the likelihood function but is roughly centered in the same region. The two obvious concerns with these strategies are that (a) the answer can still be quite sensitive to the spread of the rather arbitrarily chosen prior, and (b) centering the prior on the likelihood is a problematical double use of the data. Also, in problems with complicated likelihoods, it can be difficult to implement this strategy successfully.

The third common quasi-Bayes procedure is to write down proper (often conjugate) priors with unspecified parameters, and then treat these parameters as "tuning" parameters to be adjusted until the answer "looks nice." Unfortunately, one is sometimes not told that this has been done; that is, the choice of the parameters is, after the fact, presented as "natural."

These issues are complicated by the fact that in the hands of an expert Bayesian analyst, the quasi-Bayes procedures mentioned here can be quite reasonable, in that the expert may have the experience and skill to tell when the procedures are likely to be successful. Also, one must always consider the question: What is the alternative? I have seen many examples in which an answer was required and in which I would trust the quasi-Bayes answer more than the answer from any feasible alternative analysis.

Finally, it is important to recognize that the genie cannot be put back into the bottle. The Bayesian "machine," together with MCMC, is arguably the most powerful mechanism ever created for processing data and knowledge. The quasi-Bayes approach can rather easily create procedures of astonishing flexibility for data analysis, and its use to create such procedures should not be discouraged. However, it must be recognized that these procedures do not necessarily have intrinsic Bayesian justifications, and so must be justified on extrinsic grounds (e.g., through extensive sensitivity studies, simulations, etc.).

4. COMPUTATION AND SOFTWARE

4.1 Computational Techniques

Even 20 years ago, one often heard the refrain that "Bayesian analysis is nice conceptually; too bad it is not possible to compute Bayesian answers in realistic situations." Today, truly complex models often can only be computationally handled by Bayesian techniques. This has attracted many newcomers to the Bayesian approach and has had the interesting effect of considerably reducing discussion of "philosophical" arguments for and against the Bayesian position.

Although other goals are possible, most Bayesian computation is focused on calculation of posterior expectations, which are typically integrals of one to thousands of dimensions. Another common type of Bayesian computation is calculation of the posterior mode (as in computating MAP estimates in image processing).

The traditional numerical methods for computing posterior expectations are numerical integration, Laplace approximation, and Monte Carlo importance sampling. Numerical integration can be effective in moderate (say, up to 10) dimensional problems. Modern developments in this direction were discussed by Monahan and Genz (1996). Laplace and other saddlepoint approximations are discussed in the vignette by R. Strawderman. Until recently, Monte Carlo importance sampling was the most commonly used traditional method of computing posterior expectations. The method can work in very large dimensions and has the nice feature of producing reliable measures of the accuracy of the computation.

Today, MCMC has become the most popular method of Bayesian computation, in part because of its power in handling very complex situations and in part because it is comparatively easy to program. Because the Gibbs sampling vignette by Gelfand and the MCMC vignette by Cappé and Robert both address this computational technique, I do not discuss it here. Recent books in the area include those of Chen, Shao, and Ibrahim (2000), Gamerman (1997), Robert and Casella (1999), and Tanner (1993). It is not strictly the case that MCMC is replacing the more traditional methods listed above. For instance, in some problems importance sampling will probably always remain the computational method of choice, as will standard numerical integration in low dimensions (especially when extreme accuracy is needed).

Availability of general user-friendly Bayesian software is clearly needed to advance the use of Bayesian methods. A number of software packages exist, and these are very useful for particular scenarios. Lists and descriptions of pre-1990 Bayesian software were provided by Goel (1988) and Press (1989). A list of some of the Bayesian software developed since 1990 is given in Appendix B.

It would, of course, be wonderful to have a single general-purpose Bayesian software package, but three of the major strengths of the modern Bayesian approach create difficulties in developing generic software. One difficulty is the extreme flexibility of Bayesian analysis, with virtually any constructed model being amenable to analysis. Most classical packages need to contend with only a relatively few well-defined models or scenarios for which a classical procedure has been determined. Another strength of Bayesian analysis is the possibility of extensive utilization of subjective prior information, and many Bayesians tend to feel that software should include an elaborate expert system for prior elicitation. Finally, implementing the modern computational techniques in a software package is extremely challenging, because it is difficult to codify the "art" of finding a successful computational strategy in a complex situation.

Note that development of software implementing the objective Bayesian approach for "standard" statistical models can avoid these difficulties. There would be no need for a subjective elicitation interface, and the package could incorporate specific computational techniques suited to the various standard models being considered. Because the vast majority of statistical analyses done today use such "automatic" software, having a Bayesian version would greatly impact the actual use of Bayesian methodology. Its creation should thus be a high priority for the profession.

APPENDIX A: GENERAL BAYESIAN REFERENCES

- Historical and general monographs: Laplace (1812), Jeffreys (1961), Zellner (1971), Savage (1972), Lindley (1972), Box and Tiao (1973), de Finetti (1974, 1975), Hartigan (1983), Florens, Mouchart, and Roulin (1990)
- Graduate-level texts: DeGroot (1970), Berger (1985), Press (1989), Bernardo and Smith (1994), O'Hagan (1994), Robert (1994), Gelman, Carlin, Stern, and Rubin (1995), Poirier (1995), Schervish (1995), Piccinato (1996)
- Elementary texts: Winkler (1972), O'Hagan (1988), Albert (1996), Berry (1996), Sivia (1996), Antleman (1997), Lee (1997)
- General proceedings volumes: The International Valencia Conferences produce highly acclaimed proceedings, the last of which was edited by Bernardo et al. (1999). The Maximum Entropy and Bayesian Analysis conferences also have excellent proceedings volumes, the last of which was edited by Erickson, Rychert, and Smith (1998). The CMU Bayesian Case Studies Workshops produce unique volumes of in-depth case studies in Bayesian analysis, the last volume being edited by Gatsonis et al. (1998). The Bayesian Statistical Science Section of the ASA has an annual JSM proceedings volume, produced by the ASA.

APPENDIX B: AVAILABLE BAYESIAN SOFTWARE

- AutoClass, a Bayesian classification system (http://ic-www.arc.nasa.gov/ic/ projects/bayes-group/group/autoclass/)

- BATS, designed for Bayesian time series analysis (http://www.stat.duke.edu/~mw/bats.html)
- BAYDA, a Bayesian system for classification and discriminant analysis (http://www.cs.Helsinki.fi/research/cosco/Projects/NONE/SW/)
- BAYESPACK, etc., numerical integration algorithms (http://www.math.wsu.edu/math/faculty/genz/homepage)
- Bayesian biopolymer sequencing software (http://www-stat.stanford.edu/~jliu/)
- B/D, a linear subjective Bayesian system (http://fourier.dur.ac.uk:8000/stats/bd/)
- BMA, software for Bayesian model averaging for predictive and other purposes (http://www.research.att.com/~volinsky/bma.html)
- Bayesian regression and classification software based on neural networks, Gaussian processes, and Bayesian mixture models (http://www.cs.utoronto.ca/~radford/fbm.software.html)
- Belief networks software (http://bayes.stat.washington.edu/almond/belief.html)
- BRCAPRO, which implements a Bayesian analysis for genetic counseling of women at high risk for hereditary breast and ovarian cancer (http://www.stat.duke.edu/~gp/brcapro.html)
- BUGS, designed to analyze general hierarchical models via MCMC (http://www.mrc-bsu.cam.ac.uk/bugs/)
- First Bayes, a Bayesian teaching package (http://www.shef.ac.uk/~st1ao/1b.html)
- Matlab and Minitab Bayesian computational algorithms for introductory Bayes and ordinal data (http://www-math.bgsu.edu/~albert/)
- Nuclear magnetic resonance Bayesian software; this is the manual (http://www.bayes.wustl.edu/glb/manual.pdf)
- StatLib, a repository for statistics software, much of it Bayesian (http://lib.stat.cmu.edu/)
- Time series software for nonstationary time series and analysis with autoregressive component models (http://www.stat.duke.edu/~mw/books_software_data.html)
- LISP-STAT, an object-oriented environment for statistical computing and dynamic graphics with various Bayesian capabilities (Tierney 1991)

REFERENCES

Aitchison, J., and Dunsmore, I. R. (1975), *Statistical Prediction Analysis*, New York: Wiley.

Albert, J. H. (1996), *Bayesian Computation Using Minitab*, Belmont, CA: Wadsworth.

Andrews, R., Berger, J., and Smith, M. (1993), "Bayesian Estimation of Fuel Economy Potential due to Technology Improvements," in *Case Studies in Bayesian Statistics*, eds. C. Gatsonis et al., New York:

Springer-Verlag, pp. 1–77.

Antleman, G. (1997), *Elementary Bayesian Statistics*, Hong Kong: Edward Elgar.

Barron, A., Rissanen, J., and Yu, B. (1998), "The Minimum Description Length Principle in Coding and Modeling," *IEEE Transactions on Information Theory*, 44, 2743–2760.

Bayes, T. (1783), "An Essay Towards Solving a Problem in the Doctrine of Chances," *Philosophical Transactions of the Royal Society*, 53, 370–418.

Berger, J. (1985), *Statistical Decision Theory and Bayesian Analysis* (2nd ed.), New York: Springer-Verlag.

—— (1994), "An Overview of Robust Bayesian Analysis," *Test*, 3, 5–124.

—— (1998), "Bayes Factors," in *Encyclopedia of Statistical Sciences*, Volume 3 (update), eds. S. Kotz et al., New York: Wiley.

Berger, J., and Berry, D. (1988), "Analyzing Data: Is Objectivity Possible?," *The American Scientist*, 76, 159–165.

Berger, J., Betró, B., Moreno, E., Pericchi, L. R., Ruggeri, F., Salinetti, G., and Wasserman, L. (Eds.) (1996), *Bayesian Robustness*, Hayward, CA: Institute of Mathematical Statistics.

Berger, J., Boukai, B., and Wang, W. (1997), "Unified Frequentist and Bayesian Testing of Precise Hypotheses," *Statistical Science*, 12, 133–160.

Berger, J., and Pericchi, L. R. (1996), "The Intrinsic Bayes Factor for Model Selection and Prediction," *Journal of the American Statistical Association*, 91, 109–122.

Berliner, L. M., Royle, J. A., Wikle, C. K., and Milliff, R. F. (1999), "Bayesian Methods in the Atmospheric Sciences," in *Bayesian Statistics 6*, eds. J. M. Bernardo, J. O. Berger, A. P. Dawid, and A. F. M. Smith, London: Oxford University Press, pp. 83–100.

Bernardo, J. M. (1979), "Reference Posterior Distributions for Bayesian Inference" (with discussion), *Journal of the Royal Statistical Society*, Ser. B, 41, 113–147.

Bernardo, J. M., Berger, J. O., Dawid, A. P., and Smith, A. F. M. (Eds.) (1999), *Bayesian Statistics 6*, London: Oxford University Press.

Bernardo, J. M., and Smith, A. F. M. (1994), *Bayesian Theory*, New York: Wiley.

Berry, D. A. (1996), *Statistics: a Bayesian Perspective*, Belmont, CA: Wadsworth.

Berry, D. A., and Stangl, D. K. (Eds.) (1996), *Bayesian Biostatistics*, New York: Marcel Dekker.

—— (2000), *Meta-Analysis in Medicine and Health Policy*, New York: Marcel Dekker.

Besag, J., and Higdon, D. (1999), "Bayesian Inference for Agricultural Field Experiments" (with discussion), *Journal of the Royal Statistical Society*, Ser. B, 61, 691–746.

Bolfarine, H., and Zacks, S. (1992), *Prediction Theory for Finite Populations*, New York: Springer-Verlag.

Box, G., and Tiao, G. (1973), *Bayesian Inference in Statistical Analysis*, Reading, MA: Addison-Wesley.

Bretthorst, G. L. (1988), *Bayesian Spectrum Analysis and Parameter Estimation*, New York: Springer-Verlag.

Broemeling, L. D. (1985), *Bayesian Analysis of Linear Models*, New York: Marcel Dekker.

Brown, P. J. (1993), *Measurement, Regression, and Calibration*, Oxford, U.K.: Clarendon Press.

Buck, C. E., Cavanaugh, W. G., and Litton, C. D. (1996), *The Bayesian Approach to Interpreting Archaeological Data*, New York: Wiley.

Carlin, B. P., and Louis, T. A. (1996), *Bayes and Empirical Bayes Methods for Data Analysis*, London: Chapman and Hall.

Carlin, B., Kadane, J., and Gelfand, A. (1998), "Approaches for Optimal Sequential Decision Analysis in Clinical Trials," *Biometrics*, 54, 964–975.

Chaloner, K., and Verdinelli, I. (1995), "Bayesian Experimental Design: A Review," *Statistical Science*, 10, 273–304.

Chen, M. H., Shao, Q. M., and Ibrahim, J. G. (2000), *Monte Carlo Methods in Bayesian Computation*, New York: Springer-Verlag.

Clarotti, C. A., Barlow, R. E., and Spizzichino, F. (Eds.) (1993), *Reliability and Decision Making*, Amsterdam: Elsevier Science.

Clemen, R. T. (1996), *Making Hard Decisions: An Introduction to Decision Analysis* (2nd ed.), Belmont, CA: Duxbury.

Clyde, M. A. (1999), "Bayesian Model Averaging and Model Search Strategies," in *Bayesian Statistics 6*, eds. J. M. Bernardo, J. O. Berger, A. P. Dawid, and A. F. M. Smith, London: Oxford University Press, pp. 23–42.

Cowell, R. G., Dawid, A. P., Lauritzen, S. L., and Spiegelhalter, D. J. (1999), *Probabilistic Networks and Expert Systems*, New York: Springer.

Cyert, R. M., and DeGroot, M. H. (1987), *Bayesian Analysis in Economic Theory*, Totona, NJ: Rowman Littlefield.

Dale, A. I. (1991), *A History of Inverse Probability*, New York: Springer-Verlag.

Dawid, A. P., and Pueschel, J. (1999), "Hierarchical Models for DNA Profiling Using Heterogeneous Databases," in *Bayesian Statistics 6*, eds. J. M. Bernardo, J. O. Berger, A. P. Dawid, and A. F. M. Smith, London: Oxford University Press, pp. 187–212.

de Finetti, B. (1974, 1975), *Theory of Probability*, Vols. 1 and 2, New York: Wiley.

DeGroot, M. H. (1970), *Optimal Statistical Decisions*, New York: McGraw-Hill.

DeGroot, M. H., Fienberg, S. E., and Kadane, J. B. (1986), *Statistics and the Law*, New York: Wiley.

Dey, D., Ghosh, S., and Mallick, B. K. (Eds.) (2000), *Bayesian Generalized Linear Models*, New York: Marcel Dekker.

Dey, D., Müller, P., and Sinha, D. (Eds.) (1998), *Practical Nonparametric and Semiparametric Bayesian Statistics*, New York: Springer-Verlag.

Diaconis, P., and Freedman, D. (1986), "On the Consistency of Bayes Estimates," *The Annals of Statistics*, 14, 1–67.

Draper, D. (1995), "Assessment and Propagation of Model Uncertainty," *Journal of the Royal Statistical Society*, Ser. B, 57, 45–98.

Erickson, G., Rychert, J., and Smith, C. R. (1998), *Maximum Entropy and Bayesian Methods*, Norwell, MA: Kluwer Academic.

Fitzgerald, W. J., Godsill, S. J., Kokaram, A. C., and Stark, J. A. (1999), "Bayesian Methods in Signal and Image Processing," in *Bayesian Statistics 6*, eds. J. M. Bernardo, J. O. Berger, A. P. Dawid, and A. F. M. Smith, London: Oxford University Press, pp. 239–254.

Florens, J. P., Mouchart, M., and Roulin, J. M. (1990), *Elements of Bayesian Statistics*, New York: Marcel Dekker.

French, S., and Smith, J. Q. (Eds.) (1997), *The Practice of Bayesian Analysis*, London: Arnold.

Gamerman, D. (1997), *Markov Chain Monte Carlo: Stochastic Simulation for Bayesian Inference*, London: Chapman and Hall.

Gatsonis, C., Kass, R., Carlin, B. P., Carriquiry, A. L., Gelman, A., Verdinelli, I., and West, M. (Eds.) (1998), *Case Studies in Bayesian Statistics IV*, New York: Springer-Verlag.

Geisser, S. (1993), *Predictive Inference: An Introduction*, London: Chapman and Hall.

Gelman, A., Carlin, J. B., Stern, H. S., and Rubin, D. B. (1995), *Bayesian Data Analysis*, London: Chapman and Hall.

Geweke, J. (1999), "Using Simulation Methods for Bayesian Econometric Models: Inference, Development and Communication" (with discussion), *Econometric Reviews*, 18, 1–73.

Glymour, C., and Cooper, G. (Eds.) (1999), *Computation, Causation, and Discovery*, Cambridge, MA:

MIT Press.

Godsill, S. J., and Rayner, P. J. W. (1998), *Digital Audio Restoration*, Berlin: Springer.

Goel, P. (1988), "Software for Bayesian Analysis: Current Status and Additional Needs," in *Bayesian Statistics 3*, eds. J. M. Bernardo, J. O. Berger, A. P. Dawid, and A. F. M. Smith, London: Oxford University Press.

Good, I. J. (1983), *Good Thinking: The Foundations of Probability and Its Applications*, Minneapolis: University of Minnesota Press.

Greenland, S. (1998), "Probability Logic and Probability Induction," *Epidemiology*, 9, 322–332.

Hartigan, J. A. (1983), *Bayes Theory*, New York: Springer-Verlag.

Iversen, E. Jr., Parmigiani, G., and Berry, D. (1998), "Validating Bayesian Prediction Models: A Case Study in Genetic Susceptibility to Breast Cancer," in *Case Studies in Bayesian Statistics IV*, eds. C. Gatsonis, R. E. Kass, B. Carlin, A. Carriquiry, A. Gelman, I. Verdinelli, and M. West, New York: Springer-Verlag.

Jaynes, E. T. (1999), "Probability Theory: The Logic of Science," accessible at http://bayes.wustl.edu/etj/prob.html.

Jeffreys, H. (1961), *Theory of Probability* (3rd ed.), London: Oxford University Press.

Jensen, F. V. (1996), *An Introduction to Bayesian Networks*, London: University College of London Press.

Johnson, V. E. (1997), "An Alternative to Traditional GPA for Evaluating Student Performance," *Statistical Science*, 12, 251–278.

Johnson, V. E., and Albert, J. (1999), *Ordinal Data Models*, New York: Springer-Verlag.

Jordan, M. I. (Ed.) (1998), *Learning in Graphical Models*, Cambridge, MA: MIT Press.

Kadane, J. (Ed.) (1984), *Robustness of Bayesian Analysis*, Amsterdam: North-Holland.

—— (Ed.) (1996), *Bayesian Methods and Ethics in a Clinical Trial Design*, New York: Wiley.

Kadane, J., Schervish, M., and Seidenfeld, T. (Eds.) (1999), *Rethinking the Foundations of Statistics*, Cambridge, U.K.: Cambridge University Press.

Kadane, J., and Schuan, D. A. (1996), *A Probabilistic Analysis of the Sacco and Vanzetti Evidence*, New York: Wiley.

Kahneman, D., Slovic, P., and Tversky, A. (1986), *Judgment Under Uncertainty: Heuristics and Biases*, Cambridge, U.K.: Cambridge University Press.

Kass, R., and Raftery, A. (1995), "Bayes Factors and Model Uncertainty," *Journal of the American Statistical Association*, 90, 773–795.

Kass, R., and Wasserman, L. (1996), "The Selection of Prior Distributions by Formal Rules," *Journal of the American Statistical Association*, 91, 1343–1370.

Kim, S., Shephard, N., and Chib, S. (1998), "Stochastic Volatility: Likelihood Inference and Comparison With ARCH Models," *Review of Economic Studies*, 65, 361–194.

Kitagawa, G., and Gersch, W. (1996), *Smoothness Priors Analysis of Time Series*, New York: Springer.

Lad, F. (1996), *Operational Subjective Statistical Methods*, New York: Wiley.

Lauritzen, S. L. (1996), *Graphical Models*, London: Oxford University Press.

Laplace, P. S. (1812), *Théorie Analytique des Probabilités*, Paris: Courcier.

Leamer, E. E. (1978), *Specification Searches: Ad Hoc Inference With Nonexperimental Data*, Chichester, U.K.: Wiley.

Lee, P. M. (1997), *Bayesian Statistics: An Introduction*, London: Edward Arnold.

Lindley, D. V. (1972), *Bayesian Statistics, A Review*, Philadelphia: SIAM.

Liseo, B., Petrella, L., and Salinetti, G. (1996), "Robust Bayesian Analysis: an Interactive Approach," in *Bayesian Statistics 5*, eds. J. M. Bernardo, J. O. Berger, A. P. Dawid, and A. F. M. Smith, London:

Oxford University Press, pp. 661–666.

Liu, J., Neuwald, A., and Lawrence, C. (1999), "Markovian Structures in Biological Sequence Alignments," *Journal of the American Statistical Association*, 94, 1–15.

Monahan, J., and Genz, A. (1996), "A Comparison of Omnibus Methods for Bayesian Computation," *Computing Science and Statistics*, 27, 471–480.

Moreno, A., and Rios-Insua, D. (1999), "Issues in Service Quality Modeling," in *Bayesian Statistics 6*, eds. J. M. Bernardo, J. O. Berger, A. P. Dawid, and A. F. M. Smith, London: Oxford University Press, pp. 444–457.

Müller, P. M. (1999), "Simulation-Based Optimal Design," in *Bayesian Statistics 6*, eds. J. M. Bernardo, J. O. Berger, A. P. Dawid, and A. F. M. Smith, London: Oxford University Press, pp. 459–474.

Müller, P. M., and Rios-Insua, D. (1998), "Issues in Bayesian Analysis of Neural Network Models," *Neural Computation*, 10, 571–592.

Müller, P. M., and Vidakovic, B. (Eds.) (1999), *Bayesian Inference in Wavelet-Based Models*, New York: Springer-Verlag.

Mukhopadhyay, P. (1998), *Small Area Estimation in Survey Sampling*, New Delhi: Naroso.

Neal, R. M. (1996), *Bayesian Learning for Neural Networks*, New York: Springer.

——— (1999), "Regression and Classification Using Gaussian Process Priors," in *Bayesian Statistics 6*, eds. J. M. Bernardo, J. O. Berger, A. P. Dawid, and A. F. M. Smith, London: Oxford University Press, pp. 475–501.

O'Hagan, A. (1988), *Probability: Methods and Measurements*, London: Chapman and Hall.

——— (1994), *Kendall's Advanced Theory of Statistics, Vol. 2B—Bayesian Inference*, London: Arnold.

——— (1995), "Fractional Bayes Factors for Model Comparisons," *Journal of the Royal Statistical Society*, Ser. B, 57, 99–138.

Ó Ruanaidh, J. J. K., and Fitzgerald, W. J. (1996), *Numerical Bayesian Methods Applied to Signal Processing*, New York: Springer.

Parent, E., Hubert, P., Bobée, B., and Miquel, J. (Eds.) (1998), *Statistical and Bayesian Methods in Hydrological Sciences*, Paris: UNESCO Press.

Pearl, J. (1988), *Probabilistic Inference in Intelligent Systems*, San Mateo, CA: Morgan Kaufmann.

Perlman, M., and Blaug, M. (Eds.) (1997), *Bayesian Analysis in Econometrics and Statistics: The Zellner View*, Northhampton, MA: Edward Elgar.

Piccinato, L. (1996), *Metodi per le Decisioni Statistiche*, Milano: Springer-Verlag Italia.

Pilz, J. (1991), *Bayesian Estimation and Experimental Design in Linear Regression* (2nd ed.), New York: Wiley.

Poirier, D. J. (1995), *Intermediate Statistics and Econometrics: A Comparative Approach*, Cambridge, MA: MIT Press.

Pole, A., West, M., and Harrison, J. (1995), *Applied Bayesian Forecasting Methods*, London: Chapman and Hall.

Pollard, W. E. (1986), *Bayesian Statistics for Evaluation Research*, Beverly Hills, CA: Sage.

Press, J. (1989), *Bayesian Statistics*, New York: Wiley.

Qian, W., and Brown, P. J. (1999), "Bayes Sequential Decision Theory in Clinical Trials," in *Bayesian Statistics 6*, eds. J. M. Bernardo, J. O. Berger, A. P. Dawid, and A. F. M. Smith, London: Oxford University Press, pp. 829–838.

Racugno, W. (Ed.) (1998), *Proceedings of the Workshop on Model Selection*, special issue of Rassegna di Metodi Statistici ed Applicazioni, Bologna: Pitagora Editrice.

Regazzini, E. (1999), "Old and Recent Results on the Relationship Between Predictive Inference and Statistical Modelling Either in Nonparametric or Parametric Form," in *Bayesian Statistics 6*, eds. J.

M. Bernardo, J. O. Berger, A. P. Dawid, and A. F. M. Smith, London: Oxford University Press, pp. 571–588.

Rios-Insua, D. (1990), *Sensitivity Analysis in Multiobjective Decision Making*, New York: Springer-Verlag.

Robert, C. P. (1994), *The Bayesian Choice: A Decision-Theoretic Motivation*, New York: Springer-Verlag.

Robert, C. P., and Casella, G. (1999), *Monte Carlo Statistical Methods*, New York: Springer-Verlag.

Rubin, D. B. (1987), *Multiple Imputation for Nonresponse in Surveys*, New York: Wiley.

Savage, L. J. (1972), *The Foundations of Statistics* (2nd ed.), New York: Dover.

Sellke, T., Bayarri, M. J., and Berger, J. O. (1999), "Calibration of P Values for Precise Null Hypotheses," ISDS Discussion Paper 99-13, Duke University.

Schervish, M. (1995), *Theory of Statistics*, New York: Springer.

Sinha, D., and Dey, D. (1999), "Survival Analysis Using Semiparametric Bayesian Methods," in *Practical Nonparametric and Semiparametric Statistics*, Eds. D. Dey, P. Müller, and D. Sinha, New York: Springer, pp. 195–211.

Sivia, D. S. (1996), *Data Analysis: A Bayesian Tutorial*, London: Oxford University Press.

Smith, J. Q. (1988), *Decision Analysis: A Bayesian Approach*, London: Chapman and Hall.

Spirtes, P., Glymour, C., and Scheines, R. (1993), *Causation, Prediction, and Search*, New York: Springer-Verlag.

Stangl, D., and Berry, D. (1998), "Bayesian Statistics in Medicine: Where We are and Where We Should be Going," *Sankhya*, Ser. B, 60, 176–195.

Tanner, M. A. (1993), *Tools for Statistical Inference: Methods for the Exploration of Posterior Distributions and Likelihood Functions* (2nd ed.), New York: Springer-Verlag.

Thiesson, B., Meek, C., Chickering, D. M., and Heckerman, D. (1999), "Computationally Efficient Methods for Selecting Among Mixtures of Graphical Models," in *Bayesian Statistics 6*, eds. J. M. Bernardo, J. O. Berger, A. P. Dawid, and A. F. M. Smith, London: Oxford University Press, pp. 631–656.

Tierney, L. (1991), *Lisp-Stat, An Object-Oriented Environment for Statistical Computing and Dynamic Graphics*, New York: Wiley.

Walley, P. (1991), *Statistical Reasoning With Imprecise Probabilities*, London: Chapman and Hall.

West, M., and Harrison, J. (1997), *Bayesian Forecasting and Dynamic Models* (2nd ed.), New York: Springer-Verlag.

Winkler, R. L. (1972), *Introduction to Bayesian Inference and Decision*, New York: Holt, Rinehart, and Winston.

Wolpert, R. L., and Ickstadt, K. (1998), "Poisson/Gamma Random Field Models for Spatial Statistics," *Biometrika*, 82, 251–267.

Yang, R., and Berger, J. (1997), "A Catalogue of Noninformative Priors," ISDS Discussion Paper 97-42, Duke University.

Zellner, A. (1971), *An Introduction to Bayesian Inference in Econometrics*, New York: Wiley.

An Essay on Statistical Decision Theory

Lawrence D. Brown

1. A 1966 QUOTATION

The middle third of this century marks the summit of research in statistical decision theory. Consider this 1966 quotation from the foreword to the volume Early Statistical Papers of J. Neyman (Neyman 1967), signed "students of J. N. at Berkeley":

> The concepts of confidence intervals and of the Neyman–Pearson theory have proved immensely fruitful. A natural but far reaching extension of their scope can be found in Abraham Wald's theory of statistical decision functions. The elaboration and application of the statistical tools related to these ideas has already occupied a generation of statisticians. *It continues to be the main lifestream of theoretical statistics* [italics mine].

The italicized assertion is the focus of this vignette. Is the assertion still true today? If not, what is the current status of statistical decision theory, and what position is it likely to hold in the coming decades? Any attempt to answer these questions requires a definition of "statistical decision theory." Indeed, the answers will be largely driven by how broadly—or narrowly—the boundaries of decision theory are drawn.

2. THE SCOPE OF STATISTICAL DECISION THEORY

The term "statistical decision theory" appears to be a condensation of Wald's phrase "the theory of statistical decision functions," which occurs, for example, in the preface to his monograph (Wald 1950) as well as earlier in Wald (1942). Wald viewed his "theory" as a codification and generalization of the theory of tests and confidence intervals already developed by Neyman, often in collaboration with E. Pearson. For clear statements of Wald's view see the last two paragraphs of the introduction to Wald's pivotal paper (Wald 1939), and section 1.7 of his monograph (Wald 1950). The vignette on hypothesis testing by Marden presents an excellent review of the various manifestations of hypothesis testing. It is hard to choose a favorite among the wonderful Neyman–Pearson papers on the foundations of testing and confidence intervals, but I pick the works by Neyman and Pearson (1933) on testing and Neyman (1935) on confidence intervals.

Lawrence D. Brown is Miers Busch Professor of Statistics, Department of Statistics, University of Pennsylvania, Philadelphia, PA 19104-6302 (E-mail: lbrown@stat.wharton.upenn.edu).

The frequentist approach is the cornerstone of Neyman's statistical philosophy. Neyman (1941) provided a graphic demonstration and explanation. Thus one begins the analysis of any statistical situation with a family of possible distributions for the data. In the presence of a finite-dimensional parameterization, this can be written as $\mathcal{F} = \{F_\theta : \theta \in \Theta\}$, but the existence of such a parameterization is only an often-useful convenience, rather than a formal necessity. One then hypothesizes a possible rule for solving the problem (a test or confidence interval, or a decision function in Wald's more general terminology). The key step is to calculate the distribution of outcomes from that rule as if the true parameter were a fixed value, θ, and compare the results of such a calculation at various θ for various possible decision rules.

That "as if θ were fixed" qualification caused considerable confusion in the early years of the theory, and often continues to do so. It does not mean that θ *is* fixed. Neyman and Pearson frequently returned to emphasize that this type of calculation would guarantee validity "irrespective of the a priori truth." They denied any presumption that the statistician would be faced with a long sequence of independent repetitions of the situation, all having the same value of θ. For example, the landmark Clopper and Pearson (1934) article on confidence intervals for a binomial parameter discusses as a particular example what happens if the Clopper–Pearson prescription is used in a situation where the unknown parameter, p, can take the values 1/3, 1/2, and 2/3 with a hypothesized skewed a priori distribution. The point of this calculation is to vividly demonstrate with an example the claim that the proposed intervals have coverage probability at least the nominal value, and this fact "does not depend on any a priori knowledge regarding possible values of p." A subsidiary goal may have been to investigate by how much this nominal value would be exceeded in such a plausible example. (The answer was that the true coverage was .9676, as opposed to the nominal value of .95.) A later discussion related to this general issue appears in Neyman (1952, p. 211). Brown, Cai, and DasGupta (1999a,b) recently reexamined the problem of binomial confidence intervals and included a reassessment of the Clopper–Pearson proposal.

3. THE DECISION THEORETIC SPIRIT

According to the foregoing, the spirit of decision theory is pervasive in contemporary statistical research. Common manifestations include both mathematical and numerical attempts to check the frequentist performance of proposed procedures. This includes comparative investigations of level and power for hypothesis tests or of precision of proposed estimators as, for example, might occur in a Monte Carlo comparison of variances and biases. In particular, any presentation of statistical tests that mentions power is an embodiment of this spirit.

The frequentist interpretation of confidence intervals (and sets) relies on Neyman's previously cited articles as well as Wilson (1927) and the previously cited work of Clopper and Pearson. Note also the general formulation by Wald and Wolfowitz (1939, 1941) of nonparametric confidence bands for a cumulative distribution function, and their discovery that a fundamental probabilistic question they could not solve had earlier been settled by Kolmogorov (1933) and Smirnov (1939).

4. SEQUENTIAL ANALYSIS

Wald was justifiably proud of his formulation of sequential decision problems and his solution of fundamental issues there, as in the optimal property of the sequential probability ratio test presented by Wald and Wolfowitz (1950). Wald (1947) cited a few historical precedents in the work of others, but there is no question that this entire statistical area was his creation. The spirit of his development survives in parts of contemporary statistics and even flourishes in some, as in the methodology of early stopping for (sequential) clinical trials; for example, via group sequential testing or stochastic curtailment. Jennison and Turnbull (1990) contains a fairly recent review of clinical trials, and Siegmund (1994) provides a broader review of the current status of Wald's sequential analysis. As a prelude to my later discussion, let me already note that in this area the analytic tools originally developed by Wald (and later extended and refined by others) survive as essential building blocks in contemporary research.

5. MINIMAXITY AS A THEME AND BENCHMARK

I have argued (Brown 1994) that the minimax idea has been an essential foundation for advances in many areas of statistical research. These include general asymptotic theory and methodology, hierarchical models, robust estimation, optimal design, and nonparametric function analysis. The vignette on minimaxity by W. Strawderman clarifies and reinforces this assertion. Hence here I do not specifically pursue this crucial manifestation of the spirit of decision theory, although it is very much present in the nonparametric examples I describe.

6. BAYESIAN STATISTICS

Not so many years ago, "bayesian statistics" was frequently viewed as the antithesis of "frequentist statistics," and the feeling among some was that their favorite of these two theories would eventually triumph as the other failed ignominiously. It now is apparent that this will not happen. There is much evidence that we are currently in the midst of a productive amalgamation of these two schools of statistics.

The vignette on Bayesian analysis by Berger describes several different approaches to Bayesian analysis. Interestingly, none of these is directly the pure frequentist approach, in which the prior is a given distribution with the same frequentist validity as the family of distributions, $\mathcal{F}$. Such a situation is a conceptual and some-times realistic possibility, but modern Bayesian statistical techniques are intended to apply far beyond this possible scenario. Lehmann and Casella (1998, p. 226) provided for further discussion of this state of affairs.

According to the discussion that began this essay, a pure frequentist Bayesian approach, where realistic, is firmly within the decision-theoretic realm. Other funda-mental Bayesian approaches are not, even if they may involve loss functions and even though they may be justified, at least in part, by the decision-theoretic version of Bayes theorem that says the Bayes procedure minimizes the expected risk. However, most of these approaches are neutral rather than antithetical with respect to decision theory. Many of them involve the use of "objective" priors, such as the Jeffreys prior or the Bernardo (1979) reference priors. These approaches customarily generate decisions. As such, they can be viewed as powerful and heuristically appealing *mechanisms* for generating plausible decision rules. Having used such a device, the question remains as to how well the rule that it generates will actually perform in a suitable range of situations like the one at hand. This decision-theoretic question is implicit but equally pertinent in varieties of robust Bayesian analyses and is explicit in the realm of Γ-minimax procedures (see Hodges and Lehmann 1950 for an early decision-theoretic formulation closely related to the Γ-minimax idea). Answering such a question is increasingly (and properly) seen as requiring frequentist-style investigations of com-parative risk through simulations or theoretical calculations. Hence what is emerging is a figurative marriage of Bayes and Wald.

7. THE DECISION THEORETIC TOOLKIT

I have been arguing in general terms that Neyman and Wald (and also their collaborators and immediate students) had a particular perspective on statistics, and that this perspective is alive and thriving in contemporary statistics. Nevertheless, many statisticians apparently feel that statistical decision theory is moribund—or even already dead.

Through the decades, various technical analytical tools have been developed by avowedly decision-theoretic researchers. An assertion that decision theory is dying is probably more focused on this toolkit than it is on the decision-theoretic spirit I discussed earlier. Even here, the assertion seems to me in the main to be drastically mistaken, though one might point to certain tools that are far less broadly useful than might have been expected at the time they were being developed. I have in mind as an example of the latter the general complete-class theory involving characterization of admissible procedures. (Even here there are significant recent contributions relevant to contemporary methodology; see, e.g., Berger and Strawderman 1996 and Zhao 1999.)

This "toolkit" contains a vast array of analytical weapons for a variety of situations. Furthermore, it has continually extended and expanded from its original extent and form. It is impossible in the format of this short, broad survey to carefully trace the development of even a single important tool from its origins with Neyman or Wald (and often before them as well). So I discuss only some examples from one particular area of research, to try to emphasize the vitality and importance of that legacy.

8. NONPARAMETRIC FUNCTION ANALYSIS

8.1   Rates of Convergence

Nonparametric function analysis includes such topics as nonparametric regression, density estimation, image reconstruction, and aspects of pattern recognition. As a manifestation of the nonparametric/robust paradigm, it is already heavily infused with the influence of decision theory. For confirmation, note that on Efron's (1998) "barycentric picture of modern statistical research," the topic "robustness, nonparametrics" describes the point most heavily weighted on the "frequentist" axis, as opposed to the "Bayesian" and "Fisherian" axes. There should be no debate on this point. But I want to go further, and describe how several items in the basic toolkit are of use here.

First, the Waldian notion of loss and risk pervades the topic. A fundamental feature of this area is the presence of rates of convergence slower than the usual parametric standard of $1/\sqrt{n}$. These rates of convergence are for the *risk* under any of various loss functions. In particular, the asymptotics cannot be formulated in Fisherian terms of efficiency, because the optimal-rate risks under (integrated) quadratic loss must balance squared bias as well as variance. Nevertheless, Fisher information and the Cramer–Rao inequality occasionally can be useful (as in Brown and Low 1991), but the spirit and content of such a treatment are more directly descended from the admissibility argument of Hodges and Lehmann (1951) and the two-dimensional admissibility argument of Stein (1956b).

A precise description of these convergence rates nearly demands a minimax formulation. An exception is the early formulation of Rosenblatt (1956) and Parzen (1962), looking at optimal rates available from the use of nonnegative kernels. Brown, Low, and Zhao (1997) tried to explain why this is more emphatically so here than in the classical parametric theory.

Despite its success, there is a not-untypical shortcoming in a literal adoption of this minimax formulation—it may be unhealthily conservative. The formulation assumes that the unknown function (e.g., regression function or density) belongs to a suitably bounded class, usually explicitly involving some sort of smoothness restriction. For a basic example, in a one-dimensional setting, the assumption might be that

$$\int (f''(x))^2 \, dx \leq B$$

for a prespecified, but possibly large value of B. Corresponding to this class is an asymptotically minimax value and procedure. An adequate practical approach to this asymptotic ideal is achievable with computationally feasible procedures of various types, such as kernels, splines, and orthogonal series estimators (including wavelets). However, a procedure that is minimax in this sense, or close to it, may perform relatively poorly in practice. To see why, consider a typical simple task such as estimating an unknown probability density. One may suspect that the true density is quite "smooth"—perhaps it is thought to have a shape similar to a simple mixture of normal distributions. Various standard optimal minimax convergence rate procedures do not utilize that suspicion. Instead, they protect against the possibility that the density is as extremely wiggly as allowed by the foregoing smoothness restriction. Such extremely wiggly densities may be felt to be a priori highly unlikely and/or not very interesting. A current tidal wave of research into "adaptive" estimators is an attempt to create procedures that circumvent this shortcoming by being simultaneously nearly minimax for a broad inventory of smoothness classes.

8.2 Hardest Linear Subproblem

The minimax value just mentioned can be discovered to within a startling degree of accuracy through a very beautiful application of a fundamental device due to Wald in a form proposed by Stein (1956a) (see Donoho 1994; Donoho and Liu 1991; Donoho, Liu, and MacGibbon 1990). Consider the entirely classical parametric situation of an observation from a multivariate normal distribution with identity covariance matrix but unknown mean. Suppose that one desires to estimate some linear functional of that mean—for example, its first coordinate—and uses ordinary quadratic loss. Further, suppose the mean is known to lie in a bounded convex set, S, symmetric about the origin.

Temporarily restrict the class of available estimators to be linear. Wald's standard tool for discovering minimax values is to establish a least favorable distribution. Here that least favorable distribution turns out to be supported on a one-dimensional subset of S, say H. That is, for linear estimators, there is a "hardest one-dimensional subproblem." The minimax risk for the problem on S unrestricted to linear estimators cannot be less than that for the unrestricted problem on H. One then gets to look at an even more classical problem. Let $X \sim N(\theta, 1)$ with $\theta \in (-a, a)$. What is the minimax risk (under squared error loss), and what is its relation to the minimax risk among linear procedures? This question had been studied by Casella and Strawderman (1981), among others. Extending their result, Donoho et al. (1990) showed that the ratio of the two risks is never greater than 1.25.

This very classical (but nevertheless recent) minimax theorem about estimation of a multivariate normal mean can then be carried into the nonparametric realm with the aid of a further set of decision-theoretic tools for asymptotics largely developed by Le Cam (see, e.g., LeCam 1953, 1986).

The end result from using this decision theoretic arsenal is a powerful result enabling one to calculate the (asymptotic) minimax risk to within a factor of 1.25 and, even more important, to know how to get a (linear) procedure that comes within this factor of being minimax. Interestingly, this idea works for various loss functions in addition to squared error (see Donoho 1994).

8.3   Asymptotic Equivalence

Further utilization of Le Cam's decision theoretic ideas has proved valuable in this area. LeCam (1953, 1964, 1986) created a general concept of equivalence of statistical problems within the framework of Wald's decision theory. With the aid of additional inequalities also due at least partly to Le Cam, one can unify and greatly simplify asymptotic investigations similar to those sketched earlier.

To explain, I return briefly to a more classical parametric setting. Asymptotic questions in such settings very often can be efficiently reduced to appropriate questions involving only normal distributions. These normal distribution settings can effectively be viewed as the limit of the original problem as the sample size tends to infinity. This aspect of statistical theory has its roots well before Neyman and Wald and, I believe, lies well outside the scope of what should be considered characteristically decision theoretic. Nevertheless, both Neyman and Wald recognized its importance, made important contributions in the area, and incorporated asymptotical analyses into their basic theories. Prominent examples are Wald's (1943) proof of the asymptotic optimality of the likelihood ratio test and Neyman's (1949) explication of "best asymptotically normal" procedures.

In an analogous fashion the stochastic formulation of white noise with drift can serve as a useful, unified limit for nonparametric formulations like those already mentioned. Research on this topic is ongoing, but Brown and Low (1996) and Nussbaum (1996) have given basic results.

By all measures, wavelets provide a powerful new tool for nonparametric function analysis. The continuing development of this tool has evolved out of a combination of the function-analytic topic of wavelet bases with an intensive and extensive dose of decision theory, along with a careful evaluation of the practical problems requiring the statistical techniques being created. The debt to decision theory will be immediately evident to any reader of the fundamental works of Donoho and Johnstone and collaborators (e.g., Donoho and Johnstone 1995, 1998).

In this development there is a continuing expansion of the ideas present in the minimax treatment of Donoho et al. described previously. I will not go into further detail about that. Instead, I mention two quite different decision-theoretic aspects. One is the incorporation of Stein's unbiased estimate of the risk. Stein's powerful technical tool was developed as a more effective way of handling certain admissibility questions that arose following his surprising discovery of the inadmissibility of the usual estimator of a multivariate normal mean (see Brown 1979; James and Stein 1961; Stein 1956, 1973, 1981). Donoho and Johnstone (1995) cited this technique as the defining element of a class of wavelet estimators that they called SURE-type estimators.

A more unexpected combination of techniques is the application of Benjamini and Hochberg (1995) in the construction of wavelet estimators. Benjamini and Hochberg created an interesting new proposal for the problem of multiple comparisons. Their idea is to control what they term the "false discovery rate" (FDR). The issue of multiple comparisons is heavily decision theoretic in orientation and development. Nevertheless, the basic problem that Benjamini and Hochberg addressed would seem quite distant from the issues relevant to wavelet estimation. But Abramowich and Benjamini (1995), Abramowich, Benjamini, Donoho, and Johnstone (1999), and Johnstone (1999) described an important and productive connection between FDR and adaptive wavelet estimation.

9. MONTE CARLO INVESTIGATIONS: A CHALLENGE FOR FUTURE THEORY

I have been focusing on what I termed the decision-theoretic toolkit. I believe that this kit is largely complete and that the focus of future research will involve use of these tools, as in the foregoing stories about nonparametric functional analysis. Of course, I could be wrong. There is some interesting, continuing toolbuilding going on (as, e.g., in Eaton 1992). Then, too, maybe some brand new tool is just waiting to be discovered—perhaps a biased estimate of risk that will prove even more useful than Stein's unbiased estimate. Or maybe an accumulation of results developing out of the current toolkit will become recognized as an independently powerful tool on its own.

However, there is at least one place where I think we are lacking a general decision-theoretic tool of a new sort. As I have noted, risk comparisons of proposed procedures are often performed via simulation. (We may be simulating the performance of the respective proposals in terms of level and power or expected coverage and, sometimes, expected length or size, or in terms of bias and variance, or just to discover their sampling distribution, leaving the determination of comparative risk to the reader.)

These simulation results can provide important practical validation of an asymptotic result or of a persuasive heuristic model. However, they do not have the intellectual force of a mathematical proof. That is, in a complex situation I may be able to *convince* you with simulations that procedure A is better than procedure B, but rarely, if ever, can I *prove* it that way. This is primarily because simulations can be performed only at specific alternatives and sample sizes, and there are usually too many such alternatives of interest for an adequate simulation to be performed at each one. Hence the decision-theoretic challenge of finding a methodology for converting the simulational power of the computer into a tool able to deliver the persuasive force of a mathematical proof.

REFERENCES

Abramowich, F., and Benjamini, Y. (1995), "Threshholding of Wavelet Coefficients as Multiple Testing Procedures," in *Wavelets and Statistics*, ed. A. Antoniadis, New York: Springer-Verlag, pp. 5–14.

Abramowich, A., Benjamini, Y., Donoho, D., and Johnstone, I. M. (1999), "Sparsity and Threshholding via False Discovery Control," preprint.

Benjamini, Y., and Hochberg, Y. (1995), "Controlling the False Discovery Rate: A Practical and Powerful Approach to Multiple Testing," *Journal of the Royal Statistical Society*, Ser. B, 57, 289–300.

Berger, J. O., and Strawderman, W. E. (1996), "Choice of Hierarchical Priors: Admissibility in Estimation of Normal Means," *The Annals of Statistics*, 24, 931–951.

Bernardo, J. M. (1979), "Reference Prior Distributions for Bayesian Inference," *Journal of the Royal Statistical Society*, Ser. B, 41, 113–147.

Brown, L. D. (1979), "A Heuristic Method for Determining Admissibility of Estimators, With Applications," *The Annals of Statistics*, 7, 960–994.

—— (1994), "Minimaxity, More or Less," in *Statistical Decision Theory and Related Topics*, eds. S. S. Gupta and J. O. Berger, New York: Springer-Verlag, pp. 1–18.

Brown, L. D., Cai, T. T., and DasGupta, A. (1999a), "Interval Estimation for a Binomial Proportion," *Statistical Science*, to appear.

—— (1999b), "Confidence Intervals for a Binomial Proportion and Edgeworth Expansions," *The Annals of Statistics*, to appear.

Brown, L. D., and Low, M. G. (1991), "Information Inequality Bounds on the Minimax Risk With an Application to Nonparametric Regression," *The Annals of Statistics*, 19, 329–337.

—— (1996), "Asymptotic Equivalence of Nonparametric Regression and White Noise," *The Annals of Statistics*, 24, 2384–2398.

Brown, L. D., Low, M. G., and Zhao, L. H. (1997), "Superefficiency in Nonparametric Function Estimation," *The Annals of Statistics*, 25, 2607–2625.

Casella, G., and Strawderman, W. E. (1981), "Estimating a Bounded Normal Mean," *The Annals of Statistics*, 9, 870–878.

Clopper, C. J., and Pearson, E. S. (1934), "The Use of Confidence or Fiducial Limits Illustrated in the Case of the Binomial," *Biometrika*, 26, 404–413.

Donoho, D. L. (1994), "Statistical Estimation and Optimal Recovery," *The Annals of Statistics*, 22, 238–270.

Donoho, D. L., and Johnstone, I. M. (1995), "Adapting to Unknown Smoothness via Wavelet Shrinkage," *Journal of the American Statistical Association*, 90, 1200–1224.

—— (1998), "Minimax Estimation via Wavelet Shrinkage," *The Annals of Statistics*, 26, 879–921.

Donoho, D. L., and Liu, R. C. (1991), "Geometrizing Rates of Convergence III," *The Annals of Statistics*, 19, 668–701.

Donoho, D. L., Liu, R. C., and MacGibbon, B. (1990), "Minimax Risk for Hyperrectangles," *The Annals of Statistics*, 18, 1416–1437.

Eaton, M. L. (1992), "A Statistical Diptych: Admissible Inferences–Recurrence of Symmetric Markov Chains," *The Annals of Statistics*, 20, 1147–1179.

Efron, B. (1998), "R. A. Fisher in the 21st Century" (with discussion), *Statistical Science*, 13, 95–113.

Hodges, J. L., Jr., and Lehmann, E. L. (1950), "The Use of Previous Experience in Reaching Statistical Decisions," *Annals of Mathematical Statistics*, 23, 396–407.

——— (1951), "Some Applications of the Cramer–Rao Inequality," in *Proceedings of the Second Berkeley Symposium on Mathematical Statistics and Probability*, 1, Berkeley, CA: University of California Press, pp. 13–22.

James, W., and Stein, C. (1961), "Estimation With Quadratic Loss," in *Proceedings of the Fourth Berkeley Symposium on Mathematical Statistics and Probability*, 1, Berkeley, CA: University of California Press, pp. 311–319.

Jennison, C., and Turnbull, B. W. (1990), "Statistical Approaches to Interim Monitoring of Medical Trials: A Review and Commentary," *Statistical Science*, 5, 299–317.

Johnstone, I. M. (1999), "False Discovery Rates and Adaptive Estimation," special invited paper presented at the 1999 Joint Statistical Meetings in Baltimore.

Kolmogorov, A. (1933), "Sulla Determinazione Empirica di una Leggi di Distribuzione," *Giornale dell'Instituto Italiana degli Attuari*, 4, 83–91.

Le Cam, L. (1953), *On Some Asymptotic Properties of Maximum Likelihood Estimates and Related Bayes Procedures*, Berkeley, CA: University of California Press, pp. 277–230.

——— (1964), "Sufficiency and Approximate Sufficiency," *Annals of Mathematical Statistics*, 35, 1419–1455.

——— (1986), *Asymptotic Methods in Statistical Decision Theory*, New York: Springer-Verlag.

Lehmann, E. L., and Casella, G. (1998), *Theory of Point Estimation* (2nd ed.), New York: Springer-Verlag.

Neyman, J. (1935), "On the Problem of Confidence Intervals," *Annals of Mathematical Statistics*, 6, 111–116.

——— (1941), "Fiducial Argument and the Theory of Confidence Intervals," *Biometrika*, 32, 128–150.

——— (1949), "Contribution to the Theory of the Chi-Square Test," in *Proceedings of the Berkeley Symposium on Mathematical Statistics and Probability*, 1945, Berkeley, CA: University of California Press, pp. 239–273.

——— (1952), *Lectures and Conferences on Mathematical Statistics and Probability*, Washington, D.C.: U.S. Department of Agriculture.

——— (1967), *A Selection of Early Statistical Papers of J. Neyman*, Berkeley, CA: University of California Press.

Neyman, J., and Pearson, E. S. (1933), "The Testing of Statistical Hypotheses in Relation to Probabilities a Priori," *Proceedings of the Cambridge Philosophical Society*, 29, 492–510.

Nussbaum, M. (1996), "Asymptotic Equivalence of Density Estimation and Gaussian White Noise," *The Annals of Statistics*, 24, 2399–2430.

Parzen, E. (1962), "On the Estimation of a Probability Density and Mode," *Annals of Mathematical Statistics*, 33, 1065–1076.

Rosenblatt, M. (1956), "Remarks on Some Nonparametric Estimates of a Density Function," *Annals of Mathematical Statistics*, 27, 832–837.

Siegmund, D. (1994), "A Retrospective of Wald's Sequential Analysis—Its Relation to Changepoint Detection and Sequential Clinical Trials," in *Statistical Decision Theory and Related Topics*, eds. S. S. Gupta and J. O. Berger, New York: Springer-Verlag, pp. 19–33.

Smirnov, N. V. (1939), "On the Estimation of the Discrepancy Between Empirical Curves of Distributions for Two Independent Samples," *Bulletin of the University of Moscow*, 2(2), 3–14.

Stein, C. (1956a), "Efficient Nonparametric Testing and Estimation," in *Proceedings of the Third Berkeley Symposium on Mathematical Statistics and Probability*, Berkeley, CA: University of California Press, pp. 187–196.

———— (1956b), "Inadmissibility of the Usual Estimator for the Mean of a Multivariate Normal Distribution," in *Proceedings of the Third Berkeley Symposium on Mathematical Statistics and Probability*, Berkeley, CA: University of California Press, pp. 197–206.

———— (1973), "Estimation of the Mean of a Multivariate Distribution," in *Proceedings of the Prague Symposium on Asymptotic Statistics*, pp. 345–381.

———— (1981), "Estimation of the Mean of a Multivariate Normal Distribution," *The Annals of Statistics*, 9, 1135–1151.

Wald, A. (1939), "Contributions to the Theory of Statistical Estimation and Testing Hypotheses," *Annals of Mathematical Statistics*, 10, 299–326.

———— (1942), *On the Principles of Statistical Inference*, Notre Dame, IN.

———— (1943), "Tests of Statistical Hypotheses Concerning Several Parameters When the Number of Observations is Large," *Transactions of the American Mathematical Society*, 54, 426–482.

———— (1947), *Sequential Analysis*, New York: Wiley.

———— (1950), *Statistical Decision Functions*, New York: Wiley.

Wald, A., and Wolfowitz, J. (1939), "Confidence Limits for a Continuous Distribution Function," *Annals of Mathematical Statistics*, 10, 105–118.

———— (1941), "Note on Confidence Limits for Continuous Distribution Functions," *Annals of Mathematical Statistics*, 12, 118–119.

———— (1950), "Bayes Solution of Sequential Decision Problems," *Annals of Mathematical Statistics*, 21, 82–89.

Wilson, E. B. (1927), "Probable Inference, the Law of Succession, and Statistical Inference," *Journal of the American Statistical Association*, 22, 209–212.

Zhao, L. H. (1999), "Improved Estimators in Nonparametric Regression Problems," *Journal of the American Statistical Association*, 94, 164–173.

Markov Chain Monte Carlo: 10 Years and Still Running!

Olivier Cappé and Christian P. Robert

1. ITERATION 0: BURNIN' STEPS

Although Markov chain Monte Carlo (MCMC) techniques can be traced back to the early 1950s and are almost contemporary with independent Monte Carlo methods, the use of these techniques has only really taken off in the past 10 years, at least in "mainstream" statistics, with the work of Gelfand and Smith (1990) and the rediscovery of Metropolis's and Hastings's papers. The impact on the discipline is deep and durable, because these methods have opened new horizons in the scale of the problems that one can deal with, thus enhancing the position of statistics in most applied fields. The MCMC "revolution" has in particular boosted Bayesian statistics to new heights by providing a virtually universal tool for dealing with integration (and optimization) problems. This can be seen in, for instance, the explosion of papers dealing with complex models, hierarchical modelings, nonparametric Bayesian estimation, and spatial statistics. This trend has also created new synergies with mathematicians and probabilists, as well as econometricians, engineers, ecologists, astronomers, and others, for theoretical requests and practical implications of MCMC techniques.

It is impossible to provide here a complete perspective on MCMC methods and applications. For this, as well as for other references, we refer the reader to the books by Gilks, Richardson, and Spiegelhalter (1996) and Robert and Casella (1999). Here we focus on four particular aspects of MCMC algorithms: theoretical foundations, practical implementation, Bayesian inference, and prospects.

The Web provides many relevant sites, including the MCMC preprint server (http://www.statslab.cam.ac.uk/~mcmc/), which offers a most useful window to the current research on MCMC; Wilson's site on perfect simulation (http://dimacs.rutgers.edu/~dbwilson/exact.h), which keeps track of papers and applets on this topic; and a site on convergence control techniques (http://www.ensae.fr/crest/statistique/robert/McDiag). Even more recent developments allow for on-line computations through applets, like Breyer's (http://www.maths.lancs.ac.uk/~breyer/java.html), which impressively illustrates the behavior of different samplers for simple problems. Note

Olivier Cappé is CNRS Researcher, ENST, Paris, France (E-mail: cappe@tsi.enst.fr). Christian P. Robert is Professor, Université Paris 9–Dauphine, and Head, Laboratory of Statistics, CREST-ENSAE, INSEE, Paris, France (E-mail: robert@ensae.fr). The authors thank George Casella, Paolo Giudici, and Peter Green for comments and suggestions.

also that the BUGS and CODA software are freely available at http://www.mrc-bsu.cam.ac.uk/bu.

2. ITERATION 1: IT DOES CONVERGE!

The idea at the core of MCMC techniques is, a posteriori, quite simple: run a Markov chain $(\theta^{(t)})$ such that $(\theta^{(t)})$ is irreducible, $(\theta^{(t)})$ is ergodic, and the stationary distribution of $(\theta^{(t)})$ is the distribution of interest, π. Once the chain is on the move, all we need to do is check convergence to this distribution, as this theoretically should happen by virtue of the *ergodic theorem*. Moreover, integrals of interest can be approximated as in independent Monte Carlo studies, because

$$\lim_{T \to \infty} \frac{1}{T} \sum_{t=1}^{T} h(\theta^{(t)}) = \int h(\theta) \, d\pi(\theta).$$

This follows from basic Markov chain theory, but the formidable appeal of MCMC methods is that universal schemes exist to implement this idea: besides the Gibbs sampler (see the Gibbs sampler vignette by Gelfand), the Metropolis–Hastings algorithm allows one to pick almost any conditional density $q(\cdot|\theta)$, and to derive a transition kernel (i.e., the conditional density of the Markov chain) as

$$K(\theta^{(t)}, \theta^{(t+1)})$$

$$= q(\theta^{(t+1)}|\theta^{(t)}) \min \left(1, \frac{q(\theta^{(t)}|\theta^{(t+1)})\pi(\theta^{(t+1)})}{q(\theta^{(t+1)}|\theta^{(t)})\pi(\theta^{(t)})} \right)$$

$$+ \delta_{\theta^{(t)}}(\theta^{(t+1)}) \int \max \left(0, 1 - \frac{q(\theta^{(t)}|\xi)\pi(\xi)}{q(\xi|\theta^{(t)})\pi(\theta^{(t)})} \right)$$

$$\times q(\xi|\theta^{(t)}) \, d\xi,$$

where $\delta_{\theta^*}(\theta)$ denotes the Dirac mass in θ^*. The distribution π is then stationary under the transition K because the (detailed) balance condition holds,

$$\pi(\theta^{(t)})K(\theta^{(t)}, \theta^{(t+1)}) = \pi(\theta^{(t+1)})K(\theta^{(t+1)}, \theta^{(t)}),$$

and the chain $(\theta^{(t)})$ is irreducible and ergodic in almost every setting.

It Could Converge Faster . . .

In some cases, ergodicity simply may not be enough to guarantee convergence. The work in recent years has thus focused on more advanced properties of MCMC samplers. For instance, a property of interest is geometric ergodicity; that is, the geometric decrease (in t) of the total variation distance between the distribution at time t and π. Mengersen and Tweedie (1996) established fairly complete conditions of geometric ergodicity when $q(\xi|\theta) = g(\xi)$ (*independent samplers*) and $q(\xi|\theta) = f(\xi - \theta)$ (*random walk samplers*). Uniform ergodicity is an even stronger type of convergence, in the sense that the constant C involved in the geometric bound $C\rho^t$ $(0 < \rho < 1)$ does not depend on the starting value $\theta^{(0)}$. This property cannot hold for random walks on noncompact sets in general, but can be exhibited for some Gibbs and independent samplers.

This type of research has practical bearings, moreover, because it allows us to compare and rank samplers in specific cases (e.g., geometric vs. nongeometric). In addition, a considerable amount of work has gone into estimation of the geometric constants C and ρ. The fact that an exact rate is available for some models of interest obviously is quite a major advance, because it means that precise information is thus available on the time required by the algorithm to converge. From another point of view, it also allows the rejection of inefficient sampling schemes. So what appears at first as a theoretical game ends up after a few years and a considerable amount of work by Roberts and coauthors (Roberts and Rosenthal 1998, 1999; Roberts and Tweedie 1999; Rosenthal 1995) as a valuable tool for practical implementation.

... By Scaling the Samplers ...

Metropolis–Hastings algorithms, like their accept–reject counterparts and unlike the Gibbs sampler, have an average probability $\rho > 0$ of rejecting the proposed values. In contrast to the accept–reject setting, a high value of ρ is not a valuable feature for random walk samplers, as this indicates a propensity to avoid the tails of π. Gelman, Gilks, and Roberts (1996) characterized this feature more quantitatively by deriving optimal acceptance rates, which should get close to .234 for high-dimensional Gaussian models. Similar developments by Roberts and Rosenthal (1998) showed that this limiting rate is close to .5 for Langevin algorithms.

... Or Taking Advantage of the Posterior ...

The random walk proposal is somehow the obvious choice and is used in the early works of Metropolis and Hastings. Despite (or because of?) its universality, it does not always provide satisfactory convergence properties, particularly in large dimensions. Even though reparameterization and optimal scaling of the proposal may often improve its performance, the random walk is somehow a shot in the dark.

An improvement that takes advantage of the fact that π is known up to a constant, while keeping the universality feature, is based on discretization of the Langevin diffusion

$$dX_t = (1/2)\nabla \log \pi(X_t)\, dt + dW_t,$$

which is associated with the stationary distribution π. The discretized version, which defines a new transition q, is $\xi_{t+1} = \theta^{(t)} + (\sigma^2/2)\nabla \log \pi(\theta^{(t)}) + \sigma\varepsilon_t$. It thus adds a shift term $(\sigma^2/2)\nabla \log \pi(\theta^{(t)})$ to the random walk, which ensures that the moves are more likely directed toward regions with high values of π than toward zones of low values of π. This may be detrimental in multimodal setups, because it pulls the chain back toward the nearest mode, but it has been observed to often speed up mixing and excursions on the surface of π. This practical observation is also backed by theoretical developments on the superior convergence properties of Langevin MCMC algorithms. A last word about the universality of this scheme is that it requires very little additional programming compared to random walk, because computation of the gradient can be delegated to a numerical procedure, instead of being analytic. This allows for more universal programming and also for transparent reparameterization.

... Or Even Jumping Between Dimensions

For us, if one area must be singled out for its impact, it is the ability to deal with variable-dimension models, which appear quite naturally in, for example, change-point problems and latent variable models, as well as in model choice and variable selection. Although the Gibbs sampler cannot handle such models unless one provides an encompassing model, MCMC techniques have been devised for this purpose. The most popular solution by far is the *reversible jump* technique of Green (1995), where two models $\mathcal{M}_1$ and $\mathcal{M}_2$ are embedded in models, $\mathcal{M}_1^*$ and $\mathcal{M}_2^*$, of identical dimension, so that a bijection can be constructed between $\mathcal{M}_1^*$ and $\mathcal{M}_2^*$. Although this device requires careful implementation, in particular because of the *Jacobian* of the transformation from $\mathcal{M}_1^*$ to $\mathcal{M}_2^*$, it has been used successfully in a considerable number of settings. Alternatives like jump diffusion (Grenander and Miller 1994) and birth-and-death process (Stevens 2000) have also been implemented in realistic settings.

3. ITERATION 2: DOES IT RUN?

It Should Start on Its Own ...

The main factor in the success of MCMC algorithms is that they can be implemented with little effort in a large variety of settings. This is obviously true of the Gibbs sampler, which, provided some conditional distributions are available, simply runs by generating from these conditionals, as shown by the BUGS software. The advent of the slice sampler (for an introduction, see Robert and Casella 1999, sec. 7.1.2) even increases this automation. So far, this is not exactly the case with MCMC samplers; there is no BUMH equivalent to BUGS, for instance! This is actually rather surprising, given that some Metropolis–Hastings algorithms allow for a

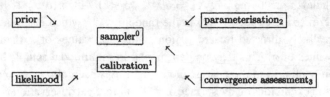

Figure 1.   Blocks Used in Modular Programming.

very modular programming (Fig. 1). This is to say, the prior, the likelihood, and the Metropolis–Hastings method can be programmed as separate blocks (procedures). For sampling schemes[0] like random walks or Langevin diffusions, the scale σ of the perturbation can be calibrated against the acceptance rate[1], and thus computed automatically. Moreover, these settings allow for straightforward changes of parameterization, which corresponds to the definition of a Jacobian procedure[2]. Finally, general convergence assessment tools[3] can be added to the blocks. This is not exactly fully automated programming, but it at least gives a hint of what the equivalent of BUGS could be. Obviously, this global strategy excludes more specific samplers, like independent samplers, because they depend more strongly on the prior/likelihood, but it works reasonably well in problems of moderate dimensions.

... And Check for Convergence ...

The main difficulty with MCMC samplers, as compared to Monte Carlo techniques, is that their validity is strictly asymptotic; that is, $\theta^{(t)}$ is converging in distribution to π. This difficulty is reflected in the impossibility of devising a general stopping rule for MCMC samplers and, correspondingly, by the emergence in the last 10 years of many partial convergence criteria that focus on specific convergence issues. Several reviews have appeared in the past (see, e.g., Brooks and Roberts 1998), and they give a rather exhaustive picture of the state of the art. A feature worth noticing is that the early debate about *parallel versus unique chains* (see, e.g., Gelman and Rubin 1992; Geyer 1992) has somehow died out, and most now agree on using signals that reflect the diversity of convergence issues and uses of the MCMC sample. This strategy is somehow conservative, as the chain keeps going until the "slowest" signal turns green, and bias results from using several stopping rules simultaneously. But these rules are not used in practice, neither with particular attention to significance level nor as exact stopping rules; one runs the sampler and checks every 1,000 or 10,000 iterations for sufficient agreement in the stopping rules. A software like CODA (Best, Cowles, and Vines 1995), which contains a series of general convergence diagnostics, actually works ex post; that is, once a batch of simulations has been done and signals whether or not simulations should be continued.

... While Improving On-Line

A given sampler will eventually fail to converge in a reasonable time for one particular problem. Solutions to this potential threat, well illustrated through pathological examples, are (a) to use several samplers simultaneously, in hybrid structures, (b) to modify temporarily the distribution to simulate, as in tempering, or (c) to adapt the proposal distribution on line; that is, use the MCMC sample (or samples) up to iteration T to modify the proposal for iterations $T + 1$ on. In (c), caution is necessary to handle the heterogeneity of the chain thus generated, but some advances have been made in this direction, establishing specific limit theorems and using renewal theory and nonparametric density estimates.

Another fruitful direction of research covers ways of accelerating MCMC algorithms; that is, to improve excursions over the support of π and/or estimation of $\mathbb{E}^\pi[h(\theta)]$. These methods include simulated tempering (Neal 1996b), simulated sintering (Liu and Sabati 1999), antithetic variates (Frigessi, Gasemyr, and Rue 2000) and Rao–Blackwellization (Robert and Casella 1999).

4. ITERATION 3: BOOSTING BAYESIAN INFERENCE

It is simply impossible to give a complete picture of the long-lasting effect of the seminal work of Gelfand and Smith (1990). We mention here some areas of inference that strongly benefit from MCMC techniques, stressing first that MCMC has clearly led to an upsurge in the use of hierarchical modeling in such areas as graphical models, generalized linear models, and image analysis.

Latent Variable Models

For the most obvious example of all, the mixture of distributions

$$\sum_{i=1}^k p_i f(x|\xi_i),$$

the advent of MCMC techniques provided a way of approximating the full posterior distribution for the first time. Since then, many other latent variable models, with distributions that are naturally expressed as marginals

$$\pi(\theta) = \int \pi^*(\theta, z) \, dz,$$

have been processed, thanks to MCMC techniques. These include hidden Markov and semi-Markov models, ARCH and switching ARMA models, stochastic volatility and discretized diffusion models, missing data and censored models, and deconvolution models (Robert 1998), with important consequences for econometrics (Kim,

Shephard, and Chib 1998), signal processing (Godsill 1997; Hodgson 1999), telecommunication, and finance.

Model Choice

The introduction of reversible jump techniques has been beneficial to the area of model choice and variable selection, as it allows for prior distributions on the dimension of a model as well as the parameter of each model. This possibility has led to a large amount of work on model choice, with new notions like *model averaging* and a renewed interest in (intrinsic and fractional) Bayes factors (Key, Perrichi, and Smith 1999). In particular, the advances in graphical models are quite impressive, as we now see setups in which the graph structures themselves can be simulated, hence allowing one to draw inference on the connections between variables (see, e.g., Madigan and York 1995). Furthermore, as this area of research allows local MCMC computations, with considerable performance improvements, very large and structured models now can be tackled properly (Giudici and Green 1999).

Nonparametric Bayes

Another area that has changed greatly over the last 10 years is Bayesian nonparametrics. The tools for density estimation, and curve fitting, and nonparametric or semiparametric regression are now quite different and improved, as shown by, for instance, Dey, Müller, and Sinha (1998). Alternatives to the Dirichlet prior, much criticized for its discrete features, have now appeared, either through the use of other processes (Beta, Levy) or of infinite mixtures. Reversible jump techniques provide an entry for density estimation via mixtures with an unknown number of components, using normal, beta, or Bernstein densities. The use of wavelets and estimation of wavelet coefficients have also been clearly enhanced by MCMC techniques (Clyde, Parmigiani, and Vidakovic 1998).

5. ITERATION 4: WHAT'S THE NEXT STEP?

The development of MCMC techniques is far from over, and we cannot predict what the next major step will be, at either the theoretical or the implementation level. Nonetheless, we can point out a few emerging, likely, or simply desirable developments.

Perfect Simulation

Although this area has been on the move since the seminal work of Propp and Wilson (1996), who showed that it is possible to use an MCMC sampler to simulate exactly from the distribution of interest by the technique of *coupling from the past,* it is yet unclear to us whether this technique will be available for statistical models on a scale larger than that for a restricted class of discrete models (besides point processes). Recent innovations like the multigamma coupler (Green and Murdoch 1999) and the perfect slice sampler (Mira, Møller, and Roberts 1999), which reduce a continuum of starting chains to a finite number of starting points, are steps in this direction, but the difficulty of implementation is considerable, and we are still very far from automated versions.

Adaptive and Sequential Algorithms

A likely development, both in theory and in practice, is increasing study of heterogeneous Markov chains and autoadaptative algorithms, as mentioned earlier (see also Holden 1998). A related area of considerable potential deals with sequential algorithms, where the distribution π_n (and possibly the dimension of the parameter space) varies with time, as in, for instance, tracking. Development of sequential importance sampling–resampling techniques (Liu and Chen 1998) and advanced importance sampling, which take advantage of the previous simulations to avoid rerunning a sample from scratch, should be able to face more and more real-time applications, such as those encountered in finance (Pitt and Shephard 1999), image analysis, signal processing, and artificial intelligence. Higher-order algorithms, like Hamiltonian schemes, could also offer a better potential to deal with these difficult setups.

Large Dimensions or Datasets

As of yet, too little innovation has been done in dealing with models involving either a large number of parameters, as in econometric or graphical models, or huge datasets, as in finance, signal processing, or teletraffic, and the MCMC samplers are currently limited in terms of the models they can handle. Neural networks, which have been investigated by Neal (1996a), are a typical example of such settings, but techniques need be devised for more classical statistical models, where completion, for instance, is too costly to be implemented. (See Mau and Newton 1997, for an illustration in genetics.)

Software

We mentioned BUGS and CODA as existing software dedicated to MCMC algorithms (more specifically to Gibbs sampling), but much remains to be done before MCMC becomes part of commercial software. This implies that each new problem requires some programming effort, most often in a basic language like Fortran or C, given that interpreted languages, like S-PLUS or Gauss, have difficulties inherent to Markov chain algorithms. An alternative would be to find ways of completely vectorizing an MCMC sampler, to allow for parallel programming, but so far little has been done in that direction. New statistical languages, like the Omega project (http://www.omegahat.org/), based on Java, may offer a solution, if they take into account MCMC requirements.

REFERENCES

Best, N. G., Cowles, M. K., and Vines, K. (1995), "CODA: Convergence Diagnosis and Output Analysis Software for Gibbs Sampling Output, Version 0.30," Technical report, MRC Biostatistics Unit, University of Cambridge.

Brooks, S. P., and Roberts, G. (1998), "Assessing Convergence of Markov Chain Monte Carlo Algorithms," *Statistics and Computing*, 8, 319–335.

Clyde, M., Parmigiani, G., and Vidakovic, B. (1998), "Multiple Shrinkage and Subset Selection in Wavelets," *Biometrika*, 85, 391–402.

Dey, D., Müller, P., and Sinha, D. (1998), *Practical Nonparametric and Semiparametric Bayesian Statistics*, New York: Springer-Verlag.

Frigessi, A., Gasemyr, J., and Rue, H. (2000), "Antithetic Coupling of Two Gibbs Sampler Chains," *Annals of Statistics*, 28, 1128–1149.

Gelfand, A. E., and Smith, A. F. M. (1990), "Sampling-Based Approaches to Calculating Marginal Densities," *Journal of the American Statistical Association*, 85, 398–409.

Gelman, A., Gilks, W. R., and Roberts, G. O. (1996), "Efficient Metropolis Jumping Rules," in *Bayesian Statistics 5*, eds. J. O. Berger, J. M. Bernardo, A. P. Dawid, D. V. Lindley, and A. F. M. Smith, London: Oxford University Press, pp. 599–608.

Gelman, A., and Rubin, D. B. (1992), "Inference From Iterative Simulation Using Multiple Sequences" (with discussion), *Statistics of Science*, 7, 457–511.

Geyer, C. J. (1992), "Practical Monte Carlo Markov Chain" (with discussion), *Statistics of Science*, 7, 473–511.

Gilks, W. R., Richardson, S., and Spiegelhalter, D. I. (1996), *Markov Chain Monte Carlo in Practice*, London: Chapman & Hall.

Giudici, P., and Green, P. J. (1999), "Decomposable Graphical Gaussian Model Determination," *Biometrika*, 86, 785–801.

Godsill, S. J. (1997), "Bayesian Enhancement of Speech and Audio Signals Which can be Modeled as ARMA Processes," *International Statistical Review*, 65, 1–21.

Green, P. J. (1995), "Reversible Jump MCMC Computation and Bayesian Model Determination," *Biometrika*, 82, 711–732.

Green, P. J., and Murdoch, D. (1999), "Exact Sampling for Bayesian Inference: Towards General Purpose Algorithms," in *Bayesian Statistics 6*, eds. J. O. Berger, J. M. Bernardo, A. P. Dawid, D. V. Lindley and A. F. M. Smith, London: Oxford University Press.

Grenander, U., and Miller, M. I. (1994), "Representation of Knowledge in Complex Systems" (with discussion), *Journal of the Royal Statistical Society*, 56, 549–603.

Hodgson, M. E. A. (1999), "A Bayesian restoration of an ion channel signal." *Journal of the Royal Statistical Society*, 61, 95–114.

Holden, L. (1998), "Adaptive Chains," Technical Report SAND/11/98, Norwegian Computing Center, Olso.

Key, J., Perrichi, L., and Smith, A. F. M. (1999), "Bayesian Model Choice: What and Why?," in *Bayesian Statistics 6*, eds. J. O. Berger, J. M. Bernardo, A. P. Dawid, and A. F. M. Smith, London: Oxford

University Press.

Kim, S., Shephard, N., and Chib, S. (1998), "Stochastic Volatility: Likelihood Inference and Comparison With ARCH Models," *Review of Economic Studies*, 65, 361–394.

Liu, J. S., and Chen, R. (1998), "Sequential Monte Carlo Methods for Dynamic Systems," *Journal of the American Statistical Association*, 93, 1032–1044.

Liu, J. S., and Sabati, C. (1999), "Simulated Sintering: Markov Chain Monte Carlo With Spaces of Varying Dimensions," in *Bayesian Statistics 6*, eds. J. O. Berger, J. M. Bernardo, A. P. Dawid, and A. F. M. Smith, London: Oxford University Press.

Madigan, D., and York, J. (1995), "Bayesian Graphical Models for Discrete Data," *International Statistical Review*, 63, 215–232.

Mau, B., and Newton, M. A. (1997), "Phylogenetic Inference on Dendograms Using Markov Chain Monte Carlo Methods," *Journal of Computing and Graphical Statistics*, 6, 122–131.

Mengersen, K. L., and Tweedie, R. L. (1996), "Rates of Convergence of the Hastings and Metropolis Algorithms," *The Annals of Statistics*, 24, 101–121.

Mira, A., Møller, J., and Roberts, G. O. (1999), "Perfect Slice Samplers," technical report, Università Degli Studi dell'Insurbia, Varese.

Neal, R. M. (1996a), *Bayesian Learning for Neural Networks*, New York: Springer-Verlag.

—— (1996b), "Sampling From Multimodal Distributions Using Tempered Transitions," *Statistics and Computing*, 4, 353–366.

Pitt, M. K., and Shephard, N. (1999), "Filtering via Simulation: Auxiliary Particle Filter," *Journal of the American Statistical Association*, 94, 590–599.

Propp, J. G., and Wilson, D. B. (1996), "Exact Sampling With Coupled Markov Chains and Applications to Statistical Mechanics," *Random Structures and Algorithms*, 9, 223–252.

Robert, C. P. (1998), "MCMC Specifics for Latent Variable Models," in *COMPSTAT 1998*, eds. R. Payne and P. J. Green, Heidelberg: Physica-Verlag. pp. 101–112.

Robert, C. P., and Casella, G. (1999), *Monte Carlo Statistical Methods*, New York: Springer-Verlag.

Roberts, G. O., and Rosenthal, J. S. (1998), "Optimal Scaling of Discrete Approximations to Langevin Diffusions," *Journal of the Royal Statistical Society*, Ser. B, 60, 255–268.

—— (1999), "Convergence of Slice Sampler Markov Chains," *Journal of the Royal Statistical Society*, Ser. B, 61, 643–660.

Roberts, G. O., and Tweedie, R. (1999), "Bounds on Regeneration Times and Convergence Rates for Markov Chains," *Stochastic Proceedings and Applications*, 80, 211–229.

Rosenthal, J. S. (1995), "Minorization Conditions and Convergence Rates for Markov Chain Monte Carlo," *Journal of the American Statistical Association*, 90, 558–566.

Stevens, M. (2000), "Bayesian Analysis of Mixtures With an Unknown Number of Components—An Alternative to Reversible Jump Methods," *The Annals of Statistics*, 28, 40–74.

Empirical Bayes: Past, Present, and Future

Bradley P. Carlin and Thomas A. Louis

1. INTRODUCTION

Despite (or perhaps because of) the enormous literature on the subject, the term *empirical Bayes* (EB) is rather difficult to define precisely. To some statisticians, it refers to a class of models; to others, a style of analysis; to still others, a philosophy for screening statistical procedures. But perhaps any definition must begin with how EB differs from an ordinary, fully Bayesian analysis. To understand this distinction, suppose that we have a distributional model $f(\mathbf{y}|\boldsymbol{\theta})$ for the observed data $\mathbf{y} = (y_1, \ldots, y_n)$ given a vector of unknown parameters $\boldsymbol{\theta} = (\theta_1, \ldots, \theta_k)$. Although the classical, or frequentist, statistician would assume that $\boldsymbol{\theta}$ is an unknown but fixed parameter to be estimated from $\mathbf{y}$, the Bayesian statistician would place a prior distribution $\pi(\boldsymbol{\theta}|\boldsymbol{\eta})$ on $\boldsymbol{\theta}$, where $\boldsymbol{\eta}$ is a vector of hyperparameters. With $\boldsymbol{\eta}$ known, the Bayesian uses Bayes's rule to compute the posterior distribution,

$$p(\boldsymbol{\theta}|\mathbf{y}, \boldsymbol{\eta}) = \frac{f(\mathbf{y}|\boldsymbol{\theta})\pi(\boldsymbol{\theta}|\boldsymbol{\eta})}{\int f(\mathbf{y}|\mathbf{u})\pi(\mathbf{u}|\boldsymbol{\eta})\,d\mathbf{u}} = \frac{f(\mathbf{y}|\boldsymbol{\theta})\pi(\boldsymbol{\theta}|\boldsymbol{\eta})}{m(\mathbf{y}|\boldsymbol{\eta})}, \tag{1}$$

where $m(\mathbf{y}|\boldsymbol{\eta})$ denotes the marginal distribution of $\mathbf{y}$.

If $\boldsymbol{\eta}$ is unknown, then information about it is captured by the marginal distribution. Moreover, if f and π form a conjugate pair of distributions (i.e., if $p(\boldsymbol{\theta}|\mathbf{y}, \boldsymbol{\eta})$ belongs to the same distributional family as π), then $m(\mathbf{y}|\boldsymbol{\eta})$ will be available in closed form. An EB analysis uses this marginal distribution to estimate $\boldsymbol{\eta}$ by $\hat{\boldsymbol{\eta}} \equiv \hat{\boldsymbol{\eta}}(\mathbf{y})$ [e.g., the marginal maximum likelihood estimator (MLE)] and then uses $p(\boldsymbol{\theta}|\mathbf{y}, \hat{\boldsymbol{\eta}})$ as the posterior distribution. In contrast, a fully Bayesian analysis (sometimes called *Bayes empirical Bayes,* or BEB) augments (1) by a hyperprior distribution, $h(\boldsymbol{\eta}|\boldsymbol{\lambda})$, and computes the posterior distribution as

Bradley P. Carlin is Professor, Division of Biostatistics, School of Public Health, University of Minnesota, Minneapolis, MN 55455. Thomas A. Louis is Senior Statistician, The RAND Corporation, Santa Monica, CA 90407. The first author's work was supported in part by National Institute of Allergy and Infectious Diseases (NIAID) grant R01-AI41966 and by National Institute of Environmental Health Sciences (NIEHS) grant 1-R01-ES07750; the second author's work was supported in part by NIEHS grant 1-R01-ES07750 and NIAID contract NO1-AI05073. The authors thank the editor and Dr. Melanie Wall for helpful comments that improved the manuscript.

$$p(\boldsymbol{\theta}|\mathbf{y}, \boldsymbol{\lambda}) = \frac{\int f(\mathbf{y}|\boldsymbol{\theta})\pi(\boldsymbol{\theta}|\boldsymbol{\eta})h(\boldsymbol{\eta}|\boldsymbol{\lambda})\,d\boldsymbol{\eta}}{\int\int f(\mathbf{y}|\mathbf{u})\pi(\mathbf{u}|\boldsymbol{\eta})h(\boldsymbol{\eta}|\boldsymbol{\lambda})\,d\mathbf{u}\,d\boldsymbol{\eta}}$$

$$= \int p(\boldsymbol{\theta}|\mathbf{y}, \boldsymbol{\eta})h(\boldsymbol{\eta}|\mathbf{y}, \boldsymbol{\lambda})\,d\boldsymbol{\eta}. \tag{2}$$

The second representation shows that the posterior is a mixture of posteriors (1) conditional on a fixed $\boldsymbol{\eta}$, with mixing via the hyperprior, updated by the data $\mathbf{y}$. Both EB and BEB use the observed data to provide information on $\boldsymbol{\eta}$ and thus "combine evidence." This combining has proven very successful in improving statistical analyses in a broad array of fields, from the estimation of false fire alarm rates in New York City (combining across neighborhoods; Carter and Rolph 1974) to modeling the decline in the CD4 counts in HIV-infected men (combining across men; DeGruttola, Lange, and Dafni 1991).

The "empirical" in EB arises from the fact that we are using the data to help determine the prior through estimation of the hyperparameter $\boldsymbol{\eta}$. Thus the basic distinction between EB and BEB is inclusion of the uppermost distribution h. (This uppermost distribution need not be precisely the third stage in the model; see the hierarchical models vignette by Hobert.) The posterior computed via (2) incorporates uncertainties associated with not knowing $\boldsymbol{\eta}$ and is generally preferred to basic EB. However, performance depends on choice of h—a difficult, important, and ongoing area of research.

The EB approach essentially replaces the integration in the rightmost part of (2) by a maximization (a substantial computational simplification), and bases inference on the estimated posterior distribution $p(\boldsymbol{\theta}|\mathbf{y}, \hat{\boldsymbol{\eta}})$. For example, any measure of the "middle" of this distribution (mean, median, mode) generally produces a suitable point estimate for $\boldsymbol{\theta}$. Interval estimates arise similarly; for instance, the upper and lower $\alpha/2$ points of the marginal estimated posterior $p(\theta_1|\mathbf{y}, \hat{\boldsymbol{\eta}})$ can be taken as a $100 \times (1-\alpha)\%$ confidence interval for θ_1. But although EB point estimates generally perform quite well, naive EB interval estimates of this type are often too narrow, as they ignore the uncertainty in estimating $\boldsymbol{\eta}$. Correcting EB intervals to have the desired (frequentist or Bayesian) coverage level has thus been the subject of significant research interest (see, e.g., Carlin and Gelfand 1990, 1991; Laird and Louis 1987).

The preceding development outlines what Morris (1983) calls *parametric* EB (PEB). Casella (1985) also provided a good introduction to this topic. The approach assumes the distribution at the penultimate level of the hierarchy, $\pi(\boldsymbol{\theta}|\boldsymbol{\eta})$, has a parametric form, so that choosing a (data-based) value for $\boldsymbol{\eta}$ is all that is required to completely specify the estimated posterior distribution. Alternatively, one can adopt a *nonparametric* EB (NPEB) approach, where the distribution at the penultimate level is known to exist, but its form is specified only generically as $\pi(\boldsymbol{\theta})$. Pioneered and championed by Robbins (1955, 1983), and further generalized and modernized by Maritz and Lwin (1989, sec. 3.4), van Houwelingen (1977), and others, this method first represents the posterior mean in terms of the unknown prior, then uses the data to estimate the Bayes rule directly. Recent advances substitute a nonparametric maximum likelihood (NPML) estimate of $\pi(\boldsymbol{\theta})$ (see Laird 1978).

EB has a long, colorful, and sometimes philosophically confused past, a vibrant present, and an uncertain future. We consider each of these in turn.

2. PAST

Somewhat ironically, the history of EB is not particularly "Bayesian," and it certainly has little in common with the traditional, subjectivist Bayesian viewpoint. As noted earlier, the earliest attempts were of the nonparametric variety and essentially represented attempts by frequentist decision theorists to use Bayesian tools to produce decision rules with good frequentist (not Bayesian) properties. For example, working in the setting where $y_i|\theta_i \overset{\text{ind}}{\sim} \text{Poisson}(\theta_i)$ and $\theta_i \overset{\text{iid}}{\sim} G(\cdot)$, Robbins (1955) showed that the NPEB estimator

$$\hat{\theta}_i^{\text{NPEB}} = \hat{E}_G(\theta_i|y_i) = (y_i + 1)\frac{\#(ys \text{ equal to } y_i + 1)}{\#(ys \text{ equal to } y_i)} \tag{3}$$

is asymptotically optimal in that as $k \to \infty$, its Bayes risk converges to the Bayes risk for the true Bayes rule where G is known. (Despite the name, "Bayes risk" is the expected value of a loss function, such as the average squared error loss $\text{ASEL}(\hat{\boldsymbol{\theta}}(\mathbf{y}), \boldsymbol{\theta}) = 1/k \sum_{i=1}^{k}(\hat{\theta}_i(\mathbf{y}) - \theta_i)^2$, where the expectation is taken *not* conditionally given the data $\mathbf{y}$, but rather *jointly* over both $\boldsymbol{\theta}$ and $\mathbf{y}$.) Asymptotically optimal estimators can perform very poorly even for fairly large sample sizes, but (3) does demonstrate the characteristic EB "borrowing strength" from data values other than y_i. Use of the NPML prior, $\hat{G}_{\text{NPML}}$, to compute the alternate estimator $E_{G_{\text{NPML}}}(\theta_i|y_i)$ produces substantial benefits over direct estimation of the Bayes rule in (3).

Parametric EB has close ties to Stein estimation, another primarily frequentist endeavor (see the vignette on minimaxity by W. Strawderman). Stein (1956) showed that in the case where $y_i|\theta_i \overset{\text{ind}}{\sim} N(\theta_i, \sigma^2), i = 1, \ldots, k$ and σ^2 is assumed known, the MLE $\hat{\boldsymbol{\theta}}^{\text{MLE}}(\mathbf{y}) = \mathbf{y}$ is inadmissible as an estimator of $\boldsymbol{\theta}$. That is, under average squared error loss there must exist another estimator with frequentist risk no larger than σ^2 for every possible $\boldsymbol{\theta}$ value. This dominating estimator was obtained by James and Stein (1961) as

$$\hat{\theta}_i^{\text{JS}}(\mathbf{y}) = \left[1 - \frac{(k-2)\sigma^2}{\|\mathbf{y}\|^2}\right]y_i,$$

where $\|\mathbf{y}\|^2 \equiv \sum y_i^2$. The connection to PEB was provided later in a celebrated series of articles by Efron and Morris (1971; 1972a,b; 1973a,b; 1975; 1977). Among many other things, these authors showed that $\hat{\boldsymbol{\theta}}^{\text{JS}}$ is exactly the PEB point estimator obtained under the assumption that $\theta_i|\tau^2 \overset{\text{iid}}{\sim} N(0, \tau^2)$ where the shrinkage factor $B \equiv \sigma^2/(\sigma^2 + \tau^2)$ is estimated by $\hat{B} = (k-2)\sigma^2/\|\mathbf{y}\|^2$.

The early EB authors' consistent use of Bayesian tools to further frequentist goals while "using the data twice" (first to help determine the prior, then again in the usual Bayesian way when computing the posterior) was not highly regarded by the (primarily subjectivist) Bayesian community of the time (e.g., de Finetti, Lindley, and Savage). In his discussion of Copas (1969), Lindley noted that "there is no one less Bayesian than an empirical Bayesian"; later in a discussion of Morris (1983), he described much of the asymptotics supporting NPEB as "technicalities out of control." In his discussion of the same work, Dempster borrowed a metaphor of Savage (1961, p. 578) in describing an empirical Bayesian as someone who "breaks the Bayesian egg but then declines to enjoy the Bayesian omelette." Still, the early EB work of the 1950s and 1960s helped further the rise of *objective* Bayesian thinking decades before the Gibbs sampler made such thinking routinely possible and acceptable. Further work by Morris (1983) and Hill (1990) encouraged a more unifying role for EB, in which evaluations of procedures could be made by averaging over *both* the data space $\mathcal{Y}$ and the parameter space Θ, with the EB position emerging as a plausible compromise between the strict Bayes and frequentist positions.

3. PRESENT

The cumulative impact of EB on statistical applications continues to be enormous. Statisticians and users of statistics, many of whom were trained to distrust Bayesian methods as overly subjective and theoretically mysterious, can nonetheless often appreciate the value of borrowing strength from similar but independent experiments. To use another metaphor, EB thus offers a way to dangle one's legs in the Bayesian water without having to jump completely into the pool. An electronic search of the latest Current Index to Statistics on "empirical Bayes" confirms its increasing popularity, yielding a median of 2.5 hits per year during 1964–1969, 11 hits per year during 1970–1979, 32 hits per year during 1980–1989, and 46 hits per year during 1990–1996. EB methods have, for example, enjoyed broad application in the analysis of longitudinal, survival, and spatially correlated data (Clayton and Kaldor 1987; Laird and Ware 1982).

The Bayes/EB formalism also ideally structures combining information from several published studies of the same research area, a scientific discipline commonly referred to as *meta-analysis* (Cooper and Hedges 1994), though in this context primary interest is in the hyperparameters (η) rather than the parameters from individual studies (θ). More generally, the hierarchical structure allows for honest assessment of heterogeneity both within and between groups, such as clinical centers or census areas. The review article by Breslow (1990) and the accompanying discussion contain an excellent summary of past and potential future application areas for Bayes and EB methods in the public health and biomedical sciences.

Important methodologic contributions also continue. For example, Efron (1996) developed an EB approach to combining likelihoods for similar but independent parameters θ_i. In keeping with the EB philosophy, little in the way of Bayesian prior information is required, and frequentist ideas (such as bias correction and robustness) are critical to the method. Other important recent EB work includes that on spatial statistics by Raghunathan (1993) and nonparametric growth curves by Altman and Casella (1995).

4. FUTURE

Returning to the theme of our opening sentence, one's view of the future of EB is indelibly tied to one's view of its past and present, as well as to one's own "upbringing." One of us (Louis) received his doctorate from Columbia University and had Herb Robbins as an advisor, and so was exposed to an NPEB framework in which achieving good frequentist properties was paramount. But Louis's view then grew to include parametric approaches and a healthy dose of purely Bayesian thinking. This leads to an optimistic view of EB's future, in which more and more applied problems are tackled using EB methods, which preserve a strong measure of frequentist validity for the resulting inferences. The other author (Carlin) came along nearly two decades later, after the appearance of the marvelous PEB discussion paper by Morris (1983) but just prior to the emergence of the Gibbs sampler and other Markov chain Monte Carlo (MCMC) tools for implementing fully Bayesian analysis. This author started from a PEB shrinkage point of view, but one in which EB methods were thought of as approximations to overly cumbersome fully Bayesian hierarchical analyses. With the widespread availability of MCMC tools such as the BUGS software (Spiegelhalter, Thomas, Best, and Gilks 1995), this produces a much more pessimistic outlook for EB, as the need for such approximations (and the corresponding restrictions on which models can be handled) has more or less vanished.

Still, fully Bayesian solutions implemented via MCMC are by no means "plug and play." Convergence of the algorithms is notoriously difficult to diagnose, with most of the usual diagnostics having well-known flaws (Cowles and Carlin 1996; Mengersen, Robert, and Guihenneuc-Jouyaux 1999). Worse, the sheer power of MCMC methods has led to the temptation to fit models larger than the data can readily support without a strongly informative prior structure—now something of a rarity in applied Bayesian work. In cases like this, an EB approach may well be superior to a full hierarchy with improper priors, as the computation will be better behaved, and thus the associated underlying theory may be better understood.

As an example of this point, consider the problem of specifying a vague hyperprior for a variance component τ^2. The most widespread current choice seems to be the gamma$(\varepsilon, \varepsilon)$ prior; that is, with mean 1 but variance $1/\varepsilon$ (see, e.g., Spiegelhalter et al. 1995). But recent work by Hodges and Sargent (2001) and Natarajan and Kass (2000) showed that such hyperpriors, though appearing "noninformative," can actually have significant impact on the resulting posterior distributions. And although this hyperprior is proper, it is "nearly improper" for suitably small ε, potentially leading to the aforementioned MCMC convergence failure—or worse, the appearance of MCMC convergence when in fact the joint posterior is also improper. Thus it might well be that reverting to an EB approach here (replacing τ^2 by $\hat{\tau}^2$) will produce a estimated posterior that, though not fully defensible from a purely Bayesian standpoint, produces improved estimates while also being safer to use and easier to obtain.

Although EB and BEB approaches have proven very effective, they are by no means a panacea, and continued research and development is needed. We urge progress in developing hierarchical analyses that are efficient and effective, but also robust with respect to prior and other specifications. If such robustness is not present, then the sources of the sensitivity must be investigated, documented, and questioned. Broadening and deepening our understanding of the influence of the hyperprior h is an important aspect of these developments. Finally, we propose that one's philosophical bent need not play a significant role in the decision to use EB or vague hyperprior BEB, as these generally produce procedures with good frequentist, Bayes, and EB properties (see, e.g., Carlin and Louis 2000, chap. 4).

REFERENCES

Altman, N. S., and Casella, G. (1995), "Nonparametric Empirical Bayes Growth Curve Analysis," *Journal of the American Statistical Association*, 90, 508–515.

Breslow, N. E. (1990), "Biostatistics and Bayes" (with discussion), *Statistical Science*, 5, 269–298.

Carlin, B. P., and Gelfand, A. E. (1990), "Approaches for Empirical Bayes Confidence Intervals," *Journal of the American Statistical Association*, 85, 105–114.

——— (1991), "A Sample Reuse Method for Accurate Parametric Empirical Bayes Confidence Intervals," *Journal of the Royal Statistical Society*, Ser. B, 53, 189–200.

Carlin, B. P., and Louis, T. A. (2000), *Bayes and Empirical Bayes Methods for Data Analysis*, 2nd ed. Boca Raton, FL: Chapman and Hall/CRC Press.

Carter, G., and Rolph, J. (1974), "Empirical Bayes Methods Applied to Estimating Fire Alarm Probabilities," *Journal of the American Statistical Association*, 69, 880–885.

Casella, G. (1985), "An Introduction to Empirical Bayes Data Analysis," *American Statistician*, 39, 83–87.

Clayton, D. G., and Kaldor, J. M. (1987), "Empirical Bayes Estimates of Age-Standardized Relative Risks for Use in Disease Mapping," *Biometrics*, 43, 671–681.

Cooper, H., and Hedges, L. (Eds.) (1994), *The Handbook of Research Synthesis*, New York: Russell Sage Foundation.

Copas, J. (1969), "Compound Decisions and Empirical Bayes," *Journal of the Royal Statistical Society*, Ser. B, 31, 397–425.

Cowles, M. K., and Carlin, B. P. (1996), "Markov Chain Monte Carlo Convergence Diagnostics: A Comparative Review," *Journal of the American Statistical Association*, 91, 883–904.

DeGruttola, V., Lange, N., and Dafni, U. (1991), "Modeling the Progression of HIV Infection," *Journal of the American Statistical Association*, 86, 569–577.

Efron, B. (1996), "Empirical Bayes Methods for Combining Likelihoods" (with discussion), *Journal of the American Statistical Association*, 91, 538–565.

Efron, B., and Morris, C. N. (1971), "Limiting the Risk of Bayes and Empirical Bayes Estimators—Part I: The Bayes Case," *Journal of the American Statistical Association*, 66, 807–815.

——— (1972a), "Limiting the Risk of Bayes and Empirical Bayes Estimators—Part II: The Empirical Bayes Case," *Journal of the American Statistical Association*, 67, 130–139.

——— (1972b), "Empirical Bayes on Vector Observations: An Extension of Stein's Method," *Biometrika*, 59, 335–347.

——— (1973a), "Stein's Estimation Rule and Its Competitors—An Empirical Bayes Approach," *Journal of the American Statistical Association*, 68, 117–130.

——— (1973b), "Combining Possibly Related Estimation Problems" (with discussion), *Journal of the Royal Statistical Society*, Ser. B, 35, 379–421.

——— (1975), "Data Analysis Using Stein's Estimator and Its Generalizations," *Journal of the American Statistical Association*, 70, 311–319.

——— (1977), "Stein's Paradox in Statistics," *Scientific American*, 236, 119–127.

Hill, J. R. (1990), "A General Framework for Model-Based Statistics," *Biometrika*, 77, 115–126.

Hodges, J. S., and Sargent, D. J. (2001), "Counting Degrees of Freedom in Hierarchical and other Richly Parameterized Models," *Biometrika*, to appear.

James, W., and Stein, C. (1961), "Estimation With Quadratic Loss," in *Proceedings of the Fourth Berkeley Symposium on Mathematical Statistics and Probability*, 1, Berkeley, CA: University of California Press, pp. 361–379.

Laird, N. M. (1978), "Nonparametric Maximum Likelihood Estimation of a Mixing Distribution," *Journal of the American Statistical Association*, 73, 805–811.

Laird, N. M., and Louis, T. A. (1987), "Empirical Bayes Confidence Intervals Based on Bootstrap Samples" (with discussion), *Journal of the American Statistical Association*, 82, 739–757.

Laird, N. M., and Ware, J. H. (1982), "Random-Effects Models for Longitudinal Data," *Biometrics*, 38, 963–974.

Maritz, J. S., and Lwin, T. (1989), *Empirical Bayes Methods* (2nd ed.), London: Chapman and Hall.

Mengersen, K. L., Robert, C. P., and Guihenneuc-Jouyaux, C. (1999), "MCMC convergence diagnostics: a revieWWW" (with discussion), in *Bayesian Statistics 6*, eds. J. M. Bernardo, J. O. Berger, A. P. Dawid, and A. F. M. Smith, London: Oxford University Press, pp. 415–440.

Morris, C. N. (1983), "Parametric Empirical Bayes Inference: Theory and Applications," *Journal of the American Statistical Association*, 78, 47–65.

Natarajan, R., and Kass, R. E. (2000), "Reference Bayesian Methods for Generalized Linear Mixed Models," *Journal of the American Statistical Association*, 95, 227–237.

Raghunathan, T. E. (1993), "A Quasi-Empirical Bayes Method for Small Area Estimation," *Journal of the American Statistical Association*, 88, 1444–1448.

Robbins, H. (1955), "An Empirical Bayes Approach to Statistics," in *Proceedings of the Third Berkeley Symposium on Mathematical Statistics and Probability*, 1, Berkeley, CA: University of California Press, pp. 157–164.

——— (1983), "Some Thoughts on Empirical Bayes Estimation," *The Annals of Statistics*, 1, 713–723.

Savage, L. J. (1961), "The Foundations of Statistics Reconsidered," in *Proceedings of the Fourth Berkeley Symposium on Mathematical Statistics and Probability*, 1, Berkeley, CA: University of California Press, pp. 575–586.

Spiegelhalter, D. J., Thomas, A., Best, N., and Gilks, W. R. (1995), "BUGS: Bayesian Inference Using Gibbs Sampling, Version 0.50," technical report, Cambridge University, Medical Research Council Biostatistics Unit, Institute of Public Health.

Stein, C. (1956), "Inadmissibility of the Usual Estimator for the Mean of a Multivariate Normal Distribution," in *Proceedings of the Third Berkeley Symposium on Mathematical Statistics and Probability*, 1, Berkeley, CA: University of California Press, pp. 197–206.

van Houwelingen, J. C. (1977), "Monotonizing Empirical Bayes Estimators for a Class of Discrete Distributions With Monotone Likelihood Ratio," *Statistica Neerlderlandia*, 31, 95–104.

Linear and Log-Linear Models

Ronald Christensen

1. INTRODUCTION

In its most general form a linear model is simply a statement that a collection of random variables $y_1, \ldots, y_n$ have mean values with linear structure. In other words, if $E(y_i) = m_i$, the model is $m_i = x_i'\beta$, where x_i is a p vector of fixed known predictor (or classification) variables and $\beta = (\beta_1, \ldots, \beta_p)'$ is a vector of unknown, unobservable parameters. Collecting terms into vectors $Y = (y_1, \ldots, y_n)'$, $m = (m_1, \ldots, m_n)'$, and collecting the x_i's into a matrix X, $E(Y) = m = X\beta$. By positing a latent vector of errors $e = (e_1, \ldots, e_n)'$ with $E(e) = 0$, we can write the model as $Y = X\beta + e$. Most often, the y_i's in a linear model are assumed to have a multivariate normal distribution (see Arnold 1981; Christensen 1996; Seber 1977).

A log-linear model is a statement that, for a collection of random variables $y_1, \ldots, y_n$ with $E(y_i) = m_i$, the model is $\log(m_i) = x_i'\beta$. If we define a multivariate log transformation to be the log applied to each element of a vector, then the log-linear model can be written as $\log[E(Y)] = \log(m) = X\beta$. In a log-linear model, the y_i's are often assumed to have a multinomial distribution, or they are partitioned into groups in which the groups have independent multinomial distributions. Alternatively, the y_i's can be assumed to be independent Poisson random variables (see Agresti 1990; Christensen 1997). Log-linear models are also used in survival analysis as accelerated failure time models, but we do not discuss such applications here.

Finally, logistic regression and logit models are a special case of log-linear models. Suppose that $y_i \sim \text{bin}(N_i, p_i)$. The logistic model has $\text{logit}(p_i) \equiv \log[p_i/(1 - p_i)] = x_i'\beta$, or using obvious vector matrix notation, $\text{logit}(p) = X\beta$. Writing $m_{1i} = E(y_i)$ and $m_{2i} = E(N_i - y_i)$ and creating corresponding vectors m_1 and m_2,

$$\text{logit}(p) = X\beta \quad \text{iff} \quad \log \begin{pmatrix} m_1 \\ m_2 \end{pmatrix} = \begin{bmatrix} I & X \\ I & 0 \end{bmatrix} \begin{bmatrix} \alpha \\ \beta \end{bmatrix},$$

where the β vectors are identical in the two models.

These three applications are all closely related to generalized linear models.

Ronald Christensen is Professor of Statistics, Department of Math and Statistics, University of New Mexico, Albuquerque, NM 87131-1141 (E-mail: fletcher@stat.unm.edu).

2. LINEAR MODELS

I think that the future of linear models is much what it was 10 years ago. The future will involve applications of the current methods (primarily for independent data) and improvements in the current methods for independent data, but the future of linear models research lies primarily in developing methods for correlated data. The basic model is

$$E(\mathbf{Y}) = \mathbf{X}\boldsymbol{\beta}, \qquad \text{cov}(\mathbf{Y}) = \mathbf{V}(\boldsymbol{\theta})$$

for some vector of parameters $\boldsymbol{\theta}$ that ensures that $\mathbf{V}(\boldsymbol{\theta})$ is nonnegative definite. Moreover, multivariate normality is often assumed; that is, $\mathbf{Y} \sim \mathbf{N}(\mathbf{X}\boldsymbol{\beta}, \mathbf{V}(\boldsymbol{\theta}))$. So the future is in developing and improving statistical procedures for evaluating $\boldsymbol{\theta}$ and also for making predictions and evaluating $\boldsymbol{\beta}$ when $\boldsymbol{\theta}$ must be estimated. Currently, to simplify the problem, the parameters in $\boldsymbol{\theta}$ are often assumed to be unrelated to the parameters in $\boldsymbol{\beta}$. If the parameters are related, then a transformation of the data may make this assumption plausible. Otherwise, methods such as those discussed for independent observations by Carroll and Ruppert (1988) are needed. In any case, real interest focuses on the special cases used in applications involving correlated data.

With spatial data, everything depends on the locations at which data are collected. If data are collected at locations $u_1, \ldots, u_n$, then we observe dependent variables $y(u_i)$ and predictor vectors $\mathbf{x}(u_i)$. The universal kriging model is simply the linear model $y(u_i) = \mathbf{x}(u_i)'\boldsymbol{\beta} + e(u_i)$. The elements of $\mathbf{V}(\boldsymbol{\theta})$ are $\text{cov}[y(u_i), y(u_j)]$. Much has been done and much remains to be done on creating good models for $\mathbf{V}(\boldsymbol{\theta})$ and developing appropriate inferential methods (see Christensen 1991; Cressie 1993; Stein 1999).

To date, the most widely used application has been mixed linear models with variance components. In a mixed model, $\mathbf{V}(\boldsymbol{\theta})$ is a linear combination of σ_s^2 parameters, $\mathbf{V}(\boldsymbol{\theta}) = \sum_{s=0}^{q} \sigma_s^2 \mathbf{Z}_s \mathbf{Z}_s'$, where the $\mathbf{Z}_s$'s are known matrices and, most often, $\mathbf{Z}_0 = \mathbf{I}$. Typically, mixed models are written as $\mathbf{Y} = \mathbf{X}\boldsymbol{\beta} + \sum_{s=1}^{q} \mathbf{Z}_s \boldsymbol{\gamma}_s + \mathbf{e}$ where the $\boldsymbol{\gamma}_s$'s are assumed to be independent with $E(\boldsymbol{\gamma}_s) = \mathbf{0}$ and $\text{cov}(\boldsymbol{\gamma}_s) = \sigma_s^2 \mathbf{I}$. (Note that the dimensions of the $\boldsymbol{\gamma}_s$'s may depend on s.) For expositions near the current state of knowledge, see Searle, Casella, and McCulloch (1992) and Khuri, Mathew, and Sinha (1998). In particular, split plot and cluster sampling models are valuable special cases of mixed models.

An interesting approach to estimating $\boldsymbol{\theta}$ in $\mathbf{V}(\boldsymbol{\theta})$ is the use of linear and nonlinear model techniques to develop quadratic estimation procedures. Briefly, for mixed models, $\mathbf{V}(\boldsymbol{\theta})$ is linear in its parameters, so one can write a linear model,

$$E[(\mathbf{Y} - \mathbf{X}\boldsymbol{\beta}) \otimes (\mathbf{Y} - \mathbf{X}\boldsymbol{\beta})] = \sum_{s=0}^{q} \sigma_s^2 \text{vec}(\mathbf{Z}_s \mathbf{Z}_s').$$

The σ_s^2's can be estimated by ordinary or weighted least squares or by various itera-tive procedures. These correspond to different versions of MINQUE and to maximum likelihood and residual maximum likelihood (REML) estimates. Of course complica-tions arise from β being unknown, but imposing a translation invariance requirement involves only simple adjustments and eliminates β from the problem. A more se-rious difficulty is that the covariance structure of the linear model depends on the same parameters (the σ_s^2's) as the linear structure. This linearization technique was initially proposed by Seely (1970). Pukelsheim (1976) provided an appropriate nota-tion. When $\mathbf{V}(\theta)$ is not linear in its parameters, methods of nonlinear regression can be applied to estimation (see Searle et al. 1992, chap. 12, and Christensen 1993 for details).

The multivariate linear model is also just a special case of the linear model. If we have q dependent variables listed in the columns of $\mathbf{Y}_{n \times q}$ with correlated columns and uncorrelated rows, with the same predictor variables $\mathbf{X}$ for each column of $\mathbf{Y}$, but separate columns of regression parameters for each dependent variable, say $\mathbf{B}_{p \times q}$, and a conformable error matrix, then we get the standard multivariate linear model $\mathbf{Y} = \mathbf{XB} + \mathbf{e}$. The equivalent standard univariate linear model is

$$\text{vec}(\mathbf{Y}) = [\mathbf{I}_q \otimes \mathbf{X}]\text{vec}(\mathbf{B}) + \text{vec}(\mathbf{e}), \qquad E[\text{vec}(\mathbf{e})] = 0, \qquad \text{cov}[\text{vec}(\mathbf{e})] = \Sigma \otimes \mathbf{I}_n.$$

Here θ is the entire matrix Σ, but estimation and inference is relatively easy. What is more interesting is modeling Σ. For instance, often the rows of $\mathbf{Y}$ indicate independent measurements on different people and the columns indicate successive measurements on the same person. In this repeated-measures (longitudinal data) setting, modeling Σ as the covariance structure for some time series is a worthwhile approach that needs continued development (Diggle 1990, chap. 5).

It is pretty obvious that linear models can be used for time series data with the linear structure modeling trend and $\mathbf{V}(\theta)$ as the covariance matrix of an appropriate stationary time series model. It is perhaps less well known that frequency domain time series is really just an application of mixed linear models. The spectral representation theorem implies that a stationary process y_i can be approximated arbitrarily closely by a mixed model based on sines and cosines (Doob 1953, p. 486). Define $\mathbf{Z}_k = [\mathbf{C}_k, \mathbf{S}_k]$ with

$$\mathbf{C}_k = \left[\cos\left(2\pi\frac{k}{n}1\right), \cos\left(2\pi\frac{k}{n}2\right), \ldots, \cos\left(2\pi\frac{k}{n}n\right) \right]'$$

and

$$\mathbf{S}_k = \left[\sin\left(2\pi\frac{k}{n}1\right), \sin\left(2\pi\frac{k}{n}2\right), \ldots, \sin\left(2\pi\frac{k}{n}n\right) \right]'.$$

Also, for simplicity, assume that n is even. The mixed model for vector $\mathbf{Y}$ is

$$\mathbf{Y} = \mathbf{J}\alpha_0 + \sum_{k=1}^{(n/2)-1} \mathbf{Z}_k\gamma_k + \mathbf{C}_{n/2}\alpha_{n/2}, \tag{1}$$

where $\mathbf{J}$ is a column of 1's, $E(y_i) = \alpha_0$, $E(\boldsymbol{\gamma}_k) = \mathbf{0}$, $E(\alpha_{n/2}) = 0$, $\mathrm{cov}(\boldsymbol{\gamma}_k) = \sigma_k^2 I_2$, $\mathrm{var}(\alpha_{n/2}) = \sigma_{n/2}^2$, $\mathrm{cov}(\boldsymbol{\gamma}_k, \boldsymbol{\gamma}_{k'}) = \mathbf{0}$ $(k \neq k')$, and $\mathrm{cov}(\boldsymbol{\gamma}_k, \alpha_{n/2}) = \mathbf{0}$. Here

$$\mathbf{V}(\boldsymbol{\theta}) = \sum_{j=1}^{[(n-1)/2]} \sigma_j^2 \mathbf{Z}_j \mathbf{Z}_j' + \sigma_{n/2}^2 \mathbf{C}_{n/2} \mathbf{C}_{n/2}'.$$

The $\mathbf{C}_k$'s and $\mathbf{S}_k$'s form an orthonormal basis for $\mathbf{R}^n$, so (1) is a saturated model. What is of more interest is finding reduced models that still explain the data; that is, identifying the important frequencies. The model is fitted using ordinary least squares, and because of the orthogonality, statistical analysis is relatively easy. For example, the periodogram is just the collection of mean squares for the different $\mathbf{Z}_k$'s (see Christensen 1991, chap. IV, for more discussion).

What I find intriguing for the next century is the similarity between this and wavelet analysis. Wavelets just use an orthonormal basis that is different from the sines and cosines basis, one in which the elements have a smaller support. In any case, wavelets fit a saturated linear model and focus on finding reduced models; that is, identifying the basis members that most help to explain the data. Although one can always use any basis to explain the data, I am unaware of a wavelet representation theorem that justifies the analysis in the same way that the spectral representation theorem justifies use of the mixed model.

These ideas can also be extended to treat nonparametric regression (Zhang 1995). Suppose that we have simple regression data $(x_i, y_i), i = 1, \ldots, n$, for which the x_i are restricted to a bounded set, say $(0, 1]$ for convenience. We can redefine a (nonorthogonal) basis to use in (1),

$$\mathbf{C}_k = [\cos(2\pi k x_1), \cos(2\pi k x_2), \ldots, \cos(2\pi k x_n)]'$$

and

$$\mathbf{S}_k = [\sin(2\pi k x_1), \sin(2\pi k x_2), \ldots, \sin(2\pi k x_n)]',$$

then fit the saturated model with ordinary least squares and identify important vectors (predictor variables) in the basis. More generally, known wavelet functions $\psi_j(x)$ can be defined on $(0, 1]$, the saturated linear model $y_i = \sum_{j=1}^n \beta_j \psi_j(x_i)$ fitted, and important wavelet functions identified. Typically, many parameters β_j will need to be nonzero to get a good fit; hence, as is often the case, the nonparametric procedure is actually a very highly parametric procedure. A relatively degenerate example involves defining functions equivalent to the Haar basis

$$\psi_j(x) = \begin{cases} 1 & \mathbf{x} \in \left(\frac{j-1}{n}, \frac{j}{n} \right] \\ 0 & \text{otherwise} \end{cases}.$$

Note the small support of each ψ_j. However, in this case, the vectors

$$\boldsymbol{\Psi}_j = [\psi_j(x_1), \psi_j(x_2), \ldots, \psi_j(x_n)]'$$

are unlikely to define a basis, and the fitting procedure is quite trivial. More interesting would be to define $\psi_j(\mathbf{x})$ as the density of a normal that essentially lives on $([(j-1)/n], j/n]$, say, $N((j/n) - (1/2n), 1/16n^2)$. With this definition of $\psi_j(x)$, the Ψ_j's should define a basis. There is great latitude in choosing the ψ_j's, but in any case, after identifying important ψ_j's and estimating their β_j coefficients, prediction for any x is trivial.

3. LOG-LINEAR MODELS

I think that the future of log-linear models is much the same as the future of linear models. The future will involve applications of the current methods and improvements in the current methods, but in terms of research, the future lies mostly in developing methods for correlated data.

Graphical log-linear models should continue to be developed to explore the relationships between discrete variables (Christensen 1997, chap. 5; Edwards 1995; Lauritzen 1996; Whittaker 1990). I expect Bayesian methods for linear, log-linear, and logistic models to be increasingly incorporated into models for directed acyclic graphs and chain graphs. In fact, I expect Bayesian logistic regression to thrive quite generally, because it gives small-sample results relatively simply (Christensen 1997, chap. 13).

Mixed models have been extended to generalized mixed models that incorporate random effects into generalized linear models such as log-linear and logistic models (McGilchrist 1994). These models should see continued development.

Currently, several methods are available for analyzing repeated measurements. These include generalized estimating equations (Zeger, Liang, and Albert 1988), generalizations of the Rasch model, (Agresti and Lang 1993; Conaway 1989), and conditional log-linear models (Gilula and Haberman 1994). This situation should get sorted out in the next century.

4. THE FUTURE OF PEDAGOGY

Linear models, log-linear models, logistic models, and all other generalized linear models share a linear structure that largely determines the nature of the model. I think that we need to start teaching regression and analysis of variance as model-based approaches to data analysis rather than as parameter-based approaches. If we do that for linear models, then the modeling skills that we teach will apply to all the other applications. For example, contrasts, including orthogonal contrasts, are wonderful for examining balanced analysis of variance problems. But I think that we would be better off teaching students how to examine the same kinds of things using model-based approaches that will apply to unbalanced situations and also to nonnormal models.

To illustrate, consider a linear regression model

$$y_i = \beta_0 + \beta_1 x_{i1} + \beta_2 x_{i2} + \beta_3 x_{i3} + \varepsilon_i, \quad \varepsilon_i\text{'s indep. N}(0, \sigma^2).$$

Suppose that we want to test $H_0 : \beta_1 + \beta_3 = 5$ and $\beta_2 = 0$. We can rewrite the hypothesis as $H_0 : \beta_1 = 5 - \beta_3$ and $\beta_2 = 0$, and create a reduced model by substitution:

$$y_i = \beta_0 + (5 - \beta_3)x_{i1} + 0x_{i2} + \beta_3 x_{i3} + \varepsilon_i, \varepsilon_i\text{'s indep. N}(0, \sigma^2).$$

Rearranging terms gives a reduced model:

$$y_i - 5x_{i1} = \beta_0 + \beta_3(x_{i3} - x_{i1}) + \varepsilon_i, \quad \varepsilon_i\text{'s indep. N}(0, \sigma^2).$$

This is just a new linear model with a new dependent variable $y_i - 5x_{i1}$ and one predictor variable $(x_{i3} - x_{i1})$. One can compute SSE and $df E$ in the usual way for this reduced model, and simply use these quantities in the usual formula for the F statistic to get the test. If the data are consistent with the reduced model, then we have a more precise model than the one we started with. Although testing provides no assurance that the reduced model is correct, the data are at least consistent with it. This more precise model can be investigated for the validity of its predictions and its usefulness in explaining the data collection process.

Reduced model procedures apply to any linear model, and, by incorporating offsets into generalized linear models, the same reduced model procedures apply to any model that uses a similar linear structure.

5. CONCLUSIONS

In the 21st century, research on linear and log-linear models will focus on models and methods for correlated data. Nonetheless, current methodologies should see increased use and refinement.

REFERENCES

Agresti, A. (1990), *Categorical Data Analysis*. New York: Wiley.

Agresti, A., and Lang, J. B. (1993), "A Proportional Odds Model With Subject-Specific Effects for Repeated Ordered Categorical Responses," *Biometrika*, 80, 527–534.

Arnold, S. F. (1981), *The Theory of Linear Models and Multivariate Analysis*, New York: Wiley.

Carroll, R. J., and Ruppert, D. (1988), *Transformation and Weighting in Regression*, New York: Chapman and Hall.

Christensen, R. (1991), *Linear Models for Multivariate, Time Series, and Spatial Data*, New York: Springer-Verlag.

—— (1993), "Quadratic Covariance Estimation and Equivalence of Predictions," *Mathematical Geology*, 25, 541–558.

—— (1996), *Plane Answers to Complex Questions: The Theory of Linear Models* (2nd ed.), New York: Springer-Verlag.

—— (1997), *Log-Linear Models and Logistic Regression* (2nd ed.), New York: Springer-Verlag.

Conaway, M. R. (1989), "Analysis of Repeated Categorical Measurements With Conditional Likelihood Methods," *Journal of the American Statistical Association*, 84, 53–62.

Cressie, N. A. C. (1993), *Statistics for Spatial Data* (rev. ed.), New York: Wiley.

Diggle, P. J. (1990), *Time Series: A Biostatistical Introduction*, Oxford, U.K.: Clarendon Press.

Doob, J. L. (1953), *Stochastic Processes*, New York: Wiley.

Edwards, D. (1995), *Introduction to Graphical Modeling*, Berlin: Springer-Verlag.

Gilula, Z., and Haberman, S. J. (1994), "Conditional Log-Linear Models for Analyzing Categorical Panel Data," *Journal of the American Statistical Association*, 89, 645–656.

Khuri, A. I., Mathew, T., and Sinha, B. K. (1998), *Statistical Tests for Mixed Linear Models*, New York: Wiley.

Lauritzen, S. L. (1996), *Graphical Models*, Oxford, U.K.: Oxford University Press.

McGilchrist, C. A. (1994), "Estimation in Generalized Mixed Models," *Journal of the Royal Statistical Society*, Ser. B, 56, 61–69.

Pukelsheim, F. (1976), "Estimating Variance Components in Linear Models," *Journal of Multivariate Analysis*, 6, 626–629.

Searle, S. R., Casella, G., and McCulloch, C. (1992), *Variance Components*, New York: Wiley.

Seber, G. A. F. (1977), *Linear Regression Analysis*, New York: Wiley.

Seely, J. (1970), "Linear Spaces and Unbiased Estimation," *Annals of Mathematical Statistics*, 41, 1725–1734.

Stein, M. (1999), *Interpolation for Spatial Data: Some Theory for Kriging*, New York: Springer-Verlag.

Whittaker, J. (1990), *Graphical Models in Applied Multivariate Statistics*, New York: Wiley.

Zeger, S., Liang, K.-Y., and Albert, P. S. (1988), "Models for Longitudinal Data: A Generalized Estimating Equation Approach," *Biometrics*, 44, 1049–1060.

Zhang, Q. (1995), "Wavelets and Regression Analysis," in *Wavelets and Statistics*, eds. A. Antoniadis and G. Oppenheim, New York: Springer-Verlag, pp. 397–407.

The Bootstrap and Modern Statistics

Bradley Efron

I once began an address to a mathematics conference with the following preposterous question: Suppose that you could buy a really fast computer, one that could do not a billion calculations per second, not a trillion, but an *infinite number*. So after you unpacked it at home, you could numerically settle the Riemann hypothesis, the Goldbach conjecture, and Fermat's last theorem (this was a while ago), and still have time for breakfast. Would this be the end of mathematics?

My question was not a very tactful one, but its intentions were honorable. I was trying to communicate the current state of statistical theory. From a pre-World War II standpoint, our current computational abilities *are* effectively infinite, at least in terms of answering many common questions that arise in statistical practice. And no, this has not spelled the end of statistical theory—though it certainly has changed (for the better, in my opinion) what constitutes a good question and a good answer.

The bootstrap provides striking verification for the "infinite" capabilities of modern statistical computation. Figure 1 shows a small but genuine example, discussed more carefully by DiCiccio and Efron (1996) and Efron (1998). Twenty AIDS patients received an experimental antiviral drug. The Pearson sample correlation coefficient between the 20 (before, after) pairs of measurements is $\hat{\theta} = .723$. What inferences can we draw concerning the true population correlation θ?

An immense amount of prewar effort, much of the best by Fisher himself, was devoted to answering this question. Most of this effort assumed a bivariate Gaussian probability model, the classic example being Fisher's z-transform for normalizing the correlation distribution. The bivariate Gaussian model, a poor fit to the AIDS data, was pushed far beyond its valid range, because there was essentially no alternative.

An exception to this statement, almost the only one, was the nonparametric delta method estimate of standard error, given in terms of sample central moments of various mixed powers by this heroic formula:

$$\widehat{SE} = \frac{\hat{\theta}}{\sqrt{n}} \left\{ \frac{\hat{\mu}_{22}}{\hat{\mu}_{11}^2} + \frac{1}{4} \left(\frac{\hat{\mu}_{40}}{\hat{\mu}_{20}^2} + \frac{\hat{\mu}_{04}}{\hat{\mu}_{02}^2} + \frac{2\hat{\mu}_{22}}{\hat{\mu}_{20}\hat{\mu}_{02}} \right) - \left(\frac{\hat{\mu}_{31}}{\hat{\mu}_{11}\hat{\mu}_{20}} + \frac{\hat{\mu}_{13}}{\hat{\mu}_{11}\hat{\mu}_{02}} \right) \right\}^{1/2} . \tag{1}$$

Formulas like (1) were an important part of the applied statistician's tool kit, heavily used for approximating standard errors, confidence intervals, and hypothesis tests.

Bradley Efron is Professor of Statistics and Biostatistics, Department of Statistics, Stanford University, Stanford, CA 94305-4065 (E-mail: brad@stat.stanford.edu).

They still are (sometimes unfortunately), even though we now are armed with more potent weaponry.

The power of modern computation is illustrated in Figure 1(b), which shows the histogram of 4,000 nonparametric bootstrap correlation coefficients $\hat{\theta}^*$. Each $\hat{\theta}^*$ was calculated by drawing 20 points at random, with replacement, from the 20 actual data points in the left panel, and then computing the Pearson sample correlation coefficient for this bootstrap data set. (A variant of this algorithm would have been used if we had wished to bootstrap the Gaussian model.) In total, about 1,000,000 elementary numerical calculations were required. This is less than 1 second of effort on a modern computer, even a small one, or perhaps 1 minute if, like me, you prefer to trade some speed for the programming ease of a high-level language like S-PLUS. The same computation on Fisher's "millionaire" mechanical calculator would have taken years of grinding human effort. Calling today's computers "infinite" is not hyperbole from this standpoint.

The sample standard deviation of the $4,000\hat{\theta}^*$ values was .0921, which is the bootstrap estimate of standard error for $\hat{\theta}$. (Here 4,000 is at least 10 times too many for a standard error, but not excessive for the confidence interval discussion to come.) This compares with .0795 from (1). An immense amount of effort has been spent justifying the theoretical basis of the bootstrap (more than 1,000 papers since Efron 1979), but the basic principle is simple, amounting in this case to an application of nonparametric maximum likelihood estimation:

1. We suppose that the data have been obtained by random sampling from some unknown probability distribution F (a bivariate distribution in the AIDS example.)

2. We are estimating the parameter of interest θ with some statistic $\hat{\theta}$.

3. We wish to know σ_F, the standard error of $\hat{\theta}$ when sampling from F.

4. We approximate σ_F with $\sigma_{\hat{F}}$, where $\hat{F}$ is the empirical distribution of the data (putting probability 1/20 on each of the 20 data points in the AIDS example).

The Monte Carlo routine for $\hat{\theta}^*$ is just a way of evaluating $\sigma_{\hat{F}}$ without going through the kind of Taylor series approximations involved in (1). In addition to being easier to use and more accurate than the Taylor series approach, it has the great advantage of being completely general. I could just as well have bootstrapped Kendall's tau, or the largest eigenvalue of the sample covariance matrix, or the ratio of 25% trimmed means. This generality allows the statistician to step fearlessly off the narrow path of prewar computational feasibility, opening the door to more flexible, realistic, and powerful data analyses. Computer-based methods, including the bootstrap, have made statisticians more useful to our scientific colleagues.

The bootstrap began life as a muscularized big brother to the Quenouille–Tukey jackknife (see Efron and Tibshirani 1993, chap. 11), with the same principal tasks in mind: routine calculation of biases and standard errors. A more ambitious goal soon pushed itself forward: automatic computation of bootstrap confidence intervals. A

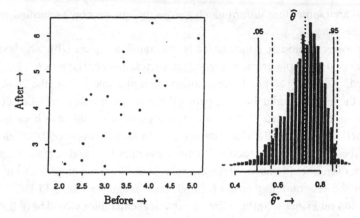

Figure 1. *Study of 20 AIDS Patients Receiving an Experimental Antiviral Drug. Left Panel shows Their CD4 counts in hundreds before and after treatment; Pearson correlation coefficient $\hat{\theta} = .723$. Right Panel Histogram of 4,000 nonparametric bootstrap replications $\hat{\theta}^*$, with solid lines indicating the central 90% BC_a interval for true correlation θ and the dashed lines indicating standard interval endpoints. The bootstrap standard error is .0921, compared to the nonparametric delta method standard error .0795.*

theory of confidence intervals more useful than the "standard intervals" $\hat{\theta} \pm 1.645\widehat{SE}$ must operate at an increased level of theoretical accuracy. Some of the deepest parts of the 1,000-paper literature concern "second-order accuracy" and how it can be obtained via the bootstrap. Many authors participated in this work, as documented in the references.

The solid lines in Figure 1 marked ".05" and ".95" indicate the endpoints of the "BC_a" bootstrap confidence interval (Efron 1987), intended to cover the true correlation θ with 90% probability. BC_a stands for bias-corrected and accelerated, enough words to suggest some difficulty in pursuing the goal of second-order accuracy. Hall (1988) verified BC_a's second-order accuracy, meaning that the actual noncoverage probabilities, intended to be 5% at each end of the interval, approach that ideal with error proportional to 1/sample size. This is an order of magnitude better than the $1/\sqrt{\text{sample size}}$ rate of the standard interval, indicated by the dashed lines in Figure 1.

The .05 and .95 lines in Figure 1 are *not* the 5th and 95th percentiles of the 4,000 bootstrap replications. In this case they are the 4.6th and 94.7th percentiles, though in other examples the disparity could be much larger. Using the "obvious" percentiles of the bootstrap distribution destroys second-order accuracy. The actual BC_a percentile depend on an automatic algorithm that takes into account the bias and changing variance of $\hat{\theta}$ as an estimator of θ.

I am going on a bit about the somewhat technical point because it reflects an important, and healthy, aspect of bootstrap research: the attempt to ground the bootstrap in the fundamental ideas of statistical theory—in this case coverage accuracy of confidence intervals. New statistical methodology is often applied promiscuously, more so if it is complicated, computer-based, and hard to check. The process of connecting it back to the basic principles of statistical inference comes later, but in the long run no methodology can survive if it flouts these principles. (The criticism process for bootstrap confidence intervals is still going strong; see Young 1994 and its discussion.) The bootstrap itself was first intended as an explanation for the success of an older methodology, the jackknife.

Fisher and his colleagues were well aware that the standard intervals gave poor results for the correlation coefficient. This was the impetus for Fisher's z-transformation. But the z-transformation only fixes up the standard intervals for the Gaussian correlation coefficient, while similar breakdowns, usually unrecognized, occur in many other contexts. Bootstrap confidence intervals automate the z-transform idea, bringing it to bear in a routine way on any estimation problem. The process of grounding the bootstrap in traditional theory has worked the other way too; quite a bit more has been learned about the theory of confidence intervals through the effort of applying it outside the traditional textbook examples.

This same two-way exchange between classic statistical theory and modern computer-based methodology is going on in other areas of research. Markov chain Monte Carlo (MCMC) offers a particularly apt example. If the bootstrap is an automatic processor for frequentist inference, then MCMC is its Bayesian counterpart. The ability to compute a posteriori distributions for almost any prior, not just mathematically convenient ones, has deepened the discussion of what those priors should be. The renewed interest in "uninformative" priors (see, e.g., Kass and Wasserman 1996), connects back to the theoretical basis of bootstrap confidence intervals. Efron (1998, secs. 6–8) speculated about these connections.

Today's computers may indeed seem infinitely fast when carrying out traditional statistical calculations. Not so though for the more ambitious data-analytic tasks suggested by modern techniques like MCMC and the bootstrap. The possibility of improved results, and the critical appraisal of just how much improvement has been achieved, create a demand for still better and inevitably more computationally intensive methodology. Bootstrap confidence intervals, usually an improvement over the traditional $\hat{\theta} \pm 1.645\widehat{SE}$, may still not give very accurate coverage in a small-sample nonparametric situation like that illustrated in Figure 1. Getting up to "third-order accuracy" seems to require bootstrapping the bootstrap, as in Beran (1987) and Loh (1987).

There is some sort of law working here whereby statistical methodology always expands to strain the current limits of computation. Our job is to make certain that the new methodology is genuinely more helpful to our scientific clientele, and not just more elaborate. I would give the statistics community a strong "A" in this regard. Here is a list (from Efron 1995) of a dozen postwar developments that have had a major effect on the practice of statistics: nonparametric and robust methods, Kaplan–Meier and proportional hazards, logistic regression and Generalized Linear Models, the jackknife and bootstrap, EM and MCMC, and empirical Bayes and James–Stein estimation.

These topics have something less healthy in common: none of them appears in most introductory statistics texts. As far as what we are teaching new students, statistics stopped dead in 1950. An obvious goal, but one that gets lost in an historical approach to our subject, is to insert intuitively simple and appealing topics like the bootstrap into the introductory curriculum.

My own education in applied statistics (a very good one in the hands of Lincoln Moses, Rupert Miller, and Byron Brown) was heavily classical. It has taken me a long time to get over the feeling that there is something magically powerful about formulas like (1) and to start trusting in the efficacy of computer-based methods like the bootstrap for routine calculations. It has been an easier transition for nonroutine analyses, where classical methods do not exist, though I still find it easy to forget that today we can answer questions that once were utterly beyond reach.

Figure 2 relates to a recent consulting experience. The figure shows data for the first five patients of an efficacy study on an experimental antiviral drug. There were 49 patients in the study, each measured on 43 predictor variables and a response, altogether creating a 49×44 data matrix $\mathbf{X}$. The goal of the study was to predict the responses from some simple function of the 43 covariates. An extensive application of step-up and step-down regression selection programs, supplemented by the scientific intuition of the investigators, resulted in a "best" model that used just three simple linear combinations of the covariates (like the sum of the "x" measurements) while giving quite accurate predictions, $R^2 = .73$. A reviewer for the medical journal asked how optimistic this R^2 might be given the amount of data mining used.

	x30	x46	x48	x54	x82	x84	x90	numo	r41	r67	r69	r70	r74	r75	r151	r184
patient.	. .	. .	. .	. .	. .	. .	. .	. .	. .	. .	. .	. .	. .	. .	. .	. .
1	0	0	0	0	0	0	0	7	1	1	0	0	0	0	0	1
2	0	0	0	0	0	1	1	7	1	0	0	0	0	0	0	1
3	0	1	0	1	1	0	1	7	1	1	0	1	0	0	0	1
4	1	0	0	0	0	0	0	4	0	0	1	1	0	1	1	1
5	0	0	1	1	1	0	0	2	0	1	1	1	0	0	0	1

	r210	r215	r219	numto	age	sex	race	aids	sqv	rtv	ind	nfv	pin	dur	tc3
1	1	1	1	10	33	0	4	3	0	0	1	0	1	28	1
2	0	1	0	4	47	0	4	3	1	0	1	1	3	48	1
3	1	1	1	7	35	0	4	3	0	1	1	0	2	60	1
4	0	0	0	5	36	0	4	1	0	0	0	1	1	12	1
5	0	0	1	9	48	0	4	2	1	0	1	0	3	88	1

	azt	d46	ddc	ddi	duron	revdur	cdb	rnb	cd4	cd12	rn4	rn12	response
1	1	1	0	1	0	82	20	6	80	120	-3	-3	1.695
2	1	1	0	1	0	76	320	5	40	40	-1	0	2.070
3	1	1	0	0	0	108	170	6	-10	-10	0	0	0.640
4	1	1	0	1	0	312	310	5	-50	0	-2	0	2.630
5	1	1	0	1	0	465	150	4	10	10	0	0	0.645

Figure 2. Data for the First Five of 49 AIDS Patients, Each Measured on 43 Predictors and Response to an Experimental Antiviral Drug. An extensive data-mining effort produced a three-variable prediction model with $R^2 = .73$. How optimistic was the R^2 value?

Our answer was based on a fundamentally straightforward bootstrap analysis:

1. Construct a bootstrap data matrix $\mathbf{X}^*$, 49×44, by resampling the rows of $\mathbf{X}$ (i.e., by resampling the patients).

2. Rerun the step-up/step-down regression selection programs on $\mathbf{X}^*$, including some allowance for the guidelines of "scientific intuition," producing a bootstrapped best prediction rule, sometimes one much different than the original rule.

3. Compute ΔR^{2^*}, the difference in predictive ability for the bootstrapped rule on its own bootstrap data set $\mathbf{X}^*$ minus its predictive ability on the original data $\mathbf{X}$.

The average value of ΔR^{2^*} over 50 bootstrap replications, which turned out to be .12, then gave a believable assessment of optimism for the original $R^2 = .73$, leaving us with a bias-corrected estimate of $R^2 = .61$. Efron and Gong (1983) discussed a more elaborate example of predictive bias correction.

The prehistory of the bootstrap is heavily involved with the jackknife. Rupert Miller's influential article "A Trustworthy Jackknife" (Miller 1964) was a successful early effort at demystifying what had seemed to be an almost magical device. Miller and I shared a sabbatical year at Imperial College in 1972–1973, and after one of Rupert's lectures David Cox asked me, in a pointed way, if I thought there was anything to this jackknife business. I took this, correctly, as a hint, and a few years later decided to make an investigation of the jackknife the subject of the 1977 Rietz lecture.

An elaborate mechanism called "the combination distribution" was to be the basis of my lecture, but the more I worked on it the less remained of the mechanism, until I was left with what seemed at the time a disappointingly simple device. One of the most helpful references for this work was a technical report by Jaeckel (1972), unfortunately unpublished, which suggested the kind of σ_F explanation given earlier.

The lecture (which became the basis of Efron 1979), was given at the Seattle joint statistical meetings, accompanied by insistent construction noise from the next room. At the end of the lecture Professor J. Wolfowitz asked me if I had any theorems to back up the bootstrap, to which I could only respond that I did not want to spoil a perfect effort. The name "bootstrap," suggested by Muenchausen's fable, was chosen for euphony with "jackknife," and I was subsequently very happy to have given up on "combination distribution." Some alternative names are reviewed in the acknowledgment of the 1979 article.

Books by Davison and Hinkley (1997), Efron and Tibshirani (1993), Hall (1992), and Shao and Tu (1995) provide different views of the bootstrap, and also extensive bibliographies. Influential articles include those of Bickel and Freedman (1981), Hall (1988), Romano (1988), and Singh (1981), but this short list excludes so many of even my personal favorites that I can only fall back on space limitations as an apology.

REFERENCES

Beran, R. (1987), "Prepivoting to Reduce Level Error of Confidence Sets," *Biometrika*, 74, 457–468.

Bickel, P., and Freedman, D. (1981), "Some Asymptotic Theory for the Bootstrap," *The Annals of Statistics*, 9, 1196–1217.

DiCiccio, T., and Efron, B. (1996), "Bootstrap Confidence Intervals" (with discussion), *Statistics of Science*, 11, 189–228.

Davison, A., and Hinkley, D. (1997), *Bootstrap Methods and Their Application*, New York: Cambridge University Press.

Efron, B. (1979), "Bootstrap Methods: Another Look at the Jackknife," *The Annals of Statistics*, 7, 1–26.

―――― (1982), *The Jackknife, the Bootstrap, and Other Resampling Plans*, Philadelphia: SIAM.

―――― (1987), "Better Bootstrap Confidence Intervals" (with discussion), *Journal of the American Statistical Association*, 82, 171–200.

―――― (1995), "The Statistical Century," *RSS News*, 22(5), 1–2.

―――― (1998), "R. A. Fisher in the 21st Century" (with discussion), *Statistics of Science*, 13, 95–122.

Efron, B., and Gong, G. (1983), "A Leisurely Look at the Bootstrap, the Jackknife, and Cross-Validation," *American Statistician*, 37, 36–48.

Efron, B., and Tibshirani, R. (1993), *An Introduction to the Bootstrap*, New York: Chapman and Hall.

Hall, P. (1988), "Theoretical Comparison of Bootstrap Confidence Intervals" (with discussion), *The Annals of Statistics*, 16, 927–985.

―――― (1992), *The Bootstrap and Edgeworth Expansion*, New York: Springer.

Jaeckel, L. (1972), "The Infinitesimal Jackknife," Technical Report MM72-1215-11, Bell Laboratories.

Kass, R., and Wasserman, L. (1996), "The Selection of Prior Distributions by Formal Rules," *Journal of the American Statistical Association*, 91, 1343–1370.

Loh, W. Y. (1987), "Calibrating Confidence Coefficients," *Journal of the American Statistical Association*, 82, 155–162.

Miller, R. (1964), "A Trustworthy Jackknife," *Annals of Mathematical Statistics*, 39, 1598–1605.

Romano, J. (1988), "A Bootstrap Revival of Some Nonparametric Distance Tests," *Journal of the American Statistical Association*, 83, 698–708.

Shao, J., and Tu, D. (1995), *The Jackknife and the Bootstrap*, New York: Springer.

Singh, K. (1981), "On the Asymptotic Accuracy of Efron's Bootstrap," *The Annals of Statistics*, 9, 1187–1195.

Young, A. (1994), "Bootstrap: More Than a Stab in the Dark?" (with discussion), *Statistical Science*, 9, 382–415.

Prospects of Nonparametric Modeling

Jianqing Fan

1. INTRODUCTION

Modern computing facilities allow statisticians to explore fine data structures that were unimaginable two decades ago. Driven by many sophisticated applications, demanded by the need of nonlinear modeling and fueled by modern computing power, many computationally intensive data-analytic modeling techniques have been invented to exploit possible hidden structures and to reduce modeling biases of traditional parametric methods. These data-analytic approaches are also referred to as nonparametric techniques. For an introduction to these nonparametric techniques, see the books by Bosq (1998), Bowman and Azzalini (1997), Devroye and Györfi (1985), Efromovich (1999), Eubank (1988), Fan and Gijbels (1996), Green and Silverman (1994), Györfi, Härdle, Sarda, and Vieu (1989), Hart (1997), Hastie and Tibshirani (1990), Müller (1988), Ogden (1997), Ramsay and Silverman (1997), Scott (1992), Silverman (1986), Simonoff (1996), Vidakovic (1999), Wahba (1990), and Wand and Jones (1995), among others.

An aim of nonparametric techniques is to reduce possible modeling biases of parametric models. Nonparametric techniques intend to fit a much larger class of models to reduce modeling biases. They allow data to search appropriate nonlinear forms that best describe the available data, and also provide useful tools for parametric nonlinear modeling and for model diagnostics.

Over the past three decades, intensive efforts have been devoted to nonparametric function estimation. Many new nonparametric models have been introduced and a vast array of new techniques invented. Many new phenomena have been unveiled, and deep insights have been gained. The field of nonparametric modeling has progressed steadily and dynamically. This trend will continue for decades to come. With the advance of information and technology, more and more complicated data mining problems emerge. The research and applications of data-analytic techniques will prove even more fruitful in the next millennium.

The field of nonparametric modeling is vast. It has taken many books to describe a part of the art. Indeed, most parametric models have their nonparametric counterpart. It

Jianqing Fan is Professor, Department of Statistics, University of California, Los Angeles, CA 90095 (E-mail: jfan@stat.ucla.edu) and Professor of Statistics, Chinese University of Hong Kong.
Fan's research was partially supported by National Science Foundation grant DMS-9804414 and a grant from the University of California at Los Angeles.

is impossible to give a complete survey of this wide field. Rather, this article highlights some of the important achievements and outlines some potentially fruitful topics of research. For a more complete review of the literature, see the aforementioned books and the references therein.

2. OVERVIEW OF DEVELOPMENTS

2.1 Density Estimation and Nonparametric Regression

Density estimation summarizes data distributions via estimating underlying densities, and nonparametric regression smooths scatterplots via estimating regression functions. They provide the simplest setup for understanding nonparametric modeling techniques and serve as useful building blocks for high-dimensional modeling. They are relatively well developed and understood.

Many useful techniques have been proposed for univariate smoothing. Among those, kernel methods (Gasser and Müller 1979; Müller 1988; Rosenblatt 1956; Wand and Jones 1995), local polynomial methods (Cleveland 1979; Fan 1993; Fan and Gijbels 1996; Stone 1977), spline methods (Eubank 1988; Green and Silverman 1994; Nychka 1995; Stone, Hansen, Kooperberg, and Truong 1997; Wahba 1977, 1990), Fourier methods (Efromovich 1999; Efromovich and Pinsker 1982) and wavelet methods (Antoniadis, 1999; Donoho and Johnstone 1994; Donoho, Johnstone, Kerkyacharian, and Picard 1995; Hall and Patil 1995; Ogden 1997; Vidakovic 1999). Different techniques have their own merits. Fan and Gijbels (1996, Chap. 2) gave an overview of these techniques.

Each nonparametric technique involves selection of smoothing parameters. Several data-driven methods have been developed. Cross-validation (Allen 1974; Rudemo 1982; Stone 1974) and generalized cross-validation (Wahba 1977) are generally applicable methods. Yet their resulting bandwidths can vary substantially (Hall and Johnstone 1992). Plug-in methods are more stable. In addition to the methods surveyed by Jones, Marron, and Sheather (1996), the preasymptotic substitution method of Fan and Gijbels (1995) and the empirical-bias method of Ruppert (1997) provide useful alternatives (see also Marron and Padgett 1987).

2.2 Multivariate Nonparametric Modeling

Univariate smoothing techniques can be extended in a straightforward manner to multivariate settings. But such extensions are not useful, due to the so-called "curse of dimensionality." Many powerful models have been proposed to avoid using "saturated" nonparametric models and hence attenuate the problems of the "curse of dimensionality." Different models incorporate different knowledge into data analyses and explore different aspects of data. Examples include additive models (Breiman and Friedman 1985; Hastie and Tibshirani 1990; Stone 1994), varying coefficient models (Cleveland, Grosse, and Shyu 1991; Hastie and Tibshirani 1993), low-dimensional interaction models (Friedman 1991; Gu and Wahba 1993; Stone et al. 1997), multiple-index models (Härdle and Stoker 1989; Li 1991), and partially linear models (Green and Silverman 1994; Speckman 1988) and their hybrids (Carroll, Fan, Gijbels, and Wand 1997; Fan, Härdle, and Mammen 1998), among others (see also semiparametric models in Bickel, Klaassen, Ritov, and Wellner 1993). Together they provide useful tool kits for processing data that arise from many scientific disciplines and for checking the adequacy of commonly used parametric models.

The area of multivariate data-analytic modeling is very dynamic. A vast array of innovative ideas has been proposed that rely on certain univariate smoothing techniques as building blocks. These include backfitting methods (Hastie and Tibshirani 1990) and average regression surface methods (Linton and Nielsen 1995; Tjøstheim and Auestad 1994) for additive modeling, sliced inverse regression method (Duan and Li 1991; Li 1991) and average derivative methods (Härdle and Stoker 1989; Samarov 1993) for multiple-index models, and combinations of these methods (Carroll et al. 1997), among others. Polynomial splines and smoothing splines can be directly applied to low-dimensional interaction models.

Tree-based regression models (Breiman, Friedman, Olshen and Stone 1993; Zhang and Signer 1999) are based on different ideas. They are also powerful tools for nonparametric multivariate regression and classification.

Nonparametric regression problems arise often from other statistical contexts, such as generalized linear models and the Cox proportional hazards model. Although some theory and methods are available, nonparametric techniques are relatively underdeveloped for likelihood and pseudolikelihood models.

2.3 Theoretical Developments

Apart from creative technological inventions, many foundational insights have been gained, and many new phenomena in infinite-dimensional spaces have been discovered. It is now well known that many nonparametric functions cannot be estimated at a root-n rate (Donoho and Liu 1991; Farrell 1972), whereas some functionals such as integrated square densities (Bickel and Ritov 1988; Fan 1991) can be estimated at a root-n rate. These optimal rates of convergence depend on the smoothness of the function classes. Adaptive procedures have been constructed so that they are nearly optimal for each given smoothness of a class of functions (see, e.g., Brown and Low 1996; Donoho et al. 1995; Lepski 1991, 1992). Adaptive estimation based on penalized least squares was discussed by Barron, Birgé, and Massart (1999). Optimal rates for hypothesis testing have also been developed (see Ingster 1993; Spokoiny 1996).

Optimal rates for multivariate analysis of variance (ANOVA) types of nonparametric models offer valuable theoretical insights into high-dimensional function estimation problems (Huang 1999; Stone 1994). It has been shown that asymptotically estimating a component in additive separable models is just as hard as the case when the other components are known (Fan, Härdle, and Mammen 1998). This property is not shared by parametric models.

3. FUTURE RESEARCH

With increasing complexity of statistical applications and the need for refinements of traditional techniques, the future of nonparametric modeling and its applications is bright and prosperous. Cross-fertilization of parametric and nonparametric techniques will be fruitful and powerful. Applications of nonparametric techniques to other scientific and engineering disciplines are increasingly demanding. Some areas of research are outlined here, but of course the list is far from exhaustive.

3.1 Nonparametric Inferences

Maximum likelihood estimation, likelihood ratio statistics, and the bootstrap offer generally applicable tools for parametric analysis. Yet there are no generally applicable principles available for nonparametric inferences. Consider the example of additive models. How can one construct simultaneous confidence bands for estimated functions? Are a set of variables significant in the models? Does a given nonlinear parametric model adequately fit the data, or is a given nonparametric model overparameterized? Although there are many collective efforts and much progress has been made, the area still requires intensive research, and widely applicable methods should be sought. Recently, Fan, Zhang, and Zhang (2001) made a start in this direction, proposing a generalized likelihood ratio method and demonstrating that it has various good statistical properties.

3.2 High-Dimensional Nonparametric Modeling

Many interesting statistical problems are multivariate and high-dimensional, with a mix of discrete and continuous variables. Although there are a number of creative nonparametric models, they cannot be expected to handle all of these statistical problems. A lack of inference tools and software has hampered their applications. High-dimensional classification problems are increasingly in demand. Applications of nonparametric modeling techniques to other statistical contexts need further development.

3.3 Functional Data Analysis

Massive datasets can now be easily collected for each individual in the form of curves or images. In ophthalmology, for example, images of a patient's cornea maps are recorded along with other demographic and ophthalmic variables. Interesting questions include studying associations between cornea shapes and demographic and ophthalmic variables, testing whether there are any differences among two or more clinical groups or treatment methods, and monitoring regression/progression of clinical surgery. (More examples and problems can be found in Capra and Müller 1997, Kneip and Gasser 1992, and Ramsay and Silverman 1997.) Feature extractions have been studied extensively (Ramsay and Silverman 1997). Yet predictions, modeling, and inferences based on functional data need substantial developments.

3.4 Applications to Other Statistical Problems

Many other statistical problems require data-analytic tools, including telecommunications and information engineering, nonlinear time series (Tong 1990; Yao and Tong 1994), and biostatistics (Hoover, Rice, Wu, and Yang 1998). They offer statisticians enormous opportunities for interdisciplinary collaboration. Context-based applications of nonparametric techniques will be fruitful. (For more discussion and details on these topics, see the expanded version of this article at http://www.stat.unc.edu/faculty/fan.html.)

3.5 Software Developments

Applications of nonparametric techniques have been hampered by lack of software. Although many nonparametric techniques have been programmed by individual researchers, these were written in many computer languages and were tested only for "in-house" use. Many modern nonparametric techniques are not available in commonly used statistical software packages. Research into fast and robust implementations of nonparametric techniques and their software developments is needed.

REFERENCES

Allen, D. M. (1974), "The Relationship Between Variable and Data Augmentation and a Method of Prediction," *Technometrics*, 16, 125–127.

Antoniadis, A. (1999), "Wavelets in Statistics: A Review" (with discussion), *Italian Journal of Statistics*, 6 (2).

Barron, A., Birgé, L., and Massart, P. (1999), "Risk Bounds for Model Selection via Penalization," *Probability Theory Related Fields*, 113, 301–413.

Bickel, P. J., Klaassen, A. J., Ritov, Y., and Wellner, J. A. (1993), *Efficient and Adaptive Inference in Semiparametric Models*, Baltimore: Johns Hopkins University Press.

Bickel, P. J., and Ritov, Y. (1988), "Estimating Integrated Squared Density Derivatives: Sharp Order of Convergence Estimates," *Sankhyā*, Ser. A, 50, 381–393.

Bosq, D. (1998), *Nonparametric Statistics for Stochastic Processes: Estimation and Prediction* (2nd ed.), Berlin: Springer-Verlag.

Bowman, A. W., and Azzalini, A. (1997), *Applied Smoothing Techniques for Data Analysis*, London: Oxford University Press.

Breiman, L., and Friedman, J. H. (1985), "Estimating Optimal Transformations for Multiple Regression and Correlation" (with discussion), *Journal of the American Statistical Association*, 80, 580–619.

Breiman, L., Friedman, J. H., Olshen, R. A., and Stone, C. J. (1993), *CART: Classification and Regression Trees*, Belmont, CA: Wadsworth.

Brown, L. D., and Low, M. (1996), "A Constrained Risk Inequality With Applications to Nonparametric Functional Estimation," *The Annals of Statistics*, 24, 2524–2535.

Capra, W. B., and Müller, H. G. (1997), "An Accelerated Time Model for Response Curves," *Journal of the American Statistical Association*, 92, 72–83.

Carroll, R. J., Fan, J., Gijbels, I., and Wand, M. P. (1997), "Generalized Partially Linear Single-Index Models," *Journal of the American Statistical Association*, 92, 477–489.

Cleveland, W. S. (1979), "Robust Locally Weighted Regression and Smoothing Scatterplots," *Journal of the American Statistical Association*, 74, 829–836.

Cleveland, W. S., Grosse, E., and Shyu, W. M. (1991), "Local Regression Models," in *Statistical Models in S*, eds. J. M. Chambers and T. J. Hastie, Pacific Grove, CA: Wadsworth & Brooks/Cole, pp. 309–376.

Devroye, L. P., and Györfi, L. (1985), *Nonparametric Density Estimation: The L_1 View*, New York: Wiley.

Donoho, D. L., and Johnstone, I. M. (1994), "Ideal Spatial Adaptation by Wavelet Shrinkage," *Biometrika*, 81, 425–455.

Donoho, D. L., Johnstone, I. M., Kerkyacharian, G., and Picard, D. (1995), "Wavelet Shrinkage: Asymptopia?," *Journal of the Royal Statistical Society*, Ser. B, 57, 301–369.

Donoho, D. L., and Liu, R. C. (1991), "Geometrizing Rate of Convergence III," *The Annals of Statistics*, 19, 668–701.

Duan, N., and Li, K-C. (1991), "Slicing Regression: A Link-Free Regression Method," *The Annals of Statistics*, 19, 505–530.

Efromovich, S. (1999), *Nonparametric Curve Estimation: Methods, Theory and Applications*, New York: Springer-Verlag.

Efromovich, S. Y., and Pinsker, M. S. (1982), "Estimation of Square-Integrable Probability Density of a Random Variable," *Problems of Information Transmission*, 18, 175–189.

Eubank, R. L. (1988), *Spline Smoothing and Nonparametric Regression*, New York: Marcel Dekker.

Fan, J. (1991), "On the Estimation of Quadratic Functionals," *The Annals of Statistics*, 19, 1273–1294.

——— (1993), "Local Linear Regression Smoothers and Their Minimax Efficiency," *The Annals of Statistics*, 21, 196–216.

Fan, J., and Gijbels, I. (1995), "Data-Driven Bandwidth Selection in Local Polynomial Fitting: Variable Bandwidth and Spatial Adaptation," *Journal of the Royal Statistical Society*, Ser. B, 57, 371–394.

——— (1996), *Local Polynomial Modelling and Its Applications*, London: Chapman & Hall.

Fan, J., Härdle, W., and Mammen, E. (1998), "Direct Estimation of Additive and Linear Components for High-Dimensional Data," *The Annals of Statistics*, 26, 943–971.

Fan, J., Zhang, C., and Zhang, J. (2001), "Generalized Likelihood Ratio Statistics and Wilks Phenomenon," *The Annals of Statistics*, 29, 153–193.

Farrell, R. H. (1972), "On the Best Obtainable Asymptotic Rates of Convergence in Estimation of a Density Function at a Point," *Annals of Mathematical Statistics*, 43, 170–180.

Friedman, J. H. (1991), "Multivariate Adaptive Regression Splines" (with discussion), *The Annals of Statistics*, 19, 1–141.

Gasser, T., and Müller, H-G. (1979), "Kernel Estimation of Regression Functions," in *Smoothing Techniques for Curve Estimation*, New York: Springer-Verlag, pp. 23–68.

Green, P. J., and Silverman, B. W. (1994), *Nonparametric Regression and Generalized Linear Models: A Roughness Penalty Approach*, London: Chapman & Hall.

Gu, C., and Wahba, G. (1993), "Smoothing Spline ANOVA With Component-Wise Bayesian 'Confidence Intervals,'" *Journal of Computing and Graphical Statistics*, 2, 97–117.

Györfi, L., Härdle, W., Sarda, P., and Vieu, P. (1989), *Nonparametric Curve Estimation from Time Series*, Berlin: Springer-Verlag.

Hall, P., and Johnstone, I. (1992), "Empirical Functionals and Efficient Smoothing Parameter Selection" (with discussion), *Journal of the Royal Statistical Society*, Ser. B, 54, 475–530.

Hall, P., and Patil, P. (1995), "Formulae for Mean Integrated Squared Error of Nonlinear Wavelet-Based Density Estimators," *The Annals of Statistics*, 23, 905–928.

Härdle, W. (1990), *Applied Nonparametric Regression*, Boston: Cambridge University Press.

Härdle, W., and Stoker, T. M. (1989), "Investigating Smooth Multiple Regression by the Method of Average Derivatives," *Journal of the American Statistical Association*, 84, 986–995.

Hart, J. D. (1997), *Nonparametric Smoothing and Lack-of Fit Tests*, New York: Springer-Verlag.

Hastie, T. J., and Tibshirani, R. (1990), *Generalized Additive Models*, London: Chapman & Hall.

—— (1993), "Varying-Coefficient Models," *Journal of the Royal Statistical Society*, Ser. B, 55, 757–796.

Hoover, D. R., Rice, J. A., Wu, C. O., and Yang, L-P. (1998), "Nonparametric Smoothing Estimates of Time-Varying Coefficient Models With Longitudinal Data," *Biometrika*, 85, 809–822.

Huang, J. (1999), "Projection Estimation in Multiple Regression With Application to Functional ANOVA Models," *The Annals of Statistics*, 26, 242–272.

Ingster, Y. I. (1993a), "Asymptotic Minimax Hypothesis Testing for Nonparametric Alternative I," *Mathematical Methods Statistics*, 2, 85–114.

—— (1993b), "Asymptotic Minimax Hypothesis Testing for Nonparametric Alternative II," *Mathematical Methods Statistics*, 3, 171–189.

—— (1993c), "Asymptotic Minimax Hypothesis Testing for Nonparametric Alternative III," *Mathematical Methods Statistics*, 4, 249–268.

Jones, M. C., Marron, J. S., and Sheather, S. J. (1996), "A Brief Survey of Bandwidth Selection for Density Estimation," *Journal of the American Statistical Association*, 91, 401–407.

Lepski, O. V. (1991), "Asymptotically Minimax Adaptive Estimation I," *Theory Probability Applications*, 36, 682–697.

—— (1992), "Asymptotically Minimax Adaptive Estimation II," *Theory Probability Applications*, 37, 3, 433–448.

Li, K-C. (1991), "Sliced Inverse Regression for Dimension Reduction" (with discussion), *Journal of the American Statistical Association*, 86, 316–342.

Linton, O. B., and Nielsen, J. P. (1995), "A Kernel Method of Estimating Structured Nonparametric Regression Based on Marginal Integration," *Biometrika*, 82, 93–101.

Kneip, A., and Gasser, T. (1992), "Statistical Tools to Analyze Data Representing a Sample of Curves," *The Annals of Statistics*, 20, 1266–1305.

Marron, J. S., and Padgett, W. J. (1987), "Asymptotically Optimal Bandwidth Selection From Randomly Right-Censored Samples," *The Annals of Statistics*, 15, 1520–1535.

Müller, H-G. (1988), *Nonparametric Regression Analysis of Longitudinal Data*.

Nychka, D. (1995), "Splines as Local Smoothers," *The Annals of Statistics*, 23, 1175–1197.

Ogden, T. (1997), *Essential Wavelets for Statistical Applications and Data Analysis*, Boston: Birkhuser.

Ramsay, J. O., and Silverman, B. W. (1997), *The Analysis of Functional Data*, Berlin: Springer-Verlag.

Rosenblatt, M. (1956), "Remarks on Some Nonparametric Estimates of a Density Function," *Annals of Mathematical Statistics*, 27, 832–837.

Rudemo, M. (1982), "Empirical Choice of Histograms and Kernel Density Estimators," *Scandinavian Journal of Statistics*, 9, 65–78.

Ruppert, D. (1997), "Empirical-Bias Bandwidths for Local Polynomial Nonparametric Regression and Density Estimation," *Journal of the American Statistical Association*, 92, 1049–1062.

Samarov, A. M. (1993), "Exploring Regression Structure Using Nonparametric Functional Estimation," *Journal of the American Statistical Association*, 88, 836–847.

Scott, D. W. (1992), *Multivariate Density Estimation: Theory, Practice, and Visualization*, New York: Wiley.

Silverman, B. W. (1986), *Density Estimation for Statistics and Data Analysis*, London: Chapman & Hall.

Simonoff, J. S. (1996), *Smoothing Methods in Statistics*, New York: Springer-Verlag.

Speckman, P. (1988), "Kernel Smoothing in Partial Linear Models," *Journal of the Royal Statistical Society*, Ser. B, 50, 413–436.

Spokoiny, V. G. (1996), "Adaptive Hypothesis Testing Using Wavelets," *The Annals of Statistics*, 24, 2477–2498.

Stone, C. J. (1977), "Consistent Nonparametric Regression," *The Annals of Statistics*, 5, 595–645.

―――― (1994), "The Use of Polynomial Splines and Their Tensor Products in Multivariate Function Estimation" (with discussion), *The Annals of Statistics*, 22, 118–184.

Stone, C. J., Hansen, M., Kooperberg, C., and Truong, Y. K. (1997), "Polynomial Splines and Their Tensor Products in Extended Linear Modeling" (with discussion), *The Annals of Statistics*, 25, 1371–1470.

Stone, M. (1974), "Cross-Validatory Choice and Assessment of Statistical Predictions" (with discussion), *Journal of the Royal Statistical Society*, Ser. B, 36, 111–147.

Tjøstheim, D., and Auestad, B. H. (1994), "Nonparametric Identification of Nonlinear Time Series: Projections," *Journal of the American Statistical Association*, 89, 1398–1409.

Tong, H. (1990), *Non-Linear Time Series: A Dynamical System Approach*, London: Oxford University Press.

Vidakovic, B. (1999), *Statistical Modeling by Wavelets*, New York: Wiley.

Wahba, G. (1977), "A Survey of Some Smoothing Problems and the Method of Generalized Cross-Validation for Solving Them," in *Applications of Statistics*, ed. P. R. Krisnaiah, Amsterdam: North-Holland, pp. 507–523.

―――― (1990), *Spline Models for Observational Data*, Philadelphia: SIAM.

Wand, M. P., and Jones, M. C. (1995), *Kernel Smoothing*, London: Chapman & Hall.

Yao, Q., and Tong, H. (1994), "On Prediction and Chaos in Stochastic Systems," *Philosophical Transactions of the Journal of the Royal Statistical Society*, Ser. A, 348, 357–369.

Zhang, H. P., and Singer, B. (1999), *Recursive Partitioning in the Health Sciences*, New York: Springer-Verlag.

Gibbs Sampling

Alan E. Gelfand

1. INTRODUCTION

During the course of the 1990s, the technology generally referred to as Markov chain Monte Carlo (MCMC) has revolutionized the way statistical models are fitted and, in the process, dramatically revised the scope of models which can be entertained.

This vignette focuses on the Gibbs sampler. I provide a review of its origins and its crossover into the mainstream statistical literature. I then attempt an assessment of the impact of Gibbs sampling on the research community, on both statisticians and subject area scientists. Finally, I offer some thoughts on where the technology is headed and what needs to be done as we move into the next millennium. The perspective is, obviously, mine, and I apologize in advance for any major omissions. In this regard, my reference list is modest, and again there may be some glaring omissions. I present little technical discussion, as by now detailed presentations are readily available in the literature. The books of Carlin and Louis (2000), Gelman, Carlin, Stern, and Rubin (1995), Robert and Casella (1999), and Tanner (1993) are good places to start. Also, within the world of MCMC, I adopt an informal definition of a Gibbs sampler. Whereas some writers describe "Metropolis steps within Gibbs sampling," others assert that the blockwise updating implicit in a Gibbs sampler is a special case of a "block at-a-time" Metropolis–Hastings algorithm. For me, the crucial issue is replacement of the sampling of a high-dimensional vector with sampling of lower-dimensional component blocks, thus breaking the so-called curse of dimensionality.

In Section 2 I briefly review what the Gibbs sampler is, how it is implemented, and how it is used to provide inference. With regard to Gibbs sampling, Section 3 asks the question "How did it make its way into the mainstream of statistics?" Section 4 asks "What has been the impact?" Finally, Section 5 asks "Where are we going?" Here, speculation beyond the next decade seems fanciful.

2. WHAT IS GIBBS SAMPLING?

Gibbs sampling is a simulation tool for obtaining samples from a nonnormalized joint density function. Ipso facto, such samples may be "marginalized," providing samples from the marginal distributions associated with the joint density.

Alan E. Gelfand is Professor, Department of Statistics, University of Connecticut, Storrs, CT 06269. (E-mail: alan@stat.uconn.edu).
His work was supported in part by National Science Foundation grant DMS 96-25383.

2.1 Motivation

The difficulty in obtaining marginal distributions from a nonnormalized joint density lies in integration. Suppose, for example, that θ is a $p \times 1$ vector and $f(\theta)$ is a nonnormalized joint density for θ with respect to Lebesgue measure. Normalizing f entails calculating $\int f(\theta) \, d\theta$. To marginalize, say for θ_i, requires $h(\theta_i) = \int f(\theta) \, d\theta_{(i)} / \int f(\theta) \, d\theta$, where $\theta_{(i)}$ denotes all components of θ save θ_i. Integration is also needed to obtain a marginal expectation or find the distribution of a function of θ. When p is large, such integration is analytically infeasible (the curse of dimensionality). Gibbs sampling offers a Monte Carlo approach.

The most prominent application has been for inference within a Bayesian framework. Here models are specified as a joint density for the observations, say $\mathbf{Y}$, and the model unknowns, say θ, in the form $h(\mathbf{Y}|\theta)\pi(\theta)$. In a Bayesian setting, the observed realizations of $\mathbf{Y}$ are viewed as fixed, and inference proceeds from the posterior density of θ, $\pi(\theta|\mathbf{Y}) \propto h(\mathbf{Y}|\theta)\pi(\theta) \equiv f(\theta)$, suppressing the fixed $\mathbf{Y}$. So $f(\theta)$ is a nonnormalized joint density, and Bayesian inference requires its marginals and expectations, as earlier. If the prior, $\pi(\theta)$, is set to 1 and if $h(\mathbf{Y}|\theta)$ is integrable over θ, then the likelihood becomes a nonnormalized density. If marginal likelihoods are of interest, then we have the previous integration problem.

2.2 Monte Carlo Sampling and Integration

Simulation-based approaches for investigating the nonnormalized density $f(\theta)$ appeal to the duality between population and sample. In particular, if we can generate arbitrarily many observations from $h(\theta) = f(\theta) / \int f(\theta)$, so-called *Monte Carlo sampling*, then we can learn about any feature of $h(\theta)$ using the corresponding feature of the sample. Noniterative strategies for carrying out such sampling usually involve identification of an importance sampling density, $g(\theta)$ (see, e.g., Geweke 1989; West 1992). Given a sample from $g(\theta)$, we convert it to a sample from $h(\theta)$, by resampling, as done by Rubin (1988) and Smith and Gelfand (1992). If one only needs to compute expectations under $h(\theta)$, this can be done directly with samples from $g(\theta)$ (see, e.g., Ripley 1987) and is referred to as *Monte Carlo integration*. Noniterative Monte Carlo methods become infeasible for many high-dimensional models of interest.

Iterative Monte Carlo methods enable us to avoid the curse of dimensionality by sampling low-dimensional subsets of the components of θ. The idea is to create a Markov process whose stationary distribution is $h(\theta)$. This seems an unlikely strategy, but, perhaps surprisingly, there are infinities of ways to do this. Then, suppose that $P(\theta \rightarrow A)$ is the transition kernel of a Markov chain with stationary distribution $h(\theta)$. (Here $P(\theta \rightarrow A)$ denotes the probability that $\theta^{(t+1)} \in A$ given $\theta^{(t)} = \theta$.) If $h^{(0)}(\theta)$ is a density that provides starting values for the chain, then, with $\theta^{(0)} \sim h^{(0)}(\theta)$, using $P(\theta \rightarrow A)$, we can develop a trajectory (sample path) of the chain $\theta^{(0)}, \theta^{(1)}, \theta^{(2)}, \ldots, \theta^{(t)}, \ldots$. If t is large enough (i.e., after a sufficiently long "burn-in" period), then $\boldsymbol{\theta}^{(t)}$ is approximately distributed according to $\mathbf{h}(\theta)$.

A bit more formally, suppose that $P(\boldsymbol{\theta} \rightarrow A)$ admits a transition density, $p(\eta|\boldsymbol{\theta})$, with respect to Lebesgue measure. Then π is an invariant density for p if $\int \pi(\boldsymbol{\theta})p(\eta|\boldsymbol{\theta})\,d\boldsymbol{\theta} = \pi(\eta)$. In other words, if $\boldsymbol{\theta}^{(t)} \sim \pi$, then $\boldsymbol{\theta}^{(t+1)} \sim \pi$. Also, Γ is a limiting (stationary, equilibrium) distribution for p if $\lim_{t\to\infty} P(\boldsymbol{\theta}^{(t)} \in A)|\boldsymbol{\theta}^{(0)} = \boldsymbol{\theta}) = \Gamma(A)$ (and thus $\lim_{t\to\infty} P(\boldsymbol{\theta}^{(t)} \in A) = \Gamma(A)$). The crucial result is that if $p(\eta|\boldsymbol{\theta})$ is aperiodic and irreducible and if π is a (proper) invariant distribution of p, then π is the unique invariant distribution; that is, π is the limiting distribution. A careful theoretical discussion of general MCMC algorithms with references was given by Tierncy (1994). Also highly recommended is the set of three Royal Statistical Society papers in 1993 by Besag and Green (1993), Gilks et al. (1993), and Smith and Roberts (1993), together with the ensuing discussion, as well as an article by Besag, Green, Higdon, and Mengersen (1996), again with discussion.

2.3   The Gibbs Sampler

The Gibbs sampler was introduced as a MCMC tool in the context of image restoration by Geman and Geman (1984). Gelfand and Smith (1990) offered the Gibbs sampler as a very general approach for fitting statistical models, extending the applicability of the work of Geman and Geman and also broadening the substitution sampling ideas that Tanner and Wong (1987) proposed under the name of data augmentation.

Suppose that we partition $\boldsymbol{\theta}$ into r blocks; that is, $\boldsymbol{\theta} = (\boldsymbol{\theta}_1, \ldots, \boldsymbol{\theta}_r)$. If the current state of $\boldsymbol{\theta}$ is $\boldsymbol{\theta}^{(t)} = (\boldsymbol{\theta}_1^{(t)}, \ldots, \boldsymbol{\theta}_r^{(t)})$, then suppose that we make the transition to $\boldsymbol{\theta}^{(t+1)}$ as follows:

$$\text{draw } \boldsymbol{\theta}_1^{(t+1)} \text{ from } h(\boldsymbol{\theta}_1|\boldsymbol{\theta}_2^{(t)}, \ldots, \boldsymbol{\theta}_r^{(t)}),$$
$$\text{draw } \boldsymbol{\theta}_2^{(t+1)} \text{ from } \mathbf{h}(\boldsymbol{\theta}_2|\boldsymbol{\theta}_1^{(t+1)}, \ldots, \boldsymbol{\theta}_3^{(t)}, \ldots, \boldsymbol{\theta}_r^{(t)}),$$
$$\vdots$$
$$\text{draw } \boldsymbol{\theta}_r^{(t+1)} \text{ from } h(\boldsymbol{\theta}_r|\boldsymbol{\theta}_1^{(t+1)}, \ldots, \boldsymbol{\theta}_{r-1}^{(t+1)}).$$

The distributions $h(\boldsymbol{\theta}_i|\boldsymbol{\theta}_1, \ldots, \boldsymbol{\theta}_{i-1}, \ldots, \boldsymbol{\theta}_{i+1}, \ldots, \boldsymbol{\theta}_r)$ are referred to as the full, or complete, conditional distributions, and the process of updating each of the r blocks as indicated updates the entire vector $\boldsymbol{\theta}$, producing one complete iteration of the Gibbs sampler. Sampling of $\boldsymbol{\theta}$ has been replaced by sampling of lower-dimensional blocks of components of $\boldsymbol{\theta}$.

2.4   How To Sample the θ_i

Conceptually, the Gibbs sampler emerges as a rather straightforward algorithmic procedure. One aspect of the art of implementation is efficient sampling of the full conditional distributions. Here there are many possibilities. Often, for some of the θ_i, the form of the prior specification will be conjugate with the form in the likelihood, so that the full conditional distribution for θ_i will be a "posterior" updating of a standard prior. Note that even if this were the case for every θ_i, $f(\theta)$ itself need not be a standard distribution; conjugacy may be more useful for Gibbs sampling than for analytical investigation of the entire posterior.

When $f(\theta_i|\theta_1, \theta_2, \ldots \theta_{i-1}, \theta_{i+1}, \ldots \theta_r)$ is nonstandard, we might consider the rejection method, as discussed by Devroye (1986) and Ripley (1987); the weighted bootstrap, as discussed by Smith and Gelfand (1992); the ratio-of-uniforms method, as described by Wakefield, Gelfand, and Smith (1992); approximate cdf inversion when θ_i is univariate, such as the griddy Gibbs sampler, as discussed by Ritter and Tanner (1992); adaptive rejection sampling, as often the full conditional density for θ_i is log concave, in which case the usual rejection method may be adaptively improved in a computationally cheap fashion, as described by Gilks and Wild (1992); and Metropolis-within-Gibbs. For the last, the Metropolis (or Hastings–Metropolis) algorithms—which, in principle, enable simultaneous updating of the entire vector θ (Chib and Greenberg 1995; Tierney 1994)—are usually more conveniently used within the Gibbs sampler for updating some of the θ_i, typically those with the least tractable full-conditional densities.

The important message here is that no single procedure dominates the others for all applications. The form of $h(\theta)$ determines which method is most suitable for a given θ_i.

2.5 Convergence

Considerable theoretical work has been done on establishing the convergence of the Gibbs sampler for particular applications, but perhaps the simplest conditions have been given by Smith and Roberts (1993). If $f(\theta)$ is lower semicontinuous at 0, if $\int f(\theta) \, d\theta_i$ is locally bounded for each i, and if the support of f is connected, then the Gibbs sampler algorithm converges. In practice, a range of diagnostic tools is applied to the output of one or more sampled chains. Cowles and Carlin (1994) and Brooks and Roberts (1998) provided comparative reviews of the convergence diagnostics literature. Also, see the related discussions in the hierarchical models vignette by Hobert and the MCMC vignette by Cappé and Robert. (In principle, convergence can never be assessed using such output, as comparison can be made only between different iterations of one chain or between different observed chains, but never with the true stationary distribution.)

2.6 Inference Using the Output of the Gibbs Sampler

The retained output from the Gibbs sampler will be a set of $\theta_j^*, j = 1, 2, \ldots, B$, assumed to be approximately iid from $h = f / \int f$. If independently started parallel chains are used, then observations from different chains are independent but observations within a given chain are dependent. "Thinning" of the output stream (i.e., taking every kth iteration, perhaps after a burn-in period) yields approximately independent observations within the chain, for k sufficiently large. Evidently, the choice of k hinges on the autocorrelation in the chain. Hence sample autocorrelation functions are often computed to assess the dependence. Given $\{\theta_j^*\}$, for a specified feature of h we compute the corresponding feature of the sample. Because B can be made arbitrarily large, inference using $\{\theta_j^*\}$ can be made arbitrarily accurate.

3. HOW DID IT HAPPEN?

The Gibbs sampler was not developed by statisticians. For at least the past half-century, scientists (primarily physicists and applied mathematicians) have sought to simulate the behavior of complex probabilistic models formulated to approximate the behavior of physical, chemical, and biological processes. Such processes were typically characterized by regular lattice structure and the joint probability distribution of the variables at the lattice points was provided through local specification; That is, the full conditional density $h(\theta_i|\theta_j, j = 1, 2, \ldots, r, j \neq i)$ was reduced to $h(\theta_i|\theta_j \in N_i)$, where N_i is a set of neighbors of location i. But then an obvious question is whether the set of densities $h(\theta_i|\theta_j \in N_i)$, a so-called Markov random field (MRF) specification, uniquely determines $h(\theta)$. Geman and Geman (1984) argued that if each full conditional distribution is a so-called Gibbs density, the answer is yes and, in fact, that this provides an equivalent definition of a MRF. The fact that each θ_i is updated by making a draw from a Gibbs distribution motivated them to refer to the entire updating scheme as Gibbs sampling.

The Gibbs sampler is, arguably, better suited for handling simulation from a posterior distribution. As noted by Gelfand and Smith (1990), $h(\theta_i|\theta_j, j \neq i) \propto f(\theta)$, where $f(\theta)$ is viewed as a function of θ_i with all other arguments fixed. Hence we always know (at least up to normalization) the full conditional densities needed to implement the Gibbs sampler. The Gibbs sampler can also be used to investigate *conditional* distributions associated with $f(\theta)$, as done by Gelfand and Smith (1990). It is also well suited to the case where $f(\theta)$ arises as the restriction of a joint density to a set S (see Gelfand, Smith, and Lee 1992).

The 1990s have brought unimaginable availability of inexpensive high-speed computing. Such computing capability was blossoming at the time of Gelfand and Smith's 1990 article. The former fueled considerable experimentation with the latter, in the process demonstrating its broad practical viability. Concurrently, the increasing computing possibilities were spurring interest in a broad range of complex modeling specifications, including generalized linear mixed models, time series and dynamic models, nonparametric and semiparametric models (particularly for censored survival data), and longitudinal and spatial data models. These could all be straightforwardly fitted as Bayesian models using Gibbs sampling.

4. WHAT HAS BEEN THE IMPACT?

Previously, within the statistical community, Bayesians, though confident in the unification and coherence that their paradigm provides, were frustrated by the computational limitations described in Section 2.1, which restricted them to "toy" problems. Though progress was made with numerical integration approaches, analytic approximation methods, and noniterative simulation strategies, fitting the rich classes of hierarchical models that provide the real inferential benefits of the paradigm (e.g., smoothing, borrowing strength, accurate interval estimates) was generally beyond the capability of these tools. The Gibbs sampler provided Bayesians with a tool to fit models previously inaccessible to classical workers. The tables were turned; if one specified a likelihood and prior, the Gibbs sampler was ready to go!

The ensuing fallout has by and large been predictable. Practitioners and subject matter researchers, seeking to explore more realistic models for their data, have enthusiastically embraced the Gibbs sampler, and Bayesians, stimulated by such receptiveness, have eagerly sought collaborative research opportunities. An astonishing proliferation of articles using MCMC model fitting has resulted. On the other hand, classical theoreticians and methodologists, perhaps feeling somewhat threatened, find intellectual vapidity in the entire enterprise; "another Gibbs sampler paper" is a familiar retort.

Though not all statisticians participate, an ideological divide, perhaps stronger than in the past, has emerged. Bayesians will argue that with a full model specification, full inference is available. And the inference is "exact" (although an enormous amount of sampling from the posterior may be required to achieve it!), avoiding the uncertainty associated with asymptotic inference. Frequentists will raise familiar concerns with prior specifications and with inference performance under experimental replication. They also will feel uncomfortable with the black box, nonanalytic nature of the Gibbs sampler. Rather than "random" estimates, they may prefer explicit expressions that permit analytic investigation.

Moreover, Gibbs sampling, as a model fitting and data-analytic technology, is fraught with the risk for abuse. MCMC methods are frequently stretched to models more complex than the data can hope to support. Inadequate investigation of convergence in high-dimensional settings is often the norm, improper posteriors surface periodically in the literature, and inference is rarely externally checked.

5. WHERE ARE WE GOING?

At this point, the Gibbs sampler and MCMC in general are well accepted and utilized for data analysis. Its use in the applied sector will continue to grow. Nonetheless, in the statistical community the frenzy over Gibbs sampling has passed, the field is now relatively stable, and future direction can be assessed. I begin with a list of "tricks of the trade," items still requiring further clarification:

- Model fitting should proceed from simplest to hardest, with fitting of simpler models providing possible starting values, mode searching, and proposal densities for harder models.
- Attention to parameterization is crucial. Given the futility of "transformation to uncorrelatedness," automatic approaches, such as that of Gelfand, Sahu, and Carlin (1995a,b), are needed. Strategies for nonlinear models are even more valuable.
- Latent and auxiliary variables are valuable devices but effective usage requires appreciation of the trade-off between simplified sampling and increased model dimension.
- When can one use the output associated with a component of θ that appears to have converged? For instance, population-level parameters, which are often of primary interest, typically converge more rapidly than individual-level parameters.
- Blocking is recognized as being helpful in handling correlation in the posterior but what are appropriate blocking strategies for hierarchical models?
- Often "hard to fit" parameters are fixed to improve the convergence behavior of a Gibbs sampler. Is an associated sensitivity analysis adequate in such cases?
- Because harder models are usually weakly identified, informative priors are typically required to obtain well behaved Gibbs samplers. How does one use the data to develop these priors and to specify them as weakly as possible?
- Good starting values are required to run multiple chains. How does one obtain "overdispersed" starting values?

With the broad range of models that can now be explored using Gibbs sampling, one naturally must address questions of model determination. Strategies that conveniently piggyback onto the output of Gibbs samplers are of particular interest as ad hoc screening and checking procedures. In this regard, see Gelfand and Ghosh (1998) and Spiegelhalter, Best, and Carlin (1998) for model choice approaches and Gelman, Meng, and Stern (1995) for model adequacy ideas.

Finally, with regard to software development, the BUGS package (Spiegelhalter, Thomas, Best, and Gilks 1995) at this point, is general and reliable enough (with no current competition) to be used both for research and teaching. CODA (Best, Cowles, and Vines 1995), is a convenient add-on to implement a medley of convergence diagnostics. The future will likely bring specialized packages to accommodate specific classes of models, such as time series and dynamic models. However, fitting cutting edge models will always require tinkering and tuning (and possibly specialized algorithms), placing it beyond extant software. But the latter can often fit simpler models before exploring harder ones and can be used to check individual code.

As for hardware, it is a given that increasingly faster machines with larger and larger storage will evolve, making feasible the execution of enormous numbers of iterations for high-dimensional models within realistic run times, diminishing convergence concerns. However, one also would expect that more capable multiprocessor machines will be challenged by bigger datasets and more complex models, encouraging parallel processing MCMC implementations.

REFERENCES

Besag, J., and Green, P. J. (1993), "Spatial Statistics and Bayesian Computation," *Journal of the Royal Statistical Society*, Ser. B, 55, 25–37.

Besag, J., Green, P., Higdon, D., and Mengersen, K. (1996), "Bayesian Computation and Stochastic Systems" (with discussion), *Statistical Science*, 10, 3–66.

Best, N. G., Cowles, M. K., and Vines, K. (1995), "CODA: Convergence Diagnostics and Output Analysis Software for Gibbs Sampling Output, Version 0.30," Medical Research Council, Biostatistics Unit, Cambridge, U.K.

Brooks, S. P., and Roberts, G. O. (1998), "Assessing Convergence of Markov Chain Monte Carlo Algorithms," *Statistics and Computing*, 8, 319–335.

Carlin, B. P., and Louis, T. A. (2000), *Bayes and Empirical Bayes Methods for Data Analysis* (2nd ed.), London: Chapman and Hall.

Chib, S., and Greenberg, E. (1995), "Understanding the Metropolis–Hastings Algorithm," *American Statistician*, 49, 327–335.

Cowles, M. K., and Carlin, B. P. (1994), "Markov Chain Monte Carlo Convergence Diagnostics: A Comparative Review," *Journal of the American Statistical Association*, 91, 883–904.

Devroye, L. (1986), *Non-Uniform Random Variate Generation*, New York: Springer-Verlag.

Gelfand, A. E., and Ghosh, S. K. (1998), "Model Choice: A Minimum Posterior Predictive Loss Approach," *Biometrika*, 85, 1–11.

Gelfand, A. E., Sahu, S. K., and Carlin, B. P. (1995a), "Efficient Parameterization for Normal Linear Mixed Effects Models," *Biometrika*, 82, 479–488.

——— (1995b), "Efficient Parameterization for Generalized Linear Mixed Models," in *Bayesian Statistics 5*, eds. J. M. Bernardo, J. O. Berger, A. P. Dawid, and A. F. M. Smith, London: Oxford University Press, pp. 47–74.

Gelfand, A. E., and Smith, A. F. M. (1990), "Sampling-Based Approaches to Calculating Marginal Densities," *Journal of the American Statistical Association*, 85, 398–409.

Gelfand, A. E., Smith, A. F. M., and Lee, T-M. (1992), "Bayesian Analysis of Constrained Parameter and Truncated Data Problems," *Journal of the American Statistical Association*, 87, 523–532.

Gelman, A., Carlin, J. B., Stern, H. S., and Rubin, D. B. (1995), *Bayesian Data Analysis*, London: Chapman and Hall.

Gelman, A., Meng, X-L., and Stern, H. S. (1995), "Posterior Predictive Assessment of Model Fitness via Realized Discrepancies" (with discussion), *Statistica Sinica*, 6, 733–807.

Geman, S., and Geman, D. (1984), "Stochastic Relaxation, Gibbs Distributions and the Bayesian Restoration of Images," *IEEE Trans. Pattern Anal. Mach. Intell.*, 6, 721–741.

Geweke, J. F. (1989), "Bayesian Inference in Econometric Models Using Monte Carlo Integration," *Econometrika*, 57, 1317–1340.

Gilks, W. R., Clayton, D. G., Spiegelhalter, D. J., Best, N. E., McNeil, A. J., Sharples, L. D., and Kirby, A. J. (1993), "Modeling Complexity: Applications of Gibbs Sampling in Medicine," *Journal of the Royal Statistical Society*, Ser. B, 55, 39–52.

Gilks, W. R., and Wild, P. (1992), "Adaptive Rejection Sampling for Gibbs Sampling," *Journal of the Royal Statistical Society*, Ser. C, 41, 337–348.

Ripley, B. D. (1987), *Stochastic Simulation*, New York: Wiley.

Ritter, C., and Tanner, M. A. (1992), "The Gibbs Stopper and the Griddy Gibbs Sampler," *Journal of the American Statistical Association*, 87, 861–868.

Robert, C. P., and Casella, G. (1999), *Monte Carlo Statistical Methods*, New York: Springer-Verlag.

Rubin, D. B. (1988), "Using the SIR Algorithm to Simulate Posterior Distributions," in *Bayesian Statistics 3*, eds. J. M. Bernardo, M. H. DeGroot, D. V. Lindley, and A. F. M. Smith, London: Oxford University Press, pp. 395–402.

Smith, A. F. M., and Gelfand, A. E. (1992), "Bayesian Statistics Without Tears," *American Statistician*, 46, 84–88.

Smith, A. F. M., and Roberts, G. O. (1993), "Bayesian Computation via the Gibbs Sampler and Related Markov Chain Monte Carlo Methods," *Journal of the Royal Statistical Society*, Ser. B, 55, 3–23.

Spiegelhalter, D., Best, N., and Carlin, B. P. (1998), "Bayesian Deviance, the Effective Number of Parameters and the Comparison of Arbitrarily Complex Models," technical report, MRC Biostatistics Unit, Cambridge, U.K.

Spiegelhalter, D. J., Thomas, A., Best, N., and Gilks, W. R. (1995), "BUGS: Bayesian Inference Using Gibbs Sampling, Version 0.50," Medical Research Council, Biostatistics Unit, Cambridge, U.K.

Tanner, M. A. (1993), *Tools for Statistical Inference* (2nd ed.), New York: Springer-Verlag.

Tanner, M. A., and Wong, W. H. (1987), "The Calculation of Posterior Distributions by Data Augmentation," *Journal of the American Statistical Association*, 82, 528–540.

Tierney, L. (1994), "Markov Chains for Exploring Posterior Distributions," *Ann. Statist.*, 22, 1701–1762.

Wakefield, J., Gelfand, A. E., and Smith, A. F. M. (1992), "Efficient Computation of Random Variates via the Ratio-of-Uniforms Method," *Statist. and Comput.*, 1, 129–133.

West, M. (1992), "Modeling With Mixtures," in *Bayesian Statistics 4*, eds. J. M. Bernardo, J. O. Berger, A. P. Dawid, and A. F. M. Smith, London: Oxford University Press, pp. 503–524.

The Variable Selection Problem

Edward I. George

The problem of variable selection is one of the most pervasive model selection problems in statistical applications. Often referred to as the problem of subset selection, it arises when one wants to model the relationship between a variable of interest and a subset of potential explanatory variables or predictors, but there is uncertainty about which subset to use. This vignette reviews some of the key developments that have led to the wide variety of approaches for this problem.

1. INTRODUCTION

Suppose that $\mathbf{Y}$, a variable of interest, and $\mathbf{X}_1, \ldots, \mathbf{X}_p$, a set of potential explanatory variables or predictors, are vectors of n observations. The problem of variable selection, or subset selection as it is often called, arises when one wants to model the relationship between $\mathbf{Y}$ and a subset of $\mathbf{X}_1, \ldots, \mathbf{X}_p$, but there is uncertainty about which subset to use. Such a situation is particularly of interest when p is large and $\mathbf{X}_1, \ldots, \mathbf{X}_p$ is thought to contain many redundant or irrelevant variables.

The variable selection problem is most familiar in the linear regression context, where attention is restricted to normal linear models. Letting γ index the subsets of $\mathbf{X}_1, \ldots, \mathbf{X}_p$ and letting q_γ be the size of the γth subset, the problem is to select and fit a model of the form

$$\mathbf{Y} = \mathbf{X}_\gamma \boldsymbol{\beta}_\gamma + \varepsilon, \tag{1}$$

where $\mathbf{X}_\gamma$ is an $n \times q_\gamma$ matrix whose columns correspond to the γth subset, $\boldsymbol{\beta}_\gamma$ is a $q_\gamma \times 1$ vector of regression coefficients, and $\varepsilon \sim N_n(0, \sigma^2 I)$. More generally, the variable selection problem is a special case of the model selection problem where each model under consideration corresponds to a distinct subset of $\mathbf{X}_1, \ldots, \mathbf{X}_p$. Typically, a single model class is simply applied to all possible subsets. For example, a wide variety of relationships can be considered with generalized linear models where $g(E(\mathbf{Y})) = \alpha + \mathbf{X}_\gamma \boldsymbol{\beta}_\gamma$ for some link function g (see the vignettes by Christensen and McCulloch). Moving further away from the normal linear model, one might instead consider relating $\mathbf{Y}$ and subsets of $\mathbf{X}_1, \ldots, \mathbf{X}_p$ with nonparametric models such as CART or MARS.

Edward I. George holds the Ed and Molly Smith Chair and is Professor of Statistics, Department of MSIS, University of Texas, Austin, TX 78712 (E-mail: egeorge@mail.utexas.edu). This work was supported by National Science Foundation grant DMS-98.03756 and Texas ARP grants 003658.452 and 003658.690.

The fundamental developments in variable selection seem to have occurred either directly in the context of the linear model (1) or in the context of general model selection frameworks. Historically, the focus began with the linear model in the 1960s, when the first wave of important developments occurred and computing was expensive. The focus on the linear model still continues, in part because its analytic tractability greatly facilitates insight, but also because many problems of interest can be posed as linear variable selection problems. For example, for the problem of nonparametric function estimation, Y represents the values of the unknown function, and $X_1, \ldots, X_p$ represent a linear basis, such as a wavelet basis or a spline basis. However, as advances in computing technology have allowed for the implementation of richer classes of models, treatments of the variable selection problem by general model selection approaches are becoming more prevalent.

One of the fascinating aspects of the variable selection problem has been the wide variety of methods that have been brought to bear on the problem. Because of space limitations, it is of course impossible to even mention them all, and so I focus on only a few to illustrate the general thrust of developments. An excellent and comprehensive treatment of variable selection methods prior to 1990 was provided by Miller (1990). As I discuss, many promising new approaches have appeared over the last decade.

2. GETTING A GRIP ON THE PROBLEM

A distinguishing feature of variable selection problems is their enormous size. Even with moderate values of p, computing characteristics for all 2^p models is prohibitively expensive, and some reduction of the model space is needed. Focusing on the linear model (1), early suggestions based such reductions on the residual sum of squares, which provided a partial ordering of the models. Taking advantage of the chain structure of subsets, branch and bound methods such as the algorithm of Furnival and Wilson (1974) were proposed to logically eliminate large numbers of models from consideration. When feasible, attention was often restricted to the "best subsets" of each size. Otherwise, reduction was obtained with variants of stepwise methods that sequentially add or delete variables based on greedy considerations (e.g., Efroymson 1966). Even with advances in computing technology, these methods continue to be the standard workhorses for reduction. Extensions beyond the linear model are straightforward; for example, in generalized linear models by substituting the deviance for the residual sum of squares.

Once attention was reduced to a manageable set of models, criteria were needed for selecting a subset model. The earliest developments of such selection criteria, again in the linear model context, were based on attempts to minimize the mean squared error of prediction. Different criteria corresponded to different assumptions about which predictor values to use, and whether they were fixed or random (see Hocking 1976; Thompson 1978 and the references therein). Perhaps the most familiar of those criteria is the Mallows $C_p = (\text{RSS}_\gamma / \hat{\sigma}^2_{\text{full}} + 2q_\gamma - n)$, where RSS_γ is the residual sum of squares for the γth model and $\hat{\sigma}^2_{\text{full}}$ is the usual unbiased estimate of σ^2 based on the full model. Motivated as an unbiased estimate of predictive accuracy of the γth model, Mallows (1973) recommended using C_p plots to help gauge subset selection (see also Mallows 1995). Although Mallows specifically warned against using minimum C_p as a selection criterion (because of selection bias), minimum C_p continues to be used as a criterion (and attributed to Mallows to boot!).

Two of the other most popular criteria, motivated from very different viewpoints, are the Akaike information criterion (AIC) and the Bayesian information criterion (BIC). Letting $\hat{l}_\gamma$ denote the maximum log-likelihood of the γth model, AIC selects the model that maximizes $(\hat{l}_\gamma - q_\gamma)$, whereas BIC selects the model that maximizes $(\hat{l}_\gamma - (\log n)q_\gamma/2)$. Akaike (1973) motivated AIC from an information theoretic standpoint (see the vignette by Soofi) as the minimization of the Kullback–Leibler distance between the distributions of Y under the γth model and under the true model. To lend further support, an asymptotic equivalence of AIC and cross-validation was shown by Stone (1977). In contrast, Schwarz (1978) motivated BIC from a Bayesian standpoint, by showing that it was asymptotically equivalent (as $n \to \infty$) to selection based on Bayes factors. BIC was further justified from a coding theory viewpoint by Rissanen (1978).

Comparisons of the relative merits of AIC and BIC based on asymptotic consistency (as $n \to \infty$) have flourished in the literature. As it turns out, BIC is consistent when the true model is fixed (Haughton 1988), whereas AIC is consistent if the dimensionality of the true model increases with n (at an appropriate rate) (Shibata 1981). Stone (1979) provided an illuminating discussion of these two viewpoints.

For the linear model (1), many of the popular selection criteria are special cases of a penalized sum of squares criterion, providing a unified framework for comparisons. Assuming σ^2 known to avoid complications, this general criterion selects the subset model that minimizes

$$(\text{RSS}_\gamma / \hat{\sigma}^2 + Fq_\gamma), \tag{2}$$

where F is a preset "dimensionality penalty." Intuitively, (2) penalizes $\text{RSS}_\gamma / \hat{\sigma}^2$ by F times q_γ, the dimension of the γth model. AIC and minimum C_p are essentially equivalent, corresponding to $F = 2$, and BIC is obtained by setting $F = \log n$. By imposing a smaller penalty, AIC and minimum C_p will select larger models than BIC (unless n is very small).

3. TAKING SELECTION INTO ACCOUNT

Further insight into the choice of F is obtained when all of the predictors are orthogonal, in which case (2) simply selects all of those predictors with t-statistics t for which $t^2 > F$. When $\mathbf{X}_1, \ldots, \mathbf{X}_p$ are in fact all unrelated to $\mathbf{Y}$ (i.e., the full model regression coefficients are all 0), AIC and minimum C_p are clearly too liberal and tend to include a large proportion of irrelevant variables. A natural conservative choice for F, namely $F = 2 \log p$, is suggested by the fact that under this null model, the expected value of the largest squared t-statistic is approximately $2 \log p$ when p is large. This choice is the risk inflation criterion (RIC) proposed by Foster and George (1994) and the universal threshold for wavelets proposed by Donoho and Johnstone (1994). Both of these articles motivate $F = 2 \log p$ as yielding the smallest possible maximum inflation in predictive risk due to selection (as $p \to \infty$), a minimax decision theory standpoint. Motivated by similar considerations, Tibshirani and Knight (1999) recently proposed the covariance inflation criterion (CIC), a nonparametric method of selection based on adjusting the bias of in-sample performance estimates. Yet another promising adjustment based on a generalized degrees of freedom concept was proposed by Ye (1998).

Many other interesting criteria corresponding to different choices of F in (2) have been proposed in the literature (see, e.g., Hurvich and Tsai 1989, 1998; Rao and Wu 1989; Shao 1997; Wei 1992; Zheng and Loh 1997 and the references therein). One of the drawbacks of using a fixed choice of F is that models of a particular size are favored; small F favors large models, and large F favors small models. Adaptive choices of F to mitigate this problem have been recommended by Benjamini and Hochberg (1995), Clyde and George (1999, 2000), George and Foster (2000), and Johnstone and Silverman (1998).

An alternative to explicit criteria of the form (2) is selection based on predictive error estimates obtained by intensive computing methods such as the bootstrap (e.g., Efron 1983; Gong 1986) and cross-validation (e.g., Shao 1993; Zhang 1993). An interesting variant of these is the little bootstrap (Brieman 1992), which estimates the predictive error of selected models by mimicking replicate data comparison. The little bootstrap compares favorably to selection based on minimum C_p or the conditional bootstrap, whose performances are seriously denigrated by selection bias.

Another drawback of traditional subset selection methods, which is beginning to receive more attention, is their instability relative to small changes in the data. Two novel alternatives that mitigate some of this instability for linear models are the nonnegative garrotte (Brieman 1995) and the lasso (Tibshirani 1996). Both of these procedures replace the full model least squares criterion by constrained optimization criteria. As the constraint is tightened, estimates are zeroed out, and a subset model is identified and estimated.

4. BAYESIAN METHODS EMERGE

The fully Bayesian approach to variable selection is as follows (George 1999). For a given set of models $M_1, \ldots, M_{2^p}$, where M_γ corresponds to the γth subset of $X_1, \ldots, X_p$, one puts priors $\pi(\beta_\gamma | M_\gamma)$ on the parameters of each M_γ and a prior on the set of models $\pi(M_1), \ldots, \pi(M_{2^p})$. Selection is then based on the posterior model probabilities $\pi(M_\gamma|Y)$, which are obtained in principle by Bayes's theorem.

Although this Bayesian approach appears to provide a comprehensive solution to the variable selection problem, the difficulties of prior specification and posterior computation are formidable when the set of models is large. Even when p is small and subjective considerations are not out of the question (Garthwaite and Dickey 1996), prior specification requires considerable effort. Instead, many of the Bayesian proposals have focused on semiautomatic methods that attempt to minimize prior dependence. Indeed, this is part of the appeal of BIC, which avoids prior specification altogether, and its properties continue to be investigated and justified (Kass and Wasserman 1995; Pauler 1998; Raftery 1996). Other examples of Bayesian treatments that avoid the prior selection difficulties in variable selection include the early proposal of Lindley (1968) to use uniform priors and a cost function for selection, the default Bayes factor criteria of Berger and Pericchi (1996a,b) and O'Hagan (1995), and the predictive criteria of Geisser and Eddy (1979), Laud and Ibrahim (1995), and San Martini and Spezzaferri (1984).

In contrast to the development of Bayesian approaches that avoid the difficulties of prior specification, the advent of Markov chain Monte Carlo (MCMC) (see the vignette by Cappé and Robert) has focused attention on Bayesian variable selection with fully specified proper parameter priors. Bypassing the difficulties of computing the entire posterior, MCMC algorithms can instead be used to stochastically search for the high-posterior probability models. The idea is that by simulating a Markov chain, which is converging to the posterior distribution, the high-probability models should tend to appear more often, and hence sooner. The resulting implementations are stepwise algorithms that are stochastically guided by the posterior, rather than by the greedy considerations of conventional stepwise methods. Such a Bayesian package is complete; it offers posterior probability as a selection criteria, associated MCMC algorithms for search, and Bayes estimates for the selected model.

The last decade has seen an explosion of research on this Bayesian variable selection approach. These developments have included proposals for new prior specifications that induce increased posterior probability on the more promising models, for new MCMC implementations that are more versatile and offer improved performance, and for extensions to a wide variety of model classes. Another closely related development in this context has been the emergence of model averaging as an alternative to variable selection. Under the Bayesian variable selection formulation, the posterior mean is an adaptive convex combination of all the individual model estimates (i.e., a model average). Although model averaging almost always improves on variable selection in terms of prediction, its drawback is that it does not lead to a reduced set of variables. Some, but by no means all, of the key developments of these Bayesian approaches to variable selection and model averaging have been discussed by Clyde (1999), Clyde, Parmigiani, and Vidakovic (1998), Draper (1995),

George and McCulloch (1993, 1997), Green (1995), and Hoeting, Madigan, Raftery, and Volinsky (1999).

5. WHAT IS NEXT

Today, variable selection procedures are an integral part of virtually all widely used statistics packages, and their use will only increase as the information revolution brings us larger datasets with more and more variables. The demand for variable selection will be strong, and it will continue to be a basic strategy for data analysis.

Although numerous variable selection methods have been proposed, plenty of work still remains to be done. To begin with, many of the recommended procedures have been given only a narrow theoretical motivation, and their operational properties need more systematic investigation before they can be used with confidence. For example, small-sample justification is needed in addition to asymptotic considerations, and frequentist justification is needed for Bayesian procedures. Although there has been clear progress on the problems of selection bias, clear solutions are still needed, especially for the problems of inference after selection (see Zhang 1992). Another intriguing avenue for research is variable selection using multiple model classes (see Donoho and Johnstone 1995). New problems will also appear as demand increases for data mining of massive datasets. For example, considerations of scalability and computational efficiency will become paramount in such a context. I suppose that all of this is good news, but there is also danger lurking ahead.

With the availability of so many variable selection procedures and so many different justifications, it has becomes increasingly easy to be misled and to mislead. Faced with too many choices and too little guidance, practitioners continue to turn to the old standards such as stepwise selection based on AIC or minimum C_p, followed by a report of the conventional estimates and inferences. The justification of asymptotic consistency will not help the naive user who should be more concerned with selection bias and procedure instability. Eventually, the responsibility for the poor performance of such procedures will fall on the statistical profession, and consumers will turn elsewhere for guidance (e.g., Dash and Liu 1997). Our enthusiasm for the development of promising new procedures must be carefully tempered with cautionary warnings of their potential pitfalls.

REFERENCES

Akaike, H. (1973), "Information Theory and an Extension of the Maximum Likelihood Principle," in *2nd International Symposium on Information Theory*, eds. B. N. Petrov and F. Csaki, Budapest: Akademia Kiado, pp. 267–281.

Benjamini, Y., and Hochberg, Y. (1995), "Controlling the False Discovery Rate: A Practical and Powerful Approach to Multiple Testing," *Journal of the Royal Statistical Society*, Ser. B, 57, 289–300.

Berger, J. O., and Pericchi, L. R. (1996a), "The Intrinsic Bayes Factor for Model Selection and Prediction," *Journal of the American Statistical Association*, 91, 109–122.

—— (1996b), "The Intrinsic Bayes Factor for Linear Models," in *Bayesian Statistics 5*, eds. J. M. Bernardo, J. O. Berger, A. P. Dawid, and A. F. M. Smith, London: Oxford University Press, pp. 25–44.

Brieman, L. (1992), "The Little Bootstrap and Other Methods for Dimensionality Selection in Regression: X-Fixed Prediction Error," *Journal of the American Statistical Association*, 87, 738–754.

—— (1995), "Better Subset Selection Using the Nonnegative Garrote," *Technometrics*, 37, 373–384.

Clyde, M. (1999), "Bayesian Model Averaging and Model Search Strategies," in *Bayesian Statistics 6*, eds. J. M. Bernardo, J. O. Berger, A. P. Dawid, and A. F. M. Smith, London: Oxford University Press.

Clyde, M., and George, E. I. (1999), "Empirical Bayes Estimation in Wavelet Nonparametric Regression," in *Bayesian Inference in Wavelet-Based Models*, eds. P. Muller and B. Vidakovic, New York: Springer-Verlag, pp. 309–322.

—— (2000), "Flexible Empirical Bayes Estimation for Wavelets," *Journal of the Royal Statistical Society*, Ser. B, 681–698.

Clyde, M., Parmigiani, G., and Vidakovic, B. (1998), "Multiple Shrinkage and Subset Selection in Wavelets," *Biometrika*, 85, 391–402.

Dash, M., and Liu, H. (1997), "Feature Selection for Classification," *Intelligent Data Analysis*, 1, 131–156.

Donoho, D. L., and Johnstone, I. M. (1994), "Ideal Spatial Adaptation by Wavelet Shrinkage," *Biometrika*, 81, 425–256.

—— (1995), "Adapting to Unknown Smoothness via Wavelet Shrinkage," *Journal of the Royal Statistical Society*, Ser. B, 90, 1200–1224.

Draper, D. (1995), "Assessment and Propagation of Model Uncertainty" (with discussion), *Journal of the Royal Statistical Society*, Ser. B, 57, 45–97.

Efron, B. (1983), "Estimating the Error Rate of a Predictive Rule: Improvement Over Cross-Validation," *Journal of the American Statistical Association*, 78, 316–331.

Efroymson, M. A. (1960), "Multiple Regression Analysis," in *Mathematical Methods for Digital Computers*, eds. A. Ralston and H. S. Wilf, New York: Wiley, pp. 191–203.

Foster, D. P., and George, E. I. (1994), "The Risk Inflation Criterion for Multiple Regression," *The Annals of Statistics*, 22, 1947–1975.

Furnival, G. M., and Wilson, R. W. (1974), "Regression by Leaps and Bounds," *Technometrics*, 16, 499–511.

Garthwaite, P. H., and Dickey, J. M. (1996), "Quantifying and Using Expert Opinion for Variable-Selection Problems in Regression" (with discussion), *Chemometrics and Intelligent Laboratory Systems*, 35, 1–34.

Geisser, S., and Eddy, W. F. (1979), "A Predictive Approach to Model Selection," *Journal of the American Statistical Association*, 74, 153–160.

George, E. I. (1999), "Bayesian Model Selection," in *Encyclopedia of Statistical Sciences, Update Vol. 3*, eds. S. Kotz, C. Read, and D. Banks, New York: Wiley, pp. 39–46.

George, E. I., and Foster, D. P. (2000), "Calibration and Empirical Bayes Variable Selection," *Biometrika*, 87, 731–747.

George, E. I., and McCulloch, R. E. (1993), "Variable Selection via Gibbs Sampling," *Journal of the American Statistical Association*, 88, 881–889.

—— (1997), "Approaches for Bayesian Variable Selection," *Statistica Sinica*, 7(2), 339–373.

Gong, G. (1986), "Cross-Validation, the Jackknife, and the Bootstrap: Excess Error Estimation in Forward Logistic Regression," *Journal of the American Statistical Association*, 393, 108–113.

Green, P. (1995), "Reversible Jump Markov Chain Monte Carlo Computation and Bayesian Model Determination," *Biometrika*, 82, 711–732.

Haughton, D. (1988), "On the Choice of a Model to Fit Data From an Exponential Family," *The Annals of Statistics*, 16, 342–355.

Hocking (1976), "The Analysis and Selection of Variables in Linear Regression," *Biometrics*, 32, 1–49.

Hoeting, J. A., Madigan, D., Raftery, A. E., and Volinsky, C. T. (1999), "Bayesian Model Averaging: A Tutorial" (with discussion), *Statistical Science*, 14, 382–417.

Hurvich, C. M., and Tsai, C. L. (1989), "Regression and Time Series Model Selection in Small Samples," *Biometrika*, 76, 297–307.

——— (1998), "A Cross-Validatory AIC for Hard Wavelet Thresholding in Spatially Adaptive Function Estimation," *Biometrika*, 85, 701–710.

Johnstone, I. M., and Silverman, B. W. (1998), "Empirical Bayes Approaches to Mixture Problems and Wavelet Regression," technical report, University of Bristol.

Kass, R. E., and Wasserman, L. (1995), "A Reference Bayesian Test for Nested Hypotheses and Its Relationship to the Schwarz Criterion," *Journal of the American Statistical Association*, 90, 928–934.

Laud, P. W., and Ibrahim, J. G. (1995), "Predictive Model Selection," *Journal of the Royal Statistical Society*, Ser. B, 57, 247–262.

Lindley, D. V. (1968), "The Choice of Variables in Multiple Regression" (with discussion), *Journal of the Royal Statistical Society*, Ser. B, 30, 31–66.

Mallows, C. L. (1973), "Some Comments on C_p," *Technometrics*, 15, 661–676.

——— (1995), "More Comments on C_p," *Technometrics*, 37, 362–372.

Miller, A. (1990), *Subset Selection in Regression*, London: Chapman and Hall.

O'Hagan, A. (1995), "Fractional Bayes Factors for Model Comparison" (with discussion), *Journal of the Royal Statistical Society*, Ser. B, 57, 99–138.

Pauler, D. (1998), "The Schwarz Criterion and Related Methods for the Normal Linear Model," *Biometrika*, 85, 13–27.

Raftery, A. E. (1996), "Approximate Bayes Factors and Accounting for Model Uncertainty in Generalized Linear Models," *Biometrika*, 83, 251–266.

Rao, C. R., and Wu, Y. (1989), "A Strongly Consistent Procedure for Model Selection in a Regression Problem," *Biometrika*, 76, 369–374.

Rissanen, J. (1978), "Modeling by Shortest Data Description," *Automatica*, 14, 465–471.

San Martini, A., and Spezzaferri, F. (1984), "A Predictive Model Selection Criterion," *Journal of the Royal Statistical Society*, Ser. B, 46, 296–303.

Schwarz, G. (1978), "Estimating the Dimension of a Model," *The Annals of Statistics*, 6, 461–464.

Shao, J. (1993), "Linear Model Selection by Cross-Validation," *Journal of the American Statistical Association*, 88, 486–494.

——— (1997), "An Asymptotic Theory for Linear Model Selection," *Statistica Sinica*, 7(2), 221–264.

Shibata, R. (1981), "An Optimal Selection of Regression Variables," *Biometrika*, 68, 45–54.

Stone, M. (1977), "An Asymptotic Equivalence of Choice of Model by Cross-Validation and Akaike's Criterion," *Journal of the Royal Statistical Society*, Ser. B, 39, 44–47.

——— (1979), "Comments on Model Selection Criteria of Akaike and Schwarz," *Journal of the Royal Statistical Society*, Ser. B, 41, 276–278.

Thompson, M. L. (1978), "Selection of Variables in Multiple Regression: Part I. A Review and Evaluation," *International Statistical Review*, 46, 1–19.

Tibshirani, R. (1996), "Regression Shrinkage and Selection via the Lasso," *Journal of the Royal Statistical Society*, Ser. B, 58, 267–288.

Tibshirani, R., and Knight, K. (1999), "The Covariance Inflation Criterion for Model Selection," *Journal of the Royal Statistical Society*, Ser. B, 61, 529–546.

Wei, C. Z. (1992), "On Predictive Least Squares Principles," *The Annals of Statistics*, 29, 1–42.

Ye, J. (1998), "On Measuring and Correcting the Effects of Data Mining and Model Selection," *Journal of the American Statistical Association*, 93, 120–131.

Zhang, P. (1992), "Inference After Variable Selection in Linear Regression Models," *Biometrika*, 79, 741–746.

——— (1993), "Model Selection via Multifold Cross-Validation," *The Annals of Statistics*, 21, 299–313.

Zheng, X., and Loh, W. Y. (1997), "A Consistent Variable Selection Criterion for Linear Models for Linear Models With High-Dimensional Covariates," *Statistica Sinica*, 7(2), 311–325.

Robust Nonparametric Methods

Thomas P. Hettmansperger, Joseph W. McKean,
and Simon J. Sheather

1. A SHORT HISTORY OF NONPARAMETRICS FOR LOCATION PROBLEMS

The terms "nonparametric statistics" and "distribution-free methods" have historically referred to a collection of statistical tests whose null distributions do not depend on the underlying distribution of the data. The area has been expanded to include estimates and confidence intervals derived from the tests and includes asymptotically distribution-free tests for complex models as well. It is now generally recognized that these statistical methods are highly efficient over large sets of possible models and are robust as well. This was not always the case. The earliest work was by Hotelling and Pabst in 1936 (references to works in this paragraph can be found in the books cited in the next paragraph) on rank correlation and by Friedman in 1937 on rank tests in a two-way design, followed in 1945 with the introduction of the signed rank and rank sum tests by Wilcoxon. In 1947, Mann and Whitney extended the ideas of Wilcoxon. All of these rank tests were considered quick and dirty but not competitive with the presumably more efficient t-tests. This perspective was about to change. In 1948, Pitman developed efficiency concepts in a set of lecture notes that, although never published, were widely disseminated. During the 1950s and 1960s, Lehmann, working with Hodges and with his students at Berkeley, showed that rank tests are surprisingly efficient and robust. In fact, the Wilcoxon rank tests are essentially efficient at the normal model and can be much more efficient than least squares methods when the underlying distribution of the data has heavy tails. In the early 1960s, Hodges and Lehmann derived estimates and confidence intervals from rank test statistics and also introduced aligned rank tests for use in the regression model. In 1967, Hájek and Šidák published their seminal book on the theory of rank tests.

It is impossible to mention all of the contributions to this early development. Hence we cite several key monographs that contain many references to this work and that extended these methods into various settings. Puri and Sen (1971) published a

Thomas P. Hettmansperger is Professor of Statistics, Department of Statistics, Pennsylvania State University, University Park, PA 16802 (E-mail: tph@stat.psu.edu). Joseph W. McKean is Professor of Statistics, Department of Mathematics and Statistics, Western Michigan University, Kalamazoo, MI 49008 (E-mail: joe@stat.wmich.edu). Simon J. Sheather is Professor, Australian Graduate School of Management, University of New South Wales, Sydney, NSW 2052, Australia (E-mail: simonsh@agsm.unsw.edu.au).

monograph on nonparametric methods in multivariate models. This work extended and applied another major work on asymptotic theory by Chernoff and Savage (1958). Lehmann's (1975) book combined applied nonparametric methods with an extensive appendix on theory. The decade closed with the publication of Randles and Wolfe's (1979) introduction to the theory of nonparametric statistics, which provided a set of tools for the development of nonparametric methods.

2. RANK-BASED PROCEDURES FOR LINEAR MODELS

The nonparametric procedures (testing and estimation) for simple location problems offer the user highly efficient and robust methods and form an attractive alternative to traditional least squares (LS) procedures. LS procedures, however, generalize easily to any linear model and to most nonlinear models. The LS procedures are not model dependent. In contrast, there are very few classical nonparametric procedures for designs other than the simple location designs, and, furthermore, these procedures vary with the problem. For instance, the ranking procedure for the Kruskal–Wallis test of treatment effect in a one-way layout is much different than the ranking procedure for the Friedman test of treatment effect in a two-way design. Further, the efficiencies of these two procedures differ widely, although both are based on linear rankings.

The generality of LS procedures can readily be seen in terms of its simple geometry. For example, suppose that a vector of responses $\mathbf{Y}$ follows a linear model of the form $\mathbf{Y} = \boldsymbol{\mu} + \mathbf{e}$, where $\mathbf{e}$ is a vector of random errors, $\boldsymbol{\mu} \in \Omega$, and Ω is a subspace of R^n. The LS estimate of $\boldsymbol{\mu}$ is the vector $\hat{\mathbf{Y}}_{LS}$ that lies "closest" to $\mathbf{Y}$ when distance is Euclidean; that is, $\|\mathbf{Y} - \hat{\mathbf{Y}}_{LS}\|_{LS} = \min \|\mathbf{Y} - \boldsymbol{\mu}\|_{LS}$ over all $\boldsymbol{\mu} \in \Omega$, where $\|\cdot\|_{LS}$ denotes the Euclidean norm. The F test of $H_0 : \boldsymbol{\mu} \in \omega$ versus $H_A : \boldsymbol{\mu} \in \Omega \cap \omega^{\perp}$ is based on the standardized difference in squared distances between $\mathbf{Y}$ and each of the subspaces $\omega \subset \Omega$. These few sentences on geometry form the essence of LS procedures for any fixed-effects linear model.

Although traditional distribution-free procedures do not generalize to any linear model, rank-based procedures based on robust estimates of regression coefficients do. Jureckova (1971) and Jaeckel (1972) obtained rank-based (R) estimates of regression coefficients for a linear model. Although their estimates are asymptotically equivalent, Jaeckel's estimates are based on minimizing a convex function and hence are easily computable. Furthermore, as discussed by McKean and Schrader (1980), Jaeckel's estimates are obtained by minimizing the norm $\|\mathbf{u}\|_{\varphi} = \sum_{i=1}^{n} a(R(u_i))u_i, \mathbf{u} \in R^n$, where $a(i) = \varphi(i/(n + 1))$ for $\varphi(u)$, a nondecreasing scores function defined on the interval $(0, 1)$, and $R(u_i)$ denotes the rank of u_i among $u_1, \ldots, u_n$. The most popular scores are the Wilcoxon (linear) scores, which are generated by $\varphi(u) = \sqrt{12}(u - (1/2))$. (More precisely, $\|\mathbf{u}\|_{\varphi}$ is a pseudonorm on R^n that is analogous to the LS-centered pseudonorm $\|\mathbf{u}\|_{LS} = \sum_{i=1}(u_i - \bar{u})^2$. This is without loss of generality, because centered designs can invariably be used in linear model problems and otherwise signed-rank norms can be used; see Hettmansperger and McKean 1983.)

Rank-based procedures for linear models are then formulated analogous to LS procedures, except that the norm $\|\mathbf{u}\|_\varphi$ is substituted for the Euclidean norm. The R estimate $\hat{\mathbf{Y}}_\varphi$ minimizes $\|\mathbf{Y} - \boldsymbol{\mu}\|_\varphi$ over all vectors $\boldsymbol{\mu} \in \Omega$; that is, $\hat{\mathbf{Y}}_\varphi$ lies "closest" to $\mathbf{Y}$ when the distance is based on $\|\mathbf{u}\|_\varphi$. If the scores are the Wilcoxon scores, then we call them the Wilcoxon estimates. The rank-based F test of $H_0 : \boldsymbol{\mu} \in \omega$ versus $H_A : \boldsymbol{\mu} \in \Omega \cap \omega^\perp$ is based on the standardized difference in distances between $\mathbf{Y}$ and each of the subspaces ω and Ω, when distance is based on the norm $\|\mathbf{u}\|_\varphi$. As with LS, these few sentences capture the essence of rank-based estimation and testing for *any* fixed-effects linear model.

Although in general these rank-based procedures are not distribution free, they are asymptotically distribution free. In practice, as with a LS analysis, a standardizing scale must be estimated. Consistent estimates are available. Small-sample studies have confirmed the validity and power of these rank-based procedures (see McKean and Sheather 1991).

The rank-based analysis generalizes simple location rank procedures and also has the same efficiency properties as rank procedures for simple location models. Depending on knowledge of the error distribution, scores can be chosen to optimize efficiency. Influence functions of the estimates and tests are bounded in response space; hence the analysis is robust in response space.

The rank-based estimates are consistent for heteroscedastic models. The analysis can be modified if the heteroscedasticity is a function of the expected response. Because the analysis is based on a norm, extensions to generalized linear and nonlinear models follow similar to the Gauss–Newton development of LS procedures for such models. Thus the analysis can be extended to logistic regression and time series.

Recently a new and interesting direction emerged for rank-based methods for multivariate data. The concept of ranking multivariate observations was developed based on ideas of invariant tests and equivariant estimates. Spatial ranks were defined that capture the properties of ranks and at the same time are the basis of tests invariant under rotations and reflections. The same has been done in the general context of affine transformations.

This work on rank-based procedures has developed over the last 25 years. Monographs include the books by Hettmansperger (1984), Koul (1992), and Puri and Sen (1985). The recent monograph by Hettmansperger and McKean (1998) developed rank-based methods for location through linear models, both univariate and multivariate. The unifying theme in this book is the geometry cited earlier and a complete set of statistical methods are developed and illustrated.

3. HIGH-BREAKDOWN, BOUNDED INFLUENCE ROBUST ESTIMATES

Although the R estimates based on minimizing the norm $\| \cdot \|_{\varphi}$ are robust in the Y-space, they are not robust in the x-space (factor space). Although this usually is not a problem for a controlled design, it can be a major problem when exploring datasets based on uncontrolled studies. Two important tools for such datasets are robust procedures with bounded influence in both spaces and diagnostic procedures that assess the quality of a fit and differentiate between fits. In this section we focus on bounded influence robust estimates. Although these estimates are robust in both spaces, they are always less efficient than the estimates discussed in Section 2.

Recent research in robust estimation has focused on the challenge of producing a regression estimator that combines both local and global stability. The local stability of an estimator is often assessed by whether its influence function is bounded in both the x and Y spaces and is continuous. The influence function of an estimator is an asymptotic notion that describes the standardized effect of a single point on the estimator. Global stability is usually measured via the breakdown point, the smallest fraction of contamination that can render the estimator meaningless, which, therefore, provides an indication of the amount of gross contamination that an estimator can tolerate.

The generalized rank-based (GR) estimate (Naranjo and Hettmansperger 1994) is a weighted Wilcoxon with weights depending on factor space. Its influence function is bounded in both spaces, so it achieves local stability at the price of a loss of efficiency to the Wilcoxon estimate. Although the GR estimate's breakdown is positive, it can be quite small. Thus it does not achieve global stability.

On the other hand, many of the high-breakdown estimators are not locally robust. In particular, the influence functions of the least trimmed squares (LTS) (Rousseeuw and Leroy 1987) and other S and generalized S estimators (Rousseeuw and Yohai 1984) are unbounded (e.g., Croux, Rousseeuw, and Hössjer 1994). Davies (1993) showed that the influence function of the least median squares (LMS) (Rousseeuw and Leroy 1987) is unbounded at the quartiles of Y. A number of authors have reported on the local instability of LMS (e.g., Ellis 1998; Hettmansperger and Sheather 1992; Sheather, McKean, and Hettmansperger 1997; Simpson and Yohai 1998). Rousseeuw, Hubert, and Van Aelst (1999) now recommend LTS over LMS.

Besides the GR estimate, a number of estimators with bounded and continuous influence functions are not globally robust. In particular, the breakdown point of a number of generalized M (GM) estimators, including the optimal Krasker–Welsch estimator (Krasker and Welsch 1982), tends to zero as the number of predictors increases.

Simpson, Ruppert, and Carroll (1992) showed how to produce a regression estimate that is both locally and globally robust. They considered one-step Newton–Raphson GM estimators of the Mallows type that use weights based on high-breakdown initial estimates of location and scatter in factor space. These weights limit the effect of extreme points in the design space on the estimates. Simpson and Yohai (1998) studied one-step Newton–Raphson estimators, including the well-known Mallows, Schweppe, and Hill–Ryan GM estimators. They showed that the choice of the weight function is crucial, as the Schweppe form and an inadequately weighted Mallows form produce a one-step estimate that has the same local instability as the initial estimate and falsely optimistic estimates of precision. On the other hand, they showed that the Hill–Ryan form is the most stable in terms of one-step estimation.

Recently, Chang, McKean, Naranjo, and Sheather (1999) proposed a high-breakdown rank-based (HBR) estimate which is a weighted Wilcoxon estimate with weights based on both high-breakdown location and scatter estimates over the x space and initial residuals from a high-breakdown estimate of the regression coefficients. This estimate has a bounded influence function over both the x and Y spaces and also has 50% breakdown, provided that the initial estimates also do. In a simulation study, its stability was a marked improvement over the stability of the initial estimates (LMS or LTS). Furthermore, it regained some of the efficiency that the GR estimate loses to the Wilcoxon estimate. Because this estimate is based on minimizing a convex function, the estimates are fully iterated. Thus, unlike GM estimates based on redescending ψ-functions (Hampel, Ronchetti, Rousseeuw, and Stahel 1986), there is no possibility that the estimates will fail to converge.

4. DIAGNOSTICS

Regression diagnostics are methods that provide information on the form of the unknown underlying regression function, the error structure, and any outliers or unusual points. Based on first-order asymptotic properties, McKean et al. (1990, 1993) showed that residual plots based on R estimates and M estimates (based on a convex loss function) can be interpreted in exactly the same manner as their LS counterparts. They also proposed numerical diagnostics, including standardized residuals that, similar to the LS standardized residuals, correct for both location in factor space and scale of the errors.

Cook (1998) described a series of graphical techniques that apply to estimates that are based on convex loss functions and not just on LS. A cornerstone for a number of these methods is a result due to Duan and Li (1991, Thm. 2.1), which in effect shows that the fit of any model can be used to obtain information on the form of the true underlying regression function. Two critical assumptions necessary for this result to hold are that the loss function is convex and that the distribution of the predictors is elliptically symmetric. The second assumption is not problematic, as methods exist for reweighting the predictors to achieve elliptical symmetry (Cook and Nachsteim 1994).

There has also been an increasing realization that residual plots based on high-breakdown robust fits cannot always be interpreted in exactly the same manner as their LS counterparts. For example, Rousseeuw, Hubert, and Van Aelst (1999) cautioned against "wearing LS-colored glasses" when looking at residual plots based on robust fits. A number of authors have reported a negative correlation in the standard plot of robust fitted values against residuals (e.g., Cook, Hawkins, and Weisberg 1992; McKean et al. 1993; Rousseeuw et al. 1999; Simpson and Chang 1997). Using first-order approximations for robust fitted values and residuals, McKean et al. (1993) showed that there can be a strong negative correlation between fits based on GM estimates and their residuals that is induced by the weights and exists whether or not the correct model has been fit. Simpson and Chang (1997, fig. 2a) provided a graphic illustration of this phenomenon with the residuals from a three-step GM estimate. In addition, Rousseeuw et al. (1999) showed that a negative correlation exists for the deepest fit, a generalization of the univariate median to regression, and the corresponding residuals.

High-breakdown regression methods ignore the tails of the data in the x-space by design. This property enables these estimates to protect against a large number of outliers from a straight line model, for example. However, methods that ignore these tails do not perform well when they contain much of the information about the form of the regression function, as is the case for polynomial models (Cook et al. 1992; McKean et al. 1994). One way to protect against outliers and still be able to detect and model curvature is to consider at least two estimates, a high-breakdown estimate and a highly efficient estimate. The high-breakdown estimate can easily detect and accommodate any outliers from a straight line model, whereas the efficient estimate can model curvature effectively. A strategy that exists in statistical folklore is to trust the highly efficient estimate when it agrees with the high-breakdown estimate and to further investigate the data when a significant difference exists between the two estimates. McKean, Naranjo, and Sheather (1996) developed graphical and numerical diagnostics to identify and further understand any such significant differences.

5. CHALLENGES

Rank-based analyses for linear models (controlled studies) offer the user an attractive highly efficient alternative to LS analyses. Many current problems, though, involve large messy datasets (uncontrolled studies), which should be fertile ground for high-breakdown, bounded influence procedures accompanied with valid diagnostic procedures to identify outliers and influential points. But here, as argued by Carroll, Ruppert, and Stefanski (1999), the field of "robust statistics lacks success stories." We view this as a challenge to develop robust methodologies for such problems. We cite three important problems in this area.

The bootstrap has become a major statistical tool for accurately assessing variability in many situations. Bootstraps are often used on large messy datasets where traditional inference based on the asymptotic distributions of either LS or robust estimation is highly questionable. A development of bootstrap procedures for high-breakdown, bounded influence estimates would further the development of valid inference procedures and diagnostic procedures for these large messy datasets.

Quasi-experimental designs and observational studies are uncontrolled studies often involving a small group of treated subjects versus large group(s) of nontreated subjects, where the grouping is nonrandomized. A methodology that is finding increasing use in this area is a matching of treated subjects versus control subjects via a propensity score analysis based on covariates (see Rubin and Thomas 1996). This methodology includes elements of logistic regression, covariance analysis, and bootstrap methods for inference. Further, such studies often result in large sample sizes for the control subjects.

Data mining deals with extremely large datasets that contain many outliers, as often the data are not clean. Such datasets need robust methods and diagnostic procedures to determine outliers and model selection.

The most important drawback to the use of robust methods for these problems is their computational expense. Procedures such as the robust QR decomposition of Ammann (1993) or the fast LTS and minimum covariance determinant algorithms of Rousseeuw and Van Driessen (1998, 1999) may be useful for quickly computing the weights for bounded influence, high-breakdown estimates. This would allow robust analyses (estimation, inference via bootstraps, and diagnostics for model checking, outlier determination and differentiation between fits) for large datasets.

REFERENCES

Ammann, L. P. (1993), "Robust Singular Value Decompositions: A New Approach to Projection Pursuit," *Journal of the American Statistical Association*, 88, 505–514.

Carroll, R. J., Ruppert, D., and Stefanski, L. A. (1999), Comment on "Regression Depth," *Journal of the American Statistical Association*, 94, 410–411.

Chang, W. H., McKean, J. W., Naranjo, J. D., and Sheather, S. J. (1999), "High-Breakdown Rank Regression," *Journal of the American Statistical Association*, 94, 205–219.

Chernoff, H., and Savage, I. R. (1958), "Asymptotic Normality and Efficiency of Certain Nonparametric Test Statistics," *Annals of Mathematical Statistics*, 39, 972–994.

Cook, R. D. (1998), *Regression Graphs*, New York: Wiley.

Cook, R. D., and Nachsteim, C. J. (1994), "Reweighting to Achieve Elliptically Contoured Covariates in Regression," *Journal of the American Statistical Association*, 89, 592–600.

Cook, R. D., Hawkins, D. M., and Weisberg, S. (1992), "Comparison of Model Misspecification Diagnostics Using Residuals From Least Mean of Squares and Least Median of Squares Fits," *Journal of the American Statistical Association*, 87, 419–424.

Croux, C., Rousseeuw, P. J., and Hössjer, O. (1994), "Generalized S-Estimators," *Journal of the American Statistical Association*, 89, 1271–1281.

Davies, P. L. (1993), "Aspects of Robust Linear Regression," *The Annals of Statistics*, 21, 1843–1899.

Duan, N., and Li, K. C. (1991), "Slicing Regression: A Link-Free Regression Method," *The Annals of Statistics*, 19, 505–530.

Ellis, S. P. (1998), "Instability of Least Squares, Least Absolute Deviation, and Least Median of Squares Linear Regression," *Statistical Science*, 13, 337–350.

Hájek, J., and Šidák, Z. (1967), *Theory of Rank Tests*, New York: Academic Press.

Hampel, F. R., Ronchetti, E. M., Rousseeuw, P. J., and Stahel, W. J. (1986), *Robust Statistics, the Approach Based on Influence Functions*, New York: Wiley.

Hettmansperger, T. P. (1984), *Statistical Inference Based on Ranks*, New York: Wiley.

Hettmansperger, T. P., and McKean, J. W. (1983), "A Geometric Interpretation of Inferences Based on Ranks in the Linear Model," *Journal of the American Statistical Association*, 78, 885–893.

———— (1998), *Robust Nonparametric Statistical Methods*, London: Arnold.

Hettmansperger, T. P., and Sheather, S. J. (1992), "A Cautionary Note on the Method of Least Median Squares," *The American Statistician*, 46, 79–83.

Jaeckel, L. A. (1972), "Estimating Regression Coefficients by Minimizing the Dispersion of Residuals," *The Annals of Mathematical Statistics*, 43, 1449–1458.

Jureckova, J. (1971), "Nonparametric Estimate of Regression Coefficients," *The Annals of Mathematical Statistics*, 42, 1328–1338.

Koul, H. L. (1992), *Weighted Empiricals and Linear Models*, Hayward, CA: Institute of Mathematical Statistics.

Krasker, W. S., and Welsch, R. E. (1982), "Efficient Bounded Influence Regression Estimation," *Journal of the American Statistical Association*, 77, 595–604.

Lehmann, E. L. (1975), *Nonparametrics: Statistical Methods Based on Ranks*, San Francisco: Holden-Day.

McKean, J. W., Naranjo, J. D., and Sheather, S. J. (1996), "Diagnostics to Detect Differences in Robust Fits of Linear Models," *Computational Statistics*, 11, 223–243.

McKean, J. W., and Schrader, R. (1980), "The Geometry of Robust Procedures in Linear Models," *Journal of the Royal Statistical Society*, Ser. B, 42, 366–371.

McKean, J. W., and Sheather, S. J. (1991), "Small-Sample Properties of Robust Analyses of Linear Models Based on R-Estimates," in *A Survey in Directions in Robust Statistics and Diagnostics*, Part II, eds. W. Stahel and S. Weisberg, New York: Springer-Verlag, pp. 1–20.

McKean, J. W., Sheather, S. J., and Hettmansperger, T. P. (1990), "Regression Diagnostics for Rank-Based Methods," *Journal of the American Statistical Association*, 85, 1018–28.

———— (1993), "The Use and Interpretation of Residuals Based on Robust Estimation," *Journal of the American Statistical Association*, 88, 1254–1263.

———— (1994), "Robust and High-Breakdown Fits of Polynomial Models," *Technometrics*, 36, 409–415.

Naranjo, J. D., and Hettmansperger, T. P. (1994), "Bounded-Influence Rank Regression," *Journal of the Royal Statistical Society*, Ser. B, 56, 209–220.

Puri, M. L., and Sen, P. K. (1971), *Nonparametric Methods in Multivariate Analysis*, New York: Wiley.

———— (1985), *Nonparametric Methods in General Linear Models*, New York: Wiley.

Randles, R. H., and Wolfe, D. A. (1979), *Introduction to the Theory of Nonparametric Statistics*, New York: Wiley.

Rousseeuw, P., Hubert, M., and Van Aelst, S. (1999), "Reply to Comments: Regression Depth," *Journal of the American Statistical Association*, 94, 419–433.

Rousseeuw, P. J., and Leroy, A. M. (1987), *Robust Regression and Outlier Detection*, New York: Wiley.

Rousseeuw, P., and Van Driessen, K. (1998), "A Fast Algorithm for the Minimum Covariance Determinant Estimator," technical report, University of Antwerp.

———— (1999), "Computing LTS Regression for Large Data Sets," technical report, University of Antwerp.

Rousseeuw, P. J., and Yohai, V. J. (1984), "Robust Regression by Means of S-Estimators," in *Robust and Nonlinear Time Series*, eds. J. Franke, W. Hardle, and R. D. Martin, New York: Springer-Verlag, pp. 256–272.

Rubin, D. B., and Thomas, N. (1996), "Matching Using Estimated Propensity Scores: Relating Theory to Practice," *Biometrics*, 52, 254–268.

Sheather, S. J., McKean, J. W., and Hettmansperger, T. P. (1997), "Finite Sample Stability Properties of the Least Median of Squares Estimator," *Journal of Statistical Computation and Simulation*, 58, 371–383.

Simpson, D. G., and Chang, Y. I. (1997), "Reweighting Approximate GM Estimators: Asymptotics and Residual-Based Graphics," *Journal of Statistical Planning and Inference*, 57, 273–293.

Simpson, D. G., Ruppert, D., and Carroll, R. J. (1992), "On One-Step GM Estimates and Stability of Inferences in Linear Regression," *Journal of the American Statistical Association*, 87, 439–450.

Simpson, D. G., and Yohai, V. J. (1998), "Functional Stability of One-Step GM Estimators in Approximately Linear Regression," *The Annals of Statistics*, 26, 1147–1169.

Hierarchical Models: A Current Computational Perspective

James P. Hobert

1. INTRODUCTION

Hierarchical models (HMs) have many applications in statistics, ranging from accounting for overdispersion (Cox 1983) to constructing minimax estimators (Strawderman 1971). Perhaps the most common use is to induce correlations among a group of random variables that model observations with something in common; for example, the set of measurements made on a particular individual in a longitudinal study. My starting point is the following general, two-stage HM.

Let $(\mathbf{Y}_1^T, \mathbf{u}_1^T)^T, \ldots, (\mathbf{Y}_k^T, \mathbf{u}_k^T)^T$ be k independent random vectors, where T denotes matrix transpose, and let $\boldsymbol{\lambda} = (\boldsymbol{\lambda}_1^T, \boldsymbol{\lambda}_2^T)^T$ be a vector of unknown parameters. The data vectors, $\mathbf{Y}_1, \ldots, \mathbf{Y}_k$, are modeled conditionally on the *unobservable* random-effects vectors, $\mathbf{u}_1, \ldots, \mathbf{u}_k$. Specifically, $f_i(\mathbf{y}_i|\mathbf{u}_i; \boldsymbol{\lambda}_1)$, the conditional density function of $\mathbf{Y}_i$ given $\mathbf{u}_i$, is assumed to be some parametric density in $\mathbf{y}_i$ whose parameters are written as functions of $\mathbf{u}_i$ and $\boldsymbol{\lambda}_1$. At the second level of the hierarchy, it is assumed that $g_i(\mathbf{u}_i; \boldsymbol{\lambda}_2)$, the marginal density function of $\mathbf{u}_i$, is some parametric density in $\mathbf{u}_i$ whose parameters are written as functions of $\boldsymbol{\lambda}_2$. (Of course, any of these random variables could be discrete, in which case we would simply replace the phrase "density function" with "mass function.") The Bayesian version of this model, in which $\boldsymbol{\lambda}$ is considered a random variable with prior $\pi(\boldsymbol{\lambda})$, is essentially the *conditionally independent hierarchical model* of Kass and Steffey (1989).

The fully specified *generalized linear mixed model* (Breslow and Clayton 1993; McCulloch 1997), which is currently receiving a great deal of attention, is a special case. Indeed, many (if not most) of the parametric HMs found in the statistical literature are special cases of this general HM. Section 2 presents three specific examples, including a model for McCullagh and Nelder's (1989) salamander data.

The likelihood function for this model is the marginal density of the *observed* data viewed as a function of the parameters; that is,

$$L(\boldsymbol{\lambda}; \mathbf{y}) = \prod_{i=1}^{k} \int f_i(\mathbf{y}_i|\mathbf{u}_i; \boldsymbol{\lambda}_1) g_i(\mathbf{u}_i; \boldsymbol{\lambda}_2) \, d\mathbf{u}_i, \tag{1}$$

James P. Hobert is Associate Professor, Department of Statistics, University of Florida, Gainesville, FL 32611 (E-mail: jhobert@stat.ufl.edu). The author is grateful to George Casella for constructive comments and suggestions and to Ranjini Natarajan for a couple of helpful conversations.

where $\mathbf{y} = (\mathbf{y}_1^T, \ldots, \mathbf{y}_k^T)^T$ denotes the observed data. As usual, Bayesian inference about λ is made through the posterior density $\pi(\lambda|\mathbf{y}) \propto L(\lambda; \mathbf{y})\pi(\lambda)$. In some applications, inferences about the $\mathbf{u}_i$'s may be of interest, but in this article I restrict attention to inference about λ.

Unfortunately, a HM is simply not useful from a statistical standpoint unless there exists a computational technique that allows for reliable approximation of the quantities necessary for inference. Indeed, it has been the ongoing development of techniques like the EM algorithm (Dempster, Laird, and Rubin 1977; McLachlan and Krishnan 1997) and Markov chain Monte Carlo (Besag, Green, Higdon, and Mengersen, 1995; Gelfand and Smith 1990; Tierney 1994) that has enabled statisticians to make use of increasingly more complex HMs over the last few decades. Section 3 contains a discussion of some key aspects of the computational techniques currently used in conjunction with Bayesian and frequentist HMs. The context of this discussion is the model for the salamander data developed in Section 2. Several areas that are ripe for new research are identified.

2. THREE EXAMPLES

An exhaustive list of all the types of data that can be modeled with the general HM of Section 1 obviously is beyond the scope of this article. However, I hope that the three examples outlined here provide the reader with a sense of how extensive such a list might be. Please note that the notation used in these specific examples is standard, but not necessarily consistent with that of Section 1.

2.1 Rat Growth Data

Gelfand, Hills, Racine-Poon, and Smith (1990) analyzed data from an experiment in which the weights of 60 young rats (30 in a control group and 30 in a treatment group) were measured weekly for 5 weeks starting on day 8. Let y_{ij} denote the weight of the ith rat on day $x_j = 7j + 1$, where $i = 1, \ldots, 60$ and $j = 1, \ldots, 5$. Suppose that $1 \leq i \leq 30$ corresponds to the control group.

The following random coefficient regression model was used by Gelfand et al. (1990). Conditional on a pair of rat-specific regression parameters, α_i and β_i, it is assumed that $y_{i1}, \ldots, y_{i5}$ are independent with

$$y_{ij}|\alpha_i, \beta_i \sim \mathrm{N}(\alpha_i + \beta_i x_j, \sigma_c^2 I(1 \leq i \leq 30) + \sigma_t^2 I(31 \leq i \leq 60)),$$

where $I(\cdot)$ is an indicator function. Thus the model allows for a difference between the variances of the error terms for the control rats and the treatment rats.

Presumably, the experimenters were not interested in the effect of treatment and control on this particular set of 60 rats, but rather on some population of rats, of which these 60 can be considered a random sample. Thus instead of taking the (α_i, β_i) pairs as fixed unknown parameters, it is assumed that $(\alpha_i, \beta_i)^T, i = 1, \ldots, 30,$ are iid bivariate normal with mean $(\alpha_c, \beta_c)^T$ and covariance matrix Σ_c, and that $(\alpha_i, \beta_i)^T, i = 31, \ldots, 60,$ are iid bivariate normal with mean $(\alpha_t, \beta_t)^T$ and covariance matrix Σ_t.

A comparison of the "c" and "t" parameters allows one to make inferences regarding treatment versus control across the relevant population of rats. Gelfand et al. (1990) performed a Bayesian analysis with conjugate priors on all of the parameters. The Gibbs sampler (see the Gibbs sampler vignette by Gelfand) was used to make inferences.

2.2   Seizure Data

Thall and Vail (1990) described an experiment in which 59 epileptics were randomly assigned to one of two treatment groups (active drug and placebo). The number of seizures experienced by each patient during four consecutive 2-week periods following treatment were recorded. Let y_{ij} denote the number of seizures experienced by the ith patient over the jth two-week period where $i = 1, \ldots, 59$ and $j = 1, 2, 3, 4$. Other available covariates are age and a pretreatment baseline seizure count.

Consider a naive generalized linear model (McCullagh and Nelder 1989) for these data in which the y_{ij}'s are assumed to be independent Poisson random variables whose means are linked to linear predictors involving the covariates. I refer to this model as naive for two reasons. First, it is probably reasonable to assume that observations on two different patients are independent, but it is surely unreasonable to assume that observations made on the same patient are independent. Second, it appears that these data are overdispersed with respect to the Poisson distribution; that is, there is more variability than would be expected under the Poisson assumption.

Booth, Casella and Hobert (2000) suggested the following HM, which allows for both dependence among an individual's observations and overdispersion. Let $u_1, \ldots, u_{59}$ be iid $N(0, \sigma^2)$ and think of these as random patient effects. Conditional on $u_i, y_{i1}, \ldots, y_{i4}$ are assumed to be iid negative binomial random variables with index α and mean $\mu_i = \exp\{\mathbf{x}_i^T \boldsymbol{\beta} + u_i\}$, where $\mathbf{x}_i^T$ is an appropriately chosen vector of covariates. More specifically,

$$P(y_{ij} = y | u_i; \boldsymbol{\beta}, \alpha) = \frac{\Gamma(y + \alpha)}{\Gamma(\alpha) y!} \left(\frac{\alpha}{\mu_i + \alpha} \right)^\alpha \left(\frac{\mu_i}{\mu_i + \alpha} \right)^y, \qquad (2)$$

where $\alpha > 0$ and $y \in \{0, 1, 2, \ldots\}$.

The random patient effects induce a positive correlation between counts from the same patient. Regarding potential overdispersion, note that $E(y_{ij}|u_i) = \mu_i$ and $\text{var}(y_{ij}|u_i) = \mu_i + \mu_i^2/\alpha$, and that (2) becomes a Poisson probability as $\alpha \to \infty$. Thus the parameter α allows for overdispersion.

2.3 The Salamander Data

The salamander data consist of three separate experiments, each performed according to the design given in table 14.3 of McCullagh and Nelder (1989) and each involving matings among salamanders in two closed groups. Thus there are a total of six closed groups of salamanders. Each group contained 20 salamanders: 5 species R females, 5 species W females, 5 species R males, and 5 species W males. (Actually, the same 40 salamanders were used in two of the experiments, but we ignore this and act as if three different sets of 40 salamanders were used because of the long time delay between these two experiments.) Within each group, only 60 of the possible 100 heterosexual crosses were observed. Let y_{hij} be the indicator of a successful mating between the ith female and jth male in group h, where $i, j = 1, 2, \ldots, 10$ and $h = 1, \ldots, 6$. Note that for fixed h, only 60 of the possible 100 (i, j) pairs are relevant. There are four possible sex–species combinations, and the experimenters collected the data hoping to answer the question: Are there differences in the four mating success probabilities?

Any two observations involving a common salamander (male or female) should obviously be modeled as correlated random variables. One way to induce such correlations is through the following HM, which is one of several models introduced by Karim and Zeger (1992). Let v_{hi} denote the random effect that the ith female in group h has across matings in which she is involved, and define w_{hj} similarly for the jth male in group h. Let $\mathbf{v}_h = (v_{h1}, \ldots, v_{h10})^T$ and $\mathbf{v} = (\mathbf{v}_1^T, \ldots, \mathbf{v}_6^T)^T$, and define $\mathbf{w}_h$ and $\mathbf{w}$ similarly. Likewise, let $\mathbf{y}_1, \ldots, \mathbf{y}_6$ be 60×1 vectors containing the binary outcomes from the six closed groups, and put $\mathbf{y} = (\mathbf{y}_1^T, \ldots, \mathbf{y}_6^T)^T$.

Conditional on $\mathbf{v}_h$ and $\mathbf{w}_h$, the elements of $\mathbf{y}_h$ are assumed independent and such that $y_{hij}|v_{hi}, w_{hj} \sim \text{Bernoulli}(\pi_{hij})$ with

$$\text{logit}(\pi_{hij}) = \mathbf{x}_{hij}^T \boldsymbol{\beta} + v_{hi} + w_{hj},$$

where $\mathbf{x}_{hij}^T$ is a 1×4 vector indicating the type of mating (i.e., it contains three 0's and a single 1) and β is an unknown regression parameter. Finally, it is assumed that the elements of $\mathbf{v}$ are iid $N(0, \sigma_v^2)$ and independent of $\mathbf{w}$ whose elements are assumed to be iid $N(0, \sigma_w^2)$. In terms of the general HM of Section 1, $\boldsymbol{\lambda} = (\boldsymbol{\beta}^T, \sigma_v^2, \sigma_w^2)^T$.

In the next section I look at Bayesian and frequentist versions of this model, and consider the computational techniques required for making inferences. There are two main reasons for using this particular dataset: (a) the models give a reasonable indication of how complex a HM statisticians are currently able to handle, and (b) I wish to take advantage of the fact that many readers will be familiar with this dataset, as it has been discussed by so many authors (e.g., Booth and Hobert 1999; Chan and Kuk 1997; Karim and Zeger 1992; Lee and Nelder 1996; McCullagh and Nelder 1989; McCulloch 1994; Vaida and Meng 1998).

3. COMPUTATION FOR THE SALAMANDER MODEL

The fact that only 60 of the possible 100 heterosexual crosses were observed in each of the six groups does not complicate the analysis of the data, but it does complicate notation. Specifically, for $h = 1, \ldots, 6$, we require

$$S_h = \{(i, j) : i\text{th female and } j\text{th male in group } h \text{ were coupled}\}.$$

The likelihood function associated with Karim and Zeger's (1992) model for the salamander data is

$$L(\boldsymbol{\beta}, \sigma_v^2, \sigma_w^2; \mathbf{y}) = \prod_{h=1}^{6} L_h(\boldsymbol{\beta}, \sigma_v^2, \sigma_w^2; \mathbf{y}_h), \tag{3}$$

where, up to a multiplicative constant,

$$L_h(\boldsymbol{\beta}, \sigma_v^2, \sigma_w^2; \mathbf{y}_h)$$

$$= \frac{1}{(\sigma_v^2 \sigma_w^2)^5} \int\!\!\int \left[\prod_{(i,j) \in S_k} \frac{\exp\{y_{hij}(\mathbf{x}_{hij}^T \boldsymbol{\beta} + v_{hi} + w_{hj})\}}{1 + \exp\{\mathbf{x}_{hij}^T \boldsymbol{\beta} + v_{hi} + w_{hj}\}} \right]$$

$$\times \exp\left\{ -\frac{1}{2\sigma_v^2} \sum_{i=1}^{10} v_{hi}^2 - \frac{1}{2\sigma_w^2} \sum_{j=1}^{10} w_{hj}^2 \right\} d\mathbf{v}_h \, d\mathbf{w}_h.$$

Thus L contains six intractable 20-dimensional integrals, which are due to the crossed nature of the design. Bayesian and frequentist computations are considered in the following two sections.

3.1 Bayesian Analysis

A consequence of the complexity of the salamander likelihood function is that, no matter what (nontrivial) prior is chosen for $\boldsymbol{\lambda} = (\boldsymbol{\beta}^T, \sigma_v^2, \sigma_w^2)^T$, the integrals defining any posterior quantities of interest will not be analytically tractable. Furthermore, the high dimension of these integrals probably rules out numerical integration. Thus, to make inferences, one is forced to use either analytical approximations (Kass and Steffey 1989; Tierney, Kass, and Kadane 1989) or Markov chain Monte Carlo (MCMC) techniques.

Suppose that the Gibbs sampler is to be used to sample from $\pi(\boldsymbol{\lambda}|\mathbf{y})$. In an attempt to be "objective" and simultaneously make the full conditionals needed for Gibbs as simple as possible, one might choose an improper conjugate prior such as $\pi(\boldsymbol{\lambda}) \propto (\sigma_v^2)^a (\sigma_w^2)^b$ for some a and b (Karim and Zeger 1992). Such priors can, however, lead to serious problems. For example, depending on the value of (a, b), it may be the case that all of the full conditionals needed to apply the Gibbs sampler are proper, whereas the posterior distribution itself is improper (Natarajan and McCulloch 1995). Compounding this potential problem is the fact that an improper posterior may not be apparent from the Gibbs output (Hobert and Casella 1996). Moreover, even if $\pi(\lambda) \propto (\sigma_v^2)^a (\sigma_w^2)^b$ does result in a proper posterior, it is not a "noninformative" prior in any formal sense for this HM. Thus it may be the case that the prior is actually driving the inferences, which is opposite of what was intended. Unfortunately, even if one is willing to suffer the (sampling) consequences of using a nonconjugate prior, standard "default" priors such as Jeffreys's prior are not available in closed form.

To guard against improper posteriors, some authors have suggested using "diffuse" proper priors, which typically means proper conjugate priors that are nearly improper. Indeed, the author is guilty of making such suggestions: as a prior for a variance component in a normal mixed model, Hobert and Casella (1996, p. 1469) suggested using $\pi(\sigma^2) \propto (\sigma^2)^{-r-1} \exp\{-1/s\sigma^2\}$ with small positive values for r and s. The problem is that these diffuse priors are still not noninformative in any formal sense and, furthermore, there is empirical evidence that such priors may lead to very slowly converging Gibbs samplers (Natarajan and McCulloch 1998).

We conclude that choosing a prior for λ is currently a real dilemma for a Bayesian with no prior information. Consequently, development of reasonable default priors for λ (of the HM in Sec. 1) seems to be a potentially rich area for future research. Natarajan and Kass (2000) discuss some recent results along these lines.

Suppose now that $\pi(\lambda)$ is a satisfactory prior for the salamander model, and that the Gibbs sampler is run to produce $(\mathbf{v}^{(t)}, \mathbf{w}^{(t)}, \beta^{(t)}, \sigma_v^{2(t)}, \sigma_w^{2(t)}), t = 0, 1, 2, \ldots, M$, where $t = 0$ corresponds to some intelligently chosen starting value. (See C. J. Geyer's web page at http://www.stat.umn.edu/~charlie/ for an interesting discussion of burn-in and the use of parallel chains in MCMC.) An obvious estimate of $E(\sigma_v^2|\mathbf{y})$ is

$$\frac{1}{M+1} \sum_{t=0}^{M} \sigma_v^{2(t)}. \tag{4}$$

Most statisticians would agree that without an associated standard error, this Monte Carlo estimate is not very useful. Despite this, estimates based on MCMC are routinely presented without reliable standard errors. The reason for this is presumably that calculating the standard error of an estimate based on MCMC is often not trivial. Indeed, establishing the existence of a central limit theorem (CLT) for a Monte Carlo estimate based on a Markov chain can be difficult (Chan and Geyer 1994; Meyn and Tweedie 1993; Tierney 1994). (Readers who are skeptical about the importance of a CLT should look at the example in sec. 4 of Roberts and Rosenthal 1998.)

The most common method of establishing the existence of CLTs is to show that the Markov chain itself is *geometrically ergodic*. Although many Markov chains that are the basis of MCMC algorithms have been shown to be geometrically ergodic (e.g., Chan 1993; Hobert and Geyer 1998; Mengersen and Tweedie 1996; Roberts and Rosenthal 1999; Roberts and Tweedie 1996), myriad complex MCMC algorithms are currently in use to which these results do not apply. An example is the Gibbs sampler for the salamander model (with a proper conjugate prior on λ). Thus another area of research that is full of possibilities is establishing geometric ergodicity (or lack thereof) of the Markov chains used in popular MCMC algorithms.

Unfortunately, even if it is known that a CLT exists, estimating the asymptotic variance may not be easy (Geyer 1992; Mykland, Tierney, and Yu 1995). For example, to apply the regenerative method of Mykland et al. (1995), a *minorization condition* (Rosenthal 1995) must be established. Depending on the complexity of the Markov chain, this can be a challenging exercise. Thus important work remains to be done on this topic as well.

3.2 Frequentist Analysis

Now the problem shifts from sampling from the posterior distribution to maximizing the likelihood function. As mentioned earlier, the likelihood function for the salamander model involves six intractable 20-dimensional integrals. Evidently, computational techniques that can be used to maximize such complex likelihood functions must make use of the Monte Carlo method. McCulloch (1997) described and compared three methods for maximizing likelihood functions based on generalized linear mixed models: Monte Carlo EM (MCEM), Monte Carlo Newton–Raphson (MCNR), and simulated maximum likelihood. He concluded that MCEM and MCNR are both generally effective (and better than simulated maximum likelihood), but noted a drawback concerning "the complications of deciding whether the stochastic versions of EM or NR have converged." Since the appearance of McCulloch's 1997 article, MCEM seems to have received more attention than MCNR. Consequently, here I focus on MCEM.

Most of the problems associated with MCEM, including the one pointed out by McCulloch, stem from an inability to quantify the Monte Carlo error introduced at each step of the algorithm. An important area for future research is the development of methods that will enable one to get a handle on this Monte Carlo error. To explain this problem in more detail, I consider using MCEM to maximize (3).

Viewing the random effects, $\mathbf{v}$ and $\mathbf{w}$, as missing data leads to the following implementation of the EM algorithm (Dempster et al. 1977). The $(r+1)$th E step entails calculation of

$$Q(\lambda|\lambda^{(r)}) = \sum_{h=1}^{6} E_h[\log\{f_h(\mathbf{y}_h|\mathbf{v}_h, \mathbf{w}_h; \lambda_1)g_h(\mathbf{v}_h, \mathbf{w}_h; \lambda_2)\}|\mathbf{y}_h; \lambda^{(r)}],$$

where $\lambda^{(r)}$ is the result of the rth iteration of EM, and E_h is with respect to the conditional distribution of $(\mathbf{v}_h, \mathbf{w}_h)$ given $\mathbf{y}_h$ with parameter value $\lambda^{(r)}$, whose density is denoted by $g_h(\mathbf{v}_h, \mathbf{w}_h|\mathbf{y}_h; \lambda^{(r)})$. Once Q has been calculated, one simply maximizes it over λ to get $\lambda^{(r+1)}$; that is, perform the $(r+1)$th M step.

There is simply no hope of computing Q in closed form, because $g_h(\mathbf{v}_h, \mathbf{w}_h | \mathbf{y}_h; \boldsymbol{\lambda}^{(r)})$ involves the same intractable integrals as those found in the likelihood function. The MCEM algorithm (Wei and Tanner 1990) avoids an intractable E step by replacing Q with a Monte Carlo approximation. For example, instead of maximizing Q, one maximizes

$$\tilde{Q}(\boldsymbol{\lambda}|\boldsymbol{\lambda}^{(r)}) = \sum_{h=1}^{6} \frac{1}{M} \sum_{l=1}^{M} \log\{f_h(\mathbf{y}_h|\mathbf{v}_h^{(l)}, \mathbf{w}_h^{(l)}; \boldsymbol{\lambda}_1) g_h(\mathbf{v}_h^{(l)}, \mathbf{w}_h^{(l)}; \boldsymbol{\lambda}_2)\},$$

where $(\mathbf{v}_h^{(l)}, \mathbf{w}_h^{(l)}), l = 1, 2, \ldots, M$, could be an iid sample from $g_h(\mathbf{v}_h, \mathbf{w}_h|\mathbf{y}_h; \boldsymbol{\lambda}^{(r)})$ or from an ergodic Markov chain with stationary density $g_h(\mathbf{v}_h, \mathbf{w}_h|\mathbf{y}_h; \boldsymbol{\lambda}^{(r)})$. (Importance sampling is another possibility.)

Of course, there is no free lunch. Although using MCEM circumvents a complicated expectation at each E step, it necessitates a method for choosing M at each MCE step. If M is too small, then the EM step will be swamped by Monte Carlo error, whereas too large an M is wasteful. Booth and Hobert (1999) argued that the ability to choose an appropriate value for M hinges on the existence of a CLT for the maximizer of $\tilde{Q}$ and one's ability to estimate the corresponding asymptotic variance. These authors provided methods for choosing M and diagnosing convergence when $\tilde{Q}$ is based on independent samples, and showed that their MCEM algorithm can be used to maximize complex likelihoods, such as the salamander likelihood described herein. However, it is clear that when the dimension of the intractable integrals in the likelihood is very large, one will be forced to construct $\tilde{Q}$ using MCMC techniques such as the Metropolis algorithm developed by McCulloch (1997). The discussion in the previous section suggests that when $\tilde{Q}$ is based on a Markov chain, verifying the existence of a CLT for the maximizer of $\tilde{Q}$ and estimating the corresponding asymptotic variance can be difficult problems (Levine and Casella 1998). This is another important and potentially rich area for new research.

REFERENCES

Besag, J., Green, P., Higdon, D., and Mengersen, K. (1995), "Bayesian Computation and Stochastic Simulation" (with discussion), *Statistical Science*, 10, 3–66.

Booth, J. G., and Hobert, J. P. (1999), "Maximizing Generalized Linear Mixed Model Likelihoods With an Automated Monte Carlo EM Algorithm," *Journal of the Royal Statistical Society*, Ser. B, 61, 265–285.

Booth, J. G., Casella, G., and Hobert, J. P. (2000), "Negative Binomial Log-Linear Mixed Models," technical report, University of Florida, Dept. of Statistics.

Breslow, N. E., and Clayton, D. G. (1993), "Approximate Inference in Generalized Linear Mixed Models," *Journal of the American Statistical Association*, 88, 9–25.

Chan, J. S. K., and Kuk, A. Y. C. (1997), "Maximum Likelihood Estimation for Probit-Linear Mixed Models With Correlated Random Effects," *Biometrics*, 53, 86–97.

Chan, K. S. (1993), "Asymptotic Behavior of the Gibbs Sampler," *Journal of the American Statistical Association*, 88, 320–326.

Chan, K. S., and Geyer, C. J. (1994), Comment on "Markov Chains for Exploring Posterior Distributions," by L. Tierney, *The Annals of Statistics*, 22, 1747–1757.

Cox, D. R. (1983), "Some Remarks on Overdispersion," *Biometrika*, 70, 269–274.

Dempster, A. P., Laird, N. M., and Rubin, D. B. (1977), "Maximum Likelihood From Incomplete Data via the EM Algorithm" (with discussion), *Journal of the Royal Statistical Society*, Ser. B, 39, 1–38.

Gelfand, A. E., Hills, S. E., Racine-Poon, A., and Smith, A. F. M. (1990), "Illustration of Bayesian Inference in Normal Data Models Using Gibbs Sampling," *Journal of the American Statistical Association*, 85, 972–985.

Gelfand, A. E., and Smith, A. F. M. (1990), "Sampling-Based Approaches to Calculating Marginal Densities," *Journal of the American Statistical Association*, 85, 398–409.

Geyer, C. J. (1992), "Practical Markov Chain Monte Carlo" (with discussion), *Statistical Science*, 7, 473–511.

Hobert, J. P., and Casella, G. (1996), "The Effect of Improper Priors on Gibbs Sampling in Hierarchical Linear Mixed Models," *Journal of the American Statistical Association*, 91, 1461–1473.

Hobert, J. P., and Geyer, C. J. (1998), "Geometric Ergodicity of Gibbs and Block Gibbs Samplers for a Hierarchical Random-Effects Model," *Journal of Multivariate Analysis*, 67, 414–430.

Karim, M. R., and Zeger, S. L. (1992), "Generalized Linear Models With Random Effects; Salamander Mating Revisited," *Biometrics*, 48, 631–644.

Kass, R. E., and Steffey, D. (1989), "Approximate Bayesian Inference in Conditionally Independent Hierarchical Models (Parametric Empirical Bayes Models)," *Journal of the American Statistical Association*, 84, 717–726.

Lee, Y., and Nelder, J. A. (1996), "Hierarchical Generalized Linear Models" (with discussion), *Journal of the Royal Statistical Society*, Ser. B, 58, 619–678.

Levine, R. A., and Casella, G. (1998), "Implementations of the Monte Carlo EM Algorithm," technical report, University of California, Davis, Intercollege Division of Statistics.

McCullagh, P., and Nelder, J. A. (1989), *Generalized Linear Models*, London: Chapman and Hall, 2nd ed.

McCulloch, C. E. (1994), "Maximum Likelihood Variance Components Estimation for Binary Data," *Journal of the American Statistical Association*, 89, 330–335.

――― (1997), "Maximum Likelihood Algorithms for Generalized Linear Mixed Models," *Journal of the American Statistical Association*, 92, 162–170.

McLachlan, G. J., and Krishnan, T. (1997), *The EM Algorithm and Extensions*, New York: Wiley.

Mengersen, K. L., and Tweedie, R. L. (1996), "Rates of Convergence of the Hastings and Metropolis Algorithms," *The Annals of Statistics*, 24, 101–121.

Meyn, S. P., and Tweedie, R. L. (1993), *Markov Chains and Stochastic Stability*, New York: Springer-Verlag.

Mykland, P., Tierney, L., and Yu, B. (1995), "Regeneration in Markov Chain Samplers," *Journal of the American Statistical Association*, 90, 233–241.

Natarajan, R., and Kass, R. E. (2000), "Reference Bayesian Methods for Generalized Linear Mixed Models," *Journal of the American Statistical Association*, 95, 227–237.

Natarajan, R., and McCulloch, C. E. (1995), "A Note on the Existence of the Posterior Distribution for a Class of Mixed Models for Binomial Responses," *Biometrika*, 82, 639–643.

――― (1998), "Gibbs Sampling with Diffuse Proper Priors: A Valid Approach to Data-Driven Inference?," *Journal of Computational and Graphical Statistics*, 7, 267–277.

Roberts, G. O., and Rosenthal, J. S. (1998), "Markov Chain Monte Carlo: Some Practical Implications of Theoretical Results" (with discussion), *Canadian Journal of Statistics*, 26, 5–31.

——— (1999), "Convergence of Slice Sampler Markov Chains," *Journal of the Royal Statistical Society*, Ser. B, 61, 643–660.

Roberts, G. O., and Tweedie, R. L. (1996), "Geometric Convergence and Central Limit Theorems for Multidimensional Hastings and Metropolis Algorithms," *Biometrika*, 83, 95–110.

Rosenthal, J. S. (1995), "Minorization Conditions and Convergence Rates for Markov Chain Monte Carlo," *Journal of the American Statistical Association*, 90, 558–566.

Strawderman, W. E. (1971), "Proper Bayes Minimax Estimators of the Multivariate Normal Mean," *The Annals of Mathematical Statistics*, 42, 385–388.

Tierney, L. (1994), "Markov Chains for Exploring Posterior Distributions" (with discussion), *The Annals of Statistics*, 22, 1701–1762.

Tierney, L., Kass, R. E., and Kadane, J. B. (1989), "Fully Exponential Laplace Approximations to Expectations and Variances of Nonpositive Functions," *Journal of the American Statistical Association*, 84, 710–716.

Vaida, F., and Meng, X-L. (1998), "A Flexible Gibbs–EM Algorithm for Generalized Linear Mixed Models With Binary Response," in *Proceedings of the 13th International Workshop on Statistical Modeling*, eds. B. Marx and H. Friedl, pp. 374–379.

Wei, G. C. G., and Tanner, M. A. (1990), "A Monte Carlo Implementation of the EM Algorithm and the Poor Man's Data Augmentation Algorithms," *Journal of the American Statistical Association*, 85, 699–704.

Hypothesis Testing: From p Values to Bayes Factors

John I. Marden

1. WHAT IS HYPOTHESIS TESTING?

Testing hypotheses involves deciding on the plausibility of two or more hypothetical statistical models based on some data. It may be that there are two hypothesis on equal footing; for example, the goal of Mosteller and Wallace (1984) was to decide which of Alexander Hamilton or James Madison wrote a number of the *Federalist Papers*. It is more common that there is a particular null hypothesis that is a simplification of a larger model, such as when testing whether a population mean is 0 versus the alternative that it is not 0. This null hypothesis may be something that one actually believes could be (approximately) true, such as the specification of an astronomical constant; or something that one believes is false but is using as a straw man, such as the efficacy of a new drug is null; or something that one hopes is true for convenience's sake, such as equal variances in the analysis of variance.

Formally, we assume that the data are represented by x, which has density $f_\theta(x)$ for parameter θ. We test the null versus the alternative hypotheses,

$$H_0 : \theta \in \Theta_0 \quad \text{versus} \quad H_A : \theta \in \Theta_A,$$

where Θ_0 and Θ_A are disjoint subsets of the parameter space. A more general view allows comparison of several models.

2. CLASSICAL HYPOTHESIS TESTING

In 1908, W. S. Gossett (Student 1908) set the stage for 20th century "classical" hypothesis testing. He derived the distribution of the Student t statistic and used it to test null hypotheses about one mean and the difference of two means. His approach is used widely today. First, choose a test statistic. Next, calculate its p *value,* which is the probability, assuming that the null hypothesis is true, that one would obtain a value of the test statistic as or more extreme than that obtained from the data. Finally, interpret this p value as the probability that the null hypothesis is true. Of course,

John I. Marden is Professor, Department of Statistics, University of Illinois at Urbana-Champaign, Champaign, IL 61820 (E-mail: marden@stat.uiuc.edu).

the p value is *not* the probability that the null hypothesis is true, but small values do cause one to doubt the null hypothesis. R. A. Fisher (1925) promoted the p value for testing in a wide variety of problems, rejecting a null hypothesis when the p value is too small: "We shall not often be astray if we draw a conventional line at .05."

Jerzy Neyman and Egon Pearson (1928, 1933) shifted focus from the p value to the fixed-level test. The null hypothesis is rejected if the test statistic exceeds a constant, where the constant is chosen so that the probability of rejecting the null hypothesis when the null hypothesis is true equals (or is less than or equal to) a prespecified level. Their work had major impact on the theory and methods of hypothesis testing.

Test Statistics Became Evaluated on Their Risk Functions. Neyman and Pearson set out to find "efficient" tests. Their famous and influential result, the *Neyman–Pearson lemma,* solved the problem when both hypotheses are simple; that is, Θ_0 and Θ_A each contain exactly one point. Letting 0 and A be those points, the *likelihood ratio test* rejects the null hypothesis when $f_A(x)/f_0(x)$ exceeds a constant. This test maximizes the power (the probability of rejecting the null when the alternative is true) among tests with its level. For situations in which one or both of the hypotheses are composite (not simple), there is typically no uniformly best test, whence different tests are compared based on their *risk functions*. The risk function is a function of the parameters in the model, being the probability of making an error: rejecting the null when the parameter is in the null or accepting the null when the parameter is in the alternative.

The Maximum Likelihood Ratio Test Became a Key Method. When a hypothesis is composite, Neyman and Pearson replace the density at a single parameter value with the maximum of the density over all parameters in that hypothesis. The *maximum likelihood ratio test* rejects the null hypothesis for large values of $\sup_{\theta \in \Theta_A} f_\theta(x) / \sup_{\theta \in \Theta_0} f_\theta(x)$. This is an extremely useful method for finding good testing procedures. It can be almost universally applied; under fairly broad conditions, it has a nice asymptotic chi-squared distribution under the null hypothesis, and usually (but not always) has reasonable operating characteristics.

2.1 Theory

Abraham Wald (1950) codified the decision-theoretic approach to all of statistics that Neyman and Pearson took to hypothesis testing. Erich Lehmann (1959) was singularly influential, setting up the framework for evaluating statistical procedures based on their risk functions. A battery of criteria based on the risk function was used in judging testing procedures, including uniformly most powerful, locally most powerful, asymptotically most powerful, unbiasedness, similarity, admissibility or inadmissibility, Bayes, minimaxity, stringency, invariance, consistency, and asymptotic efficiencies (Bahadur, Chernoff, Hodges–Lehmann, Rubin–Sethuraman—see Serfling 1980).

Authors who have applied these criteria to various problems include Birnbaum (1955), Brown, Cohen, and Strawderman (1976, 1980), Brown and Marden (1989), Cohen and Sackrowitz (1987), Eaton (1970), Farrell (1968), Kiefer and Schwartz (1965), Marden (1982), Matthes and Truax (1967), Perlman (1980), Perlman and Olkin (1980), and Stein (1956).

A topic of substantial current theoretical interest is the use of algebraic structures in statistics (see Diaconis 1988). Testing problems often exhibit certain symmetries that can be expressed in terms of groups. Such problems can be reduced to so-called *maximal invariant parameters* and *maximal invariant statistics,* which are typically of much lower dimension than the original quantities. Stein's method (Andersson 1982; Wijsman 1967) integrates the density over θ using special measures to obtain the density of the maximal invariant statistic, which can then be used to analyze the problems; for example, to find the uniformly most powerful invariant test.

The real payoff comes in being able to define entire classes of testing problems, unifying many known models and suggesting useful new ones. For example, Andersson (1975) used group symmetry to define a large class of models for multivariate normal covariance matrices, including independence of blocks of variates, intraclass correlation, equality of covariance matrices, and complex and quaternion structure. Other models based on graphs or lattices can express complicated independence and conditional independence relationships among continuous or categorical variables (see, e.g., Andersson and Perlman 1993; Lauritzen 1996; Wermuth and Cox 1992; Whitaker 1990). Such "meta-models" not only allow expression of complicated models, but also typically provide unified systematic analysis, including organized processes for implementing likelihood procedures.

2.2 Methods

As far as general methods for hypothesis testing, the granddaddy of them all has to be the chi-squared test of Karl Pearson (1900). Its extensions permeate practice, especially in categorical models. But the most popular general test is the maximum likelihood ratio test. Related techniques were developed by Rao (1947) and Wald (1943). Cox (1961) explored these methods for nonnested models; that is, models for which neither hypothesis is contained within the closure of the other. The union-intersection principle of Roy (1953) provides another useful testing method.

The general techniques have been applied in so many innovative ways that it is hard to know where to begin. Classic works that present important methods and numerous references include books by Anderson (1984) on normal multivariate analysis (notable references include Bartlett 1937; Hotelling 1931; Mahalanobis 1930); Bishop, Fienberg, and Holland (1975) on discrete multivariate analysis (see also Goodman 1968; Haberman 1974); Barlow, Bartholomew, Bremner, and Brunk (1972) and Robertson, Wright and Dykstra (1988) on order-restricted inference; and Rao (1973) on everything.

Fisher (1935) planted the seeds for another crop of methods with his *randomization* tests, in which the p value is calculated by rerandomizing the actual data. These procedures yield valid p values under much broader distributional assumptions than are usually assumed. The field of nonparametrics grew out of such ideas, the goal being to have testing procedures that work well even if all of the assumptions do not hold exactly (see Gibbons and Chakraborti 1992; Hettmansperger and McKean 1998; and the robust nonparametric methods vignette by Hettmansperger, McKean, and Sheather for overviews). Popular early techniques include the Mann–Whitney/Wilcoxon two-sample test, the Kruskal–Wallis one-way analysis-of-variance test, the Kendall τ and Spearman ρ tests for association, and the Kolmogorov–Smirnov tests on distribution functions. Later work extended the techniques. A few of the many important authors include Akritas and Arnold (1994), Patel and Hoel (1973), and Puri and Sen (1985), for linear models and analysis of variance, and Chakraborty and Chaudhuri (1999), Chaudhuri and Sengupta (1993), Choi and Marden (1997), Friedman and Rafsky (1979), Hettmansperger, Möttönen, and Oja (1997), Hettmansperger, Nyblom, and Oja (1994), Liu and Singh (1993), Puri and Sen (1971), and Randles (1989) for various approaches to multivariate one-, two-, and many-sample techniques.

A vibrant area of research today is extending the scope of nonparametric/robust testing procedures to more complicated models, such as general multivariate linear models, covariance structures, and independence and conditional independence models (see Marden 1999 for a current snapshot). More general problems benefit from the bootstrap and related methods. Davison and Hinkley (1997) provided an introduction, and Beran and Millar (1987) and Liu and Singh (1997) presented additional innovative uses of resampling.

3. BAYESIAN HYPOTHESIS TESTING

Not everyone has been happy with the classical formulation. In particular, problems arise when trying to use the level or the p value to infer something about the truth or believability of the null. Some common complaints are as follows

1. The p value is not the probability that the null hypothesis is true. Gossett (and many practitioners and students since) has tried to use the p value as the probability that the null hypothesis is true. Not only is this wrong, but it can be far from reasonable. Lindley (1957) laid out a revealing paradox in which the p value is fixed at .05, but as the sample size increases, the Bayesian posterior probability that the null hypothesis is true approaches 1. Subsequent work (Berger and Sellke 1987; Edwards, Lindeman, and Savage 1963) showed that it is a general phenomenon that the p value does not give a reasonable assessment of the probability of the null. In one-sided problems, where the null is $\mu \leq 0$ and the alternative is $\mu > 0$, the p value can be a reasonable approximation of the probability of the null. Because this is the situation that is obtained in Student 1908, we can let Gossett off the hook (see Casella and Berger

1987; Pratt 1965). Good (1987), and references therein, discussed other comparisons of p values and Bayes.

2. The p value is not very useful when sample sizes are large. Almost no null hypothesis is *exactly* true. Consequently, when sample sizes are large enough, almost any null hypothesis will have a tiny p value, and hence will be rejected at conventional levels.

3. Model selection is difficult. Instead of having just two hypotheses to choose from, one may have several models under consideration, such as in a regression model when there is a collection of potential explanatory variables. People often use classical hypothesis testing to perform step-wise algorithms. Not only are the resulting significance levels suspect, but the methods for comparing non-nested models are not widely used.

The Bayesian approach to hypothesis testing answers these complaints, at the cost of requiring prior distribution on the parameter. It is helpful to break the prior into three pieces: the density γ_A of the parameter conditional on it being in the alternative, the density γ_0 of the parameter conditional on it being in the null, and the probability π_0 that the null hypothesis is true. The prior odds in favor of the alternative are $(1 - \pi_0)/\pi_0$. Then the posterior odds in favor of the alternative are a product of the prior odds and the *Bayes factor;* that is, the integrated likelihood ratio, $\int_{\Omega_A} f_\theta(x)\gamma_A(\theta)\,d\theta / \int_{\Omega_0} f_\theta(x)\gamma_0(\theta)\,d\theta$. Thus the posterior probability that the null hypothesis is true can be legitimately calculated as $1/(1 + \text{posterior odds})$. These calculations can easily be extended to model selection.

It is obvious the posterior odds depend heavily on π_0. What is not so obvious is that the posterior odds also depend heavily on the individual priors γ_A and γ_0. For example, as the prior γ_A becomes increasingly flat, the posterior odds usually approach 0. Jeffreys (1939) recommended "reference" priors γ_A and γ_0 in many common problems. These are carefully chosen so that the priors do not overwhelm the data. Other approaches include developing priors using imaginary prior data, and using part of the data as a training sample. Berger and Pericchi (1996) found that averaging over training samples can approximate so-called "intrinsic" priors. Kass and Wasserman (1996) reviewed a number of methods for finding priors. Choosing π_0 for the prior odds is more problematic. Jeffreys's suggestion is to take $\pi_0 = \frac{1}{2}$, which means that the posterior odds equal the Bayes factor.

There is growing evidence that this Bayes approach is very useful in practice, and not just a cudgel for bashing frequentists (see Kass and Raftery 1995 and Raftery 1995 for some interesting applications and references.)

We should not discard the p value all together, but just be careful. A small p value does not necessarily mean that we should reject the null hypothesis and leave it at that. Rather, it is a red flag indicating that something is up; the null hypothesis may be false, possibly in a substantively uninteresting way, or maybe we got unlucky. On the other hand, a large p value does mean that there is not much evidence against the null. For example, in many settings the Bayes factor is bounded above by *1/p value*. Even Bayesian analyses can benefit from classical testing at the model-checking stage (see Box 1980 or the Bayesian p values in Gelman, Carlin, Stern, and Rubin 1995).

4. CHALLENGES

As we move into the next millennium, it is important to expand the scope of hypothesis testing, as statistics will be increasingly asked to deal with huge datasets and extensive collections of complex models with large numbers of parameters. (Data mining, anyone?) The Bayesian approach appears to be very promising for such situations, especially if fairly automatic methods for calculating the Bayes factors are developed.

There will continue to be plenty of challenges in the classical arena. In addition to developing systematic collections of models and broadening the set of useful non-parametric tools, practical methods for finding p values for projection-pursuit (e.g., Sun 1991) and other methods that involve searching over large spaces will be crucial.

REFERENCES

Akritas, M. G., and Arnold, S. F. (1994), "Fully Nonparametric Hypotheses for Factorial Designs, I: Multivariate Repeated Measures Designs," *Journal of the American Statistical Association*, 89, 336–343.

Anderson, T. W. (1984), *An Introduction to Multivariate Statistical Analysis* (2nd ed.), New York: Wiley.

Andersson, S. (1975), "Invariant Normal Models," *The Annals of Statistics*, 3, 132–154.

—— (1982), "Distributions of Maximal Invariants Using Quotient Measures," *The Annals of Statistics*, 10, 955–961.

Andersson, S. A., and Perlman, M. D. (1993), "Lattice Models for Conditional Independence in a Multivariate Normal Distribution," *Annals of Statistics*, 21, 1318–1358.

Barlow, R. E., Bartholomew, D. J., Bremner, J. M., and Brunk, H. D. (1972), *Statistical Inference Under Order Restrictions*, New York: Wiley.

Bartlett, M. S. (1937), "Properties of Sufficiency and Statistical Tests," *Proceedings of the Royal Society of London*, Ser. A, 160, 268–282.

Beran, R., and Millar, P. W. (1987), "Stochastic Estimation and Testing," *The Annals of Statistics*, 15, 1131–1154.

Berger, J. O., and Pericchi, L. R. (1996), "The Intrinsic Bayes Factor for Model Selection and Prediction," *Journal of the American Statistical Association*, 91, 109–122.

Berger, J. O., and Sellke, T. (1987), "Testing a Point Null Hypothesis: The Irreconcilability of p Values and Evidence" (with discussion), *Journal of the American Statistical Association*, 82, 112–122.

Birnbaum, A. (1955), "Characterizations of Complete Classes of Tests of Some Multiparametric Hypotheses, With Applications to Likelihood Ratio Tests," *Annals of Mathematical Statistics*, 26, 21–36.

Bishop, Y. M. M., Fienberg, S. E., and Holland, P. W. (1975), *Discrete Multivariate Analysis: Theory and Practice*, Cambridge, MA: MIT Press.

Box, G. E. P. (1980), "Sampling and Bayes' Inference in Scientific Modelling and Robustness" (with discussion), *Journal of the Royal Statistical Society*, Ser. A, 143, 383–430.

Brown, L. D., Cohen, A., and Strawderman, W. E. (1976), "A Complete Class Theorem for Strict Monotone Likelihood Ratio With Applications," *The Annals of Statistics*, 4, 712–722.

—— (1980), "Complete Classes for Sequential Tests of Hypotheses," *The Annals of Statistics*, 8, 377–398, Corr. 17, 1414–1416.

Brown, L. D., and Marden, J. I. (1989), "Complete Class Results for Hypothesis Testing Problems With Simple Null Hypotheses," *The Annals of Statistics*, 17, 209–235.

Casella, G., and Berger, R. L. (1987), "Reconciling Bayesian and Frequentist Evidence in the One-Sided Testing Problem," *Journal of the American Statistical Association*, 82, 106–111.

Chakraborty, B., and Chaudhuri, P. (1999), "On Affine Invariant Sign and Rank Tests in One- and Two-Sample Multivariate Problems," in *Multivariate Analysis, Design of Experiments, and Survey Sampling*, ed. S. Ghosh, New York, Marcel Dekker, pp. 499–522.

Chaudhuri, P., and Sengupta, D. (1993), "Sign Tests in Multidimension: Inference Based on the Geometry of the Point Cloud," *Journal of the American Statistical Association*, 88, 1363–1370.

Choi, K. M., and Marden, J. I. (1997), "An Approach to Multivariate Rank Tests in Multivariate Analysis of Variance," *Journal of the American Statistical Association*, 92, 1581–1590.

Cohen, A., and Sackrowitz, H. B. (1987), "Unbiasedness of Tests for Homogeneity," *The Annals of Statistics*, 15, 805–816.

Cox, D. R. (1961), "Tests of Separate Families of Hypotheses," *Proceedings of the Fifth Berkeley Symposium on Mathematical Statistics and Probability*, 1, 105–123.

Davison, A. C., and Hinkley, D. V. (1997), *Bootstrap Methods and Their Application*, Cambridge, U.K.: Cambridge University Press.

Diaconis, P. (1988), *Group Representations in Probability and Statistics*, Heyward, CA: Institute for Mathematical Statistics.

Eaton, M. L. (1970), "A Complete Class Theorem for Multidimensional One-Sided Alternatives," *Annals of Mathematical Statistics*, 41, 1884–1888.

Edwards, W., Lindeman, H., and Savage, L. J. (1963), "Bayesian Statistical Inference for Psychological Research," *Psychological Review*, 70, 193–242.

Farrell, R. H. (1968), "Towards a Theory of Generalized Bayes Tests," *Annals of Mathematical Statistics*, 39, 1–22.

Fisher, R. A. (1925), *Statistical Methods for Research Workers*: London: Oliver and Boyd.

—— (1935), *Design of Experiments*, London: Oliver and Boyd.

Friedman, J. H., and Rafsky, L. C. (1979), "Multivariate Generalizations of the Wald–Wolfowitz and Smirnov Two-Sample Tests," *The Annals of Statistics*, 7, 697–717.

Gelman, A., Carlin, J., Stern, H., and Rubin, D. (1995), *Bayesian Data Analysis*, London: Chapman Hall.

Gibbons, J. D., and Chakraborti, S. (1992), *Nonparametric Statistical Inference*, New York: Marcel Dekker.

Good, I. J. (1987), Comments on "Testing a Point Null Hypothesis: The Irreconcilability of p Values and Evidence" by Berger and Sellke and "Reconciling Bayesian and Frequentist Evidence in the One-Sided Testing Problem" by Casella and Berger, *Journal of the American Statistical Association*, 82, 125–128.

Goodman, L. A. (1968), "The Analysis of Cross-Classified Data: Independence, Quasi-Independence, and Interactions in Contingency Tables With or Without Missing Entries," *Journal of the American Statistical Association*, 63, 1091–1131.

Haberman, S. J. (1974), *The Analysis of Frequency Data*, Chicago: University of Chicago Press.

Hettmansperger, T. P., and McKean, J. W. (1998), *Robust Nonparametric Statistical Methods*, London: Arnold.

Hettmansperger, T. P., Möttonen, J., and Oja, H. (1997), "Affine-Invariant Multivariate One-Sample Signed-Rank Tests," *Journal of the American Statistical Association*, 92, 1591–1600.

Hettmansperger, T. P., Nyblom, J., and Oja, H. (1994), "Affine Invariant Multivariate One-Sample Sign Tests," *Journal of the Royal Statistical Society*, Ser. B, 56, 221–234.

Hotelling, H. (1931), "The Generalization of Student's Ratio," *Annals of Mathematical Statistics*, 2, 360–378.

Jeffreys, H. (1939), *Theory of Probability*, Oxford, U.K.: Clarendon Press.

Kass, R. E., and Raftery, A. E. (1995), "Bayes Factors," *Journal of the American Statistical Association*, 90, 773–795.

Kass, R. E., and Wasserman, L. (1996), "The Selection of Prior Distributions by Formal Rules," *Journal of the American Statistical Association*, 91, 1343–1370.

Kiefer, J., and Schwartz, R. (1965), "Admissible Bayes Character of T^2-, R^2-, and Other Fully Invariant Tests for Classical Multivariate Normal Problems," *Annals of Mathematical Statistics*, 36, 747–770. Corr. 43, 1742.

Lauritzen, S. L. (1996), *Graphical Models*, Oxford, U.K.: Clarendon Press.

Lehmann, E. L. (1959), *Testing Statistical Hypotheses*, New York: Wiley.

Lindley, D. V. (1957), "A Statistical Paradox," *Biometrika*, 44, 187–192.

Liu, R. Y., and Singh, K. (1993), "A Quality Index Based on Data Depth and Multivariate Rank Tests," *Journal of the American Statistical Association*, 88, 252–260.

——— (1997), "Notions of Limiting p Values Based on Data Depth and Bootstrap," *Journal of the American Statistical Association*, 92, 266–277.

Mahalanobis, P. C. (1930), "On Tests and Measures of Group Divergence," *Journal and Proceedings of the Asiatic Society of Bengal*, 26, 541–588.

Marden, J. I. (1982), "Minimal Complete Classes of Tests of Hypotheses With Multivariate One-Sided Alternatives," *The Annals of Statistics*, 10, 962–970.

——— (1999), "Multivariate Rank Tests," in *Multivariate Analysis, Design of Experiments, and Survey Sampling*, ed. S. Ghosh, New York: Marcel Dekker, pp. 401–432.

Matthes, T. K., and Truax, D. R. (1967), "Test of Composite Hypotheses for the Multivariate Exponential Family," *Annals of Mathematical Statistics*, 38, 681–697. Corr. 38, 1928.

Mosteller, F., and Wallace, D. L. (1984), *Applied Bayesian and Classical Inference: The Case of the Federalist Papers*, New York: Springer.

Neyman, J., and Pearson, E. (1928), "On the Use and Interpretation of Certain Test Criteria for Purposes of Statistical Inference. Part I," *Biometrika*, 20A, 175–240.

——— (1933), "On the Problem of the Most Efficient Tests of Statistical Hypotheses," *Philosophical Transactions of the Royal Society*, Ser. A, 231, 289–337.

Patel, K. M., and Hoel, D. G. (1973), "A Nonparametric Test for Interaction in Factorial Experiments," *Journal of the American Statistical Association*, 68, 615–620.

Pearson, K. (1900), "On the Criterion That a Given System of Deviations From the Probable in the Case of a Correlated System of Variables is Such That it can be Reasonably Supposed to Have Risen From Random Sampling," *Philosophical Magazine*, 1, 157–175.

Perlman, M. D. (1980), "Unbiasedness of Multivariate Tests: Recent Results," in *Multivariate Analysis V*, Amsterdam: North-Holland/Elsevier, pp. 413–432.

Perlman, M. D., and Olkin, I. (1980), "Unbiasedness of Invariant Tests for MANOVA and Other Multivariate Problems," *The Annals of Statistics*, 8, 1326–1341.

Pratt, J. W. (1965), "Bayesian Interpretation of Standard Inference Statements," *Journal of the Royal Statistical Society*, Ser. B, 27, 169–203.

Puri, M. L., and Sen, P. K. (1971), *Nonparametric Methods in Multivariate Analysis*, New York: Wiley.

——— (1985), *Nonparametric Methods in General Linear Models*, New York: Wiley.

Raftery, A. E. (1995), "Bayesian Model Selection in Social Research," in *Sociological Methodology 1995*, ed. P. V. Marsden, Oxford, U.K.: Blackwells, pp. 111–196.

Randles, R. H. (1989), "A Distribution-Free Multivariate Sign Test Based on Interdirections," *Journal of the American Statistical Association*, 84, 1045–1050.

Rao, C. R. (1947), "Large-Sample Tests of Statistical Hypotheses Concerning Several Parameters With Applications to Problems of Estimation," *Proceedings of the Cambridge Philosophical Society*, 44, 50–57.

——— (1973), *Linear Statistical Inference and its Applications*, New York: Wiley.

Robertson, T., Wright, F. T., and Dykstra, R. L. (1988), *Order-Restricted Statistical Inference*, New York: Wiley.

Roy, S. N. (1953), "On a Heuristic Method of Test Construction and Its Use in Multivariate Analysis," *Annals of Mathematical Statistics*, 24, 220–238.

Serfling, R. (1980), *Approximation Theorems of Mathematical Statistics*, New York: Wiley.

Stein, C. (1956), "The Admissibility of Hotelling's T^2 Test," *Annals of Mathematical Statistics*, 27, 616–623.

Student (1908), "The Probable Error of a Mean," *Biometrika*, 6, 1–25.

Sun, J. (1991), "Significance Levels in Exploratory Projection Pursuit," *Biometrika*, 78, 759–769.

Wald, A. (1943), "Tests of Statistical Hypotheses Concerning Several Parameters When the Number of Observations is Large," *Transactions of the American Mathematical Society*, 54, 426–482.

——— (1950), *Statistical Decision Functions*, New York: Wiley.

Wermuth, N., and Cox, D. R. (1992), "Graphical Models for Dependencies and Associations," *Proceedings of the Tenth Symposium on Computational Statistics*, 1, 235–249.

Whitaker, J. (1990), *Graphical Models in Applied Multivariate Statistics*, New York: Wiley.

Wijsman, R. A. (1967), "Cross-Sections of Orbits and Their Application to Densities of Maximal Invariants," *Proceedings of the Fifth Berkeley Symposium on Mathematical Statistics and Probability*, 1, 389–400.

Generalized Linear Models

Charles E. McCulloch

1. INTRODUCTION AND SOME HISTORY

What is the difference, in absolute value, between logistic regression and discriminant analysis? I will not make you read this entire article to find the answer, which is 2. But you will have to read a bit further to find out why.

As most statisticians know, logistic regression and probit regression are commonly used techniques for modeling a binary response variable as a function of one or more predictors. These techniques have a long history, with the term "probit" traced by David (1995) back to Bliss (1934), and Finney (1952) attributing the origin of the technique itself to psychologists in the late 1800s. In its earliest incarnations, probit analysis was little more than a transformation technique; scientists realized that the sigmoidal shape often observed in plots of observed proportions of successes versus a predictor x could be rendered a straight line by applying a transformation corresponding to the inverse of the normal cdf.

For example, Bliss (1934) described an experiment in which nicotine is applied to aphids and the proportion killed is recorded. (How is that for an early antismoking message?) Letting $\Phi^{-1}(\cdot)$ represent the inverse of the standard normal cdf, and $\hat{p}_i$ the observed proportion killed at dose d_i of the nicotine, Bliss showed a plot of $\Phi^{-1}(\hat{p}_i)$ versus $\log d_i$. The plot seems to indicate that a two-segment linear regression model in $\log d_i$ is the appropriate model. In an article a year later, Bliss (1935) explained the methodology in more detail as a weighted linear regression of $\Phi^{-1}(\hat{p}_i)$ on the predictor x_i using weights equal to $[n_i\phi(p_i)^2]/\{\Phi(p_i)[1 - \Phi(p_i)]\}$, where $\phi(\cdot)$ represents the standard normal pdf and n_i is the sample size for calculating $\hat{p}_i$. These weights can be easily derived as the inverse of the approximate variance found by applying the delta method to $\Phi^{-1}(\hat{p}_i)$.

This approach obviously has problems if an observed proportion is either 0 or 1. As a brief appendix to Bliss's work, Fisher (1935) outlined the use of maximum likelihood to obtain estimates using data in which $\hat{p}_i$ is either 0 or 1. Herein lies a subtle change: Fisher was no longer describing a model for the transformed proportions, but instead was directly modeling the mean of the binary response. Users of generalized

Charles E. McCulloch is Professor, Division of Biostatistics, University of California, San Francisco, CA 94143 (E-mail: chuck@biostat.ucsf.edu). This is BU-1449-M in the Biometrics Unit Technical Report Series at Cornell University and was supported by National Science Foundation grant DMS-9625476.

linear models (GLMs) will recognize the distinction between a transformation and a link.

This technique of maximum likelihood is suggested only as a method of last resort when the observed proportions that are equal to 0 or 1 must be incorporated in the analysis. The computational burdens were simply too high for it to be used on a regular basis in that era.

So, by the 1930s, models of the following form were being posited, and fitting with the method of maximum likelihood was at least being entertained. With p_i denoting the probability of a success for the ith observation, the model is given by

$$Y_i \sim \text{indep. Bernoulli}(p_i)$$

and

$$p_i = \Phi(\mathbf{x}_i'\boldsymbol{\beta}), \tag{1}$$

where $\mathbf{x}_i'$ denotes the ith row of the matrix of predictors. With a slight abuse of notation, and to make this look similar to a linear model, we can rewrite (1) as

$$\mathbf{Y} \sim \text{indep. Bernoulli}(\mathbf{p})$$

and

$$\mathbf{p} = \Phi(\mathbf{X}\boldsymbol{\beta}) \tag{2}$$

or, equivalently,

$$\Phi^{-1}(\mathbf{p}) = \mathbf{X}\boldsymbol{\beta}. \tag{3}$$

By 1952, this had changed little. In that year Finney more clearly described the use of maximum likelihood for fitting these models in an appendix titled "Mathematical Basis of the Probit Method" and spent six pages in another appendix laying out the recommended computational method. This includes steps such as "34. Check steps 30–33" and the admonishment to the computer (a person!) that "A machine is not a complete safeguard against arithmetical errors, and carelessness will lead to wrong answers just as certainly as in nonmechanized calculations." This is clearly sage advice against overconfidence in output even from today's software.

Finney was practically apologetic about the effort required: "The chief hindrances to the more widespread adoption of the probit method . . . (is) . . . the apparent laboriousness of the computations." He recognized that his methods must be iterated until convergence to arrive at the maximum likelihood estimates but indicated that "with experience the first provisional line may often be drawn so accurately that only one cycle of the calculations is needed to give a satisfactory fit . . . ."

With computations so lengthy, what iterative method of fitting was used? Finney recommended using "working probits," which he defined as (ignoring the shift of five units historically used to keep all the calculations positive):

$$z_i = \mathbf{x}_i'\boldsymbol{\beta} + \frac{y_i - \Phi(\mathbf{x}_i'\boldsymbol{\beta})}{\phi(\mathbf{x}_i'\boldsymbol{\beta})}. \tag{4}$$

The working probits for a current value of β were regressed on the predictors using weights the same as suggested by Bliss, namely $[\phi(p_i)^2]/\{\Phi(p_i)[1 - \Phi(p_i)]\}$ to get the new value of β.

When I first learned about the EM algorithm (Dempster, Laird, and Rubin 1977), I was struck by its similarity to Finney's algorithm. A common representation of (1) is via a threshold model. That is, hypothesize a latent variable W_i such that

$$W_i \sim \text{ indep } N(\mathbf{x}_i'\boldsymbol{\beta}, 1). \tag{5}$$

Then, using $Y_i = I_{\{W_i > 0\}}$ yields (1). To implement the EM algorithm, it is natural to regard the W_i as missing data and fill them in. Once the W_i are known, ordinary least squares can be used to get the new estimate of β. The E step fills in the W_i using the formula

$$E[W_i|Y_i] = \mathbf{x}_i'\boldsymbol{\beta} + \phi(\mathbf{x}_i'\boldsymbol{\beta})\frac{y_i - \Phi(\mathbf{x}_i'\boldsymbol{\beta})}{\Phi(\mathbf{x}_i'\boldsymbol{\beta})[1 - \Phi(\mathbf{x}_i'\boldsymbol{\beta})]}, \tag{6}$$

and the M step estimates β as $(\mathbf{X}'\mathbf{X})^{-1}\mathbf{X}'\mathbf{W}$.

Thus the term added to $\mathbf{x}_i'\boldsymbol{\beta}$ in the EM algorithm is the same as the term added using working probits, once they are multiplied by the weight. Practical usage of EM and working probits, however, shows that working probits invariably converge much more quickly than EM!

So as early as 1952 many of the key ingredients of GLMs are seen: using "working variates" and link functions, fitting using a method of iteratively weighted fits, and using likelihood methods. But lack of computational resources simply did not allow widespread use of such techniques.

Logistic regression was similarly hampered. Over a decade later, Cox (1966) stated that "since the maximization of a function of many variables may not be straightforward, even with an electronic computer, it is worth having 'simple' methods for solving maximum likelihood equations, especially for use when there are single observations at each x value, so that the linearizing transformation is not applicable." Note the need for simple methods despite the fact that "computers" in 1966 are now machines.

For the logistic regression model akin to (2), namely

$$\mathbf{Y} \sim \text{ indep. Bernoulli}(\mathbf{p})$$

and

$$\mathbf{p} = 1/(1 + \exp[-\mathbf{X}\boldsymbol{\beta}]) \tag{7}$$

or

$$\log[\mathbf{p}/(1-\mathbf{p})] = \mathbf{X}\beta,$$

it is straightforward to show that the maximum likelihood equations are given by

$$\mathbf{X}'\mathbf{Y} = \mathbf{X}'\mathbf{p}. \tag{8}$$

Because β enters $\mathbf{p}$ in a nonlinear fashion in (8), it is not possible to analytically solve this equation for β. However, using the crude approximation (Cox 1966), $1/(1 + \exp[-t]) \approx 1/2 + t/6$, which clearly is applicable only for the midrange of the curve, (8) can be rewritten approximately as

$$\begin{aligned} \mathbf{X}'\mathbf{Y} &= \mathbf{X}'\left(\tfrac{1}{2}\mathbf{1} + \tfrac{1}{6}\mathbf{X}\beta\right) \\ &= \tfrac{1}{2}\mathbf{X}'\mathbf{1} + \tfrac{1}{6}\mathbf{X}'\mathbf{X}\beta. \end{aligned} \tag{9}$$

This leads to

$$\mathbf{X}'\left(\mathbf{Y} - \frac{1}{2}\mathbf{1}\right) = \frac{1}{6}\mathbf{X}'\mathbf{X}\beta, \tag{10}$$

which we can solve as

$$\hat{\beta} = (\mathbf{X}'\mathbf{X})^{-1}\mathbf{X}'6\left(\mathbf{Y} - \frac{1}{2}\mathbf{1}\right) = (\mathbf{X}'\mathbf{X})^{-1}\mathbf{X}'\mathbf{Y}^*, \tag{11}$$

where Y_i^* is equal to 3 for a success and -3 for a failure. That is, we can approximate the logistic regression coefficients in a crude way by an ordinary least squares regression on a coded Y.

Logistic regression is often used as an alternate method for two-group discriminant analysis (Cox and Snell 1989), by using the (binary) group identifier as the "response" and the multivariate vectors as the "predictors." This is a useful alternative when the usual multivariate normality assumption for the multivariate vectors is questionable; for example, when one or more variables are binary or categorical.

When it *is* reasonable to assume multivariate normality, the usual Fisher discriminant function is given by $\mathbf{S}^{-1}(\bar{\mathbf{X}}_1 - \bar{\mathbf{X}}_2)$, where $\bar{\mathbf{X}}_i$ is the mean of the vectors for the ith group. If we code the successes and failures as 1 and -1, then $\bar{\mathbf{X}}_1 - \bar{\mathbf{X}}_2 = \mathbf{X}'\mathbf{Y}$. Thus we see that the difference between logistic regression and discriminant function analysis is 2, in absolute value.

2. ORIGINS

GLMs appeared on the statistical scene in the path-breaking article of Nelder and Wedderburn (1972). Even though virtually all of the pieces had previously existed, these authors were the first to put forth a unified framework showing the similarities between seemingly disparate methods, such as probit regression, linear models, and contingency tables. They recognized that fitting a probit regression by iterative fits using the "working probits," namely (4), could be generalized in a straightforward way to unify a whole collection of maximum likelihood problems. Replacing $\Phi^{-1}(\cdot)$ with a general "link" function, $g(\cdot)$, and defining a "working variate" via

$$z \equiv g(\mu) + (y - \mu)g'(\mu) \tag{12}$$

gave, via iterative weighted least squares, a computational method for finding the maximum likelihood estimates. More formally, we can write the model as

$$Y_i \sim \text{ indep. } f_{Y_i}(y_i),$$

$$f_{Y_i}(y_i) = \exp\{(y_i\boldsymbol{\theta}_i - b(\boldsymbol{\theta}_i))/a(\phi) - c(y_i, \phi)\},$$

$$E[Y_i] = \mu_i,$$

and

$$g(\mu_i) = \mathbf{x}_i'\boldsymbol{\beta}, \tag{13}$$

where $\boldsymbol{\theta}_i$ is a known function of $\boldsymbol{\beta}$ and $g(\cdot)$ is a known function that transforms (or links) the mean of y_i (not y_i itself!) to the linear predictor. The iterative algorithm is used to give maximum likelihood estimates of $\boldsymbol{\beta}$.

More important, it made possible a style of thinking that freed the data analyst from the need to look for a transformation that simultaneously achieved linearity in the predictors and normality of the distribution (as in Box and Cox 1964).

I think of building GLMs by making three decisions:

1. What is the distribution of the data (for fixed values of the predictors and possibly after a transformation)?
2. What function of the mean will be modeled as linear in the predictors?
3. What will the predictors be?

What advantages does this approach have? First, it unifies what appear to be very different methodologies, which helps to understand, use, and teach the techniques. Second, because the right side of the equation is a linear model after applying the link, many of the concepts of linear models carry over to GLMs. For example, the issues of full-rank versus overparameterized models are similar.

The application of GLMs became a reality in the mid 1970s, when GLMs were incorporated into the statistics package GENSTAT and made available interactively in the GLIM software. Users of these packages could then handle linear regression, logistic and probit regression, Poisson regression, log-linear models, and regression with skewed continuous distributions, all in a consistent manner. Both packages are still widely used and are currently distributed by the Numerical Algorithms Group (www.nag.com). Of course, by now, most major statistical packages have facilities for GLMs; for example, SAS Proc GENMOD.

GLMs received a tremendous boost with the development of quasi-likelihood by Wedderburn in 1974. Using only the mean-to-variance relationship, Wedderburn showed how statistical inference could still be conducted. Perhaps surprisingly, given the paucity of assumptions, these techniques often retain full or nearly full efficiency (Firth 1987). Further, the important modification of *overdispersion* is allowed; that is, models with variance proportionally larger than predicted by the nominal distribution, say, a Poisson distribution. Such situations arise commonly in practice. Quasi-likelihood was put on a firmer theoretical basis by McCullagh (1983).

That year also saw the publication of the first edition of the now-classic book *Generalized Linear Models* (McCullagh and Nelder 1983). With a nice blend of theory, practice, and applications this text made GLMs more widely used and appreciated. A colleague once asked me what I thought of the book. I replied that it was absolutely wonderful and that the modeling and data analytic philosophy that it espoused was visionary. After going on for several minutes, I noticed that he looked perplexed. When I inquired why, he replied "I think it's terrible—it has no theorems." Perhaps that was the point.

3. MAJOR DEVELOPMENTS

GLMs are now a mature data-analytic methodology (e.g., Lindsey 1996) and have been developed in numerous directions. There are techniques for choosing link functions and diagnosing link failures (e.g., Mallick and Gelfand 1994; Pregibon 1980) as well as research on the consequences of link misspecification (e.g., Li and Duan 1989; Weisberg and Welsh 1994). There are techniques for outlier detection and assessment of case influence for model checking (e.g., Cook and Croos-Dabrera 1998; Pregibon 1981). There are methods of modeling the dispersion parameters as a function of covariates (e.g., Efron 1986) and for accommodating measurement error in the covariates (e.g., Buzas and Stefanski 1996; Stefanski and Carroll 1990), as well as ways to handle generalized additive models (Hastie and Tibshirani 1990).

An extremely important extension of GLMs is the approach pioneered by Liang and Zeger (Liang and Zeger 1986; Zeger and Liang 1986) known as generalized estimating equations (GEEs). In my opinion, GEEs made two valuable contributions: the accommodation of a wide array of correlated data structures and the popularization of the "robust sandwich estimator" of the variance–covariance structure. Current software implementations of GEEs are designed mostly to accommodate longitudinal data structures; that is, ones in which the data can be arranged as repeat measurements on a series of independent "subjects" (broadly interpreted, of course). The use of the "robust sandwich estimator," which basically goes back to Huber (1967) and Royall (1986), allows the specification of a "working" covariance structure. That is, the data analyst must specify a guess as to the correct covariance structure, but inferences remain asymptotically valid even if this structure is incorrectly specified (as it always is to some degree). Not surprisingly, the efficiency of inferences can be affected if the "working" structure is far from truth (e.g., Fitzmaurice 1995).

Distribution theory for modifications of exponential families for use in GLMs has been developed further by, for example, Jorgensen (1997), and the theory of quasi-likelihood is detailed in the book-length treatment of Heyde (1997).

4. LOOKING FORWARD

Anyone making predictions runs the risk of someone actually checking later to see whether the predictions are correct. So I am counting on the "Jean Dixon effect," defined by the *Skeptic's Dictionary* (http://skeptic.com) as "the tendency of the mass media to hype or exaggerate a few correct predictions by a psychic, guaranteeing that they will be remembered, while forgetting or ignoring the much more numerous incorrect predictions." Because likelihood and quasi-likelihood methods are based on large-sample approximations, an important area of development is the construction of tests and confidence intervals that are accurate in small- and moderate-sized samples. This may be through "small-sample asymptotics" (e.g., Jorgensen, 1997; Skovgaard, 1996) or via computationally intensive methods like the bootstrap (Davison and Hinkley 1997; Efron and Tibshirani 1993). The extension of GLMs to more complex correlation structures has been an area of active research and will see more development. Models for time series (e.g., Chan and Ledolter 1995), random-effects models (e.g., Breslow and Clayton 1993; Stiratelli, Laird, and Ware 1984), and spatial models (e.g., Heagerty and Lele 1999) have all been proposed. Unfortunately, likelihood analysis of many of the models has led to intractable, high-dimensional integrals. So, likewise, computing methods for these models will continue to be an ongoing area of development. Booth and Hobert (1999) and McCulloch (1997) used a Monte Carlo EM approach; Quintana, Liu, and del Pino (1999) used a stochastic approximation algorithm; and Heagerty and Lele (1998) took a composite likelihood tack.

Attempts to avoid likelihood analysis via techniques such as penalized quasi-likelihood (for a description see Breslow and Clayton 1993) have not been entirely successful. Approaches based on working variates (e.g., Schall 1991) and Laplace approximations (e.g., Wolfinger 1994) generate inconsistent estimates (Breslow and Lin 1994) and can be badly biased for distributions far from normal (i.e., the important case of Bernoulli-distributed data). Clearly, reliable, well-tested, general-purpose fitting algorithms need to be developed before these models will see regular use in practice.

The inclusion of random effects in GLMs raises several additional questions: What is the effect of misspecification of the random-effects distribution (e.g., Neuhaus, Hauck, and Kalbfleisch 1992) and how can it be diagnosed? What is the best way to predict the random effects and how can prediction limits be set, especially in small- and moderate-sized samples (e.g., Booth and Hobert 1998)? How can outlying random effects be identified or downweighted? All of these are important practical questions that must be thoroughly investigated for regular data analysis.

The whole idea behind GLMs is the development of a strategy and philosophy for approaching statistical problems, especially those involving nonnormally distributed data, in a way that retains much of the simplicity of linear models. Areas in which linear models have been heavily used (e.g., simultaneous equation modeling in econometrics) have and will see adaptations for GLMs. As such, GLMs will continue in broad use and development for some time to come.

REFERENCES

Bliss, C. (1934), "The Method of Probits," *Science*, 79, 38–39.

———— (1935), "The Calculation of the Dose–Mortality Curve," *Annals of Applied Biology*, 22, 134–167.

Booth, J. G., and Hobert, J. P. (1998), "Standard Errors of Prediction in Generalized Linear Mixed Models," *Journal of the American Statistical Association*, 93, 262–272.

———— (1999), "Maximizing Generalized Linear Mixed Model Likelihoods With an Automated Monte Carlo EM Algorithm," *Journal of the Royal Statistical Society*, Ser. B, 61, 265–285.

Box, G. E. P., and Cox, D. R. (1964), "An Analysis of Transformations" (with discussion), *Journal of the Royal Statistical Society*, Ser. B, 26, 211–252.

Breslow, N. E., and Clayton, D. G. (1993), "Approximate Inference in Generalized Linear Mixed Models," *Journal of the American Statistical Association*, 88, 9–25.

Breslow, N. E., and Lin, X. (1994), "Bias Correction in Generalized Linear Mixed Models With a Single Component of Dispersion," *Biometrika*, 82, 81–91.

Buzas, J. S., and Stefanski, L. A. (1996), "Instrumental Variable Estimation in Generalized Linear Measurement Error Models," *Journal of the American Statistical Association*, 91, 999–1006.

Chan, K. S., and Ledolter, J. (1995), "Monte Carlo EM Estimation for Time Series Models Involving Counts," *Journal of the American Statistical Association*, 90, 242–252.

Cook, R. D., and Croos-Dabrera, R. (1998), "Partial Residual Plots in Generalized Linear Models," *Journal of the American Statistical Association*, 93, 730–739.

Cox, D. R. (1966), "Some Procedures Connected With the Logistic Qualitative Response Curve," in *Research Papers in Statistics*, ed. F. N. David, New York: Wiley, pp. 55–72.

Cox, D. R., and Snell, E. J. (1989), *Analysis of Binary Data*, London: Chapman and Hall.

David, H. A. (1995), "First (?) Occurrence of Common Terms in Probability and Statistics," *The American Statistician*, 49, 121–133.

Davison, A. C., and Hinkley, D. V. (1997), *Bootstrap Methods and Their Application*, Cambridge, U.K.: Cambridge University Press.

Dempster, A. P., Laird, N. M., and Rubin, D. B. (1977), "Maximum Likelihood From Incomplete Data via the EM Algorithm" (with discussion), *Journal of the Royal Statistical Society*, Ser. B, 39, 1–38.

Efron, B. (1986), "Double Exponential Families and Their Use in Generalized Linear Regression," *Journal of the American Statistical Association*, 81, 709–721.

Efron, B., and Tibshirani, R. J. (1993), *An Introduction to the Bootstrap*, New York: Chapman and Hall.

Finney, D. J. (1952), *Probit Analysis*, Cambridge, U.K.: Cambridge University Press.

Firth, D. (1987), "On the Efficiency of Quasi-Likelihood Estimation," *Biometrika*, 74, 233–245.

Fisher, R. A. (1935), Appendix to "The Calculation of the Dose–Mortality Curve" by C. Bliss, *Annals of Applied Biology*, 22, 164–165.

Fitzmaurice, G. M. (1995), "A Caveat Concerning Independence Estimating Equations With Multivariate Binary Data," *Biometrics*, 51, 309–317.

Hastie, T. J., and Tibshirani, R. J. (1990), *Generalized Additive Models*, London: Chapman and Hall.

Heagerty, P., and Lele, S. (1998), "A Composite Likelihood Approach to Binary Spatial Data," *Journal of the American Statistical Association*, 93, 1099–1111.

Heyde, C. C. (1997), *Quasi-Likelihood and Its Application*, New York: Springer.

Huber, P. J. (1967), "The Behaviour of Maximum Likelihood Estimators Under Nonstandard Conditions," in *Proceedings of the Fifth Berkeley Symposium on Mathematical Statistics and Probability*, eds. L. M. LeCam and J. Neyman, pp. 221–233.

Jorgensen, B. (1997), *The Theory of Dispersion Models*, London: Chapman and Hall.

Li, K-C., and Duan, N. (1989), "Regression Analysis Under Link Violation," *The Annals of Statistics*, 17, 1009–1052.

Liang, K-Y., and Zeger, S. L. (1986), "Longitudinal Data Analysis Using Generalized Linear Models," *Biometrika*, 73, 13–22.

Lindsey, J. K. (1996), *Applying Generalized Linear Models*, New York: Springer.

Mallick, B. K., and Gelfand, A. E. (1994), "Generalized Linear Models With Unknown Link Functions," *Biometrika*, 81, 237–245.

McCullagh, P. (1983), "Quasi-Likelihood Functions," *The Annals of Statistics*, 11, 59–67.

McCullagh, P., and Nelder, J. (1983), *Generalized Linear Models*, London: Chapman and Hall.

McCulloch, C. (1997), "Maximum Likelihood Algorithms for Generalized Linear Mixed Models," *Journal of the American Statistical Association*, 92, 162–170.

Nelder, J. A., and Wedderburn, R. W. M. (1972), "Generalized Linear Models," *Journal of the Royal Statistical Society*, Ser. A, 135, 370–384.

Neuhaus, J. M., Hauck, W-W., and Kalbfleisch, J. D. (1992), "The Effects of Mixture Distribution Misspecification When Fitting Mixed Effects Logistic Models," *Biometrika*, 79, 755–762.

Pregibon, D. (1980), "Goodness of Link Tests for Generalized Linear Models," *Applied Statistics*, 29, 15–24.

——— (1981), "Logistic Regression Diagnostics," *The Annals of Statistics*, 9, 705–724.

Quintana, R., Liu, J., and del Pino, G. (1999), "Monte Carlo EM With Importance Reweighting and Its Application in Random Effects Models," *Computational Statistics and Data Analysis*, 29, 429–444.

Royall, R. (1986), "Model Robust Inference Using Maximum Likelihood Estimators," *International Statistical Review*, 54, 221–226.

Schall, R. (1991), "Estimation in Generalized Linear Models With Random Effects," *Biometrika*, 78, 719–727.

Skovgaard, I. M. (1996), "An Explicit Large-Deviation Approximation to One-Parameter Tests," *Bernoulli*, 2, 145–165.

Stefanski, L. A., and Carroll, R. J. (1990), "Score Tests in Generalized Linear Measurement Error Models," *Journal of the Royal Statistical Society*, Ser. B, 52, 345–359.

Stiratelli, R., Laird, N., and Ware, J. H. (1984), "Random-Effects Models for Serial Observations With Binary Response," *Biometrics*, 40, 961–971.

Wedderburn, R. W. M. (1974), "Quasi-Likelihood Functions, Generalized Linear Models, and the Gauss–Newton Method," *Biometrika*, 61, 439–447.

Weisberg, S., and Welsh, A. H. (1994), "Adapting for the Missing Link," *The Annals of Statistics*, 22, 1674–1700.

Wolfinger, R. W. (1994), "Laplace's Approximation for Nonlinear Mixed Models," *Biometrika*, 80, 791–795.

Zeger, S., and Liang, K-Y. (1986), "Longitudinal Data Analysis for Discrete and Continuous Outcomes," *Biometrics*, 42, 121–130.

Missing Data: Dial M for ???

Xiao-Li Meng

The question mark is common notation for the missing data that occur in most applied statistical analyses. Over the past century, statisticians and other scientists not only have invented numerous methods for handling missing/incomplete data, but also have invented many forms of missing data, including data augmentation, hidden states, latent variables, potential outcome, and auxiliary variables. Purposely constructing unobserved/unobservable variables offers an extraordinarily flexible and powerful framework for both scientific modeling and computation and is one of the central statistical contributions to natural, engineering, and social sciences. In parallel, much research has been devoted to better understanding and modeling of real-life missing-data mechanisms; that is, the unintended data selection process that prevents us from observing our intended data. This article is a very brief and personal tour of these developments, and thus necessarily has much missing history and citations. The tour consists of a number of M's, starting with a historic story of the mysterious method of McKendrick for analyzing an epidemic study and its link to the EM algorithm, the most popular and powerful method of the 20th century for fitting models involving missing data and latent variables. The remaining M's touch on theoretical, methodological, and practical aspects of missing-data problems, highlighted with some common applications in social, computational, biological, medical, and physical sciences.

1. McKENDRICK, A MYSTERY, AND EM

Table 1, adopted from Meng (1997), tells a fascinating story of missing data from the early part of the 20th century. The first two rows describe an epidemic of cholera in an Indian village, where x represents the number of cholera cases within a household and n_x is the corresponding observed number of such households. Prior to presenting this example, McKendrick (1926) derived a Poisson model for such data. However, the direct Poisson fit, reported in the third row, is so poor that any goodness-of-fit method that fails to reject the Poisson model must itself be rejected.

Had McKendrick (1926) settled for the simple Poisson model, it would not have

Xiao-Li Meng is Professor, Department of Statistics, University of Chicago, IL 60637 (E-mail: meng@galton.uchicago.edu). This research is supported in part by National Science Foundation grant DMS 96-26691 and National Security Agency grant MDA 904-99-1-0067. The author thanks George Casella for the invitation to write this Y2K vignette and for comments, and also thanks Radu Craiu, Dan Heitjan, Mary Sara McPeek, Dan Nicolae, William Rosenberger, Jay Servidea, and David van Dyk for comments and/or proofreading.

Table 1.   Data and Fitted Values For McKendrick's Problem

x	0	1	2	3	4	≥ 5	Total
n_x	168	32	16	6	1	0	223
Direct Poisson fit	151.64	58.48	11.28	1.45	0.00	.01	223
McKendrick's fit	36.89	34.11	15.77	4.86	1.12	.24	93
MLE fit	33.46	32.53	15.81	5.12	1.25	.29	88.46

been the earliest citation in the seminal paper on the EM algorithm by Dempster, Laird, and Rubin (1977), nor would it have been reprinted in *Breakthroughs in Statistics*, Vol. III (Kotz and Johnson 1997). McKendrick's approach seems rather mysterious, especially because he did not provide a derivation. He first calculated $s_1 = \sum_x x n_x = 86$, $s_2 = \sum_x x^2 n_x = 166$, and

$$\hat{n} = \frac{s_1^2}{s_2 - s_1} = 92.45. \tag{1}$$

Next, he treated $\hat{n} \approx 93$ as the Poisson sample size, and thus estimated the Poisson mean by $\tilde{\lambda} = s_1/\hat{n} = .93$. The fitted counts were then calculated via $\hat{n}\tilde{\lambda}^x \exp(-\tilde{\lambda})/x!$, as given in the fourth row of Table 1. The fit is evidently very good for $x \geq 1$, but exhibits an astonishingly large discrepancy for $x = 0$.

This large discrepancy gives a clue to McKendrick's approach—there were too many 0's for the simple Poisson model to fit. Earlier a lieutenant-colonel in the Indian Medical Service and then a curator of the College of Physicians at Edinburgh, McKendrick had astute insight into the excess 0's, as he wrote that "[T]his suggests that the disease was probably water borne, that there were a number of wells, and that the inhabitants of 93 out of 223 houses drank from one well which was infected." In other words, a household can have no cases of cholera either because it was never exposed to cholera or because it was exposed but no member of it was infected. The existence of these unexposed households complicates the analysis, for without external information, one cannot distinguish an unexposed from an exposed but uninfected household. To a modern statistician, this immediately suggests using the binomial/Poisson mixture model, also known as the zero-inflated Poisson (ZIP) model (see, e.g., Böhning, Dietz, and Schlattmann 1999), which models a binomial indicator for the exposure status and, conditional on being exposed, a Poisson variable as before. Although McKendrick (1926) was not explicit, he fit a zero-truncated Poisson (ZTP) model; that is, he ignored the observed $n_0 = 168$ zero-class count and used remaining data under the Poisson model to *impute* the unobserved zero-class count from the exposed population. Once he had the imputed total, the rest is history. The ingenious part of McKendrick's approach is his imputation of the total Poisson size n via (1), which equates the sample variance with the sample mean. Neither s_1 nor s_2 is affected by the actual value of n_0, yet $\lim_{n \to \infty} \hat{n}/n = \lambda^2/(\lambda^2 + \lambda - \lambda) = 1$, and thus $\hat{n}$ is a consistent imputation/estimate of the true Poisson sample size n. The mystery is then unfolded.

The key ingredients of McKendrick's approach are to first identify a missing data structure, perhaps constructed, then impute the missing data, and finally analyze the completed data set as if there were no missing data. This procedure is a predecessor of many modern missing-data methods. A key advance of modern methods, thanks to enormously improved computing power, is iterative repetition of such types of processes, as in the EM algorithm, or multiple repetitions, as with multiple imputation (Rubin 1987). The need for this iteration/repetition was recognized by Irwin (1963), who noted that once an estimator of λ was obtained via McKendrick's approach, n can be reimputed by $n^{(t+1)} = n^{(t)} \exp(-\lambda^{(t)}) + n_{obs}$, where $n_{obs} = \sum_{x \geq 1} n_x$ and t indexes iteration, and in turn λ can be reestimated via $\lambda^{(t+1)} = s_1/n^{(t+1)}$. Irwin's method, though not a special case, resembles the two-step EM algorithm. In the Expectation step, the complete-data log-likelihood $l(\theta|Y_{com})$ is imputed by its conditional expectation $Q(\theta|\theta^{(t)}) = E[l(\theta|Y_{com})|\theta^{(t)}, Y_{obs}]$, where Y_{obs} is the observed data. In the Maximization step, $Q(\theta|\theta^{(t)})$ is maximized as a function of θ to determine $\theta^{(t+1)}$. For the ZTP model, $\theta = \lambda$, and the M step is the same as McKendrick's or Irwin's; that is, $\lambda^{(t+1)} = s_1/n^{(t+1)}$. The E step is an improved version of Irwin's imputation; that is, $n^{(t+1)} = n_{obs}/(1 - \exp(-\lambda^{(t)}))$. Combining the E and M steps yields $\lambda^{(t+1)} = (s_1/n_{obs})(1 - \exp(-\lambda^{(t)}))$, which, like Irwin's method, converges to the maximum likelihood estimate, $\hat{\lambda} = .972$, as long as $\lambda^{(0)} > 0$. The fifth row of Table 1 gives the corresponding fit.

McKendrick's problem also highlights the celebrated idea of data augmentation when one adopts the binomial/Poisson mixture model. Because a complete sample is available from this model, there are no missing data in the traditional sense. Nonetheless, we can view the mixture/exposure indicator as missing data and construct the corresponding EM algorithm (Meng 1997). Purposely constructing missing data, such as mixture indicators, random effects, and latent factors, is a key contribution of Dempster et al. (1977) and has seen an enormous number of applications in statistical and scientific studies, as illustrated in Sections 4–6. This, along with a large number of recent improvements and extensions of EM (see Liu, Rubin, and Wu 1998; McLaughlan and Krishnan 1997; Meng and van Dyk 1997; and the references therein), has served to substantially increase the applicability and speed of EM-type algorithms.

2. MISSING-DATA MECHANISM

A profound difficulty in dealing with real-life missing-data problems is to reasonably understand and model the *missing-data mechanism* (MDM), namely the process that prevents us from observing our intended data. This process is a data selection process, like a sampling process; yet because it is typically not controlled by or even unknown to the data collector, it can be subject to all kinds of (hidden) biases, known collectively as *nonresponse bias*. Although the general theoretical foundation of sampling processes existed in the early part of the 20th century (e.g., Neyman 1934) and the impact of selection bias (e.g., from a purposive selection) has long been understood, the corresponding foundation for MDM was not formally developed until much later, starting with Rubin (1976a). Two key mechanisms introduced by Rubin, namely missing at random (MAR) and missing completely at random (MCAR), now appear in most statistical articles that contain analyses of incomplete data, often even without citation. These concepts have also been extended (see, e.g., Heitjan 1997, 1999; Heitjan and Rubin 1991).

Assuming MCAR basically means that we believe the observed data are a random subsample of the intended sample, and thus we can analyze it just as we analyze the intended sample, only with reduced size. Because this assumption is generally very far from the truth, common convenient approaches such as ignoring any case with missing values can be strongly biased (see, e.g., Little and Rubin 1987). MAR is a much weaker assumption, which allows the MDM to depend on observed quantities, but not on unobserved quantities. Under MAR, we can ignore the MDM in a likelihood inference based on the observed data without inducing nonresponse bias (but possibly inducing inefficiency when there is a priori dependence between the estimand and parameters governing the MDM, i.e., when the *parameter distinctness* assumption of Rubin (1976a) is violated; see Shih 1994). However, for sampling-based inference, it generally requires MCAR to ignore the MDM (see Heitjan and Basu (1996) for illustrations.)

When the MDM is not MAR (and thus not MCAR), the probability of missingness depends on the unobserved values themselves. The MDM is then generally not ignorable, meaning that the validity of our inference depends crucially on the particular model of the MDM we adopt. Because ignorability is fundamentally untestable from the observed data alone, one must exercise great caution when drawing substantive conclusions from any inference under a nonignorable model. Sensitivity analysis to the specification of an MDM model is a necessity, and subjective knowledge can play a critical role, as illustrated by Molenberghs, Goetghebeur, Lipsitz, and Kenward (1999). Modeling nonignorable MDMs is currently a very active research area with many open problems (see, e.g., Heitjan 1999; Ibrahim, Lipsitz, and Chen 1999; Mohlenberghs, Kenward, and Lesaffre 1997; Scharfstein, Rotnitzky, and Robins 1999).

3. MULTIPLE IMPUTATION AND UNCONGENIALITY

The common usage of *nonresponse bias* for general biases induced by an MDM reflects the historical fact that nonresponse in sample surveys is the most visible missing-data problem in general practice, especially in social sciences. Thanks to the efforts made by many statisticians and social scientists throughout the 20th century, we are seeing fewer and fewer articles using convenient missing-data "methods" such as mean imputation and complete case analyses without acknowledging their potential serious flaws. On the other hand, the simplicity of these "methods" is so attractive that preventing practitioners from being seduced requires scientifically and statistically more defensible methods with comparable simplicity. Multiple imputation (Rubin 1987) was motivated by this need. In the context of public-use or shared databases, the first step of Rubin's multiple imputation is to have the data collector build a sensible imputation model given available data and knowledge about the MDM, which are typically far more comprehensive than what could possibly be available to an average user (e.g., Barnard and Meng 1999; Meng 1994a; Rubin 1996). The data collector then draws M (e.g., 5–10) independent samples of all the missing values, as a set, from the imputation model, thereby creating M completed-data sets and thus permitting general users to directly assess and account for the increased variability/uncertainties due to nonresponse. For a user, analyzing a multiply imputed dataset means conducting M separate complete-data analyses, one for each of the M completed-data sets, and then combining these M completed-data analysis outputs using a few general rules. Readers are referred to Schafer (1999) for an updated tutorial; Gelman, King, and Liu (1998) for a recent application in public opinion polls; and Schafer (1997) for a comprehensive treatment of the practical implementation of Rubin's multiple imputation, including software.

In the context of public-use data files, there is a crucial separation between the data collector/imputer and general users. The two parties typically have different goals, data, information, and assessments and thus often adopt different models or even different modes of inference. Consequently, the imputation model is usually *uncongenial* to the user's analysis procedure; that is, the latter cannot be embedded into a (Bayesian) model that is compatible with the imputation model (Meng 1994a). One way to reduce this uncongeniality, of course, is to encourage more information exchange, such as having the imputer provide additional imputation quantities beyond the imputed datasets (e.g., Meng 1994a; Robins and Wang 2000; Schafer and Schenker 2000). Although this is clearly a direction for more research, the practical constraints (e.g., confidentiality, a user's choice of inferential mode) ensure the issue of uncongeniality will always remain. Consequently, much research is needed to establish a more flexible "multiparty" paradigm for comparing and evaluating statistical procedures, a paradigm that promotes most effective procedures given resource and practical constraints, rather than those that are misguided by impossible idealizations (see Rubin 1996), even if such idealizations are sensible in a congenial environment.

4. MCMC AND PERFECT SIMULATION

Constructing unobserved variables—namely, the method of data augmentation (e.g., Tanner and Wong 1987) or of auxiliary variables (e.g., Besag and Green 1993; Edwards and Sokal 1988)—has played a critical role in the development of efficient

Markov chain Monte Carlo (MCMC) algorithms. Some recent findings (e.g., Liu and Wu 1999; Meng and van Dyk 1999; van Dyk and Meng 2001) demonstrate the seemingly limitless potential of this method. Here I briefly describe one of its uses for perfect simulation, a rapidly growing area of MCMC—the list of references in *http://dimacs.rutgers.edu/~dbwilson/exact* is updated constantly.

Perfect simulation, or exact sampling, refers to a class of MCMC algorithms that in finite time provide genuine and independent draws from the limiting (i.e., stationary) distribution of a Markov chain. This seemingly impossible task was made possible by the *backward coupling* method of Propp and Wilson (1996) which, in a very rough sense, is a clever stochastic counterpart of the deterministic method for finding the optimizer by comparing the value of the objective function at each point. Consequently, this class of methods is most effective with finite-state chains, though they are by no means restricted to such chains (e.g., Green and Murdoch 1999; Murdoch 2000; Murdoch and Green 1998).

Indeed, data augmentation can help us to transform a continuous state-space problem into a finite one. For example, suppose that we are interested in simulating from $p(X)$ and we can augment this model to $p(X, Y)$ such that both $p(X|Y)$ and $p(Y|X)$ are easy to sample and the augmented variable Y is discrete. We can then implement a two-step Gibbs sampler, which induces a marginal Markov chain on Y. Because Y has a finite state space, in some cases we can directly implement the backward coupling method with this discrete chain to obtain iid draws from $p(Y)$. The desired iid draws from $p(X)$ are then obtained easily by drawing from $p(X|Y)$ given the draws of Y's. If the state space of Y is too large for the direct backward coupling method, then one can try multistage backward coupling (Meng 2000). An immediate application of this approach is to Bayesian finite mixtures (joint work with D. Murdoch), where Y is the subpopulation indicator. Readers are referred to Møller and Nicholls (1999) and the references therein for other methods of using discrete hidden variables to make perfect simulation effective for routine applications in statistics.

5. MENDELIAN LIKELIHOODS AND RELATIVE INFORMATION

It was exactly one century ago when Mendel's basic theory of heredity was re-discovered and gained general recognition (e.g., McPeek 1996; Thompson 1996), marking the real birth of modern genetics. The theory of Mendel (1866) provides general principles for probabilistic modeling of the inheritance of genes from parents to offspring. However, in common pedigree analyses, we typically miss some data on genotype information (e.g., allele types at some genetic markers), on the genealogi-cal tree (e.g., whether an allele was from the paternal side or the maternal side), and even on phenotype (e.g., a disease status of an ancestor). Consequently, Mendelian modeling and the associated computation are intrinsically problems of missing data, typically of very high dimension with exceedingly complex structures. Monte Carlo simulation is an effective general approach for such problems, but its efficiency depends on the choice of an underlying data augmentation scheme. Two common, and sometime competing, choices are genotypes and inheritance vectors/meiosis indicators, which indicate whether the origin of the gene is grandpaternal or grandmaternal (see Lander and Green 1987; Thompson 1994, 2000). Once we obtain draws of the missing data Y_{mis} from $p(Y_{\text{mis}}|Y_{\text{obs}}, \theta)$ (e.g., via MCMC), the computation of the observed-data Mendelian likelihood ratio $L(\theta_1|Y_{\text{obs}})/L(\theta_2|Y_{\text{obs}})$ can be dealt with effectively via bridge sampling (Bennett 1976; Jensen and Kong 1999; Meng and Wong 1996) and the reweighted mixture method of Geyer (1991). A currently challenging problem is to make such methods more accurate for computing the likelihood ratio when θ_1 and θ_2 belong to different marker regions (Thompson 2000), and the warp bridge sampling method (Meng 1999; Meng and Schilling 1999) is a possible direction to explore, because it increases efficiency by increasing the overlaps of the underlying densities by warping their shapes. The use of bridge sampling for assessing the convergence of Monte Carlo EM (Wei and Tanner 1990), which is useful for genetic linkage analysis (Guo and Thompson 1992), was detailed by Meng and Schilling (1996).

Another important missing-data problem in genetic linkage analysis is estimating the amount of information in the observed data *relative* to the total amount of information that would have been available had there been no missing data. The statistical literature on the fraction of missing information has been largely focused on its theoretical properties (e.g., Dempster et al. 1977; Liu 1994; Meng 1994b) and methodological uses (e.g., Meng and Rubin 1991) in computation and estimation. However, the focus here is more on design, with the aim of directly guiding follow-up strategies; for example, using more genetic markers with existing DNA samples versus collecting DNA samples from additional families, by assessing how much more information could be obtained if, say, we add more markers. An additional difference is that, because hypothesis testing is a useful screening tool for linkage studies (e.g., Thompson 1996), we need to measure the relative information in the context of hypothesis testing. This requires considering the roles of both the null hypothesis (i.e., no linkage) and the alternative hypothesis (as specified by a trait model). Although this issue does not arise in the estimation context, the basic identities given by Dempster et al. (1977) are fundamental for establishing a general theoretical framework for studying relative information in the context of hypothesis testing; details will be given elsewhere (as a joint paper with A. Kong and D. Nicolae).

6. MAPPING THE BRAIN AND THE UNIVERSE

Image reconstruction, a critical component in many medical and physical studies, is fundamentally another class of missing data problems. In the medical imaging context, perhaps the best-known example to statisticians is positron emission tomography (PET), for which the use of the EM algorithm signifies statisticians' direct involvement in the developing stage of the technique (e.g., Lange and Carson 1984; Shepp and Vardi 1982). The overview given by Vardi, Shepp, and Kaufman (1985), using brain mapping as an example, showed the intrinsic missing-data nature of PET, for we cannot directly observe the count of photons emitted from each pixel (i.e., a location in the brain). In addition, we face missing-data problems such as attenuation by the body's tissues and the escape of photons that travel along lines that do not intersect with any detector. As with linkage analysis, the choice of data augmentation schemes, or hidden-data spaces, has a direct impact on the speed of computation. As an example, Fessler and Hero (1995) discussed clever choices of hidden-data spaces that have made EM-type reconstructions more practical, overcoming the slowness of early EM reconstructions that use individual pixel counts as hidden states. An algorithmic analysis of the method of Fessler and Hero was given by Meng and van Dyk (1997) in the framework of the AECM algorithm.

Similar imaging techniques also play an important role in astrophysics, where the use of EM-type algorithms, such as the Richardson–Lucy algorithm (Lucy 1974; Richardson 1972), predates the publication of Dempster et al. (1977), though the development of fast EM and related Bayesian imaging algorithms has just begun (e.g., van Dyk 1999; van Dyk, Connors, Kashyap, and Siemiginowska 2001). The Poisson spectral imaging model, designed to analyze data from the Chandra observatory (launched on the space shuttle *Columbia,* July 1999) and other upcoming detectors, is an example of needing efficient methodologies for handling data from the new generation of high-resolution satellite telescopes. The Poisson model here is designed to summarize the relative frequency of photon energies (x-ray or γ-ray), collected as counts in a number of bins, arriving at a detector. The detected photons originate from many sources (e.g., a "continuum" and a number of "line profiles") and have been subject to background contamination, instrument response, and stochastic absorption. Each of these distortions requires a layer of modeling (e.g., Poisson, multinomial), forming an overall *multilevel* hierarchical model for the observed binned energies, a typical situation with real-data latent-variable modeling. Each of these levels, as well as any combination or function of them, is a candidate for data augmentation in fitting the model. An efficient choice can substantially improve the computational speed; van Dyk (1999) and van Dyk and Meng (1999) gave details and empirical evidences.

7. MILLENNIUM WISHES

The topic of missing data is as old and as extensive as statistics itself—after all, statistics is about knowing the unknowns. It is thus impossible in a few pages to discuss all of the main areas of past and present research. Areas not discussed here include, among many others, noniterative methods (e.g., Baker, Rosenberger, and DerSimonian 1992; Rubin 1976b), direct maximization of observed-data likelihoods (e.g., Molenberghs and Goetghebeur 1997), pattern-mixture models (e.g., Little 1993), bootstrap methods (e.g., Efron 1994), estimating equation approaches (e.g., Heyde and Morton 1996; Robins, Rotnitzky, and Zhao 1994; Lipsitz, Ibrahim, and Zhao 1999), and potential outcome in causal inferences (e.g., Barnard, Du, Hill, and Rubin 1998; Rubin 1978). Consequently, the 82 references listed in this article are really just the tip of the iceberg—even with many missing articles, Meng and Pedlow (1992) found more than 1,000 EM-related articles, about 85% of which were in nonstatistical journals. The number must have doubled by now.

Much remains to be done, however. The most pressing task, in my opinion, is placing further emphasis on the general recognition and understanding, at a conceptual level, of the necessity of properly dealing with the missing-data mechanism, as part of our ongoing emphasis on the importance of the data collection process in any meaningful statistical analysis. The missing-data mechanism is in the blood of statistics, and it is the nastiest and the most deceptive cell, especially for nonstatisticians—why on earth should anyone be concerned with data that one does not even have? I conclude with an excerpt from a referee's report of Tu, Meng, and Pagano (1994), to make one of my wishes for the new millennium. Reports like this will soon be of great value, but only on auction.

> The statement, "The naive approach of ignoring the missing data and using only the observed portion could provide very misleading conclusions" is nonsense to me (and I think the authors should also recognize it as nonsense in the real world). Similarly, what does it mean, "When analyzing such missing data, . . ."; if the data are missing, you can't analyze them. Except for old, rigid, demanding, clunky data treatment methods like the Yates algorithm (and except for the ridge problems discussed), it is unlikely that ". . . the analysis could still be very complicated due to the unbalanced structure of the observed data . . ." (page 4). Does any chemometrician every (sic) worry about making "it possible to utilize computer routines already developed for complete-data maximization"? I don't think any chemometricians every (sic) use data-specific data-treatment methods.

(To purchase a copy of this referee report, please dial M for Meng!)

REFERENCES

Baker, S. G., Rosenberger, W. F., and DerSimonian, R. (1992), "Closed-Form Estimates for Missing Counts in Two-Way Contingency Tables," *Statistics in Medicine*, 11, 643–657.

Barnard, J., Du, J., Hill, J. L., and Rubin, D. B. (1998), "A Broader Template for Analyzing Broken Randomized Experiments," *Sociological Methods and Research*, 27, 285–317.

Barnard, J., and Meng, X-L. (1999), "Applications of Multiple Imputation in Medical Studies: From AIDS to NHANES," *Statistical Methods in Medical Research*, 8, 17–36.

Bennett, C. H. (1976), "Efficient Estimation of Free Energy Differences From Monte Carlo Data," *Journal of Computational Physics*, 22, 245–268.

Besag, J., and Green, P. J. (1993), "Spatial Statistics and Bayesian Computation," *Journal of the Royal Statistical Society*, Ser. B, *Methodological*, 55, 25–37.

Böhning, D., Dietz, E., and Schlattmann, P. (1999), "The Zero-Inflated Poisson Model and the Decayed, Missing and Filled Teeth Index in Dental Epidemiology," *Journal of the Royal Statistical Society*, Ser. B, 162, 195–209.

Dempster, A. P., Laird, N. M., and Rubin, D. B. (1977), "Maximum Likelihood From Incomplete Data via the EM Algorithm" (with discussion), *Journal of the Royal Statistical Society*, Ser. B, 39, 1–37.

Edwards, R., and Sokal, A. (1988), "Generalization of the Fortuin–Kasteleyn–Swendsen–Wang Representation and Monte Carlo Algorithm," *Physical Review Letters*, 28, 2009–2012.

Efron, B. (1994), "Missing Data, Imputation, and the Bootstrap," *Journal of the American Statistical Association*, 89, 463–475.

Fessler, J. A., and Hero, A. O. (1995), "Penalized Maximum-Likelihood Image Reconstruction Using Space-Alternating Generalized EM Algorithm," *IEEE Transactions on Image Processing*, 4, 1417–1438.

Gelman, A., King, G., and Liu, C. (1998), "Not Asked and Not Answered: Multiple Imputation for Multiple Surveys" (with discussion), *Journal of the American Statistical Association*, 93, 846–874.

Geyer, C. (1991), "Reweighting Monte Carlo Mixtures," Technical Report 568, University of Minnesota, School of Statistics.

Green, P. J., and Murdoch, D. J. (1999), "Exact Sampling for Bayesian Inference: Towards General Purpose Algorithms," in eds. J. M. Bernardo, J. O. Berger, A. P. Dawid, and A. F. M. Smith, Bayesian Statistics 6, London: Oxford University Press.

Guo, S. W., and Thompson, E. A. (1992), "A Monte Carlo Method for Combined Segregation and Linkage Analysis," *American Journal of Human Genetics*, 51, 1111–1126.

Heitjan, D. F. (1994), "Ignorability in General Incomplete-Data Models," *Biometrika*, 81, 701–708.

—— (1997), "Ignorability, Sufficiency and Ancillarity," *Journal of the Royal Statistical Society*, Ser. B, 59, 375–381.

—— (1999), "Causal Inference in Clinical Trials, A Comparative Example," *Controlled Clinical Trials*, 20, 309–318.

Heitjan, D. F., and Basu, S. (1996), "Distinguishing 'missing at random' and 'missing completely at random'," *The American Statistician*, 50, 207–213.

Heitjan, D. F., and Rubin, D. B. (1991), "Ignorability and Coarse Data," *The Annals of Statistics*, 19, 2244–2253.

Heyde, C. C., and Morton, R. (1996), "Quasi-likelihood and Generalizing the EM Algorithm," *Journal of the Royal Statistical Society*, Ser. B, 58, 317–327.

Ibrahim, J. G., Lipsitz, S. R., and Chen, M-H. (1999), "Missing Covariates in Generalized Linear Models When the Missing Data Mechanism is Nonignorable," *Journal of the Royal Statistical Society*, Ser. B, 61, 173–190.

Irwin, J. O. (1963), "The Place of Mathematics in Medical and Biological Statistics," *Journal of the Royal Statistical Society*, Ser. B, 126, 1–45.

Jensen, C. S., and Kong, A. (1999), "Blocking Gibbs Sampling for Linkage Analysis in Large Pedigrees With Many Loops," *American Journal of Human Genetics*, 65, 885–901.

Kotz, S., and Johnson, N. L. (1997), *Breakthroughs in Statistics, Vol. III*, New York: Springer.

Lander, E. S., and Green, P. (1987), "Construction of Multilocus Genetic Linkage Maps in Humans," *Proceedings of the National Academy of Sciences*, 84, 2363–2367.

Lange, K., and Carson, R. (1984), "EM Reconstruction Algorithms for Emission and Transmission Tomography," *Journal of Computer Assisted Tomography*, 8, 306–316.

Lipsitz, S. R., Ibrahim, J. G., and Zhao, L. P. (1999), "A Weighted Estimating Equation for Missing Covariate Data With Properties Similar to Maximum Likelihood," *Journal of the American Statistical Association*, 94, 1147–1160.

Little, R. J. (1993), "Pattern-Mixture Models for Multivariate Incomplete Data," *Journal of the American Statistical Association*, 88, 125–134.

Little, R. J., and Rubin, D. B. (1987), *Statistical Analysis With Missing Data*, New York: Wiley.

Liu, C., Rubin, D. B., and Wu, Y. N. (1998), "Parameter Expansion for EM Acceleration—the PXEM Algorithm," *Biometrika*, 75, 755–770.

Liu, J. S. (1994), "Fraction of Missing Information and Convergence Rate of Data Augmentation," in *Computationally Intensive Statistical Methods: Proceedings of the 26th Symposium Interface*, pp. 490–497.

Liu, J. S., and Wu, Y. N. (1999), "Parameter Expansion Scheme for Data Augmentation," *Journal of the American Statistical Association*, 94, 1264–1274.

Lucy, L. B. (1974), "An Iterative Technique for the Rectification of Observed Distributions," *Astronomical Journal*, 79, 745–754.

McKendrick, A. G. (1926), "Applications of Mathematics to Medical Problems," *Proceedings of the Edinburgh Mathematics Society*, 44, 98–130.

McLaughlan, G. J., and Krishnan, T. (1997), *The EM Algorithm and Extensions*, New York: Wiley.

McPeek, M. S. (1996), "An Introduction to Recombination and Linkage Analysis," in *Genetic Mapping and DNA Sequencing*, eds. T. Speed and M. Waterman, New York: Springer, pp. 1–14.

Mendel, G. J. (1866), *Experiments in Plant Hybridisation*, English trans., Edinburgh: Oliver and Boyd, 1965.

Meng, X-L. (1994a), "Multiple Imputation Inferences With Uncongenial Sources of Input" (with discussion), *Statistical Sciences*, 9, 538–573.

——— (1994b), "On the Rate of Convergence of the ECM Algorithm," *The Annals of Statistics*, 22, 326–339.

——— (1997), "The EM Algorithm and Medical Studies: A Historical Link," *Statistical Methods in Medical Research*, 6, 3–23.

——— (1999), "Invited discussions of Matther Stephens's and Simon Tavaré's papers on Statistical and Computational Approaches to Genetic Evolution," in *Bulletin of the International Statistical Institute; 52nd Session, Helsinki*. Available at http://www.stat.fi/isi99/proceedings.html

——— (2000), "Toward a More General Propp–Wilson Algorithm: Multistage Backward Coupling," *Fields Institute Communications Series Vol. 26: Monte Carlo Methods*, American Mathematical Society.

Meng, X-L., and Pedlow, S. (1992), "EM: A Bibliographic Review With Missing Articles," in *Proceedings of the Statistical Computing Section, American Statistical Association*, pp. 24–27.

Meng, X-L., and Rubin, D. B. (1991), "Using EM to Obtain Asymptotic Variance-Covariance Matrices: The SEM Algorithm," *Journal of the American Statistical Association*, 86, 899–909.

Meng, X.-L., and Schilling, S. (1996), "Fitting Full-Information Item Factor Models and an Empirical Investigation of Bridge Sampling," *Journal of the American Statistical Association*, 91, 1254–1267.

——— (1999), "Warp Bridge Sampling," *Revised for The Journal of Computational and Graphical Statistics*.

Meng, X.-L., and van Dyk, D. A. (1997), "The EM Algorithm—An Old Folk Song Sung to a Fast New Tune" (with discussion), *Journal of the Royal Statistical Society*, Ser. B, 59, 511–567.

——— (1999), "Seeking Efficient Data Augmentation Schemes via Conditional and Marginal Augmentation," *Biometrika*, 86, 301–320.

Meng, X.-L., and Wong, W. H. (1996), "Simulating Ratios of Normalizing Constants via a Simple Identity: A Theoretical Explanation," *Statistica Sinica*, 6, 831–860.

Molenberghs, G., and Goetghebeur, E. (1997), "Simple Fitting Algorithms for Incomplete Categorical Data," *Journal of the Royal Statistical Society*, Ser. B, 59, 401–414.

Molenberghs, G., Goetghebeur, E. J., Lipsitz, S. R., and Kenward, M. G. (1999), "Nonrandom Missingness in Categorical Data: Strengths and Limitations," *The American Statistician*, 53, 110–118.

Molenberghs, G., Kenward, M. G., and Lesaffre, E. (1997), "The Analysis of Longitudinal Ordinal Data With Nonrandom Drop-Out," *Biometrika*, 84, 33–44.

Møller, J., and Nicholls, G. K. (1999), "Perfect Simulation for Sample-Based Inference," Preprint.

Murdoch, D. J. (2000), "Exact Sampling for Bayesian Inference: Unbounded State Space," *Fields Institute Communications Series Vol. 26: Monte Carlo Methods*, American Mathematical Association.

Murdoch, D. J., and Green, P. J. (1998), "Exact Sampling From a Continuous State Space," *Scandinavian Journal of Statistics*, 25, 483–502.

Neyman, J. (1934), "On the Two Different Aspects of the Representative Method: The Method of Stratified Sampling and the Method of Purposive Selection" (with discussion), *Journal of the Royal Statistical Society*, Ser. A, 97, 558–625.

Propp, J. G., and Wilson, D. B. (1996), "Exact Sampling With Coupled Markov Chains and Applications to Statistical Mechanics," *Random Structures and Algorithms*, 9, 1, 2, 223–252.

Richardson, W. H. (1972), "Bayesian-Based Iterative Methods of Image Restoration," *Journal of the Optical Society of America*, 62, 55–59.

Robins, J. M., Rotnitzky, A., and Zhao, L. P. (1994), "Estimation of Regression Coefficients When Some Regressors Are Not Always Observed," *Journal of the American Statistical Association*, 89, 846–866.

Robins, J. M., and Wang, N. (2000), "Inference for Imputation Estimators," *Biometrika*, 87, 113–124.

Rubin, D. B. (1976a), "Inference and Missing Data," *Biometrika*, 63, 581–592.

——— (1976b), "Noniterative Least Squares Estimates, Standard Errors and F Tests for Analyses of Variance With Missing Data," *Journal of the Royal Statistical Society*, Ser. B, 38, 270–274.

——— (1978), "Bayesian Inference for Causal Effects: The Role of Randomization," *The Annals of Statistics*, 6, 34–58.

——— (1987), *Multiple Imputation for Nonresponse in Surveys*, New York: Wiley.

——— (1996), "Multiple Imputation After 18+ Years" (with discussion), *Journal of the American Statistical Association*, 91, 473–489.

Schafer, J. L. (1997), *Analysis of Incomplete Multivariate Data*, London: Chapman and Hall.

——— (1999), "Multiple Imputation: A Primer," *Statistical Methods in Medical Research*, 8, 3–15.

Schafer, J. L., and Schenker, N. (2000), "Inference With Imputed Conditional Means," *Journal of the American Statistical Association*, 95, 144–154.

Scharfstein, D. O., Rotnitzky, A., and Robins, J. M. (1999), "Adjusting for Nonignorable Drop-out Using Semiparametric Nonresponse Models" (with discussion), *Journal of the American Statistical Association*, 94, 1096–1146.

Shepp, L. A., and Vardi, Y. (1982), "Maximum Likelihood Reconstruction for Emission Tomography," *IEEE Transactions on Image Processing*, 2, 113–122.

Shih, W. J. (1994), Discussion of "Informative Drop-Out in Longitudinal Data Analysis," by Diggle and Kenward, *Applied Statistics*, 43, 87.

Tanner, M. A., and Wong, W. H. (1987), "The Calculation of Posterior Distributions by Data Augmentation" (with discussion), *Journal of the American Statistical Association*, 82, 528–550.

Thompson, E. A. (1994), "Monte Carlo Likelihood in Genetic Mapping," *Statistical Science*, 9, 355–366.

——— (1996), "Likelihood and Linkage: From Fisher to the Future," *The Annals of Statistics*, 24, 449–465.

——— (2000), *Statistical Inference From Genetic Data*, Heyward, CA: Institute of Mathematical Statistics.

Tu, X. M., Meng, X-L., and Pagano, M. (1994), "On the Use of Conditional Maximization in Chemometrics," *Journal of Chemometrics*, 8, 365–370.

van Dyk, D. A. (1999), "Fast New EM-Type Algorithms With Applications in Astrophysics," technical report, Department of Statistics, Harvard University.

van Dyk, D. A., Connors, A., Kashyap, V. L., and Siemiginowska, A. (2001), "Analysis of Energy Spectrum With Low Photon Counts via Bayesian Posterior Simulation," *Astrophysical Journal*, to appear.

van Dyk, D. A., and Meng, X-L. (1999a), "Algorithms Based on Data Augmentation: A Graphical Representation and Comparison," in *Models, Predictions, and Computing: Proceedings of the 31st Symposium on the Interface*, eds. M. Pourahmadi and K. Berk, pp. 230–239.

——— (2001), "The Art of Data Augmentation" (with discussion), *Journal of Computational and Graphical Statistics*, to appear.

Vardi, Y., Shepp, L. A., and Kaufman, L. (1985), "A Statistical Model for Positron Emission Tomography," *Journal of the American Statistical Association*, 80, 8–19.

Wei, G., and Tanner, M. A. (1990), "A Monte Carlo Implementation of the EM Algorithm and the Poor Man's Data Augmentation Algorithm," *Journal of the American Statistical Association*, 85, 699–704.

A Robust Journey in the New Millennium

Stephen Portnoy and Xuming He

A brief search through *Current Index to Statistics* found more than 3,000 entries with "robust" or "robustness" as key words. That this number exceeded the number of appearances for "density," "rank," "bootstrap," "censored," or "smoothing" suggests that the name of this field might also be an appropriate appellation for those doing the research. Here we present a very personal view of a few of the most fundamental insights offered by the study of robustness and suggest some promising areas for future robust travel.

1. A PAGE OF HISTORY IS WORTH A VOLUME OF LOGIC (Oliver Wendell Holmes, 1929)

A common view sees robustness as providing alternatives to least squares methods and Gaussian theory. Indeed, this approach may be justified by noting that mathematical and computational convenience formed the basis for the Gaussian assumptions of normality and quadratic loss. In fact, a statistical procedure derived from the least squares principle may be highly nonrobust against deviations from normality. For example, Tukey (1960) showed that contamination by just two observations from a $N(0, 9)$ distribution among 1,000 standard normal observations suffices for the mean absolute deviation estimator to be more efficient than the sample standard deviation, which is asymptotically optimal for the Gaussian scale parameter. What we learn from such examples goes a long way in explaining the extent of research in robustness mentioned in the abstract.

More generally, robustness can be viewed from many different perspectives, and its implications have never been constant in history. Box and Tiao (1962) distinguished two types of robustness: criterion robustness and inference robustness. The former selects a criterion for statistical optimality and then investigates its variation as the parent distribution deviates from the form assumed. The latter concerns changes in

Stephen Portnoy is Professor and Xuming He is Associate Professor, Department of Statistics, University of Illinois at Urbana-Champaign, Champaign, IL 61820. Stephen Portnoy's research was partially supported by National Science Foundation grant DMS97-03758. Xuming He's research was partially supported by National Science Foundation grant SBR96-17278. This vignette is a personal and subjective view of what the authors believe makes robustness enjoyable and useful. Like all since Newton, they note that they have stood on the shoulders of giants, the large majority of whom they have not formally referenced here. They hope those whom they have slighted will accept this general acknowledgment as sincere thanks, and beg forgiveness for including so many of their own works among a few seminal papers and some others that they feel should be better known.

quantities leading to inference (significance levels, coverage probability, etc.). Later researchers have studied efficiency robustness, qualitative robustness, bias robustness, Bayesian robustness, and so on.

It is noteworthy that the use of least absolute deviations in a regression problem predates least squares by half a century. Stigler (1986) discussed this approach of Boscovich (ca. 1755) and other early alternatives to least squares. More modern attempts to bypass the strict normality assumption include Pearson curves and other super-models. The importance of using robust statistical methods was recognized by eminent statisticians like Pearson, Student, Box, and Tukey. However, the modern theory of robust statistics emerged more recently, led by Huber's (1964) classic minimax approach and Hampel's (1968, 1974) introduction of influence functions. Since then, several areas of high growth have revolutionized the way statisticians understand the power of robustness.

Huber (1964) might be considered the first to formalize robustness as "approximate validity of a parametric model." Almost all parametric models used in statistical analysis are approximations to reality. A statistical procedure that is optimal at one single model is easy to come by but not always trustworthy. Huber introduced a contamination model and several distance-based neighborhoods, arguing that a good method of estimation should behave well in a full nonparametric neighborhood of the assumed model. This obviates the need to construct a finite parameter super-model and provides protection against small departures in arbitrary directions from a nominal model. Furthermore, Huber introduced specific optimality criteria (e.g., minimax asymptotic variance) to develop robust procedures with good finite-sample properties (see Andrews et al. 1972).

Hampel took a different approach by studying the properties of statistical functionals. A statistical functional maps from the space of probability distribution functions to a parameter space of interest. Typically an estimator or test statistic can be viewed as a statistical functional evaluated at the empirical distribution. This perspective was not totally new, but it provided the springboard for the development of three important robustness concepts: qualitative robustness (continuity), influence function (derivative), and breakdown point. Hampel also proposed finding estimators with optimal efficiency given an upper bound on the influence function (see Hampel, Ronchetti, Rousseeuw, and Stahel 1986).

Early robustness studies focused on generalized maximum likelihood estimators, commonly referred to as M estimators. They are obtained either by minimizing a loss function or solving the associated first-order conditions. Robust choices of the loss function can be motivated from both Huber's and Hampel's approaches for the location-scale models. In the regression setting, He (1991) showed that obtaining local robustness requires using information beyond the residuals. Generalized M estimators use design-based weights on the residuals to achieve this goal.

Although Huber's original focus was on variance, a combination of the neighborhood concept and Hampel's functional perspective has led to a more systematic study of bias robustness. For instance, Donoho and Liu (1988) and He and Simpson (1993) provided global lower bounds for the bias of statistical functionals under a variety of neighborhoods and showed that minimum distance estimators are often close to achieving minimax bias. He and Simpson also showed that for high-dimensional parameters, globally minimax bias functionals have to compromise for lack of local linearity, suggesting that no optimality criteria (including robustness ones) should be pushed to their limits. Nevertheless, the use of bias curves for estimating functionals has led to a number of new proposals (see, e.g., Maronna, Yohai, and Zamar 1993), and the breakdown functions of He, Simpson, and Portnoy (1990) provided a means for global robustness assessment of statistical tests.

A robust procedure is expected to work well when there are outliers in the data that are hard to detect. Traditional methods for outlier detection and diagnostics often suffer from masking; that is, the appearance of several outliers whose similarity frustrates the ability of "delete-k" methods to identify them. As a formal notion in robustness, the breakdown point was made popular at least in part by Donoho and Huber (1983). Roughly speaking, it is the smallest proportion of outliers that would in the worst case drive the estimator beyond all bounds. Researchers were intrigued by the fact that all M estimators have low breakdown points in problems with multiple parameters. Rousseeuw (1984) opened the door for finding affine-equivariant estimators with a breakdown point that does not decrease with dimension. A number of high-breakdown estimators for regression and for multivariate location-scatter models have been proposed since then. Estimators that combine local stability, good efficiency at the central model, and a high breakdown point are generally believed to be superior. Theoretically, a one-step Newton iteration toward a highly efficient generalized M estimator starting from a high breakdown fit would do the job (see Jurečková and Portnoy 1987).

Most people view robust statistics as parametric. That is, robust statistics is concerned with deviations from a nominal parametric model. Closely related to this is what we call nonparametric methods. They include distribution-free methods, such as rank tests, that are correct for a large class of distributions (see the vignette on robust nonparametric methods by Hettmansperger, McKean, and Sheather). We use the following example to illustrate the difference between parametric robustness and nonparametric statistics. Suppose that a random sample is drawn from a mixture model with 90% from N(0, 1) and the remaining 10% from N(t, 1) for some large value of t. If we use a central model of N(θ, 1), then a robust estimate of location would aim at the center of the majority; that is, $\theta = 0$. Of course, the estimate will be biased from the contamination, but a robust estimate tries to control the bias regardless of the size of t. A nonparametric estimate of location would aim at the center of the mixture distribution, which is not 0 in this case. Therefore, it is important to know what we are seeking. If we believe in a parametric model that approximates reality and wish to estimate the parameters associated with the model, then robust estimates are our best bet. On the other hand, if we consider the data as a sample from a single population and are interested in a population quantity, then a nonparametric estimate is more appropriate (see Portnoy and Welsh 1992).

It is not unusual for a nonparametric procedure to be robust at certain models. For example, the median is a nonparametric statistic, but it is also a highly robust estimate for estimating the center of a symmetric distribution as a central model. The same statistic plays two different roles.

Under our view of nonparametrics, we would like to emphasize estimators that are consistent for their population parameters under a large class of distributions. Such estimators may or may not be robust for parametric estimation, but there are well-defined and well understood population quantities being estimated. In linear models, robust estimators are typically consistent for the slope parameters when the errors are iid. For more general error structures, the regression quantiles of Koenker and Bassett (1978) provide consistent estimates of conditional quantiles in a semi-parametric sense. Though most useful for analyzing general forms of variability (e.g., heteroscedasticity), regression quantiles also provide a natural ordering of structured data that may be used to construct robust (trimmed) estimates (see, e.g., Ruppert and Carroll 1980; Koenker and Portnoy 1987). Note that regression quantiles are robust against gross errors in the response but not in the design.

For multivariate data, efforts have been made to construct the ranks and estimate the median and other depth contours. Strictly speaking, ranking itself is a univariate concept. In the multivariate case, it is possible to generalize the concept of ranking by starting from the center and moving outward. A number of proposals on multivariate ranks and on data depth are now available, and further research is sure to lead to consolidation and new standards in this area.

In historical summary, robustness has provided at least three major insights into statistical theory and practice:

1. Relatively small perturbations from nominal models can have very substantial deleterious effects on many commonly used statistical procedures and methods (as in Tukey's example cited at the beginning of this article).

2. There is a gain in both clarity and generality in considering nonparametric neighborhoods of a nominal model, and such considerations often suggest optimal procedures that are especially natural and simple. Huber's minimax M estimator serves as a good example in this regard.

3. Robust methods are needed for detecting or accommodating outliers in the data. (A number of such successful applications are mentioned in the rejoinder to Rousseeuw and Hubert 1999.)

The discussions in and of Stigler (1977) marked an era of anxiety when we ask whether robust methods really work with real data. By now, robust methods of various forms have been in use in almost all areas of statistics and biostatistics and in a broad variety of scientific fields, including recent ones like computer vision. The question today is how they can be better used and more widely appreciated.

2. THE FUTURE LIES AHEAD
(Attributed to Governor Brown, 1966)

The best advice for prognosticators is the proverbial, "buy low, sell high." However, in the statistical market, mature stocks have gone through their period of explosive growth. In a robust market, some such stocks may continue to pay high dividends. Some candidates for the advice, "buy high, sell higher," are the following.

Computational issues, especially for moderately large problems (i.e., with sample size $n > 1,000$ and parametric dimension $p > 10$), are sure to remain a major focus of attention. Computation of an estimator maximizing a multimodal objective function or solving an equation with multiple roots is generally problematic. When an extensive search is not practical, approximate algorithms are often used. Some progress has been made in developing general algorithms (e.g., simulated annealing, genetic algorithms) and in finding faster algorithms to compute specific robust estimators. As a result, a number of high-breakdown estimators, such as the least trimmed squares and the minimum covariance determinant estimator (see Rousseeuw and Van Driessen 1999), can now be approximately computed much faster than before. However, if the search is based on random subsamples instead of on all possible candidates, then the resulting estimator may not share the statistical properties of the mathematical solution. For example, the idea of a "deepest" point has been used as a robust descriptive means of identifying the center of a set of observations. Recent work of Rousseeuw and Hubert (1999) not only has generalized depth to the linear regression setting, but also has provided a way of defining an estimator of a conditional quantile having high breakdown and being root-n consistent. There are computational difficulties, especially if one is to maintain asymptotic distributional properties (as we believe is necessary). Though it seems possible to obtain a consistent pseudomaximal depth quantile estimator by searching over ever larger finite subsets of elemental solutions, the asymptotics developed by Bai and He (1999) would not hold unless the size of the subsets grows rapidly with the sample size (faster than the order of $n^{p/2}$). This growth surely precludes computation in moderately large problems. Clearly, new approaches need to be developed in the future. For estimators of this type, it seems more hopeful to find computational methods that make good use of the specific properties of the estimators than to rely on advances in general-purpose optimization algorithms.

Even for convex objective functions, computation in very large samples can be challenging. Traditionally, computation in such large problems has been restricted to least squares methods because of timing considerations. Portnoy and Koenker (1997) recently showed that a combination of stochastic preprocessing and modern interior point methods can provide an algorithm for L_1 estimators in statistical linear models that is faster than least squares for problems where $np > 10^6$ (if p is not too large compared with n). In general, stochastic preprocessing steps can be extremely useful both in terms of computation speed, and the ability to analyze the resulting estimator in statistical terms. Improved computation should permit robust procedures to be used in computer-intensive areas like data mining.

Asymptotics for increasing parametric dimension also seems a good bet to remain an active area of research. As suggested by substantial anecdotal evidence and shown empirically by Koenker (1988), parametric dimension tends to grow inexorably as the sample size increases. A fairly satisfactory theory for certain M estimators in regression settings was developed by Portnoy (1985) and many others subsequently, and some other special cases have been considered (e.g., contingency tables, exponential families, and item response models). However, similar results for more general procedures in more general models remain to be developed. Present results suggest that if p is the number of parameters and n is the sample size, then consistency might be anticipated if $p/n \to 0$. However, asymptotic normality seems to require at least $p^2/n \to 0$, except in cases of somewhat special symmetry. Results for nonsmooth estimating equations (like those for regression quantiles) appear especially difficult, and may require $p^3/n \to 0$. Nonetheless, questions of large p asymptotics arise naturally in a wide variety of multivariate situations, and further work, though theoretically difficult, will certainly be needed.

Certain minimum distance estimators are known for their robustness without compromising on asymptotic efficiency. For example, the minimum Hellinger distance estimator is first-order equivalent to the maximum likelihood estimator but enjoys a bounded influence function and a high breakdown point at a variety of parametric models. This is particularly useful for models with discrete data (see Simpson 1987). For models with continuous data, a smooth density estimate appears to be needed to compute the Hellinger distance (see Beran 1977; Lindsay 1994). We expect minimum distance methods to continue to play a role in the development of robust statistics.

Research on depth and ranks for multivariate data is expected to continue. There is evidence that work in this area will lead to the development of rank-based multivariate distribution-free tests and to the understanding of multivariate regression quantiles. Although they do not fall into the parametric framework of robust statistics, their interplay with robust statistics must not be ignored. In pursuit of robust procedures, statisticians have traditionally taken the view of robustifying Gaussian likelihood. It is also possible to use likelihood directly at a non-Gaussian model. The t-likelihood has been used in constructing robust estimates in regression (see He, Simpson, and Wang 2000) and in mixed models (see Welsh and Richardson 1997). This approach offers some distinctive advantages, including the common appeal of likelihood and the availability of an objective function (as compared to an estimating equation with multiple roots). It is highly likely that a shift from super-models to robust models will offer applied statisticians a more satisfactory solution to robustness needs.

Extensions and applications of insights from robustness will continue to be made in all areas of statistical research. Clearly, work will continue in such areas as time series and multivariate analysis seeking procedures robust to distributional perturbation. The problem of outliers has also been recognized recently in the data mining arena. Some other less studied areas include sequential analysis and design of experiments. More fundamentally, as nominal models become more complex, robustness to other model departures becomes more important. For example, in nonstationary models, each observation has its own distribution to be perturbed. Some work on local bias has application here, and certain models for heteroscedasticity have been proposed (see Koenker and Zhao 1996) for a recent application to heteroscedastic time series models). However, it seems extremely hard to imagine how to deal with a fully nonparametric neighborhood over which all distributions are permitted to vary. Dependent situations provide a similar problem: Though it is possible to vary the marginal distribution in stationary time series models, the analysis of fully nonparametric neighborhoods in the space of n-dimensional distributions seems remote even for stationary nominal models, and may well be hopeless in the nonstationary case. Nonetheless, the importance and ubiquity of such models suggest the need for further research and lead us to our final, more speculative prognostications.

As noted earlier, there has been considerable work on finding procedures robust to marginal distributional perturbation in various dependent models. However, since the early work of Gastwirth and Rubin (1971) and Portnoy (1977), relatively little research has been done on robustness to dependent departures from a nominal model. The problem is defining an appropriate and analyzable neighborhood of dependent models. One approach from the 1980s considers stationary procedures whose spectra lie in certain "bandwidth-bounded" sets (e.g., Kassam and Poor 1985). The least favorable distributions are the standard time-domain autoregressive moving average (ARMA) models. However, in most statistical applications, it is extremely difficult to relate intuitive properties of reasonable dependent perturbations to the bandwidth bounds on the spectra. In fact, these bandwidth-bounded processes seem so remarkably special as to be inappropriate for most robustness considerations. In our opinion, fundamentally unsolved problems remain in developing appropriate families of alternative dependent models and phrasing appropriate questions whose answers would clarify robustness. Progress in this area might provide substantial insights and improvements in the development of statistical analyses that are not sensitive to the presence of small but otherwise unknown dependence. Random effects and mixed models also present a class of models where remarkably little progress has been made on the robustness front. The difficulty seems clear: Once normality is relaxed, the choice of an appropriate scale parameter is unclear. Furthermore, nice models for scale parameters mesh poorly with the additive model for components of variance. For example, in "repeated measurement" (or "panel data") models with the effect of the observation unit being random, a linear model for conditional quantiles at each time point will no longer be linear across time, because of convolution with the random effect (except for normal models). Some authors have robustified estimates of variance components or studied the robustness of standard methods (see, e.g., Richardson and Welsh 1995), but variances have no special place in nonnormal models. Thus, in addition to the problem of identifying appropriate neighborhoods, there is the more fundamental problem of finding and making sense of appropriate parameters. It is likely that a particular definition must be tailored to each application.

This incomplete list of potential research suggests productive and exciting times ahead for the robust statistician. We eagerly anticipate the continuing journey.

REFERENCES

Andrews, D. F., Bickel, P. J., Hampel, F. R., Huber, P. J., Rogers, W. H., and Tukey, J. W. (1972), *Robust Estimates of Location*, Princeton, NJ: Princeton University Press.

Bai, Z. D., and He, X. (1999), "Asymptotic Distributions of the Maximal Depth Regression and Multivariate Location," *The Annals of Statistics*, 27, 1616–1637.

Beran, R. (1977), "Minimum Hellinger Distance Estimates for Parametric Models," *The Annals of Statistics*, 5, 445–463.

Box, G. E. P., and Tiao, G. C. (1962), "A Further Look at Robustness via Bayes Theorem," *Biometrika*, 49, 419–432.

Donoho, D. L., and Huber, P. J. (1983), "The Notion of Breakdown Point," in *A Festschrift for Erich L. Lehmann*, Belmont, CA: Wadsworth, Inc., pp. 157–184.

Donoho, D., and Liu, R. (1988), "The 'Automatic' Robustness of Minimum Distance Functionals," *The Annals of Statistics*, 21, 552–586.

Gastwirth, J., and Rubin, H. (1971), "Effect of Dependence on the Level of Some One-Sample Tests," *Journal of the American Statistical Association*, 66, 816–820.

Hampel, F. R. (1968), "Contributions to the Theory of Robust Estimation," Ph. D. thesis, University of California, Berkeley.

———— (1974), "The Influence Curve and Its Role in Robust Estimation," *Journal of the American Statistical Association*, 69, 383–393.

Hampel, F. R., Ronchetti, E. M., Rousseeuw, P. J., and Stahel, W. A. (1986), *Robust Statistics, The Approach Based on Influence Functions*, New York: Wiley.

He, X. (1991), "A Local Breakdown Property of Robust Tests in Linear Regression," *Journal of Multivariate Analysis*, 38, 294–305.

He, X., and Simpson, D. G. (1993), "Lower Bounds for Contamination Bias: Globally Minimax Versus Locally Linear Estimation," *The Annals of Statistics*, 21, 314–337.

He, X., Simpson, D. G., and Portnoy, S. (1990), "Breakdown Robustness of Tests," *Journal of the American Statistical Association*, 85, 446–452.

He, X., Simpson, D. G., and Wang, G. Y. (2000), "Breakdown Points of t-Type Regression Estimators," *Biometrika*, 87, 675–687.

Huber, P. (1964), "Robust Estimation of a Location Parameter," *The Annals of Mathematical Statistics*, 35, 73–101.

Jurečková, J., and Portnoy, S. (1987), "Asymptotics for One-Step M-Estimators in Regression With Application to Combining Efficiency and High Breakdown Point," *Communication of Statistics, Theory and Methods*, 16, 2187–2200.

Kassam, S., and Poor, H. V. (1985), "Robust Techniques for Signal Processing: A Survey," *Proceedings of the IEEE*, 73, 433–481.

Koenker, R. (1988), "Asymptotic Theory and Econometric Practice," *Journal of Applied Econometrics*, 3, 139–147.

Koenker, R., and Bassett, G. (1978), "Regression Quantiles," *Econometrica*, 46, 33–50.

Koenker, R., and Portnoy, S. (1987), "L-Estimation for Linear Models," *Journal of the American Statistical Association*, 82, 851–857.

Koenker, R., and Zhao, Q. (1996), "Conditional Quantile Estimation and Inference for ARCH Models," *Econometric Theory*, 12, 793–813.

Lindsay, B. G. (1994), "Efficiency Versus Robustness: The Case for Minimum Hellinger Distance and Related Methods," *The Annals of Statistics*, 22, 1081–1114.

Maronna, R., Yohai, V., and Zamar, R. (1993), "Bias-Robust Regression Estimation: A Partial Survey," in *New Directions in Statistical Data Analysis and Robustness*, Boston: Birkhauser, pp. 157–176.

Portnoy, S. (1977), "Robust Estimation in Dependent Situations," *The Annals of Statistics*, 5, 22–43.

—— (1985), "Asymptotic Behavior of M-Estimators of p Regression Parameters When p^2/n is Large; II. Normal Approximation," *The Annals of Statistics*, 13, 1403–1417.

Portnoy, S., and Koenker, R. (1997), "The Gaussian Hare and the Laplacian Tortoise: Computability of Squared-Error and Absolute-Error Estimation," *Statistical Science*, 12, 279–300.

Portnoy, S., and Welsh, A. (1992), "Exactly What is Being Modelled by the Systematic Component in a Heteroscedastic Linear Regression," *Statistical Probability of Letters*, 13, 253–258.

Richardson, A. M., and Welsh, A. H. (1995), "Robust Restricted Maximum Likelihood in Mixed Linear Models," *Biometrics*, 51, 1429–1439.

Rousseeuw, P. J. (1984), "Least Median of Squares Regression," *Journal of the American Statistical Association*, 79, 871–880.

Rousseeuw, P. J., and Hubert, M. (1999), "Regression Depth" (with discussion), *Journal of the American Statistical Association*, 94, 388–433.

Rousseeuw, P. J., and Van Driessen, K. (1999), "A Fast Algorithm for the Minimum Covariance Determinant Estimator," *Technometrics*, 41, 212–223.

Ruppert, D., and Carroll, R. (1980), "Trimmed Least Squares Estimation in the Linear Model," *Journal of the American Statistical Association*, 75, 828–838.

Simpson, D. G. (1987), "Minimum Hellinger Distance Estimation for the Analysis of Count Data," *Journal of the American Statistical Association*, 82, 802–807.

Stigler, S. M. (1977), "Do Robust Estimators Work With Real Data?" (with discussion), *The Annals of Statistics*, 5, 1055–1077.

—— (1986), *The History of Statistics*, Cambridge, MA: Belknap Press.

Tukey, J. (1960), "A Survey of Sampling from Contaminated Distributions," in *Contributions to Probability and Statistics*, ed. I. Olkin, Stanford, CA: Stanford University Press.

Welsh, A. H., and Richardson, A. M. (1997), "Approaches to the Robust Estimation of Mixed Models. Robust Inference," in *Handbook of Statistics*, 15, Amsterdam: North-Holland.

Likelihood

N. Reid

1. INTRODUCTION

In 1997 a study conducted at the University of Toronto concluded that the risk of a traffic accident increased by four-fold when the driver was using a cellular telephone (Redelmeier and Tibshirani 1997a). The report also stated that such a large increase was very unlikely to be due to chance, or to unmeasured confounding variables, although the latter could not be definitively ruled out. Figure 1(a) shows the *likelihood function* for the important parameter in the investigators' model, the relative risk of an accident. Figure 1(b) shows the log of the likelihood function plotted against the log of the relative risk. The likelihood function was the basis for the inference reported. (The point estimate of relative risk from the likelihood function is actually 6.3, although 4.0 was the reported value. The maximum likelihood estimate was downweighted by a method devised to accommodate some complexities in the study design.) As with most real life studies, there were a number of decisions related first to data collection, and then to modeling the observed data, that involved considerable creativity and a host of small but important decisions relating to details of constructing the appropriate likelihood function. A nontechnical account of some of these was given by Redelmeier and Tibshirani (1997c), and a more statistically oriented version was given by Redelmeier and Tibshirani (1997b). In this vignette I am simply using the data to provide an illustration of the likelihood function.

Assume that one is considering a parametric model $f(\mathbf{y}; \boldsymbol{\theta})$, which is the probability density function with respect to a suitable measure for a random variable $\mathbf{Y}$. The parameter is assumed to be k-dimensional and the data are assumed to be n-dimensional, often representing a sequence of iid random variables: $\mathbf{Y} = (Y_1, \ldots Y_n)$. The *likelihood function* is defined to be a function of $\boldsymbol{\theta}$, proportional to the model density,

$$L(\boldsymbol{\theta}) = L(\boldsymbol{\theta}; \mathbf{y}) = cf(\mathbf{y}; \boldsymbol{\theta}), \tag{1}$$

where c can depend on $\mathbf{y}$ but not on $\boldsymbol{\theta}$. Within the context of the given parametric model, the likelihood function measures the relative plausibility of various values of $\boldsymbol{\theta}$, for a given observed data point $\mathbf{y}$. The notation for the likelihood function emphasizes that the parameter $\boldsymbol{\theta}$ is the quantity that varies, and that the data value

N. Reid is Professor, Department of Statistics, University of Toronto, Ontario, Canada M5S 3G3 (E-mail: reid@utstat.utoronto.ca).

is considered fixed. The constant of proportionality in the definition is needed, for example, to accommodate one-to-one transformations of the random variable Y that do not involve θ, as these clearly should have no effect on our inference about θ. Another way to say the same thing is that the likelihood function is not calibrated in θ, or that only relative values $L(\theta_1)/L(\theta_2)$ are well determined.

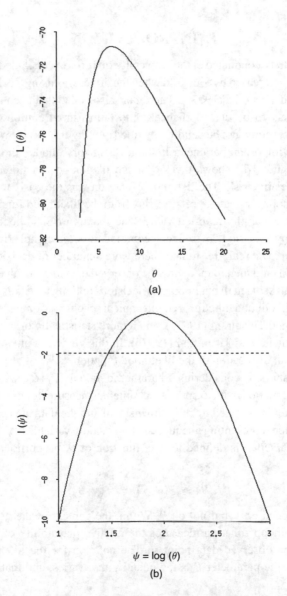

Figure 1. Likelihood and Log-Likelihood Function for the Relative Risk; Based on Redelmeier and Tibshirani (1997c).

The likelihood function was proposed by Fisher (1922) as a means of measuring the relative plausibility of various values of θ by comparing their likelihood ratios. When θ is one- or two-dimensional, the likelihood function can be plotted and provides a visual assessment of the set of likelihood ratios. Several authors, beginning with Fisher, suggested that ranges of plausible values for θ can be directly determined from the likelihood function, first by determining the *maximum likelihood estimate,* $\hat{\theta} = \hat{\theta}(\mathbf{y})$, the value of θ that maximizes $L(\theta; \mathbf{y})$, and then using as a guideline

$$L(\hat{\theta})/L(\theta) \in (1, 3), \quad \text{very plausible;}$$

$$L(\hat{\theta})/L(\theta) \subset (3, 10), \quad \text{somewhat implausible;}$$

and

$$L(\hat{\theta})/L(\theta) \in (10, \infty) \quad \text{highly implausible.}$$

The ranges suggested here are taken from Kass and Raftery (1995), attributed to Jeffreys. Other authors have suggested different cutoff points; for example, Fisher (1956, p. 71) suggested using 2, 5, and 15, and Royall (1997) suggested 4, 8, and 32. General introductions to the definition of the likelihood function and its informal use in inference were given by Fisher (1956), Edwards (1972), Kalbfleisch (1985), Azzalini (1996), and Royall (1997).

2. LIKELIHOOD FUNCTION AND INFERENCE

2.1 Bayesian Inference

Although the use of likelihood as a plausibility scale is sometimes of interest, probability statements are usually preferred in applications. The most direct way to obtain these is by combining the likelihood with a *prior probability* function for θ, to obtain a posterior probability function,

$$\pi(\theta|\mathbf{y}) \propto \pi(\theta)L(\theta; \mathbf{y}), \tag{2}$$

where the constant of proportionality is $\int \pi(\theta)L(\theta; \mathbf{y}) \, d\theta$.

This leads directly to Bayesian inferences of this sort: Using the prior density $\pi(\theta)$, we conclude that values of θ greater than θ_U have posterior probability less than .05 and are hence inconsistent with the model and the prior. Jeffreys (1961) emphasized this use of the likelihood function, and investigated the possibility of using "flat" or "noninformative" priors. He also suggested the plausibility range described earlier, in the context of Bayesian inference with a flat prior.

One difficulty in applying Bayesian inference is in constructing a suitable prior, and interest has been renewed in the construction of noninformative priors, which lead to posterior probability intervals that are in one way or another minimally affected by the prior density. One example of a noninformative prior is one for which the posterior probability limit θ_U described in the previous paragraph does in fact lead to an interval that, when considered as a confidence interval, has (at least approximately) coverage equal to its posterior probability content. If θ is a scalar parameter, then the appropriate prior is Jeffreys's prior $\pi(\theta) \propto \{i(\theta)\}^{1/2}$, where $i(\theta)$ is the Fisher information in the model $f(y; \theta)$,

$$ i(\theta) = E\left\{ \frac{\partial L(\theta; Y)}{\partial \theta} \right\}^2 = \int \left\{ \frac{\partial L(\theta; y)}{\partial \theta} \right\}^2 f(y; \theta)\, dy. \tag{3} $$

This result was derived by Welch and Peers (1963) in response to a question raised by Lindley (1958). Unfortunately, there is no satisfactory general prescription for such a *probability matching prior* when $\boldsymbol{\theta}$ is multidimensional. Another type of noninformative prior, motivated rather differently, is the *reference prior* of Bernardo (Berger and Bernardo 1992). Kass and Wasserman (1996) provided an excellent review of noninformative priors.

Another difficulty in applying Bayesian inference with multidimensional parameters, or in more complex situations, is the high-dimensional integration needed either to evaluate the normalizing constant in (2) or to compute marginal posterior densities for particular parameters of interest from the multidimensional posterior. These difficulties have largely been solved by the introduction of a number of numerical methods, including importance sampling and Markov chain Monte Carlo (MCMC) methods. An introduction to Gibbs sampling was given by Casella and George (1992); see also the vignettes on Gibbs sampling and MCMC methods in this issue.

Bayesian inference respects the so-called *likelihood principle,* which states that inference from an experiment should be based only on the likelihood function for the observed data. Any inference that uses the sampling distribution of the likelihood function, as described in the next section, does not obey the likelihood principle. The discovery by Birnbaum (1962) that the principles of sufficiency and conditionality imply the likelihood principle led to considerable discussion in the 1960s and 1970s on various aspects of the foundations of inference. A good overview was provided by Berger and Wolpert (1984). More recently, there has been less interest in these foundational issues.

2.2   Classical Inference

Frequentist probability statements can be constructed from the likelihood function by considering the sampling distribution of the likelihood function and derived quantities. In fact, this is practically necessary from a frequentist standpoint, because the likelihood map is sufficient, which in particular implies that the minimal sufficient statistic in any model is determined by the likelihood map $L(\boldsymbol{\theta}; \cdot)$. This is why, for example, the Neyman–Pearson lemma concludes that the most powerful test depends on the likelihood ratio statistic.

The conventional derived quantities for a parametric likelihood function are the score function

$$l'(\boldsymbol{\theta}) = \partial \log L(\boldsymbol{\theta})/\partial \boldsymbol{\theta}, \tag{4}$$

the maximum likelihood estimate

$$\sup_{\boldsymbol{\theta}} l(\boldsymbol{\theta}) = l(\hat{\boldsymbol{\theta}}), \tag{5}$$

and the observed Fisher information

$$\mathbf{j}(\hat{\boldsymbol{\theta}}) = -\partial^2 l(\boldsymbol{\theta})/\partial \boldsymbol{\theta}^2|_{\boldsymbol{\theta}-\hat{\boldsymbol{\theta}}}, \tag{6}$$

where $l(\boldsymbol{\theta}) = \log L(\boldsymbol{\theta})$ is the log-likelihood function.

In the case where $\mathbf{Y} = (Y_1, \ldots, Y_n)$ is a sample of iid random variables, the log-likelihood function is a sum of n iid quantities, and under some conditions on the model a central limit theorem can be applied to the score function (4). More general sampling, such as $(Y_1, \ldots, Y_n)$ independent, but not identically distributed, or weakly dependent, can be accommodated if the model satisfies enough regularity conditions to ensure a central limit theorem for a suitably standardized version of the score function. Under many types of sampling, the score function is a martingale, and the martingale central limit theorem can be applied. Thus for a wide class of models, the following results can be derived:

$$l'(\boldsymbol{\theta})^T \{\mathbf{j}(\hat{\boldsymbol{\theta}})\} l'(\boldsymbol{\theta}) \xrightarrow{d} \chi_p^2, \tag{7}$$

$$(\hat{\boldsymbol{\theta}} - \boldsymbol{\theta})^T \{\mathbf{j}(\hat{\boldsymbol{\theta}})\}^{-1}(\hat{\boldsymbol{\theta}} - \boldsymbol{\theta}) \xrightarrow{d} \chi_p^2, \tag{8}$$

and

$$2\{l(\hat{\boldsymbol{\theta}}) - l(\boldsymbol{\theta})\} \xrightarrow{d} \chi_p^2, \tag{9}$$

where χ_p^2 is the chi-squared distribution on p degrees of freedom and p is the dimension of $\boldsymbol{\theta}$.

Similar results are available for inference about component parameters: writing $\boldsymbol{\theta} = (\boldsymbol{\psi}, \boldsymbol{\lambda})$, and letting $\hat{\boldsymbol{\lambda}}_{\psi}$ denote the *restricted maximum likelihood estimate* of $\boldsymbol{\lambda}$ for $\boldsymbol{\psi}$ fixed,

$$\sup_{\boldsymbol{\lambda}} l(\boldsymbol{\psi}, \boldsymbol{\lambda}; \mathbf{y}) = l(\boldsymbol{\psi}, \hat{\boldsymbol{\lambda}}_{\psi}; \mathbf{y}) = l_p(\boldsymbol{\psi}), \tag{10}$$

one has, for example,

$$2\{l(\hat{\boldsymbol{\psi}}, \hat{\boldsymbol{\lambda}}) - l(\boldsymbol{\psi}, \hat{\boldsymbol{\lambda}}_{\psi})\} \xrightarrow{d} \chi_q^2, \tag{11}$$

where q is the dimension of ψ. The function $l_p(\psi)$ defined in (10) is called the *profile log-likelihood* function.

These limiting results are taken as the size of the sample, n, in an independent sampling context, increases, with the dimension of θ held fixed. More generally, limit statements can be derived for the limit as the amount of Fisher information in $\mathbf{Y}$ increases.

The approximations suggested by these limiting results, such as

$$\hat{\theta} \overset{\cdot}{\sim} \mathrm{N}\{\theta, \mathbf{j}(\hat{\theta})\}, \tag{12}$$

called first-order approximations, are widely used in practice for inference about θ. The development of high-speed computers throughout the last half of the 20th century has enabled accurate and fast computation of maximum likelihood estimators in a wide variety of models, and most statistical packages have general-purpose routines for calculating derived likelihood quantities. This has meant in particular that development of alternative methods of point and interval estimation derived in the first half of the century is less important for applied work than it once was.

2.3   Likelihood as Pivotal

A major development in likelihood-based inference of the past 20 years is the discovery that the likelihood function can be used directly to provide an approximate sampling distribution for derived quantities that is more accurate than approximations like (12). The main result, usually called Barndorff-Nielsen's approximation, was initially developed in a series of articles in the August 1980 issue of *Biometrika* (Barndorff-Nielsen; Cox; Durbin; Hinkley), all of which derived in one version or another that

$$f(\hat{\theta}; \theta | \mathbf{a}) \overset{\cdot}{=} c|\mathbf{j}(\hat{\theta})|^{1/2} \exp\{l(\hat{\theta}) - l(\theta)\}. \tag{13}$$

The right side of (13) is often called Barndorff-Nielsen's p^* approximation. This formula generalizes an exact result for location models due to Fisher (1934). The renormalizing constant c is equal to $(2\pi)^{-p/2}\{1 + O(n^{-1})\}$. In some generality, (13) is a *third-order* approximation, meaning the ratio of the right side to the true sampling density of $\hat{\theta}$ (given $\mathbf{a}$) is $1 + O(n^{-3/2})$. Despite its importance, a rigorous proof of (13) is not yet available, although Skovgaard (1990) gave a very careful and helpful derivation. It is necessary to condition on a statistic $\mathbf{a}$ so that (13) is meaningful, because the likelihood function appearing on the right side depends on the data $\mathbf{y}$, yet it is being used as the sampling distribution for $\hat{\theta}$. The role of $\mathbf{a}$ is to complete a one-to-one transformation from $\mathbf{y}$ to $(\hat{\theta}, \mathbf{a})$. For (13) to be useful for inference, $\mathbf{a}$ must have a distribution either exactly or approximately free of θ; otherwise, we have lost information about θ in reducing to the conditional model.

The importance of (13) for the theory of inference is that it shows that the distribution of the maximum likelihood estimator (and other derived quantities) is obtained to a very high order of approximation directly from the likelihood function, as it is in a location model.

A result related to (13) and more directly useful for inference is the approximation of the cumulative distribution function for $\hat{\theta}$. In the case where θ is a scalar, this is expressed as

$$\Pr(\hat{\Theta} \leq \hat{\theta}; \theta | \mathbf{a}) = F(\hat{\theta}; \theta | \mathbf{a}) \doteq \Phi(r) + \phi(r)\left(\frac{1}{r} - \frac{1}{q}\right), \tag{14}$$

where

$$r = \text{sign}(q)\sqrt{[2\{l(\hat{\theta}) - l(\theta)\}]} \tag{15}$$

and

$$q = \{l_{;\hat{\theta}}(\hat{\theta}) - l_{;\hat{\theta}}(\theta)\}\{j(\hat{\theta})\}^{-1/2}, \tag{16}$$

where $l_{;\hat{\theta}}(\theta) = \partial l(\theta; \hat{\theta}, \mathbf{a})/\partial \hat{\theta}$. As with (13), this is an approximation with relative error $O(n^{-3/2})$. Two advantages of (14) over (13) are that it gives tail areas or p-values directly, and that it depends on $\mathbf{a}$ rather weakly, through a first derivative on the sample space. Approximation (14) shows that the first-order approximation to the likelihood ratio statistic [the scalar parameter version of (9)] provides the leading term in an asymptotic expansion to its distribution, that the next term in the expansion is easily computed directly from the likelihood function, and that in frequentist-based inference, the sample space derivative of the log-likelihood function plays an essential role. This last result has the potential to clarify (and also narrow) the difference between frequentist and Bayesian inference. Approximation (14) is often called the Lugannani and Rice approximation, as a version for exponential families was first developed by Lugannani and Rice (1980). There are analogous versions of (14), (15), and (16) for inference about a scalar component of θ in the presence of a nuisance parameter; a partial review was given by Reid (1996), and more recent work was presented by Barndorff-Nielsen and Wood (1998), Fraser, Reid, and Wu (1999), and Skovgaard (1996). (See also the approximations vignette by R. Strawderman.)

3. PARTIAL LIKELIHOOD AND ALL THAT

3.1 Nuisance Parameters

I defined at (10) the profile log-likelihood function $l_p(\psi)$, which is often used in problems in which the parameter of the model θ is partitioned into a parameter of interest ψ and a nuisance parameter λ. Typically λ is introduced into the model to make it more realistic. More generally, one can define

$$l_p(\psi) = \sup_{\psi = \psi(\theta)} l(\theta). \tag{17}$$

The profile likelihood is not a real likelihood function, in that it is not proportional to the sampling distribution of an observable quantity. However, there are limiting results analogous to (7)–(9), such as (11), that continue to provide first-order approximations. These approximations are expected to be poor if the dimension of the nuisance parameter λ is large relative to n, as it is known that the results break down if the dimension of θ increases with n. More intuitively, because no adjustment is made for errors of estimation of the nuisance parameter in (10) or (17), it is likely that the apparent precision of (10) or (17) is overstated. Several methods have been suggested for constructing a likelihood function better suited to problems with nuisance parameters. Some models may contain a conditional or marginal distribution that contains all the information about the parameter of interest, or is at least free of the nuisance parameter, and this density provides a true conditional or marginal likelihood. In fact, Figure 1 is a plot of the conditional likelihood of a component of the minimal sufficient statistic for the model, this likelihood depending only on the relative risk of an accident and not on nuisance parameters describing the background risk. More precisely, that model has the factorization

$$f(\mathbf{y}; \psi, \lambda) \propto f(s|\mathbf{t}; \psi) f(\mathbf{t}; \psi, \lambda), \tag{18}$$

and Figure 1 shows $L_c(\psi) \propto f(s|\mathbf{t}; \psi)$. The justification for ignoring the term $f(\mathbf{t}; \psi, \lambda)$ is not entirely clear and not entirely agreed on, although the claim is usually made that this component contains "little" information about ψ in the absence of knowledge of λ. A review of some of this work was given by Reid (1995).

In models where a conditional or marginal likelihood is not available, a natural alternative is a suitably defined approximate conditional or marginal likelihood, and approximation (13) has led to several suggestions for *modified profile likelihoods*. These typically have the form

$$l_m(\psi) = l_p(\psi) - \frac{1}{2} |\mathbf{j}_{\lambda\lambda}(\psi, \hat{\lambda}_\psi)| + B(\psi) \tag{19}$$

for some choice of $B(\cdot)$ of the same asymptotic order as the second term in (18), typically $O_p(1)$. The original modified profile likelihood is due to Barndorff-Nielsen (1983); Cox and Reid (1987) suggested using (18) with $B(\psi) = 0$, and several other versions have been proposed. Brief overviews were given by Mukerjee and Reid (1999) and Severini (1998).

3.2 Partial Likelihood

In more complex models there is often a partition analogous to (18), say

$$L(\theta; \mathbf{y}) = L_1(\psi; \mathbf{y}) L_2(\psi, \lambda; \mathbf{y}) \tag{20}$$

where it seems intuitively obvious that the second component cannot provide information about ψ in the absence of knowledge of λ. The most famous model for which this is the case is Cox's proportional hazards model for failure time data, where L_1 depends on the observed failure times and L_2 depends on the failure process between observed failure times. Cox (1972) proposed basing inference about the parameters of interest on L_1, which he called a conditional likelihood, later changed to partial likelihood (Cox 1975). Cox (1972) also showed that a martingale central limit theorem could be applied to the score statistic computed from L_1, leading to asymptotic normality for derived quantities such as the partial maximum likelihood estimate.

There are many related models where a partial likelihood leads to an adequate first-order approximation (Andersen, Borgan, Gill, and Keiding 1993; Murphy and van der Vaart 1997). There is not yet a theory of higher-order approximations in this setting, however. Likelihood partitions, such as (20), were discussed in some generality by Cox (1999).

3.3 Pseudolikelihood

One interpretation of partial likelihood is that the probability distribution of only part of the observed data is modeled, as this makes the problem tractable and with luck provides an adequate first-order approximation. A similar construction was suggested for complex spatial models by Besag (1977), using the conditional distribution of the nearest neighbors of any given point, and using the product of these conditional distributions as a pseudolikelihood function. A more direct approach to likelihood inference in spatial point processes was described by Geyer (1999).

3.4 Quasi-Likelihood

The last 30 years have also seen the development of an approach to modeling that does not specify a full probability distribution for the data, but instead specifies the form of, for example, the mean and the variance of each observation. This viewpoint is emphasized in the development of generalized linear models (McCullagh and Nelder 1989) and is central to the theory of generalized estimating equations (Diggle, Liang, and Zeger 1994). A *quasi-likelihood* is a function that is compatible with the specified mean and variance relations. Although it may not exist, when it does, it has in fairly wide generality the same asymptotic distribution theory as a likelihood function (Li and McCullagh 1994; McCullagh 1983).

3.5 Likelihood and Nonparametric Models

Suppose that we have a model in which we assume that $Y_1, \ldots Y_n$ are iid from a completely unknown distribution function $F(\cdot)$. The natural estimate of $F(\cdot)$ is the empirical distribution function,

$$F_n(y) = \frac{1}{n} \sum_{i=1}^{n} 1\{Y_i \leq y\}. \tag{21}$$

Although it is not immediately clear what the likelihood function or likelihood ratio is in a nonparametric setting, for a suitably defined likelihood $F_n(\cdot)$ is the maximum likelihood estimator of $F(\cdot)$. This was generalized to much more complex sampling, including censoring, by Andersen et al. (1993).

The empirical distribution function plays a central role in two inferential techniques closely connected to likelihood inference: the bootstrap and empirical likelihood. The nonparametric bootstrap uses samples from F_n for constructing an inference, usually by Monte Carlo resampling. The parametric bootstrap uses samples from $F(\cdot; \hat{\theta})$, where $\hat{\theta}$ is the maximum likelihood estimator. There is a close connection between the parametric bootstrap and the asymptotic theory of Section 2.3, although the precise relationship is still elusive. A good review was given by DiCiccio and Efron (1996).

An alternative to the nonparametric bootstrap is the empirical likelihood function, a particular type of profile likelihood function for a parameter of interest, treating the distribution of the data otherwise as the nuisance "parameter." The empirical likelihood was developed by Owen (1988), and has been shown to have an asymptotic theory similar to that for parametric likelihoods.

Empirical likelihood and likelihoods related to the bootstrap were described by Efron and Tibshirani (1993).

4. CONCLUSION

Whether from a Bayesian or a frequentist perspective, the likelihood function plays an essential role in inference. The maximum likelihood estimator, once regarded on an equal footing among competing point estimators, is now typically the basis for most inference and subsequent point estimation, although some refinement is needed in problems with large numbers of nuisance parameters. The likelihood ratio statistic is the basis for most tests of hypotheses and interval estimates. The emergence of the centrality of the likelihood function for inference, partly due to the large increase in computing power, is one of the central developments in the theory of statistics during the latter half of the 20th century.

REFERENCES

Andersen, P. K., Borgan, O., Gill, R. D., and Keiding, N. (1993), *Statistical Models Based on Counting Processes*, New York: Springer-Verlag.

Azzalini, A. (1996), *Statistical Inference*, London: Chapman & Hall.

Barndorff-Nielsen, O. E. (1980), "Conditionality Resolutions," *Biometrika*, 67, 293–310.

—— (1983), "On a Formula for the Distribution of the Maximum Likelihood Estimator," *Biometrika*, 70, 343–365.

Barndorff-Nielsen, O. E., and Wood, A. T. (1998), "On Large Deviations and Choice of Ancillary for p^* and r^*," *Bernoulli*, 4, 35–63.

Berger, J. O., and Bernardo, J. (1992), "On the Development of Reference Priors" (with discussion), in *Bayesian Statistics IV*, eds. J. M. Bernardo, J. O. Berger, A. P. Dawid, and A. F. M. Smith, London: Oxford University Press, pp. 35–60.

Berger, J. O., and Wolpert, R. (1984), *The Likelihood Principle*, Hayward, CA: Institute of Mathematical Statistics.

Besag, J. (1977), "Efficiency of Pseudo-Likelihood Estimation for Simple Gaussian Fields," *Biometrika*, 64, 616–618.

Birnbaum, A. (1962), "On the Foundations of Statistical Inference," *Journal of the American Statistical Association*, 57, 269–306.

Casella, G., and George, E. I. (1992), "Explaining the Gibbs Sampler, *American Statistician*, 46, 167–174.

Cox, D. R. (1972), "Regression Models and Life Tables" (with discussion), *Journal of the Royal Statistical Society*, Ser. B, 34, 187–220.

—— (1975), "Partial Likelihood," *Biometrika*, 62, 269–276.

—— (1980), "Local Ancillarity," *Biometrika*, 67, 279–286.

—— (1999), "Some Remarks on Likelihood Factorization," in *State of the Art in Probability and Statistics*, eds. M. C. M. de Gunst, C. A. J. Klaussen, A. W. van der Waart, Hayward: Institute of Mathematical Statistics.

Cox, D. R., and Reid, N. (1987), "Parameter Orthogonality and Approximate Conditional Inference" (with discussion), *Journal of the Royal Statistical Society*, Ser. B, 49, 1–39.

DiCiccio, T. J., and Efron, B. (1996), "Bootstrap Confidence Intervals" (with discussion), *Statistical Science*, 11, 189–228.

Diggle, P. J., Liang, K. Y., and Zeger, S. L. (1994), *The Analysis of Longitudinal Data*, Oxford, U.K.: Oxford University Press.

Durbin, J. (1980), "Approximations for Densities of Sufficient Statistics," *Biometrika*, 67, 311–333.

Edwards, A. F. (1972), *Likelihood*, Cambridge, U.K.: Cambridge University Press.

Efron, B., and Tibshirani, R. J. (1993), *An Introduction to the Bootstrap*, London: Chapman & Hall.

Fisher, R. A. (1922), "On the Mathematical Foundations of Theoretical Statistics," *Philosophical Transactions of the Royal Society*, Ser. A, 222, 309–368.

—— (1934), "Two New Properties of Mathematical Likelihood," *Proceedings of the Royal Society*, Ser. A, 144, 285–307.

—— (1956), *Statistical Methods and Scientific Inference*, Edinburgh: Oliver and Boyd.

Fraser, D. A. S., Reid, N., and Wu, J. (1999), "A Simple General Formula for Tail Probabilities for Frequentist and Bayesian Inference," *Biometrika*, 86, 249–264.

Geyer, C. (1999), "Likelihood Inference for Spatial Point Processes," in *Stochastic Geometry: Likelihood and Computation*, eds. O. E. Barndorff-Nielsen, W. S. Kendall, and M. N. M. van Lieshout, London: Chapman and Hall/CRC, pp. 79–140.

Hinkley, D. V. (1980), "Likelihood as Approximate Pivotal," *Biometrika*, 67, 287–292.

Jeffreys, H. (1961), *Theory of Probability*, Oxford, U.K.: Oxford University Press.

Kalbfleisch, J. G. (1985), *Probability and Statistics*, Vol. II (2nd ed.), New York: Springer-Verlag.

Kass, R. E., and Raftery, A. E. (1995), "Bayes Factors," *Journal of the American Statistical Association*, 90, 773–795.

Kass, R. E., and Wasserman, L. (1996), "Formal Rules for Selecting Prior Distributions: A Review and Annotated Bibliography," *Journal of the American Statistical Association*, 91, 1343–1370.

Li, B., and McCullagh, P. (1994), "Potential Functions and Conservative Estimating Functions," *The Annals of Statistics*, 22, 340–356.

Lindley, D. V. (1958), "Fiducial Distributions and Bayes's Theorem," *Journal of the Royal Statistical Society*, Ser. B, 20, 102–107.

Lugannani, R., and Rice, S. (1980), "Saddlepoint Approximation for the Distribution of the Sum of Independent Random Variables," *Advances in Applications and Probabilities*, 12, 475–490.

McCullagh, P. (1983), "Quasi-Likelihood Functions," *Annals of Statistics*, 11, 59–67.

McCullagh, P., and Nelder, J. A. (1989), *Generalized Linear Models* (2nd ed.), London: Chapman & Hall.

Mukerjee, R., and Reid, N. (1999), "On Confidence Intervals Associated With the Usual and Adjusted Likelihoods," *Journal of the Royal Statistical Society*, Ser. B, 61, 945–953.

Murphy, S. A., and van der Vaart, A. (1997), "Semiparametric Likelihood Ratio Inference," *The Annals of Statistics*, 25, 1471–1509.

Owen, A. (1988), "Empirical Likelihood Ratio Confidence Intervals for a Single Functional," *Biometrika*, 75, 237–249.

Redelmeier, D., and Tibshirani, R. J. (1997a), "Association Between Cellular Phones and Car Collisions," *New England Journal of Medicine*, February 12, 1997, 1–5.

—— (1997b), "Is Using a Cell Phone Like Driving Drunk?," *Chance*, 10, 5–9.

—— (1997c), "Cellular Telephones and Motor Vehicle Collisions: Some Variations on Matched Pairs Analysis," *Canadian Journal of Statistics*, 25, 581–593.

Reid, N. (1995), "The Roles of Conditioning in Inference," *Statistics in Science*, 10, 138–157.

—— (1996), "Likelihood and Higher-Order Approximations to Tail Areas: A Review and Annotated Bibliography," *Canadian Journal of Statistics*, 24, 141–166.

Royall, R. (1997), *Statistical Evidence*, London: Chapman & Hall.

Severini, T. A. (1998), "An Approximation to the Modified Profile Likelihood Function," *Biometrika*, 82, 1–23.

Skovgaard, I. M. (1990), "On the Density of Minimum Contrast Estimators," *The Annals of Statistics*, 18, 779–789.

—— (1996), "An Explicit Large-Deviation Approximation to One-Parameter Tests," *Bernoulli*, 2, 145–165.

Welch, B. L., and Peers, H. W. (1963), "On Formulae for Confidence Points Based on Integrals of Weighted Likelihoods," *Journal of the Royal Statistical Society*, Ser. B, 25, 318–329.

Conditioning, Likelihood, and Coherence: A Review of Some Foundational Concepts

James Robins and Larry Wasserman

1. INTRODUCTION

Statistics is intertwined with science and mathematics but is a subset of neither. The "foundations of statistics" is the set of concepts that makes statistics a distinct field. For example, arguments for and against conditioning on ancillaries are purely statistical in nature; mathematics and probability do not inform us of the virtues of conditioning, but only on how to do so rigorously. One might say that foundations is the study of the fundamental conceptual principles that underlie statistical methodology. Examples of foundational concepts include ancillarity, coherence, conditioning, decision theory, the likelihood principle, and the weak and strong repeated-sampling principles. A nice discussion of many of these topics was given by Cox and Hinkley (1974).

There is no universal agreement on which principles are "right" or which should take precedence over others. Indeed, the study of foundations includes much debate and controversy. An example, which we discuss in Section 2, is the likelihood principle, which asserts that two experiments that yield proportional likelihood functions should yield identical inferences. According to Birnbaum's theorem, the likelihood principle follows logically from two other principles: the conditionality principle and the sufficiency principle. To many statisticians, both conditionality and sufficiency seem compelling yet the likelihood principle does not. The mathematical content of Birnbaum's theorem is not in question. Rather, the question is whether conditionality and sufficiency should be elevated to the status of "principles" just because they seem compelling in simple examples. This is but one of many examples of the type of debate that pervades the study of foundations.

This vignette is a selective review of some of these key foundational concepts. We make no attempt to be complete in our coverage of topics. In Section 2 we discuss the likelihood function, the likelihood principle, the conditionality principle,

James Robins is Professor, Department of Epidemiology, Harvard School of Public Health, Boston, MA 02115. Larry Wasserman is Professor, Department of Statistics, Carnegie Mellon University, Pittsburgh, PA 15213. This research was supported by National Institutes of Health grants R01-CA54852-01 and R01-A132475-07 and National Science Foundation grants DMS-9303557 and DMS-9357646. The authors thank David Cox, Phil Dawid, Sander Greenland, Erich Lehmann, and Isabella Verdinelli for many helpful suggestions.

and the sufficiency principle. In Section 3 we briefly review conditional inference. In Section 4 we discuss Bayesian inference and coherence. In Section 5 we provide a brief look at some newer foundational work that suggests that in complex problems, the conditionality principle, the likelihood principle, and coherence arguments may be less compelling than in the low-dimensional problems, where they are usually discussed.

2. THE LIKELIHOOD FUNCTION AND THE LIKELIHOOD PRINCIPLE

Consider a random variable Y and a model $\mathcal{M} = \{p_\theta(\cdot); \theta \in \Theta\}$ for the distribution of Y. Here each $p_\theta(\cdot)$ is a density for Y, θ is an unknown parameter, and the parameter space Θ is a subset of $\mathcal{R}^k$. Assume that we have n iid replicates of Y denoted by $Y^n = (Y_1, \ldots, Y_n)$, generated from p_{θ_0}, where $\theta_0 \in \Theta$ denotes the true value of the unknown parameter θ. Fisher (1921, 1925, 1934) defined the likelihood function as

$$\mathcal{L}_n(\theta) = \prod_{i=1}^{n} p_\theta(Y_i).$$

Of course, the likelihood function appeared implicitly much earlier in the work of Bayes and Laplace, who used what we now call Bayesian inference to solve statistical problems. But it was Fisher who first emphasized the essential role of the likelihood function itself in inference. (See Aldrich 1997, Edwards 1997, and Fienberg 1997, for discussions on the history of likelihood.)

From a Bayesian perspective, inferences are based on the posterior $p(\theta|Y^n) \propto \mathcal{L}_n(\theta)\pi(\theta)$, where $\pi(\theta)$ is a prior distribution for θ. In the Bayesian framework, the data enter the inferences only through the likelihood function.

From a frequentist perspective, the likelihood is a source of point and interval estimators. For example, the maximum likelihood estimator (MLE) $\hat{\theta}_n$—the point at which $\mathcal{L}_n(\theta)$ takes its maximum—is known, under weak conditions, to be asymptotically normal with the smallest possible asymptotic variance. The set $C_n(c) = \{\theta; \mathcal{L}_n(\theta)/\mathcal{L}_n(\hat{\theta}_n) \leq c\}$ gives the asymptotically shortest, $1 - \alpha$ confidence set if c is chosen appropriately, if θ is scalar.

The likelihood function is central to many statistical analyses. More controversial is the role of the likelihood principle (LP). The LP says that when two experiments yield proportional likelihoods they should yield identical inferences. This principle of inference is accepted as a guiding principle by some statisticians but is considered unreasonable by others. Most forms of frequentist inference, such as significance testing and confidence intervals, violate the LP and so are ruled out if one wishes to follow the LP. For example, consider the likelihood-based confidence interval $C_n(c)$ defined earlier. Suppose that two different experiments yielded proportional likelihoods. Then the form of the interval $C_n(c)$ would be the same in the two experiments, but the coverage probability ascribed to $C_n(c)$ could be different, thus violating the LP.

A famous example that illustrates the likelihood principle involves binomial versus negative binomial sampling. Suppose that we want to estimate the probability θ of "heads" for a coin. In the binomial experiment, we toss the coin n times and observe the number of "heads" Y. Here n is fixed and Y is random. In the negative binomial experiment, we observe the number of tosses N required to obtain y "heads." Here N is random and y is fixed. Suppose that we observe 3 heads in 5 tosses. Under binomial sampling, these data yield the likelihood function

$$\mathcal{L}_1(\theta) = \left(\begin{array}{c} 5 \\ 3 \end{array} \right) \theta^3 (1-\theta)^2.$$

Under negative binomial sampling, the likelihood function is

$$\mathcal{L}_2(\theta) = \left(\begin{array}{c} 4 \\ 2 \end{array} \right) \theta^3 (1-\theta)^2.$$

As functions of θ, these likelihood functions are proportional, so any method that obeys the likelihood principle should yield identical inferences about θ regardless of whether the data were obtained by binomial or negative binomial sampling. Frequentist confidence intervals violate the likelihood principle, because the coverage of an interval is evaluated under hypothetical repetitions of the experiment. The set of possible outcomes in these hypothetical repetitions will differ, depending on whether binomial or negative binomial sampling is used. An interesting discussion of this problem was provided by Lindley and Phillips (1976).

Birnbaum (1962) showed that the likelihood principle follows logically from two other principles: the conditionality principle (CP) and the sufficiency principle (SP). A clear exposition of the details of Birnbaum's theorem and its implications was given by Berger and Wolpert (1984). This and the next section draw heavily from that monograph (see also Cox and Hinkley 1974).

The conditionality principle (Cox 1958) says that if we decide which of two experiments to do by the flip of a fair coin, then the final inference should be the same as if the experiment had been chosen without flipping the coin. More formally, the (weak) form of CP can be described as follows. Suppose that we consider two experiments, E_1 and E_2, for the same parameter θ. We flip a fair coin and perform E_1 if the coin is "heads" and perform E_2 if the coin is "tails." This is called a "mixed experiment." CP asserts that if we obtain "heads" and perform E_1, then our inferences should be identical to the inferences that we would make if E_1 were performed without first flipping the coin (and similarly for E_2). The outcome of the coin flip is an example of an ancillary statistic; that is, a statistic whose distribution does not depend on the unknown parameter. The sufficiency principle says that two outcomes of an experiment that yield the same value of a sufficient statistic should yield identical inferences. Many statisticians find CP and SP quite appealing, yet they do not find LP appealing, despite the fact the LP follows logically from these two principles. In fact, Evans, Fraser, and Monette (1986) since showed that LP follows from a slightly stronger version of CP alone.

The LP and CP have many supporters and detractors. Often, statisticians who find Bayesian methods appealing favor the LP. A non-Bayesian approach that obeys the LP was given by Edwards (1972). Those who prefer frequentist methods, as developed by Neyman, Wald, and others, find LP less compelling. Let us add our own point of view. The CP seems compelling in simple examples, provided that the experiment performed gives no additional information about the parameter beyond that contained in the data. In Section 5 we show that the CP is less compelling in high-dimensional models.

3. CONDITIONING, ANCILLARITY, AND RELEVANT SUBSETS

In the preceding section, when we discussed the conditionality principle, it seemed natural that inferences in the mixed experiment should be made conditionally on the value of the coin flip. Such inferences would then obey CP. This raises a more general question: Should inferences be carried out conditionally on an appropriate statistic? Inferences made conditional on some statistic go under the rubric of "conditional inference." Conditional inference can be traced back to Fisher (1956), Cox (1958), and others. The appeal of conditioning is evident from the following simple example (example 1 of Berger and Wolpert 1984). We observe two iid random variables Y_1 and Y_2, where $P_\theta(Y_i = \theta - 1) = P_\theta(Y_i = \theta + 1) = 1/2, i = 1, 2$. Here θ is an unknown real number. Let $C = \{(Y_1 + Y_2)/2\}$ if $Y_1 \neq Y_2$ and $C = \{Y_1 - 1\}$ otherwise. C is a 75% confidence set, although, unlike the typical confidence set, it contains only a single point. Thus $P_\theta(\theta \in C) = .75$ for all θ. But when $Y_1 \neq Y_2$, we are certain that $\theta \in C$ and when $Y_1 = Y_2, C$ contains θ 50% of the time; that is, $\Pr(\theta \in C | Y_1 = Y_2) = 1/2$. This suggests partitioning the sample space into $\{B, B^c\}$, where $B = \{Y_1 \neq Y_2\}$, and then reporting different inferences depending on whether B occurs or does not occur. This is equivalent to reporting inferences conditional on the statistic $S = |Y_1 - Y_2|$. The example is meant to suggest that inferences will be more intuitively plausible if they are performed conditional on an appropriate statistic.

If we accept that inferences should sometimes be conditional on something, then that raises the question of what we should condition on. The most common choices to condition on are ancillary statistics and relevant subsets. Both concepts were discussed by Fisher (1956).

An ancillary statistic is a statistic whose distribution does not depend on θ. In the foregoing example, $S = |Y_1 - Y_2|$ is ancillary, and it seems quite reasonable to report a confidence of 1 when $S = 1$ and a confidence of .50 when $S = 0$. Specifically, we report $P_\theta(\theta \in C|S = 1) = 1$ and $P_\theta(\theta \in C|S = 0) = .5$. The coin flip in the mixed experiment described in Section 2 is another example of an ancillary statistic.

Consider a $1 - \alpha$ confidence set $C(Y)$; that is, $P_\theta(\theta \in C(Y)) = 1 - \alpha$ for all θ. We say that B is a relevant subset if there is some $\varepsilon > 0$ such that either $P_\theta(\theta \in C(Y)|Y \in B) \le (1-\alpha)-\varepsilon$ for all θ or $P_\theta(\theta \in C(Y)|Y \in B) \ge (1-\alpha)+\varepsilon$ for all θ. If a relevant subset exists, then it seems tempting to report different inferences depending on whether $Y \in B$ or $Y \notin B$. Buehler and Fedderson (1963) and Brown (1967) showed that there exists a relevant subset even for the familiar Student t intervals for the mean of a normal. Thus the existence of relevant subsets is far from pathological.

Bayesian inference is an extreme form of conditional inference, because the posterior conditions on the data itself, as opposed to a conditioning on some statistic. There have been attempts to build formal, non-Bayesian theories of conditional inference. The best-known attempt is probably that of Kiefer (1977). Other contributions have come from Brown (1967, 1978), Buehler (1959), Casella (1987), Cox (1958, 1980), Fraser (1977), and Robinson (1976, 1979). The main idea is to partition the sample space as $\mathcal{Y} = \cup_s \mathcal{Y}_s$ and report conditional confidence $P_\theta(\theta \in C(Y)|Y \in \mathcal{Y}_s)$ when $Y \in \mathcal{Y}_s$.

Conditional inference is appealing because it seems to solve the apparently counterintuitive results like those in the simple example presented at the beginning of this section. Nevertheless, in many models there is no known ancillary or relevant subset on which to condition, or there may be many ancillaries, in which case it is not clear on which to condition. One way to extend the applicability of conditional inference when there is no exact ancillary is to use approximate conditional inference, in which one conditions on a statistic that is asymptotically ancillary (see, e.g., Amari 1982; Barndorff-Nielsen 1983; Cox 1988; Cox and Reid 1987; DiCiccio 1986; Efron and Hinkley 1978; Robins and Morgenstern 1987; Severini 1993; Sweeting 1992).

Brown (1990) and Foster and George (1996) gave examples in which an estimator is admissible conditional on an ancillary statistic but is unconditionally inadmissible. Such examples show explicitly that procedures that obey the CP can have poor unconditional properties. We discuss this point further in Section 5.

Another situation where conditioning has received attention is in the problem of estimating a common odds ratio ψ in a series of 2×2 tables. Specifically, we observe independent binomial random variables $X_{0k} \sim \text{bin}(n_{0k}, p_{0k})$ and $X_{1k} \sim \text{bin}(n_{1k}, p_{1k})$ $k = 1, \ldots, K$, where $p_{1k} = p_{0k}/(p_{0k} + \psi(1 - p_{0k}))$. The likelihood is $L_1(\psi)L_2(\psi, \mathbf{p}_0)$, where

$$L_1(\psi) = f(\mathbf{X}_1|\mathbf{X}_+; \psi) = \prod_{k=1}^{K} f(X_{1k}|X_{+k}; \psi)$$

and

$$L_2(\psi, \mathbf{p}_0) = f(\mathbf{X}_+; \psi, \mathbf{p}_0) = \prod_{k=1}^{K} f(X_{+k}; \psi, p_{0k}),$$

where $\mathbf{p}_0 = (p_{01}, \ldots, p_{0K})$, $\mathbf{X}_1 = (X_{11}, \ldots, X_{1k})$, $\mathbf{X}_+ = (X_{+1}, \ldots, X_{+K})$, and $X_{+k} = X_{0k} + X_{1k}$. Now the set of row totals $\mathbf{X}_+$ is not S ancillary for ψ; that is, there is no global reparameterization (ψ, θ) such that ψ and θ are variation independent (i.e., the parameter space is a product space) and such that $f(\mathbf{X}_+; \psi, \theta) = f(\mathbf{X}_+; \theta)$. Thus the CP does not imply that inference for ψ should be performed conditional on $\mathbf{X}_+$. However, it has been argued that inference for ψ (conditional or unconditional) should be based on the conditional likelihood $L_1(\psi)$ if the marginal likelihood $L_2(\psi, \mathbf{p}_0)$ contains no independent information about ψ when $\mathbf{p}_0$ is unknown. The conditional MLE maximizing $L_1(\psi)$ is asymptotically efficient for ψ both in large-stratum asymptotics, in which $n_{0k} \to \infty$ and $n_{1k}/n_{0k} \to c_k$ for each k and in a sparse data asymptotics, in which $K \to \infty, n_{1k}$ and n_{0k} are bounded and the p_{0k} are drawn independently from a common distribution (Bickel, Klaassen, Ritov, and Wellner 1993; Lindsay 1980). Thus, asymptotically, $L_2(\psi, \mathbf{p}_0)$ contains no additional information about ψ. However, Sprott (1975) provided an interesting example to show that $L_2(\psi, \mathbf{p}_0)$ can contain some information about ψ. He considered the special case where $n_{1k} = n_{0k} = 1$. Suppose that only $\mathbf{X}_+$ is observed and $X_{+k} = 1$ for all k. Then he argued that for large K, the hypothesis $\psi = 1$ can be rejected, because for any p_{0k}, the probability that $X_{+k} = 1$ when $\psi = 1$ can never exceed $1/2$. Of course, the information about ψ contained in the marginal law of $\mathbf{X}_+$ is asymptotically negligible as $K \to \infty$ compared to that in $L_1(\psi)$, because the conditional MLE is efficient.

Finally, we should add that conditioning is used for other reasons as well; for example, in the construction of similar tests (Cox and Hinkley 1974, chap. 5; Lehmann 1986, chap. 4). A general discussion of conditional inference was provided by Lehmann (1986, chap. 10).

4. COHERENCE AND BAYESIAN INFERENCE

Some researchers have attempted to create a foundationally sound method of inference by stating axioms for inference and then characterizing all inferential methods that satisfy these axioms. Often, such axioms are called axioms of coherence, as they are meant to capture what a coherent (i.e., self-consistent) inference is. Usually, these arguments lead to conclusions of the form that inferences are coherent if and only if they are Bayesian. It may appear that this line of research has had a greater impact on statistical practice than conditioning arguments, because Bayesian methods have become increasingly popular in practice, whereas conditional inference has not. However, we believe the increasing use of Bayesian methods has more to do with their conceptual simplicity as well as advances in computing than with the coherence arguments. Still, these arguments do add interesting insight into inferential issues.

There are many versions of coherence arguments (see, e.g., de Finetti 1974, 1975; Freedman and Purves 1969; Heath and Sudderth 1978, 1989; Jeffreys 1961; Ramsey 1930; Regazzini 1987; Savage 1954). Here we describe the Heath–Sudderth approach. We begin with a model $\{P_\theta; \theta \in \Theta\}$, where each P_θ is a probability distribution for a random variable Y. An inference Q is a map that assigns a (possibly finitely additive) probability measure Q_y over Θ to each outcome y. The function Q_y is regarded as a set of probabilities from which bets are made about θ after observing $Y = y$. An inference Q is called "coherent" if it is impossible to place a finite number of bets on subsets of Θ on observing $Y = y$, which have a strictly positive expected payoff. Heath and Sudderth proved that an inference Q is coherent if and only if it is a posterior for some, possibly finitely additive, prior π over Θ. Formally, Q is a posterior distribution for the prior π, if for every bounded, measurable function $\phi(\theta, y)$,

$$\int \int \phi(\theta, y) P_\theta(dy) \pi(d\theta) = \int \int \phi(\theta, y) Q_y(d\theta) m(dy).$$

Here, m is the marginal distribution for Y induced by the model and prior; that is, $\int g(y) m(dy) = \int \int g(y) P_\theta(dy) \pi(d\theta)$ for every bounded, measurable function $g(y)$.

The implication is that inferences are coherent if and only if they are Bayesian. Such results increase the appeal of Bayesian methods to many statisticians. Of course, the results are only as compelling as the axioms. Given the choice between a method that is coherent and a method that has, say, correct frequentist coverage, many statisticians would choose the latter. The issue is not mathematical in nature. The question is under which circumstances one finds coherence or correct coverage more important.

In some cases it is possible both to be coherent and to have good frequentist properties, in large samples. Indeed, in a finite-dimensional model, with appropriate regularity conditions, the following facts are known. Let $\hat{\theta}_n$ be the MLE, let $\bar{\theta}$ be the posterior mean, and let $Q(d\theta|Y^n)$ be the posterior based on n iid observations $Y^n = (Y_1, \ldots, Y_n)$. Then the following hold:

1. $\bar{\theta}_n - \hat{\theta} = O_P(n^{-1})$.
2. There exist regions C_n such that both $\int_{C_n} Q(d\theta|Y^n) = 1 - \alpha$, and C_n has frequentist coverage $1 - \alpha + O(n^{-1})$.
3. If N is any open, fixed Euclidean neighborhood of the true value θ_0, then $\int_N Q(d\theta|Y^n)$ tends to 1 almost surely.
4. The posterior concentrates around the true value θ_0 at rate $n^{-1/2}$; that is, $Q(\{\theta; |\theta - \theta_0| \geq a_n n^{-1/2}\}|Y^n) = o_P(1)$ for any sequence $a_n \to \infty$.

In words, (1) the MLE and posterior mean are asymptotically close, (2) Bayesian posterior intervals and confidence intervals agree asymptotically, (3) the posterior is consistent, and (4) the posterior converges at the same rate as the maximum likelihood estimate. Facts 1, 3, and 4 follow from standard asymptotic arguments (see, Schervish

1995). Fact 2 was shown by Welch and Peers (1963), for the more difficult one-sided case.

In infinite-dimensional models, the situation is less clear. Consistency is sometimes attainable and sometimes not (see, e.g., Barron 1988; Barron, Schervish, and Wasserman 1999; Diaconis and Freedman 1986, 1990, 1993, 1997; Doob 1949; Freedman 1963, 1965; Freedman and Diaconis 1983; Ghosal, Ghosh, and Ramamoorthi 1999; Schwartz 1960, 1965) for example. Similarly, good rates of convergence are sometimes possible (see Ghosal, Ghosh, and van der Vaart 1998; Shen and Wasserman 1998; Zhao 1993, 1998). The issue of matching posterior probability and frequentist coverage has received less attention. While some negative results were reported by Cox (1993), this topic remains mostly unexplored territory.

5. A LOOK TO THE FUTURE: FOUNDATIONS IN INFINITE-DIMENSIONAL MODELS

For the most part, foundational thinking has been driven by intuition based on low-dimensional parametric models. But in modern statistical practice, it is routine to use high-dimensional or even infinite-dimensional (nonparametric or semiparametric) methods. There is some danger in extrapolating our intuition from finite-dimensional to infinite-dimensional models. Should we rethink foundations in light of these methods? Here we argue that the answer is "yes." We summarize an example that was discussed in great detail by Robins and Ritov (1997). To keep things brief and simple, we give a telegraphic version and omit the details; see the Robins and Ritov article for a full discussion. Although it will not be immediately obvious, the example stems from a real problem—the analysis of treatment effects in randomized trials. See also Robins, Rotnitzky, and Van der Laan (2000).

Let $(X_1, Y_1), \ldots, (X_n, Y_n)$ be n iid copies of a random vector (X, Y), where X is continuous taking values in the k-dimensional unit cube $\mathcal{X} = (0, 1)^k$ and Y given $X = x$ is normal with mean $\theta_0(x)$ and variance 1. The conditional mean function θ_0: $(0, 1)^k \to \mathcal{R}$ is assumed to be continuous and to satisfy $\sup_{x \in (0,1)^k} |\theta(x)| \leq M$ for some known positive constant M. Let Θ denote all such functions. The density $f_0(x)$ of X is assumed to belong to the set of densities

$$\mathcal{F}_X = \{f; c < f(x) < 1/c \text{ for } x \in \mathcal{X}\},$$

where $c \in (0, 1)$ is a fixed constant. Our goal is to estimate the parameter $\psi_0 = \int_{\mathcal{X}} \theta_0(x)\, dx$. A pair (θ, f) completely determines a law of (X, Y). The likelihood function is

$$\mathcal{L}(\theta, f) = \mathcal{L}_1(\theta)\mathcal{L}_2(f)$$

$$= \left\{\prod_{i=1}^{n} \phi(Y_i - \theta(X_i))\right\}\left\{\prod_{i=1}^{n} f(X_i)\right\},$$

where $\phi(\cdot)$ denotes the standard normal density. Notice that in this likelihood, the parameters θ and f are functions. The model is infinite dimensional because the set Θ cannot be put into a smooth, one-to-one correspondence with a finite-dimensional Euclidean space. Let $\mathbf{X} = \{X_i; i = 1, \ldots, n\}$ denote the observed X_i's. When f_0 is known, $\mathbf{X}$ is ancillary. When f_0 is unknown, $\mathbf{X}$ is still ancillary but in a slightly different sense. Technically, $\mathbf{X}$ is S ancillary for ψ, because the conditional likelihood given $\mathbf{X}$ is a function of θ alone and the marginal likelihood of $\mathbf{X}$ is a function of f alone, θ and f are variation independent (i.e., the parameter space is a product space), and ψ is a function of θ only (Barndorff-Nielsen 1978; Cox and Hinkley 1974).

Now we shall see that whether or not we know the distribution f_0 of the ancillary $\mathbf{X}$ has drastic implications for inference, in contrast to the usual intuition about ancillarity. When f_0 is unknown, Robins and Ritov (1997) showed that no uniformly consistent estimator of ψ_0 exists. But when f_0 is known, there do exist uniformly consistent estimators of ψ_0. In fact, there exist estimators that are $\sqrt{n}$-consistent uniformly over all $\theta \times f \in \Theta \times \mathcal{F}_X$. For example, define the random variable $V = Y/f_0(X)$. Then $\bar{V} = n^{-1} \sum_{i=1}^{n} V_i = n^{-1} \sum_{i=1}^{n} Y_i/f_0(X_i)$ is uniformly $\sqrt{n}$-consistent. (Uniformity is important because it links asymptotic behavior to finite-sample behavior. This is especially important in high-dimensional examples; i.e., when k is large.)

This result has implications for many common inferential methods. In particular, standard likelihood-based and Bayesian estimator methods will fail to be uniformly consistent. To see this, note that maximum likelihood inference, profile likelihood inference, and Bayesian inference with independent priors on θ and f all share the following property: They provide the same inferences for ψ whatever the known $f_0 \in \mathcal{F}_X$ that generated the data. We call such methods *strict factorization-based* (SFB) methods. Indeed, in the model with f_0 known, any inference method that satisfies the likelihood principle is SFB. Robins and Ritov (1997) showed that no SFB estimator can be consistent for ψ_0 uniformly over $(\theta_0, f_0) \in \Theta \times \mathcal{F}_X$.

The deficiencies in SFB methods extend to interval estimation. Any interval estimator that is not a function of f_0 will not be "valid." By valid, we mean that under all $(\theta_0, f_0) \in \Theta \times \mathcal{F}_X$ the coverage is at least $(1 - \alpha)$ at each sample size n and the expected length goes to 0 with increasing sample size. There are valid interval estimators for ψ, but these depend on f_0 and hence are not SFB, and they violate LP. An example of such an interval estimator is $\bar{V} \pm dn^{-1/2}$, where

$$d^2 = \frac{M^2 + 1}{1 - \alpha} \int_{\mathcal{X}} \frac{dx}{f_0(x)}.$$

That this has coverage exceeding $1 - \alpha$ follows from Chebyshev's inequality. Note that this interval is not only valid, but its length shrinks at rate $n^{-1/2}$. However, even with f_0 known, there is no interval estimator that has expected length tending to 0, with conditional coverage at least $1 - \alpha$ given $\mathbf{X}$, on a set of $\mathbf{X}$ with f_0 probability 1, for all $\theta_0 \in \Theta$. Our example is connected to Godambe and Thompson's (1976) criticism of likelihood-based inference in the context of finite-population inference from sample survey data and to the "ancillarity paradoxes" of Brown (1990) and Foster and George (1996) mentioned earlier. Indeed, the development of Brown (1990) suggests that any estimator that is unconditionally admissible for squared error loss will fail to be SFB and hence will violate the LP.

It is often stated that Bayesian inference always satisfies the likelihood principle. This is correct only when the prior does not depend on the experiment. Consider, for example, an observer with a prior π that makes θ and f dependent. Now suppose that this observer learns the true value f_0 of f. Then his posterior distribution of θ and ψ will depend on f_0 and thus violate the LP. Note that f_0 indexes the experiment being performed. Since, for any two experiments f_0 and f_0^*, the likelihood ratio $\mathcal{L}(\theta, f_0)/\mathcal{L}(\theta, f_0^*)$ is not a function of θ, any estimator that depends on f_0 violates LP. This result does not contradict Birnbaum's theorem, because the observer's inferences also violate the CP because with this prior, knowledge of which experiment was actually performed (i.e., the true f_0 that generated the data) contains information concerning θ. To see this, consider the extreme case where the observer gets no data but learns f_0. Then observer's posterior for θ will be the conditional prior $\pi(\theta|f_0)$. However, if the process determining which experiment was chosen had been superseded and the experiment had instead been chosen by a coin flip, then the posterior for θ would be the marginal prior $\pi(\theta)$. In other words, the observer's inferences are (correctly) influenced by how the observer got to learn f_0.

The example we presented in this section is important for two reasons. First, as we noted earlier, a version of this problem actually arises in the problem of estimating treatment effects in randomized trials when the randomization probabilities depend on observed covariates. Second, the example illustrates the general point that good frequentist performance and the LP can be in severe conflict in the sense that any procedure with good frequentist properties must violate the LP. In the future, we believe that more attention should be directed to examining foundational principles in infinite-dimensional settings.

REFERENCES

Aldrich, J. (1997), "R. A. Fisher and the Making of Maximum Likelihood 1912–1922," *Statistical Science*, 12, 162–176.

Amari, S. (1982), "Geometrical Theory of Asymptotic Ancillarity and Conditional Inference," *Biometrika*, 69, 1–17.

Barndorff-Nielsen, O. (1978), *Information and Exponential Families in Statistical Theory*, New York: Wiley.

———— (1983), "On a Formula for the Distribution of the Maximum Likelihood Estimator," *Biometrika*, 70, 343–365.

Barron, A. (1988), "The Exponential Convergence of Posterior Probabilities With Implications for Bayes Estimators of Density Functions," unpublished manuscript.

Barron, A., Schervish, M., and Wasserman, L. (1999), "Consistency of Posterior Distributions in Nonparametric Problems," *The Annals of Statistics*, 27, 536–561.

Berger, J., and Wolpert, R. (1984), *The Likelihood Principle*, Hayward, CA: Institute for Mathematical Statistics.

Bickel, P., Klaassen, C., Ritov, Y., and Wellner, J. (1993), *Efficient and Adaptive Estimation for Semiparametric Models*, Baltimore: Johns Hopkins University Press.

Birnbaum, A. (1962), "On the Foundations of Statistical Inference" (with discussion), *Journal of the American Statistical Association*, 57, 269–306.

Brown, L. D. (1967), "The Conditional Level of Student's t Test," *The Annals of Mathematical Statistics*, 38, 1068–1071.

——— (1978), "A Contribution to Kiefer's Theory of Conditional Confidence Procedures," *The Annals of Statistics*, 6, 59–71.

——— (1990), "An Ancillary Paradox which Appears in Multiple Linear Regression," *The Annals of Statistics*, 8, 471–493.

Buehler, R. J. (1959), "Some Validity Criteria for Statistical Inference," *The Annals of Mathematical Statistics*, 30, 845–863.

Buehler, R. J., and Feddersen, A. P. (1963), "Note on a Conditional Property of Student's t," *The Annals of Mathematical Statistics*, 34, 1098–1100.

Casella, G. (1987), "Conditionally Acceptable Recentered Set Estimators," *The Annals of Statistics*, 15, 1363–1371.

Cox, D. R. (1958), "Some Problems Connected With Statistical Inference," *The Annals of Mathematical Statistics*, 29, 357–372.

——— (1980), "Local Ancillarity," *Biometrika*, 67, 279–286.

——— (1988), "Some Aspects of Conditional and Asymptotic Inference: A Review," *Sankhya*, 50, 314–337.

——— (1993), "An Analysis of Bayesian Inference for Nonparametric Regression," *The Annals of Statistics*, 21, 903–923.

Cox, D. R., and Hinkley, D. (1974), *Theoretical Statistics*, London: Chapman and Hall.

Cox, D. R., and Reid, N. (1987), "Parameter Orthogonality and Approximate Conditional Inference," *Journal of the Royal Statistical Society*, Ser. B, 49, 1–18.

de Finetti, B. (1974, 1975), *Theory of Probability, Vols. I and II*, trans. by A. F. M. Smith and A. Machi, New York: Wiley.

Diaconis, P., and Freedman, D. (1986), "On the Consistency of Bayes Estimates," *The Annals of Statistics*, 14, 1–26.

——— (1990), "On the Uniform Consistency of Bayes Estimates for Multinomial Probabilities," *The Annals of Statistics*, 18, 1317–1327.

——— (1993), "Nonparametric Binary Regression: A Bayesian Approach," *The Annals of Statistics*, 21, 2108–2137.

——— (1997), "Consistency of Bayes Estimates for Nonparametric Regression: A Review," in *Festschrift for L. Le Cam*, 157–165.

DiCiccio, T. (1986), "Approximate Conditional Inference for Location Families," *Canadian Journal of Statistics*, 14, 5–18.

Doob, J. L. (1949), "Application of the Theory of Martingales," in *Le Calcul des Probabilités et ses Applications*, Colloques Internationaux du Centre National de la Recherche Scientifique, Paris, pp. 23–27.

Edwards, A. W. F. (1972), *Likelihood*, Cambridge, U.K.: Cambridge University Press.

—— (1997), "What did Fisher Mean by 'Inverse Probability?'," *Statistical Science*, 12, 177–184.

Efron, B., and Hinkley, D. (1978), "Assessing the Accuracy of the Maximum Likelihood Estimator: Observed Versus Expected Fisher Information," *Biometrika*, 65, 457–482.

Evans, M., Fraser, D. A. S., and Monette, G. (1986), "On Principles and Arguments to Likelihood" (with discussion), *Canadian Journal of Statistics*, 14, 181–199.

Fienberg, S. (1997), "Introduction to R. A. Fisher on Inverse Probability and Likelihood," *Statistical Science*, 12, 161.

Fisher, R. A. (1921), "On the 'Probable Error' of a Coefficient of Correlation Deduced From a Small Sample," *Metron*, I. part 4, 3–32.

—— (1925), "Theory of Statistical Estimation," *Proceedings of the Cambridge Philosophical Society*, 22, 700–725.

—— (1934), "Two New Properties of Mathematical Likelihood," *Proceedings of the Royal Society*, Ser. A, 144, 285–307.

—— (1956), *Statistical Methods and Scientific Inference*, Edinburgh: Oliver and Boyd.

Foster, D. P., and George, E. I. (1996), "A Simple Ancillarity Paradox," *Scandinavian Journal of Statistics*.

Fraser, D. (1977), "Confidence, Posterior Probability and the Buchler Example," *The Annals of Statistics*, 5, 892–898.

Freedman, D. (1963), "On the Asymptotic Behavior of Bayes's Estimates in the Discrete Case," *The Annals of Mathematical Statistics*, 34, 1386–1403.

—— (1965), "On the Asymptotic Behavior of Bayes Estimates in the Discrete Case, II," *The Annals of Mathematical Statistics*, 36, 454–456.

Freedman, D., and Diaconis, P. (1983), "On Inconsistent Bayes Estimates in the Discrete Case," *The Annals of Statistics*, 11, 1109–1118.

Freedman, D. A., and Purves, R. A. (1969), "Bayes Methods for Bookies," *The Annals of Mathematical Statistics*, 40, 1177–1186.

Ghosal, S., Ghosh, J. K., and Ramamoorthi, R. V. (1999), "Posterior Consistency of Dirichlet Mixtures in Density Estimation," *The Annals of Statistics*, 27, 143–158.

Ghosal, S., Ghosh, J. K., and van der Vaart, A. (1998), "Rates of Convergence of Posteriors," technical report, Free University, Amsterdam.

Godambe, V. P., and Thompson, M. E. (1976), "Philosophy of Survey-Sampling Practice" (with discussion), in *Foundations of Probability Theory, Statistical Inference and Statistical Theories of Science, Vols. I, II, and III*, eds. W. L. Harper and A. Hooker, Dordrecht: D. Reidel.

Heath, D., and Sudderth, W. (1978), "On Finitely Additive Priors, Coherence, and Extended Admissibility," *The Annals of Statistics*, 6, 333–345.

—— (1989), "Coherent Inference From Improper Priors and From Finitely Additive Priors," *The Annals of Statistics*, 17, 907–919.

Jeffreys, H. (1961), *The Theory of Probability*, Oxford, U.K.: Clarendon Press.

Kiefer, J. (1977), "Conditional Confidence Statements and Confidence Estimators" (with discussion), *Journal of the American Statistical Association*, 72, 789–827.

Lehmann, E. L. (1986), *Testing Statistical Hypotheses* (2nd ed.), New York: Wiley.

Lindley, D. V., and Phillips, L. D. (1976), "Inference for a Bernoulli Process (A Bayesian view)," *The American Statistician*, 30, 112–119.

Lindsay, B. (1980), "Nuisance Parameters, Mixture Models, and the Efficiency of Partial Likelihood Estimators," *Philosophical Transactions of the Royal Society*, 296, 639–665.

Ramsey, F. (1931), *The Foundations of Mathematics and Other Logical Essays*, Paterson, NJ: Littlefield-Adams.

Regazzini, E. (1987), "De Finetti's Coherence and Statistical Inference," *The Annals of Statistics*, 15, 845–864.

Robins, J. M., and Morgenstern, H. (1987), "The Foundations of Confounding in Epidemiology," *Computers and Mathematics with Applications*, 14, 869–916.

Robins, J. M., and Ritov, Y. (1997), "A Curse of Dimensionality Appropriate (CODA) Asymptotic Theory for Semiparametric Models," *Statistics in Medicine*, 16, 285–319.

Robins, J. M., Rotnitzky, A., and van der Laan, M. (2000), Comment on "On Profile Likelihood," by S. A. Murphy and A. W. van der Vaart, *Journal of the American Statistical Association*, 95, 477–482.

Robinson, G. K. (1976), "Properties of Student's t and of the Behrens–Fisher Solution to the Two Means Problem," *The Annals of Statistics*, 5, 963–971.

——— (1979), "Conditional Properties of Statistical Procedures," *The Annals of Statistics*, 7, 742–755.

Savage, L. J. (1954), *The Foundations of Statistics*, New York: Wiley.

Schervish, M. (1995), *Theory of Statistics*, New York: Springer-Verlag.

Schwartz, L. (1960), "Consistency of Bayes Procedures," Ph.D. dissertation, University of California.

——— (1965), "On Bayes Procedures," *Z. Wahrsch. Verw. Gebiete*, 4, 10–26.

Severini, T. (1993), "Local Ancillarity in the Presence of a Nuisance Parameter," *Biometrika*, 80, 305–320.

Shen, X., and Wasserman, L. (1998), "Rates of Convergence of Posterior Distributions," Technical Report 678, Carnegie Mellon University, Statistics Dept.

Sprott, D. A. (1975), "Marginal and Conditional Sufficiency," *Biometrika*, 62, 599–606.

Sweeting, T. (1992), "Asymptotic Ancillarity and Conditional Inference for Stochastic Processes," *The Annals of Statistics*, 20, 580–589.

Welch, B. L., and Peers, H. W. (1963), "On Formulae for Confidence Points Based on Integrals of Weighted Likelihoods," *Journal of the Royal Statistical Society*, Ser. B, 25, 318–329.

Zhao, L. (1993), "Frequentist and Bayesian Aspects of Some Nonparametric Estimation Problems," Ph.D. thesis, Cornell University.

——— (1998), "A Hierarchical Bayesian Approach in Nonparametric Function Estimation," technical report, University of Pennsylvania, Wharton School.

The End of Time Series

V. Solo

1. INTRODUCTION

Just as physicists have spent the last three-quarters of a century working out the consequences of the quantum mechanics paradigm established by the late 1920s, so have statisticians spent the last half century working out the consequences of the statistical paradigm assembled by the end of World War II.

As a branch of statistics, time series has always occupied a special place. Not only is it mostly practiced by nonstatisticians, but, as a consequence, most of the innovation has come from outside statistics. But statisticians have nevertheless played a disproportionate role given their relatively small numbers. A list of areas making a significant use of time series methods includes communications (Haykin 1983), control engineering (Ljung 1987), econometrics (Hamilton 1994), geophysics (Aki and Richards 1980), meteorology (Daley 1991), optical signal processing (Papoulis 1968), signal processing (Kay 1988), and radio astronomy (Steward 1987). In fact, any applied discipline that makes heavy use of Fourier methods is likely to have some need for time series.

Before plunging into the details, I outline the dimension of my discussion space. Some elementary partitions are obvious: discrete and continuous time, stationary and nonstationary, scalar and vector, linear and nonlinear, time domain and frequency domain, and parametric and nonparametric. Finally, I observe, less obviously, that the problems of time series analysis resolve mostly into modeling and/or filtering. Because each of the resulting 128 cells will have several subheadings, it is clear that the discussion must be selective! Some of these cells have been well worked out, others have been barely touched, and others, though much studied, have thrown up problems that have long resisted solution. Some other topics, such as wavelets (being time-frequency methods) and semiparametrics, require partitioning to allow interaction; this results in 2,187 cells!

2. A BRIEF HISTORY OF TIME (SERIES)

The period of interest may be roughly divided into two subperiods: 1945–1975 and 1975–1999. In the first subperiod, the following developments can be identified:

V. Solo is Professor of Electrical Engineering, University of New South Wales, Sydney, NSW 2052, Australia (E-mail: vsolo@syscon.ee.unsw. edu.au). This work was completed while the author was Professor of Statistics at Macquarie University, Sydney, Australia.

1. Development of basic filtering and smoothing algorithms, including fast versions. Thus we have Levinson's scalar algorithm (Bartlett 1955; Durbin 1960, who found a variance update missed by Levinson; Wiener 1949, appendix); the vector version of Levinson's algorithm (Robinson 1967; Whittle 1963a); the Kalman filter (Kalman 1960) based on state-space representations. Spectral factorization for prediction filter design is closely related—it is the infinite-dimensional version of the finite-data Cholesky factorization (of the covariance matrix inverse) achieved by the Kalman filter. It is quite remarkable that even today, some textbook authors and at least one distinguished figure do not seem to understand that the Kalman filter can, for example, track a unit root process stably. The basic frequency domain smoother is based on two-sided Wiener filtering (Whittle 1963b). The state-space-based smoothers were not, despite a common conception, developed by Kalman, but rather by other control engineers (Bryson and Ho 1969).

2. Development of a thorough-going modeling paradigm, stimulated by Box and Jenkins (1970) and continuing through the application of the Kalman filter to likelihood construction (Harvey 1989); development of nonparametric methods of spectral estimation (Jenkins and Watts 1969).

3. Development of vector time series methods: Brillinger (1981) gave extensions of classical multivariate analysis methods in a frequency domain context; Hannan (1970) gave an asymptotic theory as well as the beginnings of a resolution of the identifiability problem for vector time series. Fundamental insight and associated methods were provided by Akaike (1974). The identifiability problem was ultimately fully resolved by the combined efforts of control engineers (Hannan and Deistler 1988; Kailath 1980).

4. Development of a paradigm of asymptotic analysis especially for stationary time series (Hannan 1970).

By the end of this period, time series had reached a considerable level of maturity. The second subperiod brought the following:

1. Development of methods and asymptotics for integrated time series, starting with unit root techniques by Fuller and Dickey (Fuller 1976) and continuing with the development of cointegration methods (Engle and Granger 1987; Hamilton 1994). Martingale methods and weak convergence came to the fore here (Hall and Heyde 1980). On the smoothing side, methods based on smoothing splines became important (Wahba 1990) and are closely related to nonparametric function estimation, as well as to stochastic estimation with Brownian motion priors.

2. Development of nonparametric methods based on selection of model dimension. This starts with Akaike (1969), but real understanding develops later (Hannan and Deistler 1988; Shibata 1976, 1980). Other important developments are those of Rissanen (1978, 1986) and Schwarz (1978).

3. Slow but accelerating development of nonlinear modeling methods (Tong 1990), aided most recently by the advent of new kinds of simulation methods (Gordon,

Salmond, and Smith 1993; Kitagawa and Gersch 1996).

4. Development of methods for fractional time series modeling (Beran 1994).

5. Further development of vector time series, especially relating to integrated time series. But other important topics had received substantial treatment, including linear causality (Caines 1988; Granger 1969) and errors in variables (Anderson and Deistler 1984; Deistler and Anderson 1989).

6. Development of reliable and easy-to-use software packages and platforms consequent on the huge gains in computing power in this period.

At the end of this period, time series of integrated processes stood on a firmer footing and some solid headway had been made on the difficult problem of inference for fractional time series. Also at last, the possibility for real progression on nonlinear time series modeling appears. An unfortunate development has been the mistaken belief in some quarters that frequency domain methods are no longer important.

3. THE LAST FRONTIER

This section discusses some current research directions that hold promise for the future. The list is not exhaustive and might have included more topics; for example, wavelets and chaos.

3.1 Parametric Nonlinear Modeling

The problem of fitting nonlinear state-space models to observational data has been an important open problem for a very long time. Although nonlinear filtering algorithms were written down (in continuous time) in the 1960s (Jaszwinski 1970), they were computationally intractable. Recently, however, the development of Markov chain Monte Carlo simulation methods has opened the possibility for progress in this area. Significant advances have been made by Gordon et al. (1993), Kitagawa (1993), and Kitagawa and Gersch (1996), which opens the possibility for progress on the modeling problem. Some preliminary work in this direction has been done by Chib, Nardari, and Shephard (1998) who also provided a good set of references. A very interesting approach has also recently been developed by Gallant and Tauchen (1998). The potential here is just beginning to be exploited, and basic questions, such as the convergence of the Monte Carlo filters, are open.

3.2 Nonparametric Nonlinear Modeling

In the last few years, interest has developed in the financial econometrics community (Campbell, Lo, and MacKinlay 1997) in modeling (e.g., stock prices and interest rates) with nonlinear stochastic differential equations. Such models are necessary for more accurate forecasting, and also for the valuation of derivative financial instruments. There is a growing literature here, and the mixture of nonlinearity, non-parametrics, and stochastic differential equations makes for challenging problems (Ait-Sahalia 1996; Bosq 1998; Gallant and Tauchen 1998).

3.3 High-Dimensional Vector Time Series

Steady improvements in econometric data collection and storage over the last two decades have led to the emergence of huge time series datasets. A typical example might involve several thousand short time series (with, say, 10 years of monthly observations) all relating to an economic phenomenon of interest. The vector time series methodology so far developed is incapable of dealing with this kind of dataset, and yet data of this kind will only grow in occurrence, size, and complexity. Recent work in this area (Stock and Watson 1998) showed that remarkable forecasting gains can be made even with simple principal components analyses.

3.4 Variable (Lag) Selection

Results on order selection were mentioned earlier, but the problem of variable selection is perhaps the major unresolved problem in statistics (Breiman 1992). Given the widespread use of regression methods, its resolution would have enormous impact in many disciplines. The results cited earlier cover only restrictive order-selection problems. Recently, promising new variable selection methods based on least squares fitting with nonquadratic penalties have appeared (Alliney and Ruzinsky 1994; Chen, Donoho, and Saunders 1996; Tibshirani 1996). These methods produce solutions with exactly zeroed coefficients. A deep understanding of their abilities must be developed from both computational and theoretical perspectives. When the regressors are orthogonal, these methods collapse to methods that are currently used informally (i.e., inspection of t values) and also go under the name thresholding when used for wavelet signal or function estimation (Donoho and Johnstone 1995).

3.5 Functional Data Analysis

Huge time series datasets are appearing not only in econometrics, but also in areas such as the biomedical sciences. Here the data typically consist of a small number of very long time series (e.g., 50 independent electrocardiograms each of length 5,000). There has been some time series work on these singular longitudinal data analysis problems (Brillinger 1973, 1980; Shumway 1970, 1987), but this work was not followed up, perhaps because of limited computing facilities. Recently, these kinds of problems have gained attention under the name "functional data analysis" (Besse and Ramsay 1986; Ramsay and Silverman 1997). The difference between the new methodology and the older time series methods is that stationarity is not assumed; rather, a nonstationary covariance kernel is estimated nonparametrically.

4. THE END

Like all other sciences, time series has been affected by the computing revolution as well as the growth in data collection capability of the last two decades. These two factors together have already helped generate new solutions to old problems and produce quite new and challenging problems. Some have thought that the rise of great computing power means the demise of theory. The last 150 years of the history of science and technology includes many examples of marvelous new developments that displaced old ways and of disciplines (like optics) that have waxed and waned (several times). But through all this, theory and (now computer) experiments have continued to go hand in hand. Fluid mechanics and condensed matter physics are good examples here, as computational power is hugely important in both areas. And so it is with time series.

Although some parts of the theory have reached considerable maturity, there remain outstanding problems as well as the constant appearance of new problems. Furthermore, new tools pop up from time to time; wavelets is the most celebrated recent example. An end to time series, then, whether in theory or practice, does not seem to be in sight.

REFERENCES

Ait-Sahalia, Y. (1996), "Testing Continuous Time Models of the Spot Interest Rate," *Review of Financial Studies*, 9, 385–426.

Akaike, H. (1969), "Fitting Autoregressive Models for Prediction," *Annals of the Institute of Statistical Mathematics*, 21, 243–247.

Akaike, N. (1974), "Stochastic Theory of Minimal Realization," *IEEE Transactions in Automated Control*, 19, 667–674.

Aki, K., and Richards, P. (1980), *Quantitative Seismology: Theory and Practice*, New York: W.H. Freeman.

Alliney, S., and Ruzinsky, S. (1994), "An Algorithm for the Minimization of Mixed l_1 and l_2 Norms With Application to Bayesian Estimation," *IEEE Transactions in Signal Processing*, 42, 618–627.

Anderson, B., and Deistler, M. (1984), "Identifiability of Dynamic Errors in Variables Models," *Journal of Time Series Analysis*, 5, 1–13.

Bartlett, M. (1955), *An Introduction to Stochastic Processes*, Cambridge, U.K.: Cambridge University Press.

Beran, J. (1994), *Statistics for Long-Memory Processes*, London: Chapman and Hall.

Besse, P., and Ramsay, J. (1986), "Principal Components Analysis of Sampled Functions," *Psychometrika*, 51, 285–311.

Bosq, D. (1998), *Nonparametric Statistics for Stochastic Processes*, New York: Springer-Verlag.

Box, G., and Jenkins, G. (1970), *Time Series Analysis: Forecasting and Control*, San Francisco: Holden-Day.

Breiman, L. (1992), "The Little Bootstrap and Other Methods for Dimensionality Reduction in Regression: X-Fixed Prediction Error," *Journal of the American Statistical Association*, 87, 738–752.

Brillinger, D. (1973), "The Analysis of Time Series Collected in an Experimental Design," in *Multivariate Analysis III*, New York: Academic Press, pp. 241–256.

———— (1980), "Analysis of Variance and Problems Under Time Series Models," in *Handbook of Statistics*, Vol. I, Amsterdam: North-Holland, pp. 237–278.

———— (1981), *Time Series: Data Analysis and Theory*, San Francisco: Holden-Day.

Bryson, A., and Ho, Y. (1969), *Applied Optimal Control*, Cambridge, MA: Blaisdell.

Caines, P. (1988), *Linear Stochastic Systems*, New York: Wiley.

Campbell, J., Lo, A., and MacKinlay, A. (1997), *The Econometrics of Financial Markets*, Princeton, NJ: Princeton University Press.

Chen, S., Donoho, D., and Saunders, M. (1996), "Atomic Decomposition by Basis Pursuit," technical report, Stanford University.

Chib, S., Nardari, F., and Shephard, N. (1998), "Markov Chain Monte Carlo Methods for Generalised Stochastic Volatility Models," technical report, Oxford University, Dept. of Economics.

Daley, R. (1991), *Atmospheric Data Analysis*, Cambridge, U.K.: Cambridge University Press.

Deistler, M., and Anderson, B. (1989), "Linear Dynamic Errors in Variables, Some Structure Theory," *Journal of Econometrics*, 41, 39–63.

Donoho, D., and Johnstone, I. (1995), "Adapting to Unknown Smoothness via Wavelet Shrinkage," *Journal of the American Statistical Association*, 90, 1200–1224.

Durbin, J. (1960), "The Fitting of Time Series Models," *R.W. I.S.I*, 28, 233–244.

Engle, R., and Granger, C. (1987), "Co-Integration and Error Correction: Representation, Estimation and Testing," *Econometrica*, 55, 251–276.

Fuller, W. (1976), *Introduction to Statistical Time Series*, New York: Wiley.

Gallant, M., and Tauchen, G. (1998), "Reprojecting Partially Observed Systems With Application to Interest Rate Diffusions," *Journal of the American Statistical Association*, 93, 10–25.

Gordon, N., Salmond, D., and Smith, A. (1993), "Novel Approach to Nonlinear/Nongaussian Bayesian State Estimation," *IEE: Proc. F*, 140, 107–113.

Granger, C. (1969), "Investigating Causal Relations by Econometric Models and Cross-Spectral Methods," *Econometrica*, 37.

Hall, P., and Heyde, C. (1980), *Martingale Limit Theory and Its Application*, New York: Academic Press.

Hamilton, J. (1994), *Time Series Analysis*, Princeton, NJ: Princeton University Press.

Hannan, E. (1970), *Multiple Time Series*, New York: Wiley.

Hannan, E., and Deistler, M. (1988), *Statistical Theory of Linear Systems*, New York: Wiley.

Harvey, A. (1989), *Forecasting, Structural Time Series Models and the Kalman Filter*, Cambridge, U.K.: Cambridge University Press.

Haykin, S. (1983), *Communication Systems*, New York: Wiley.

Jaszwinski, A. (1970), *Stochastic Processes and Filtering Theory*, New York: Academic Press.

Jenkins, G., and Watts, D. (1969), *Spectral Analysis and Its Applications*, San Francisco: Holden-Day.

Kailath, J. (1980), *Linear Systems*, New York: Wiley.

Kalman, R. (1960), "A New Approach to Linear Filtering and Prediction Problems," *Journal of Basic Engineering*, 82, 35–45.

Kay, S. (1988), *Modern Spectral Estimation*, Englewood Cliffs, NJ: Prentice-Hall.

Kitagawa, G. (1993), "A Monte Carlo Filtering and Smoothing Method for Non-Gaussian Nonlinear State-Space Models," in *Proceedings of the Second U.S.–Japan Joint Seminar on Time Series*, pp. 110–131. Institute of Statistical Mathematics.

Kitagawa, G., and Gersch, W. (1996), *Smoothness Priors Analysis of Time Series*, Berlin: Springer-Verlag.

Ljung, L. (1987), *System Identification: Theory for the User*, Englewood Cliffs, NJ: Prentice-Hall.

Papoulis, A. (1968), *Systems and Transforms With Applications in Optics*, Malabar, FL: Krieger.

Ramsay, J., and Silverman, B. (1997), *Functional Data Analysis*, New York: Springer-Verlag.

Rissanen, J. (1978), "Modeling by Shortest Date Description," *Automatica*, 14, 465–471.

——— (1986), "Stochastic Complexity and Modeling," *The Annals of Statistics*, 14, 1080–1100.

Robinson, E. (1967), *Multichannel Time Series*, New York: Prentice-Hall.

Schwarz, G. (1978), "Estimating the Dimension of a Model," *The Annals of Statistics*, 6, 461–464.

Shibata, R. (1976), "Selection of the Order of an Autoregressive Model by AIC," *Biometrika*, 63, 117–126.

——— (1980), "Asymptotically Efficient Selection of the Order of the Model for Estimating Parameters of a Linear Process," *The Annals of Statistics*, 8, 147–164.

Shumway, R. (1970), "Applied Regression and Analysis of Variance for Stationary Time Series," *Journal of the American Statistical Association*, 65, 1527–1546.

——— (1987), *Applied Statistical Time Series Analysis*, Englewood Cliffs, NJ: Prentice-Hall.

Steward, E. (1987), *Introduction to Fourier Optics*, Sussex, U.K.: Ellis Horwood.

Stock, J., and Watson, M. (1998), "Diffusion Indexes," technical report, Harvard University.

Tibshirani, R. (1996), "Regression Shrinkage and Selection via the Lasso," *Journal of the Royal Statistical Society*, Ser. B, 58, 267–288.

Tong, H. (1990), *Non-Linear Time Series*, Oxford, U.K.: Oxford University Press.

Wahba, G. (1990), *Spline Models for Observational Data*, Philadelphia: SIAM.

Whittle, P. (1963a), "On the Fitting of Multivariate Autoregression and the Approximate Canonical Factorisation of the Spectral Density Matrix," *Biometrika*, 50, 129–134.

——— (1963b), *Prediction and Regulation by Linear Least Squares Methods*, Princeton, NJ: van Nostrand.

Wiener, M. (1949), *Extrapolation, Interpolation and Smoothing of Stationary Time Series*, New York: Wiley.

Principal Information Theoretic Approaches

Ehsan S. Soofi

1. INTRODUCTION

Among the impacts of World War II on the scientific endeavors was the development of information entropy in the field of communication engineering by Shannon (1948). No more than a decade was needed for developing information theoretic principles of inference and methodologies based on the entropy and its generalizations in statistics and physics by Kullback and Leibler (1951), Kullback (1954), Lindley (1956), and Jaynes (1957).

The fundamental contribution of information theory to statistics has been to provide a unified framework for dealing with notion of information in a precise and technical sense in various statistical problems. As Kullback (1959) stated, "We shall use information in the technical sense to be defined, and it should not be confused with our semantic concept, though it is true that the properties of the measure of information following from the technical definition are such as to be reasonable according to our intuitive notion of information." In the seminal book, Kullback (1959), a unification of "heterogeneous development of statistical procedures scattered through the literature" was attained by "a consistent application of the concepts and properties of information theory."

During the 1960s and 1970s, information theoretic methods and principles were further developed by the continuous endeavors of Kullback and his associates, Lindley and his students, and Jaynes, with important contributions by Zellner and Akaike. During the 1980s and 1990s, information theoretic methods became integral parts of statistics and were developed for testing, estimation, prediction, classification, association, modeling, and diagnostics with applications in various branches of statistics and related fields (see, e.g., Alwan, Ebrahimi, and Soofi 1998; Bozdogan 1994; Brockett 1991; Brockett, Charnes, Cooper, Learner, and Phillips 1995; Burnham and Anderson 1998; Cover and Thomas 1991; Fomby and Hill 1997; Golan, Judge, and Miller 1996; Kapur 1989; Maasoumi 1993; Pourahmadi and Soofi 2000; Soofi 1994; Soofi, Ebrahimi, and Habibullah 1995; Theil and Fiebig 1984; Zellner 1997).

Ehsan S. Soofi is Professor of Business Statistics, School of Business Administration, University of Wisconsin-Milwaukee, P.O. Box 742, Milwaukee, WI 53201 (E-mail: esoofi@csd.uwm.edu). The author acknowledges with thanks the comments received from Arnold Zellner and George Casella on the first draft of this article.

2. INFORMATION FRAMEWORK

Information, in a technical sense, in many seemingly diverse statistical problems is quantified in a unified manner by using a suitably chosen *discrimination information* (Kullback-Leibler, cross-entropy, relative entropy) function

$$K(f : g) \equiv \int \log \frac{f(x)}{g(x)} dF(x) \geq 0, \tag{1}$$

where $f(x) = dF(x)$ is a probability density (mass) function, absolutely continuous with respect to g. Equality holds if and only if $f(x) = g(x)$ almost everywhere. $K(f : g)$ is the principal information function introduced by Kullback and Leibler (1951) as a generalization of the information functions developed by Shannon (1948). The symmetric version of (1), $J(f, g) = K(f : g) + K(g : f)$, was used by Jeffreys (1946) as a measure of divergence between two distributions, but not as an information function. The log-ratio in (1) was used by Good (1950) as the *weight of evidence* for $f(x)$, given x.

Two information functions of Shannon (1948) are special cases of (1). Shannon's entropy is

$$\begin{aligned} H(X) &\equiv H(f) = -\int \log f(x) \, dF(x) \\ &= H(U) - K(f : U), \end{aligned} \tag{2}$$

with U being the uniform distribution. Shannon derived the discrete version of $H(X)$ based on a set of axioms of information and named it *entropy* because of its similarity with thermodynamics entropy. The continuous version was defined by analogy. In the discrete case with a finite number of points, $H(X)$ is nonnegative and measures the expected information in a signal x transmitted without noise from the source X according to a distribution $f(x)$. The entropy measures uniformity of $f(x)$ and is used in various fields as measures of disorder, diversity, and uncertainty in predicting an outcome x. The negative entropy, $-H(X) = E_f[\log f(X)]$, is the average log-height of the density and provides a measure of concentration of the distribution and is used as a measure of information about x. A distribution f_2 is more informative than f_1 if $H(f_1) - H(f_2) > 0$.

Shannon's mutual information function is defined by the expected entropy difference,

$$\begin{aligned} \vartheta(Y \wedge X) &\equiv H(Y) - E_x[H(Y|x)] \\ &= K[f(x, y) : f(x)f(y)], \end{aligned} \tag{3}$$

where $H(Y|x)$ is the entropy of the conditional distribution $f(y|x)$. The mutual information is Shannon's measure of the expected information about Y transmitted through a noisy channel. As is apparent in (3), $\vartheta(Y \wedge X)$ is nonnegative and measures the extent of functional dependency between X and Y; $\vartheta(Y \wedge X) = 0$ if and only if the two variables are independent.

Kullback and Leibler (1951), based on the Bayes theorem, interpreted (1) as the expected *information* in x for discrimination between f and g, studied its mathematical properties, showed its relationship with Fisher information (Fisher 1921), and introduced the notion of *discrimination sufficiency*. Let $Y = T(X)$ be a transformation, and let $f_Y(y)$ and $g_Y(y)$ denote the distributions induced by T on $f_X(x)$ and $g_X(x)$, where f is absolutely continuous with respect to g. Then $K(f_Y : g_Y) \leq K(f_X : g_X)$ with equality if and only if T is sufficient for discrimination, i.e.,

$$\frac{f_Y(T(x))}{g_Y(T(x))} = \frac{f_X(x)}{g_X(x)}, \tag{4}$$

for almost all x. When $f = f(x|\theta)$ and $g = g(x|\theta)$, $K(f_Y : g_Y) \leq K(f_X : g_X)$ with equality if and only if T is a sufficient statistic for θ. Discrimination sufficiency is considered to be a generalization of sufficiency in the parametric case.

Kullback (1954) established a number of important inequalities for (1) through which the basic form of the distribution f^* that minimizes $K(f : g)$ subject to some constraints on f was found. Kullback also developed the notion of *discrimination efficiency* in terms of $K(f_Y : g_Y)/K(f_X : g_X) \leq 1$, with equality if and only if (4) holds. Information efficiency provides a generalization of parametric estimation efficiency in terms of the Cramer–Rao inequality.

Kullback (1959) fully developed the theoretical grounds for various applications of the minimum discrimination information (MDI) statistics. Many of the known statistical results and procedures to date were shown to be related to (1), and new results were found. In a very general and formal framework, Kullback (1959) showed that "grouping, condensation, or transformation of observations by a statistic will, in general, result in loss of information," and that "information theory provides a unification of known results, and leads to natural generalizations and the derivation of new results." The MDI model relative to a reference distribution g was obtained by minimizing $K(f : g)$ with respect to f subject to the information constraint of the form $E_f[T_j(X)] = \theta_j, j = 1, \ldots, J$. The MDI model, if it exists, is in the form of

$$f^*(x|\boldsymbol{\theta}, g) = Cg(x) \exp\left[\sum_{j=1}^{J} \eta_j T_j(x)\right], \tag{5}$$

where $\eta_j = \eta_j(\boldsymbol{\theta}), j = 1, \ldots, J$ are Lagrange multipliers and $C = C(\boldsymbol{\theta})$ is the normalizing factor. Various MDI procedures, their applications, and computational algorithms developed by Kullback and his associates for analysis of categorical data were given by Gokhale and Kullback (1978).

An *information processing rule* for inference was formulated by Zellner (1988). The rule is defined in terms of the change in predata and postdata information due to statistical processing by

$$\Delta[f_p(\theta|x)] \equiv [\text{output information}] - [\text{input information}], \quad (6)$$

where the input information is the sum of information in an *antedata* density $f_a(\theta)$ and in the conditional data distribution $f(x|\theta)$, and the output information is the sum of information in the *postdata* density $f_p(\theta|x)$ and the predictive density $f(x)$. Zellner used various *logarithmic* measures of information in (6) such that

$$\Delta[f_p(\theta|x)] = \int f_p(\theta|x) \log \left[\frac{f_p(\theta|x)f(x)}{f_a(\theta)f(x|\theta)} \right] d\theta. \quad (7)$$

Zellner defined the *information conservation principle* as $\Delta[f_p(\theta|x)] = 0$, and showed that the Bayes rule is the optimal information processing rule, in that $\Delta[f_p(\theta|x)] = 0$ if and only if $f_p(\theta|x) = f_b(\theta|x)$, where $f_b(\theta|x)$ is the posterior distribution obtained by updating $f_a(\theta)$ via the Bayes rule. Kullback, in his discussion of Zellner's article, noted that the information processing rule (6) is a discrimination information function,

$$\Delta[f_p(\theta|x)] = K[f_p(\theta|x) : f_b(\theta|x)]. \quad (8)$$

However, Zellner (1988) maintained that formulation (7) "can be readily generalized to apply to other information processing problems." Jaynes, in his discussion of Zellner's article, emphasized a noteworthy aspect of the information framework by noting that "the logarithmic measures of information might appear arbitrary at first glance; yet as Kullback showed, this is not the case."

3. INFORMATION ABOUT PARAMETER

The most celebrated information measure in statistics is the one developed by Fisher (1921) for the purpose of quantifying information in $f(x|\theta)$ about the parameter, given by

$$\mathcal{F}(\theta) \equiv \int \left[\frac{\partial \log f(x|\theta)}{\partial \theta} \right]^2 dF(x|\theta). \quad (9)$$

$\mathcal{F}(\theta)$ is a measure of information in the sense that it quantifies "the ease with which a parameter can be estimated" by x (Lehmann 1983, p. 120).

Kullback and Leibler (1951) showed that when $f = f_\theta$ and $g = f_{\theta+\Delta\theta}$ belong to the same parametric family where θ and $\theta + \Delta\theta$ are two neighboring points in the parameter space Θ, then $K(f_\theta : f_{\theta+\Delta\theta}) \approx 2(\Delta\theta)^2 \mathcal{F}(\theta)$. Thus $\mathcal{F}(\theta)$ can be interpreted in terms of the expected information in x for discrimination between the neighboring points in Θ. Along this line, Rao (1973) reiterated that information about a parameter should be measured in terms of a discrepancy measure between f_θ and

$f_{\theta+\Delta\theta}$ and showed that $\mathcal{F}(\theta)$ is an approximation to the Hellinger distance between f_θ and $f_{\theta+\Delta\theta}$. At the conceptual level, Rao characterized information as follows: "By information on an unknown parameter θ contained in a random variable or its distribution, we mean the extent to which uncertainty regarding the unknown value of θ is reduced as a consequence of an observed value of the random variable" (Rao 1973, p. 331). In Rao's characterization of information, the *existence* of uncertainty about the parameter is presumed, but is not mapped by a probability distribution.

Lindley (1956) was the first to develop a measure of information in data x about a parameter θ that ranges over the parameter space Θ with the prior distribution $f(\theta)$. Lindley adopted Shannon's mutual information for measuring the expected information in data x about θ as

$$\vartheta(\Theta \wedge X) = H(\Theta) - E_w[H(\Theta|x)]$$
$$= E_x\{K[f(\theta|x) : f(\theta)]\}. \tag{10}$$

Lindley (1961) showed that ignorance between two neighboring values θ and $\Delta\theta$ in the parameter space implies that $\vartheta(\Theta \wedge X) \approx 2(\Delta\theta)^2 \mathcal{F}(\theta)$. Bernardo (1979a), based on the second expression, provided an expected utility interpretation of $\vartheta(\Theta \wedge X)$ that is now prevalent in Bayesian literature.

Lindley's measure has been successfully applied in developing information loss (gain) diagnostics for experimental design, collinearity, dimension reduction, and censoring problems. This is an active and promising line of research.

In Lindley's adoption of Shannon's mutual information, the expected reduction of uncertainty about θ is computed by averaging the entropy of the posterior distribution in (10) with respect to the marginal distribution $f(x)$. Zellner (1971) defined an information function as the difference between the prior entropy and the entropy of the sampling distribution (likelihood) $H[f(x|\theta)]$, averaged with respect to the prior distribution,

$$G(\Theta) \equiv H(\Theta) - E_\theta[H(X|\theta)]$$
$$= E_\theta\{K[f(x|\theta) : f(\theta)]\}. \tag{11}$$

The second expression gives an interpretation of $G(\Theta)$ in terms of the discrimination information function between the likelihood and the prior. Zellner (1997) gave a new interpretation of $G(\Theta)$ in terms of "the *total* information provided by an experiment over and above the prior," defined as $I(\exp) \equiv H(\Theta) - H(\Theta, X) = -E_\theta[H(X|\theta)]$, where $H(\Theta, X)$ is the entropy of the joint distribution.

The information functions (9)–(11) have played major roles in developing data-based prior distributions for θ (see, e.g., Yang and Berger 1996). Jeffreys's invariant prior is proportional to the square root of Fisher information. The *reference* priors proposed by Bernardo (1979b) are obtained by maximizing $\vartheta(\Theta \wedge X)$ with respect to $f(\theta)$ (see Berger 2000). In general, maximization of $\vartheta(\Theta \wedge X)$ does not give an explicit solution, and a reference prior is obtained as an approximate solution. The maximal data information prior (MDIP) maximizes $G(\Theta)$, provides solutions in the form of $p(\theta) \propto \exp\{-H[f(x|\theta)]\}$, and is capable of incorporating side information. Currently, developing sample-based (objective) priors is an active area of research, and information theory has much to offer in this endeavor.

4. MAXIMUM ENTROPY

Parallel to the developments of the MDI in statistics by Kullback, Jaynes (1957) developed the *maximum entropy* (ME) principle of scientific inference in physics. The ME distribution is the one that maximizes (2) with respect to f subject to a set of constraints that reflect some partial knowledge about f. The ME model subject to moment constraints is given by (5) with $g(x) = 1$. In the general case, the MDI is referred to as the *minimum cross-entropy* principle.

The ME is a formalized approach to developing probability distributions based on partial knowledge and is considered a generalization of Laplace's "principle of insufficient reason" for assigning probabilities (Jaynes 1968). Axiomatic justifications for the MDI and ME approaches were given by Csiszar (1991) and Shore and Johnson (1980).

The range of applications of ME is quite broad and provides ample research opportunities. Some examples are as follows. Many well-known parametric families of distributions are characterized as ME subject to specific constraints. Developing ME distributional fit diagnostics requires nonparametric entropy estimation, which is an active research area (see, e.g., Soofi et al. 1995). In the categorical case, various logit and log-linear models are derived as MDI and ME solutions subject to various forms of constraints on covariates and indicators of categories (Gokhale and Kullback 1978; Soofi and Gokhale 1997). The ME inversion technique is shown to be remarkably powerful for image recovery (see, e.g., Gull 1989). An entropy-moment inequality given by Wyner and Ziv (1969) provides a sharp lower bound for the prediction error variance in terms of the entropy, in the same spirit as the inverse of Fisher information in the Cramer–Rao inequality (see Pourahmadi and Soofi 2000). Recently, Zellner proposed the Bayesian method-of-moments (BMOM) procedure, in which data-based moments for parameters are used for producing postdata (posterior and predictive) distributions by the ME (see Zellner 1997). The BMOM procedure is a versatile alternative when a likelihood is not available for application of the Bayes theorem.

5. MINIMUM DISCRIMINATION INFORMATION LOSS ESTIMATION

A parametric model, $f^*(x|\theta)$, $\theta = (\theta_1, \ldots, \theta_J)'$, derived by the MDI, ME, or simply assumed, is a convenient mathematical formula that we utilize in practice and hope that it is a reasonable approximation to the unknown true data-generating distribution $f(x)$. Given a sample $\mathbf{x} = (x_1, \ldots, x_n)$ from $f(\mathbf{x})$, it is then natural to estimate the model in such a way that the model approximation to the data-generating distribution is improved. The MDI or minimum relative entropy loss estimation procedure serves this purpose.

The loss of approximating $f(x)$ by an estimated model $f^*(x|\tilde{\boldsymbol{\theta}})$ is measured by the information discrepancy $K[f(\mathbf{x}) : f^*(\mathbf{x}|\tilde{\boldsymbol{\theta}})]$. The MDI or minimum relative entropy loss estimate of $\boldsymbol{\theta}$ is defined by

$$\tilde{\boldsymbol{\theta}}_{\text{MDI}} = \arg \min_{\boldsymbol{\theta}} K[f(\mathbf{x}) : f^*(\mathbf{x}|\boldsymbol{\theta})]. \tag{12}$$

The entropy loss has been used with frequentist and Bayesian risk functions in various parametric estimation problems and for model selection (see Sooti 1997 and references therein).

Akaike (1973) showed that "choice of the information theoretic loss function is a very natural and reasonable one to develop a unified asymptotic theory of estimation." He developed an MDI approach for estimating the parameter $\boldsymbol{\theta}_J$ of the family of models $f^*(\mathbf{x}|\boldsymbol{\theta}_J)$ when the dimension $J, J = 1, \ldots, L$ is unknown. Akaike (1974) observed that decomposing the log-ratio in (1) gives

$$K[f(\mathbf{x}) : f^*(\mathbf{x}|\tilde{\boldsymbol{\theta}}_J)] = H_f[f^*(\mathbf{x}|\tilde{\boldsymbol{\theta}}_J)] - H[f(\mathbf{x})], \tag{13}$$

where

$$H_f[f^*(\mathbf{x}|\tilde{\boldsymbol{\theta}})] \equiv -E_f[\log f^*(\mathbf{X}|\tilde{\boldsymbol{\theta}})]. \tag{14}$$

The entropy of the data-generating distribution is free of $\tilde{\boldsymbol{\theta}}_J$, so $H[f(\mathbf{x})]$ in (13) can be ignored in the minimization. By this clever observation, the cumbersome problem of minimizing the information discrepancy between the unknown data-generating distribution and the model is reduced to the simpler problem of maximizing the expectation in (14). Akaike proposed estimating (14) by the mean log-likelihood function, which gives

$$\tilde{\boldsymbol{\theta}}_{J_{\text{MDI}}} = \arg \max_{\boldsymbol{\theta}_{\mathbf{J}}} \frac{1}{n} \log f^*(\mathbf{x}|\boldsymbol{\theta}_{\mathbf{J}}) = \hat{\boldsymbol{\theta}}_J,$$

where $\hat{\boldsymbol{\theta}}_J$ is the maximum likelihood estimate (MLE) of $\boldsymbol{\theta}_J$ under the model $f^*(\mathbf{x}|\boldsymbol{\theta}_{\mathbf{J}})$. Thus the MLE minimizes an estimate of the information discrepancy between the data-generating distribution $f(\mathbf{x})$ and the model $f^*(\mathbf{x}|\boldsymbol{\theta}_{\mathbf{J}})$ over the parameter space. When a set of models with various J or in different families is under consideration, first the parameters of each candidate model are estimated by the MLE, and then the optimal MDI model in the set is selected. Akaike interpreted this approach as an extension of the maximum likelihood principle.

Akaike (1973), under the assumption of $f(x) = f^*(x|\boldsymbol{\theta}_L)$ with $\boldsymbol{\theta}_L = (\theta_1, \ldots, \theta_J,$ $\theta_{J+1}, \ldots, \theta_L)'$, computed an approximate frequentist risk of selecting a submodel of $f^*(\mathbf{x}|\hat{\boldsymbol{\theta}}_L)$ as

$$2E_{\hat{\boldsymbol{\theta}}_J} K[f^*(\mathbf{x}|\boldsymbol{\theta}_L) : f^*(\mathbf{x}|\hat{\boldsymbol{\theta}}_J)]$$
$$\approx -\frac{2}{n} \log \frac{f^*(\mathbf{x}|\hat{\boldsymbol{\theta}}_J)}{f^*(\mathbf{x}|\hat{\boldsymbol{\theta}}_L)} + \frac{2J}{n} - \frac{L}{n}, \qquad J \le L. \tag{15}$$

In a problem that n, L, and the likelihood function $f^*(\mathbf{x}|\hat{\boldsymbol{\theta}}_L)$ remain constant, the minimum risk model is the one that minimizes the information criterion

$$\text{AIC}(J) = -2 \log f^*(\mathbf{x}|\hat{\boldsymbol{\theta}}_J) + 2J, \qquad J = 1, 2, \ldots, L.$$

Akaike's work popularized application of an information function among users of statistics and generated vast interest in developing model selection diagnostics, which has been a very active line of research for the last quarter of the 20th century.

REFERENCES

Akaike, H. (1973), "Information Theory and an Extension of the Maximum Likelihood Principle," *2nd International Symposium on Information Theory*, 267–281.

——— (1974), "A New Look at the Statistical Model Identification," *IEEE Transactions of Automated Control*, AC-19, 716–723.

Alwan, L. C., Ebrahimi, N., and Soofi, E. S. (1998), "Information-Theoretic Framework for Statistical Process Control," *European Journal of Operational Research*, 111, 526–541.

Berger, J. O. (2000), "Bayesian Analysis: A Look at Today and Thoughts for Tomorrow," *Journal of the American Statistical Association*, 95, 1269–1276.

Bernardo, J. M. (1979a), "Expected Information as Expected Utility," *The Annals of Statistics*, 7, 686–690.

——— (1979b), "Reference Posterior Distributions for Bayesian Inference" (with discussion), *Journal of the Royal Statistical Society*, Ser. B, 41, 113–147.

Bozdogan, H. (1994), *Proceedings of the First U.S./Japan Conference of the Frontiers of Statistical Modeling*, Vols. 1–3, Amsterdam: Kluwer.

Brockett, P. L. (1991), "Information Theoretic Approach to Actuarial Science: A Unification and Extension of Relevant Theory and Applications," *Transactions of Society of Actuaries*, 43, 73–135.

Brockett, P. L., Charnes, A., Cooper, W. W., Learner, D., and Phillips, F. Y. (1995), "Information Theory as a Unifying Statistical Approach for Use in Marketing Research," *European Journal of Operational Research*, 84, 310–329.

Burnham, K. P., and Anderson, D. R. (1998), *Model Selection and Inference: A Practical Information-Theoretic Approach*, New York: Springer-Verlag.

Cover, T. M., and Thomas, J. A. (1991), *Elements of Information Theory*, New York: Wiley.

Csiszar, I. (1991), "Why Least Squares and Maximum Entropy? An Axiomatic Approach to Inference in Linear Inverse Problems," *The Annals of Statistics*, 19, 2032–2066.

Fisher, R. A. (1921), "On Mathematical Foundations of Theoretical Statistics," *Philosophical Transactions of the Royal Society of London*, Ser. A, 222, 309–368.

Fomby, T. B., and Hill, R. C. (1997), *Advances in Econometrics: Applying Maximum Entropy to Econometric Problems*, Vol. 12, Greenwich, CT: JAI Press.

Gokhale, D. V., and Kullback, S. (1978), *The Information in Contingency Tables*, New York: Marcel Dekker.

Golan, A., Judge, G. G., and Miller, D. (1996), *Maximum Entropy Econometrics: Robust Estimation with Limited Data*, New York: Wiley.

Good, I. J. (1950), *Probability and Weighting of Evidence*, London: Griffin.

Gull, S. F. (1989), "Developments in Maximum Entropy Data Analysis," in *Maximum Entropy and Bayesian Methods*, ed. J. Skilling, Boston: Kluwer.

Jaynes, E. T. (1957), "Information Theory and Statistical Mechanics," *Physical Review*, 106, 620–630.

——— (1968), "Prior Probabilities," *IEEE Transactions Systems in Science and Cybernetics*, SSC-4, 227–241.

Jeffreys, H. (1946), "An Invariant Form for the Prior Probability in Estimation Problems," *Proceedings of the Royal Statistical Society (London)*, Ser. A, 186, 453–461.

Kapur, J. N. (1989), *Maximum Entropy Models in Science and Engineering*, New York: Wiley.

Kullback, S. (1954), "Certain Inequalities in Information Theory and the Cramer–Rao Inequality," *The Annals of Mathematical Statistics*, 25, 745–751.

——— (1959), *Information Theory and Statistics*, New York: Wiley.

Kullback, S., and Leibler, R. A. (1951), "On Information and Sufficiency," *Annals of Mathematical Statistics*, 22, 79–86.

Lehmann, E. L. (1983), *Theory of Point Estimation*, New York: Wiley.

Lindley, D. V. (1956), "On a Measure of Information Provided by an Experiment," *Annals of Mathematical Statistics*, 27, 986–1005.

——— (1961), "The Use of Prior Probability Distributions in Statistical Inference and Decision," in *Proceedings of the Fourth Berkeley Symposium*, 1, Berkeley, CA: University of California Press, pp. 436–468.

Maasoumi, E. (1993), "A Compendium to Information Theory in Economics and Econometrics," *Econometric Reviews*, 12(2), 137–181.

Pourahmadi, M., and Soofi, E. S. (2000), "Predictive Variance and Information Worth of Observations in Time Series," *Journal of Time Series Analysis*, 21, 413–434.

Rao, C. R. (1973), *Linear Statistical Inference and Its Applications* (2nd ed.), New York: Wiley.

Shannon, C. E. (1948), "A Mathematical Theory of Communication," *Bell System Technical Journal*, 27, 379–423.

Shore, J. E., and Johnson, R. W. (1980), "Axiomatic Derivation of the Principle of Maximum Entropy and the Principle of Minimum Cross-Entropy," *IEEE Transactions on Informational Theory*, IT-26, 26–37.

Soofi, E. S. (1994), "Capturing the Intangible Concept of Information," *Journal of the American Statistical Association*, 89, 1243–1254.

——— (1997), "Information Theoretic Regression Methods," in *Advances in Econometrics: Applying Maximum Entropy to Econometric Problems*, 12, eds. T. B. Fomby and R. C. Hill, Greenwich, CT: JAI Press, pp. 25–83.

Soofi, E. S., Ebrahimi, N., and Habibullah, M. (1995), "Information Distinguishability With Application to Analysis of Failure Data," *Journal of the American Statistical Association*, 90, 657–668.

Soofi, E. S., and Gokhale, D. V. (1997), "Information Theoretic Methods for Categorical Data," in *Advances in Econometrics: Applying Maximum Entropy to Econometrics Problems*, 12, eds. T. B. Fomby and R. C. Hill, Greenwich, CT: JAI Press, pp. 107–134.

Theil, H., and Fiebig, D. Z. (1984), *Exploiting Continuity: Maximum Entropy Estimation of Continuous Distributions*, Cambridge, MA: Ballinger.

Wyner, A. D., and Ziv, J. (1969), "On Communication of Analog Data From Bounded Source Space," *Bell System Technical Journal*, 48, 3139–3172.

Yang, R., and Berger, J. (1997), "A Catalog of Noninformative Priors," ISDS Discussion Paper 97-42, Duke University.

Zellner, A. (1971), *An Introduction to Bayesian Inference in Econometrics*, New York: Wiley.

—— (1988), "Optimal Information Processing and Bayes's Theorem" (with discussion), *The American Statistician*, 42, 278–284.

—— (1997), *Bayesian Analysis in Econometrics and Statistics: The Zellner View and Papers*, Cheltenham, U.K.: Edward Elgar.

Measurement Error Models

L. A. Stefanski

1. INTRODUCTION

Following an invited paper session on errors-in-variables models at the 1983 American Statistical Association Annual Meeting in Toronto, two comments foretold of change in the study of regression models in which predictor variables are measured with error. The first was an opinion that the negative connotation of the name "errors-in-variables models" was a hindrance to the impact of the research on such models. The second was a complaint that the scope of models studied was not keeping pace with the complexity of problems encountered in practice.

With regard to names and key words, *errors-in-variables models, regression with errors in x*, and *measurement error models* are still in use. However, publication of the book *Measurement Error Models* (Fuller 1987) established the latter as the phrase associated with the study of regression models in which the independent variables are measured with error and explains why this vignette shares the same title.

As for the second comment, the fact that statisticians are moving toward more complex models in general is clear from reading any current issue of *JASA*. At the time of the Toronto meetings, the field of measurement error modeling was already broadening to include more realistic models required for certain applications. In the period since then, the breadth and complexity of the models studied under the heading of measurement error models has grown at an ever-increasing rate. This expansion is straining the terminology, notation, and assumptions that were adequate for describing linear errors-in-variables models, and also is creating new problems for research.

Despite the rapid developments in the field and the increasing use of measurement error models in practice over the last 15–20 years, it is likely that many readers are not familiar with measurement error problems. The sections that follow provide an introduction to measurement error problems and models and a necessarily cursory review of the area, emphasizing certain key contributions and current and future research trends.

L. A. Stefanski is Professor, Department of Statistics, North Carolina State University, Raleigh, NC 27695 (E-mail: stefansk@stat.ncsu.edu). The author thanks *JASA* for the opportunity to write this vignette and acknowledges the editor's welcomed encouragement, the financial support of the National Science Foundation, and the career-long support as well as the specific suggestions of R. J. Carroll and David Ruppert.

2. MEASUREMENT ERROR PROBLEMS

In many areas of application, statistically meaningful models are defined in terms of variables X that for some reason are not directly observable. In such situations, it is common for substitute variables W to be observed instead. The substitution of W for X complicates the statistical analysis of the observed data when the purpose of the analysis is inference about a model defined in terms of X. Problems of this nature are commonly called measurement error problems, and the statistical models and methods for analyzing such data are called measurement error models.

The terminology belies the generality of the class of problems and models connoted by the phrase "measurement error." Its origins lie in the special case in which the substitute variable W is a measurement (in the usual sense of the word) of the true value of X. For example, the measured perpendicular distance W that an animal lies from a transect line can differ significantly from the true distance X depending on the method of measurement. Similarly, a person's measured systolic blood pressure W differs from his or her long-term average systolic blood pressure X because of significant temporal variation as well as instrument and reader error. These cases are fundamentally different from the situation in which the ambient concentrations of NO_2 in the bedroom and kitchen of a child's home $\mathbf{W} = (W_1, W_2)$ are substitutes for the child's personal exposure to NO_2 X (which is directly measurable via a personal monitor but not done for reasons of cost or inconvenience). The latter situation is similar to that in which airborne particulate matter concentration measured daily at a central location W is used as a substitute for daily average (over a population of individuals) personal exposure to particulate matter X in time series studies of the relationship between daily air pollution and daily mortality.

These four examples have in common the feature that for certain statistical analyses, the variable X is preferred over W: Methods for estimating animal abundance from line-transect data assume that distances are measured without error; for assessing the relationship between heart disease and blood pressure, long-term average blood pressure is the meaningful risk factor; personal exposure is the relevant risk factor in studies of the effects of NO_2 exposure on respiratory illness in children; and excess daily mortality due to exposure to particulate matter is more directly related to average personal exposure than to concentrations measured at central monitoring stations. These examples also have in common the feature that X is not observable for each observational unit, whereas W is. The problem is to fit models described in terms of X (and other observed variables) given data on W (and the other observed variables).

Other, more detailed descriptions of measurement error problems have been given by Carroll, Ruppert, and Stefanski (1995) and Fuller (1987).

2.1 Measurement Error in Simple Linear Regression

Consider the usual simple linear regression model $Y_i = \alpha + \beta X_i + \varepsilon_i, i = 1, \ldots, n$. Suppose that the observed $W_i = X_i + U_i$, where U_i represents measurement error. Let $\hat{\beta}_{Y|W}$ denote the slope estimator from the least squares regression of Y on W. Similarly let $\hat{\beta}_{Y|X}$ denote the slope estimator obtained by least squares regression of Y on X. Then in the seemingly benign case in which the measurement errors U_i have mean $\mu_U = 0$ and constant variance σ_U^2 and are independent of the true predictors X_i and the equation errors ε_i, it transpires that $\hat{\beta}_{Y|W} = \hat{\lambda}\hat{\beta}_{Y|X} + o_p(1)$, where $\hat{\lambda} = s_X^2 / (s_X^2 + s_U^2)$, where s^2 denotes the sample variance of the subscript variable. Because $0 < \hat{\lambda} < 1$, a consequence of ignoring measurement error is attenuation (bias toward 0) in the slope estimator.

This simple model illustrates the key features of measurement error in regression models. The parameters of interest (α, β) appear in the regression model for $E(Y|X)$, which depends on the unobserved variable X; the second component of the model relates the observed substitute variable W to X (in this case via the additive error model $W = X + U$, note that in this case $E(W|X) = X$); and the model fit to the observed data results in biased estimators of the parameters of interest. A less obvious feature that it shares with other measurement error models is a lack of identifiability. The regression parameters (α, β) cannot be consistently estimated without additional data (e.g., replicate measurements) or additional information in the form of distributional or moment restrictions on the error distributions or the distribution of X. An often-studied, early version of the model known as the *classical errors-in-variables model* (Fuller 1987, sec. 1.3) incorporates the assumption that the ratio of error variances $\eta = \sigma_\varepsilon^2 / \sigma_U^2$ is known. Under this identifiability assumption, the regression parameters are consistently estimated using the method of orthogonal least squares, formulated and described more than 100 years ago by R. J. Adcock (1877, 1878). However, the historical importance of the model in which the ratio of error variances is assumed known is greater than its practical significance (Carroll et al. 1995; Carroll and Ruppert 1996). Finally, it is important to distinguish between the *classical errors-in-variables model* and the *classical error model*. The latter phrase refers only to the measurement error component of a model and describes the additive error model $W = X + U$, with U and X independent (or uncorrelated).

3. LINEAR ERRORS-IN-VARIABLES MODELS

Generalizations and variations of linear error-in-variables models were studied extensively in the 100 years following Adcock's description of orthogonal least squares. The major generalizations include multiple and multivariate multiple linear regression, in which one or more predictor variables are measured with error and some predictor variables are error free. The major variations depend on assumptions about the unobserved predictors and the type of data or distributional (moment) information available, ensuring identifiability of the regression parameters.

In *functional* errors-in-variables models, the unobserved X_i are modeled as unknown, nonrandom constants (parameters), whereas in *structural* errors-in-variables models, the observables and unobservables jointly vary in repeated sampling. For example, consider modeling the relationship between aquatic species diversity Y and acid-neutralizing capacity X, given measurements of (Y, X) from each of n lakes. If the only lakes of interest are those represented in the sample, then it is appropriate to model $X_i, i = 1, \ldots, n$, as unknown constants. Alternatively, if the lakes represented in the data are a random sample from a large population of lakes, then it is appropriate to model $X_i, i = 1, \ldots, n$, as iid random variables.

Parameters are identified under a wide variety of combinations of error distributional or moment restrictions and the existence of additional data (e.g., replicate measurements), each giving rise to a unique and interesting inference problem. Discussion of the numerous variations of the model, the methods of analysis adapted to each, and the early contributions of many well-known statisticians to the field were provided by Madansky (1959). Sprent (1989) provided a more recent overview of the field, including references to other key review articles.

The traditional distinction in the literature between functional models and structural models is, for some applications, not as relevant as the distinction between *functional modeling and structural modeling* as defined by Carroll et al. (1995, sec. 1.2), which facilitates classification of estimation methods based on the strength of the assumptions made about the latent variables $X_1, X_2, \ldots$. However, functional models played an important role in the study of measurement error models and in statistics more generally.

In a functional errors-in-variables model, the unobserved latent variables are unknown parameters. Consequently, for a functional model with sample size n, and hence measurements on n latent variables, the unknown parameter vector includes $(X_1, X_2, \ldots, X_n)$ and so has dimension increasing linearly with sample size. Functional errors-in-variables models are practical and classical examples of models with infinitely many nuisance parameters, for which the deficiencies of maximum likelihood estimation are now well known. The pioneering work of Neyman and Scott (1948) and Kiefer and Wolfowitz (1956) on estimation in the presence of numerous nuisance parameters had significant impact not only on the direction of research in the errors-in-variables literature, but also on the course of research in statistics in general.

4. MEASUREMENT ERROR IN NONLINEAR MODELS

In the 1980s, the research emphasis in measurement error models shifted to problems in which the model of interest defined in terms of X is something other than a linear model. With increasing frequency, articles appeared on nonlinear regression models with errors in both variables, generalized linear models with predictor variables measured with error, and nonparametric distribution/density/regression estimation in the presence of measurement error. But these are not the first investigations of measurement error in nonlinear models. Eddington (1913) studied distribution estimation, and Trumpler and Weaver (1953, sec. 1.56) described an interesting approach to nonlinear regression errors-in-variables problems; however, these researches were the exceptions, whereas by the late 1980s, the focus of measurement error modeling research was on nonlinear models.

The early research in errors-in-variables models was driven by applications in the physical sciences, especially astronomy, and soon thereafter also by econometric applications. Much of the current research in nonlinear measurement error models is motivated by applications in the health sciences; this is especially true of research in generalized linear measurement error models. The article by Carroll, Spiegelman, Lan, Bailey, and Abbott (1984) on measurement error in binary regression marks the shift in emphasis from linear to nonlinear measurement error models, and is noteworthy for breaking ground in the application of measurement error modeling in the health sciences and in the study of generalized linear models with measurement error.

Greater variety in the types of error model (the model relating the observed W and unobserved X) studied also characterizes the recent research in measurement error modeling. The so-called Berkson error model, wherein the observed values W are fixed in repeated sampling and the X values vary (Berkson 1950), which is little studied in linear models because of the lack of rectifiable problems that it causes, plays a more significant role in the study of nonlinear measurement error models. The distinction between Berkson error and classical error models is subtle but important because of the very different problems associated with each (Carroll et al. 1995, sec. 1.3; Fuller 1987, sec. 1.6.4). A frequently cited example due to Fuller (1987, sec. 1.6.4) illustrates the key feature of Berkson error. Consider an experiment in which the quality of cement, Y, is to be studied as a function of the amount of water in the mixture. The amount of water is controlled by setting a metered valve to specified values. Because of fluctuations in water pressure and inaccuracies in the metered valve, the true amount of water in a mixture, X, differs from the prescribed amount W. Note that W is controlled as part of the experimental design and is not random. In replication of the experiment at the same value of W, the true amount X will vary. If the valve is correctly calibrated, then a reasonable model is $X = W + U$, where U is a mean-0 error regardless of the value of W. Thus for unbiased Berkson error, $E(X|W) = W$, whereas for unbiased classical error, $E(W|X) = X$.

There are experimental and sampling designs/studies in which the pure forms of error models (classical and Berkson) arise naturally, and hence their importance in applied and theoretical work. However, there are applications where neither error model is directly applicable (measuring a child's exposure to NO_2 and measuring average exposure to particulate matter are two such examples), although either may be appropriate after transformation of the raw measurements (error calibration and regression calibration; Carroll et al. 1995, sec. 1.3). An incorrect assumption about the error model has the potential of causing problems as great as those created by ignoring measurement error completely. Thus correct error model identification is crucial to the successful use of measurement error models.

4.1 A Logistic Regression Measurement Error Model

Logistic regression is a common, important nonlinear model in which measurement error is frequently a concern. The simple version of the model described in this section illustrates key features of measurement error in nonlinear models.

The model discussed herein is not unlike that used by MacMahon et al. (1990) to investigate attenuation due to measurement in their study of blood pressure, stroke, and coronary heart disease. MacMahon et al. (1990) described the practical consequences of failing to adjust for attenuation (which they refer to as regression-dilution bias) and emphasized the need to account for measurement in the analysis of their data.

Consider the logistic regression model for the dependence of the binary response Y on the scalar predictor X in which $\Pr(Y = 1|X) = H(\alpha + \beta X)$, where $H(t) = \{1 + \exp(-t)\}^{-1}$. Given data $(X_i, Y_i), i = 1, \ldots, n$, maximum likelihood estimation requires numerical maximization. Now suppose that X_i is not observed, but the measurement $W_i = X_i + U_i$ with additive independent error is observed.

Because the maximum likelihood estimates have no closed-form expression, the effect of substituting W_i for X_i in logistic regression is not easily determined. Although it is generally true that the estimate of β is attenuated as in the case of linear regression, and by approximately the same factor, this is not always the case (Stefanski and Carroll 1985). The logistic model is typical in this regard in that assessing the effects of measurement error on estimates in nonlinear models is intrinsically more difficult than in linear models (Carroll et al. 1995, sec. 2.5).

Estimation in the logistic model with normally distributed measurement error usually proceeds under the assumption that the error variance is known, or that it is independently estimable, say from replicate measurements or validation data. The model is identified more generally, but in this case identifiability does not imply the ability to obtain estimators with acceptable finite-sample properties (Carroll et al. 1995, sec. 7.1.1). A number of estimation methods have been studied for this model and were described by Carroll et al. (1995). One such method, chosen because of its link to the work of Neyman and Scott (1948), is described next.

Consider the functional version of the logistic measurement error model with errors, U_i, that are normally distributed with known variance σ_U^2. In this case, the density of (Y_i, W_i) is

$$f_{YW}(y, w|\beta_0, \beta_x, X_i)$$

$$= \{H(\alpha + \beta X_i)\}^y \{1 - H(\alpha + \beta X_i)\}^{1-y} \frac{1}{\sigma_U} \phi\left(\frac{w - X_i}{\sigma_U}\right),$$

where $\phi(\ )$ is the standard normal density function. The functional maximum likelihood has $n + 2$ parameters $X_1, \ldots, X_n, \alpha, \beta$, and its maximization does not produce consistent estimators of the logistic regression parameters (Stefanski and Carroll 1985).

However, examination of the foregoing density reveals that the parameter-dependent statistic $\Delta_i = W_i + Y_i \sigma_U^2 \beta$ is sufficient for the unknown X_i in the sense that the conditional distribution of (Y_i, W_i) given Δ_i does not depend on the nuisance parameter X_i. This fact can be exploited to obtain unbiased estimating equations for the regression parameters (α, β) using either conditional likelihood methods or mean-variance function models (based on the conditional moments of Y_i given Δ_i) and quasi-likelihood methods. Estimating equations derived in these ways is called *conditional scores,* and the corresponding estimators are called *conditional score estimators.*

Conditioning on sufficient statistics is a common strategy for eliminating nuisance parameters, and it applies more generally to measurement error models. The key features of the logistic model with normal measurement error are that both component models are exponential family densities. Thus there are a number of other generalized linear measurement error models to which conditioning approach applies. Details and examples of the method for logistic regression and generalizations to other generalized linear models with measurement error have been provided by Carroll et al. (1995, sec. 6.4) and Stefanski and Carroll (1987).

This section closes with some numerical results providing an empirical illustration of attenuation and the conditional score estimation method described earlier. The data used are a subset of the data from the Framingham Heart Study and consist of two measurements (from two exams 2 years apart) of systolic blood pressure (SBP) and an indicator of coronary heart disease for each of 1,615 individuals. Blood pressure measurements are transformed so that the variables used in the analyses are Y, W_1, and W_2, where Y is the binary indicator and W_i is the natural logarithm of the measured SBP from the ith exam, $i = 1, 2$. The model assumed is that W_1 and W_2 are iid normal replicate measurements of X, defined here as an individual's long-term, mean log-transformed SBP. In other words $W_i = X + U_i, i = 1, 2$, where U_i is normally distributed with mean 0 and variance σ_U^2. The error variance estimated from the replicates is $\hat{\sigma}_U^2 = .006$. Define $W = (W_1 + W_2)/2$ and note that W is the preferred measurement of X with an error variance of $\sigma_U^2/2$ (with estimate .003).

The three unbiased measurements of $X (W_1, W_2, W)$ permit an empirical demonstration of attenuation due to measurement error. The measurement error variances of W_1 and W_2 are equal and are twice as large as the measurement error variance of their average W. Thus it would be expected that the attenuation due to measurement error in the regressions of Y on W_1 and Y on W_2 is equal, whereas the regression of Y on W should manifest less attenuation.

The estimates of slope from the logistic regressions of Y on W_1 and Y on W_2 are 2.98 and 2.86, with an average of 2.92. The estimate of slope from the logistic regression of Y on W is 3.30. In the absence of appropriate standard errors, it is unreasonable to claim that these estimates provide proof of attenuation due to measurement error; however, the relationships among the estimates, $2.98 \approx 2.86 < 3.30$, are consistent with what is known about measurement error and attenuation—the larger the measurement error, the greater the attenuation.

The conditional score estimate of slope using W as the measurement of X and assuming that the error variance is known and equal to .003 (a reasonable assumption in light of the large degrees of freedom associated with this variance estimate) is 3.75. Note that this estimate exceeds 3.30. This is as expected, because the latter estimate is attenuated, whereas the conditional score estimate completely corrects (at least in theory) for the attenuation due to measurement error.

The reasonableness of the conditional score estimate can be argued using a simple jackknife-like bias adjustment. Note that for the measurements with error variance of .006, the average estimated slope is 2.92 (corresponding to the average of the leave-one-out estimates in a typical jackknife analysis—in the present case, one of the two measurements is left out). For the measurement with error variance of .003, the estimated slope is 3.30 (corresponding to the full-data estimate in the usual jackknife setting—in the present case, no measurements are left out). A linear, jackknife-like bias correction extrapolates this trend to the case of zero measurement error resulting in the jackknife, bias-adjusted estimate $2(3.30) - 2.92 = 3.68$, whose closeness to the conditional score estimate (3.75) is not unexpected.

The reason that the jackknife-like estimate works is because bias (attenuation) due to measurement error is a function of the measurement error variance in much the same way that for nonlinear estimators in general, finite-sample bias is a function of inverse sample size. This fact is the basis of a jackknife-like method for reducing measurement error bias that uses simulated measurement error as an alternative to deleting measurements (Carroll et al. 1995; Cook and Stefanski 1994; Stefanski and Cook 1995).

5. FUTURE RESEARCH TRENDS

Carroll et al. (1995) described a number of viable methods of inference in non-linear measurement error models, ranging from simple moment-based methods for reducing measurement error bias to fully specified parametric-model maximum likelihood, including semiparametric and flexible-parametric maximum likelihood. Further methodological developments are inevitable in light of the increasing variety of problems tackled under the heading of measurement error modeling.

However, the long-term viability of the field will be determined by how measurement error models and methods are integrated into the sciences from which they evolved. The recent methodological developments form a foundation for inference in measurement error models, and what is needed now are strategies for implementing these methods in the commonly encountered, practical situations in which model formulation is a necessary part of the data analysis. Some relevant areas of research are model selection, model robustness, and variable selection in the presence of measurement error.

Structural measurement error modeling is appealing because of its simplicity. Once a model is specified, maximum likelihood estimation is an obvious alternative. However, full specification of a structural measurement error model requires a model for the response variable given X and other observable predictors, a model relating W and X possibly depending on other observables, and a model for the distribution of X. Thus model selection and robustness in measurement error models are multifaceted problems that are further complicated by the fact that parameter identifiability is often achieved through some combination of supplementary data and error model assumptions. In the worst of cases in which neither replicate measurements nor validation data (observational units on which both X and W are observed) are available, model parameter estimates are largely determined by assumptions about the type of error model and magnitude of the error variance. In the best of cases in which internal validation data are available (X and W observed on a subsample of the main-study observational units—in which case the problem can also be viewed as a missing-data problem; Carroll et al. 1995, sec. 1.8) there are data for estimating error-model parameters, but seldom are the validation data sufficient for error-model selection.

The fundamental inference problems on which model building and variable selection procedures depend are more complicated in measurement error models, especially with multiple correlated predictors measured with error. Known results on the adverse effects of measurement error on hypothesis tests (Carroll et al. 1995, sec. 11.3) indicate the need to account for the effects of measurement error when using (W, Y) to select variables or build models for (X, Y). In recent investigations of the health effects of particulate matter (Lipfert and Wyzga 1995; Schwartz, Dockery, and Neas 1996), wherein both coarse and fine particulates are measured with error, measurement error has been invoked to explain patterns of statistical (non) significance in the analysis of the observed data (coarse and fine particulate measured at central sites). The fact that measurement error affects variable selection and model building is not in dispute, although the severity of its effect in multivariable models is not known. Model building and variable selection in the presence of measurement error is a potentially interesting and useful topic of study.

REFERENCES

Adcock, R. J. (1877), "Note on the Method of Least Squares," *Analyst*, 4, 183–184.

——— (1878), "A Problem in Least Squares," *Analyst*, 5, 53–54.

Berkson, J. (1950), "Are There Two Regressions?," *Journal of the American Statistical Association*, 45, 164–180.

Carroll, R. J., and Ruppert, D. (1996), "The Use and Misuse of Orthogonal Regression in Linear Errors-in-Variables Models," *The American Statistician*, 50, 1–6.

Carroll, R. J., Ruppert, D., and Stefanski, L. A. (1995), *Measurement Error in Nonlinear Models*, London: Chapman and Hall.

Carroll, R. J., Spiegelman, C., Lan, K. K., Bailey, K. T., and Abbott, R. D. (1984), "On Errors-in-Variables for Binary Regression Models," *Biometrika*, 71, 19–26.

Cook, J., and Stefanski, L. A. (1994), "A Simulation Extrapolation Method for Parametric Measurement Error Models," *Journal of the American Statistical Association*, 89, 1314–1328.

Eddington, A. S. (1913), "On a Formula for Correcting Statistics for the Effects of a Known Probable Error of Observation," *Mon. Not. R. Astrom. Soc.*, 73, 359–360.

Fuller, W. A. (1987), *Measurement Error Models*, New York: Wiley.

Kiefer, J., and Wolfowitz, J. (1956), "Consistency of the Maximum Likelihood Estimator in the Presence of Infinitely Many Incidental Parameters," *Annals of Mathematical Statistics*, 27, 887–906.

Lipfert, F. W., and Wyzga, R. E. (1995), "Uncertainties in Identifying Responsible Pollutants in Observational Epidemiology Studies," *Inhalation Toxicology*, 7, 671–689.

MacMahon, S., Peto, R., Cutler, J., Collins, R., Sorlie, P., Neaton, J., Abbott, R., Godwin, J., Dyer, A., and Stamler, J. (1990), "Blood Pressure, Stroke and Coronary Heart Disease: Part 1, Prolonged Differences in Blood Pressure: Prospective Observational Studies Corrected for the Regression Dilution Bias," *Lancet*, 335, 765–774.

Madansky, A. (1959), "The Fitting of Straight Lines When Both Variables are Subject to Error," *Journal of the American Statistical Association*, 54, 173–205.

Neyman, J., and Scott, E. L. (1948), "Consistent Estimates Based on Partially Consistent Observations," *Econometrica*, 16, 1–32.

Schwartz, J., Dockery, D. W., and Neas, L. M. (1996), "Is Daily Mortality Associated Specifically With Fine Particles?," *Journal of the Air and Waste Management Association*, 46, 927–939.

Sprent, P. (1989), "Some History of Functional and Structural Relationships," in *Statistical Analysis of Measurement Error Models and Applications*, eds. P. J. Brown and W. A. Fuller, Providence, RI: American Mathematical Society.

Stefanski, L. A., and Carroll, R. J. (1985), "Covariate Measurement Error in Logistic Regression," *The Annals of Statistics*, 13, 1335–1351.

——— (1987), "Conditional Scores and Optimal Scores in Generalized Linear Measurement Error Models," *Biometrika*, 74, 703–716.

Stefanski, L. A., and Cook, J. (1995), "Simulation Extrapolation: The Measurement Error Jackknife," *Journal of the American Statistical Association*, 90, 1247–1256.

Trumpler, R. J., and Weaver, H. F. (1953), *Statistical Astronomy*, Berkeley, CA: University of California Press.

Higher-Order Asymptotic Approximation: Laplace, Saddlepoint, and Related Methods

Robert L. Strawderman

1. INTRODUCTION

Approximation is ubiquitous in both statistical theory and practice. Many approximations routinely used in statistics can be derived by approximating certain integrals and are asymptotic in nature. That is, such approximations are typically developed under the assumption that some quantity connected to the amount of information available (e.g., sample size) becomes infinitely large. In the case of approximating density and distribution functions, one example here is "local linearization" (i.e., the delta method combined with the central limit theorem). This method of approximation has a long history, with roots that can be traced back to the time of Laplace and Gauss (i.e., early 1800s). In general, local linearization leads to approximations that are accurate to "first order." Methods of "higher-order" asymptotic approximation, in principle more accurate than those of first order, include Edgeworth expansion, tilted Edgeworth expansion, and saddlepoint approximations for density and distribution functions. A related technique for approximating integrals known as Laplace's method has been a popular tool in recent work in Bayesian inference and random- and mixed-effects models.

In the last 20 years or so, the statistical literature on higher-order asymptotic approximation has grown significantly. As this article is not meant as a comprehensive review of the area, many useful and interesting contributions have necessarily been left out. For example, the vast literature on Edgeworth expansion is essentially ignored. The significant recent growth has not been limited to applications of Laplace's method and saddlepoint approximation. However, much of the work reviewed in Section 3 is connected to these basic methods of integral approximation at some level, and these works have arguably had the most significant influence on development in the area of higher-order asymptotic approximation to date. A parallel motivation for writing this article, emphasized primarily in Section 2, is that Laplace and saddlepoint approximations exist independently of statistics and have substantially broader applicability than is often recognized.

Robert L. Strawderman is Associate Professor, Department of Biometrics, Cornell University, 434 Warren Hall, Ithaca NY 14853 (E-mail: rls54@cornell.edu).

2. ASYMPTOTIC APPROXIMATIONS
FOR INTEGRALS

2.1 Laplace's Method

It is well known that the initial development of an asymptotic theory now central
to both statistical theory and practice originated with Laplace's seminal work on the
central limit theorem, a result first read before the Paris Academy in 1810 (Stigler
1986). Perhaps less well known is that Laplace's original development of the central
limit theorem actually made use of Laplace's method for approximating integrals and
Fourier inversion. Indeed, Laplace's method was actually conceived earlier in 1774
as part of the solution to a problem left unsolved by Thomas Bayes that involved
finding the posterior distribution of a binomial probability under a uniform prior.
This led to an asymptotic approximation to the (now) well-known incomplete beta
function. In 1785, Laplace refined and used his method of approximation in devising
a Fourier-type inversion formula for point probabilities derived from the generating
function for a sum of independent discrete random variables, solving a problem that
had stymied the likes of Abraham De Moivre, Thomas Simpson, and Joseph Lagrange.
These asymptotic approximations also provided the impetus for his development of
the central limit theorem. Laplace's work here grew into a considerably more general
analysis of the behavior of sums of independent random variables by the time his
monograph *Théorie analytique de probabilités* was published in 1812.

Laplace's method is designed to approximate integrals of the form

$$I(\lambda) = \int_a^b f(t)e^{\lambda\phi(t)}\,dt \tag{1}$$

as $\lambda \to \infty$, where $[a,b] \subseteq \mathbb{R}$ and $f(t)$ and $\phi(t)$ are real-valued functions with
certain properties. The intuition underlying Laplace's method is very simple and is as
follows. Suppose that $f(t)$ and $\phi(t)$ are smooth functions independent of λ and that
$t_{\max} = \max_{t \in [a,b]} \phi(t)$ maximizes $e^{\lambda\phi(t)}$ for any $\lambda > 0$. Suppose further that $e^{\lambda\phi(t)}$
becomes very strongly peaked at $t = t_{\max}$ and decreases rapidly away from $t = t_{\max}$
on $[a,b]$ as $\lambda \to \infty$; that is, $\phi(t_{\max}) < 0$. Then, as $\lambda \to \infty$, it is generally true that
the major contribution to $I(\lambda)$ comes from the behavior of the integrand in a small
neighborhood about $t_{\max}$. One common form of this result, valid when $\phi(\cdot)$ has a
unique maximum on (a,b) and generally stated under strong smoothness conditions,
is

$$I(\lambda) \approx \left(\frac{2\pi}{\lambda|\phi''(t_{\max})|} \right)^{1/2} f(t_{\max})e^{\lambda\phi(t_{\max})}. \tag{2}$$

The right side is usually referred to as the (first-order) *Laplace approximation* to
$I(\lambda)$. A standard heuristic derivation of (2) from (1) proceeds by Taylor expansion,

and eventually involves evaluating the integral of the kernel of a normal density function (see, e.g., Bleistein and Handelsman 1975, chap. 5).

The form of the expansion (2) is directly tied to the presence of the normal density function, and the usual heuristic derivation based on Taylor expansion arises because of the inherent smoothness assumption on $\phi(\cdot)$ in a neighborhood about $t = t_{\max}$. A considerably more general approximation to (1), requiring no smoothness conditions on $f(\cdot)$ or $\phi(\cdot)$, has been given by Fulks (1960). Basically, in line with the foregoing intuition, all that is really required is that the integral eventually exists and that $f(\cdot)$ and $\phi(\cdot)$ admit asymptotic expansions of a certain form in a neighborhood of the critical point $t_{\max}$. A parallel version of this result is available for multivariate integrals (Fulks and Sather 1961). One important consequence of these results is that locally *quadratic* behavior of $\phi(\cdot)$ near $t_{\max}$ (i.e., "asymptotically normal" behavior) is not a required assumption for the successful application of Laplace's method. However, in cases where such locally quadratic behavior does not hold, the form of the resulting approximation generally differs from (2).

2.2 Saddlepoint Approximations

Laplace's method for approximating integrals can be viewed as a special case of a more general class of methods for approximating integrals known as saddlepoint methods. The saddlepoint method and an important special case known as the method of steepest descent are designed to approximate integrals of the Laplace type in which both the integrand and contour of integration are allowed to be complex valued. The origin of the method of steepest descent has been traced back to a posthumous work of Riemann (1863) and was placed on a more rigorous foundation by Debye (1909). The exact origin of the saddlepoint method is less clear. These methods of approximation are probably the most well known of all procedures for determining the asymptotic behavior of integrals. Excellent introductions to the basic principles underlying the method of steepest descent and saddlepoint approximations were provided by de Bruijn (1961), Bleistein and Handelsman (1975), and Ablowitz and Fokas (1997). I now summarize these methods, combining the important points made by these three authors.

Consider the problem of approximating

$$I(\lambda) = \int_C f(z) e^{\lambda \phi(z)} \, dz \tag{3}$$

as $\lambda \to \infty$, where $z = a + ib$ is complex valued, C is a curve (i.e., contour) in the complex plane $\mathcal{C}$, and $\phi(z)$ and $f(z)$ are analytic functions of z in a region $D \subset \mathcal{C}$ that contains C. The goal of both the saddlepoint method and the method of steepest descent is to derive an approximation to (3) as $\lambda \to \infty$. Evidently, (1) is a special case of (3) in which $C, \phi(\cdot), f(\cdot)$, and z are all real valued. However, the asymptotic approximation of (3) brings with it a number of additional considerations and difficulties that do not present themselves in the approximation of (1). For example, because $\phi(z)$ is complex valued, the integrand in (3) can oscillate very rapidly as $\lambda \to \infty$, instead of decaying to 0 exponentially fast.

The basic idea behind steepest descent approximation is to deform the contour C in such a way so as to pass through important critical points of $\phi(z)$ and ensure that the imaginary portion of $\phi(z)$ is constant on this new contour. The deformation of contours is a common technique in complex analysis that is justified by Cauchy's closed-curve theorem (e.g., Bak and Newman 1996). The ability to deform the contour C into a new contour C^* on which the imaginary portion of $\phi(z)$ is constant is quite important. In particular, it eliminates the rapid oscillation of the integrand for large λ, leaving an integral that is more amenable to approximation via Laplace's method.

It is a well-known fact from complex analysis that the paths on which the imaginary portion of $\phi(z)$ are constant are also paths for which the real part of $\phi(z)$, say $u(z)$, either decreases or increases the fastest (e.g., Bleistein and Handelsman 1975, sec. 7.1). The paths on which $u(z)$ decreases the fastest are known as paths of *steepest descent;* hence the name of the method. Typically, steepest descent paths pass through a point (or points) z_0 for which $\phi'(z) = 0$ (i.e., saddlepoints), and, much like Laplace's method, the major contributions to $I(\lambda)$ come from the behavior of the integrand in a neighborhood of these points. In the case where $u(z)$ is maximized at a point interior to the path C, the following expansion often holds (Watson 1944, sec. 8.3):

$$I(\lambda) = e^{\lambda\phi(z_0)}\left[f(z_0)\left(-\frac{2\pi}{\lambda\phi''(z_0)}\right)^{1/2} + O(\lambda^{-3/2})\right]. \qquad (4)$$

The method of steepest descent is a special case of the more commonly used saddlepoint method. In the saddlepoint method, the original steepest descent contour C^* is replaced by a new contour C^{**} that coincides with C^* over only a small segment, say S^*. The coincidental segment S^* is chosen to contain a relevant saddlepoint; that is, a point for which $\phi'(z) = 0$. The remainder of the contour C^{**} is set to be a conveniently parameterized descent contour. As before, deformation of the original contour C into C^{**} is justified by Cauchy's closed-curve theorem and is generally done because (a) the steepest descent contour C^*, though often easy to characterize locally, is difficult to characterize globally; and (b) under sufficient smoothness conditions, the main contribution to the integral as $\lambda \to \infty$ comes from the behavior of the integrand about the saddlepoint. The resulting saddlepoint approximation is obtained by approximating the integral over S^* via the method of steepest descent and showing that the contributions to the integral over $C^{**} - S^*$ are asymptotically negligible.

3. SOME KEY DEVELOPMENTS IN STATISTICS

In my opinion, eight articles have had the most significant influence on the development of higher-order asymptotic approximations in statistics. In order of publication, these are the articles of Daniels (1954), Barndorff-Nielsen and Cox (1979), Barndorff-Nielsen (1980), Lugannani and Rice (1980), Barndorff-Nielsen (1983), Barndorff-Nielsen (1986), Tierney and Kadane (1986), and Skovgaard (1987). As remarked earlier, I do not mean to imply that these are the only important works in the area of asymptotic approximation, for this is clearly not the case. However, I think that it can be successfully argued that these works contain the key developments from which most, if not all, of today's statistical literature on the topic of higher-order asymptotic approximation has been generated. I should note that, except for the articles of Daniels (1954) and Barndorff-Nielsen and Cox (1979), the order in which the articles are discussed here should not necessarily be taken as an implication of relative importance.

Daniels (1954) introduced the saddlepoint method to statisticians in the context of approximating the density of a sample mean of n iid random variables $X_1 \ldots X_n$. This was accomplished by using the saddlepoint method described in Section 2.2 to derive an approximation to the Fourier inversion integral representation of the density of the sample mean; that is, to

$$f_{\bar{x}}(x) = \frac{n}{2\pi i} \int_{\tau-i\infty}^{\tau+i\infty} e^{n[K(z)-zx]} \, dz,$$

where $K(t)$ is the cumulant-generating function for X_1. This integral is exactly of the form (3) with a particularly simple choice of the contour C, $f(z) \equiv 1, \lambda = n$, and $\psi(z) = K(z) - zx$. Applying (4) leads to the relation

$$f_{\bar{x}}(x) = \left(\frac{n}{2\pi K''(t_0)}\right)^{1/2} e^{n[K(t_0)-t_0 x]} \times (1 + O(n^{-1})), \tag{5}$$

where t_0 uniquely solves $K'(t_0) = x$ and thus depends on x. Daniels demonstrated to the statistical community that substantial improvements over approximations provided by both the central limit theorem and Edgeworth expansions could be obtained with this method of approximation. Daniels also connected the saddlepoint approximation to an "exponentially tilted" version of the Edgeworth expansion, the latter being a particularly statistical interpretation of the more general technique of saddlepoint approximation. These important results were largely overlooked by statisticians for the next 25 years or so. For example, although the article of Daniels (1954) has probably been referenced more than any other article in statistical work on higher-order asymptotics, most of these references appear in works published in 1979 or later.

Although Daniels (1954) is generally credited with introducing saddlepoint methods (and tilted Edgeworth expansion) to statistics, the work of Barndorff-Nielsen and

Cox (1979) really represents the cornerstone of development for much of the future work in the area of higher-order asymptotic approximation in statistics. Many of the key ideas that now permeate the literature were given by Barndorff-Nielsen and Cox (1979), including (a) a statistically oriented approach to devising higher-order asymptotic expansions (i.e., through tilted Edgeworth expansion); (b) further development of the "double saddlepoint" approximation for conditional distributions, a technique originated by Daniels (1958); (c) introduction of the "single-saddlepoint" approximation to conditional distributions; and (d) insights into the value and use of these methods in unconditional and conditional likelihood inference. The ideas and observations presented in the discussion to this article also provide a window on useful future developments; see in particular the comments of Durbin, Peers, Lindley, Bickel, Hampel, and Hinkley.

In work on higher-order approximations to tail probabilities, it is a virtual guarantee that either one or both of the articles by Lugannani and Rice (1980) and Skovgaard (1987) will be referenced. According to Jensen (1995, p. 88), Lugannani and Rice (1980) were the first to derive a saddlepoint approximation to the tail probability $P\{\bar{X} \geq x\}$ that was uniformly valid throughout the range of x. Previous attempts by others generally led to approximations that performed poorly either for x near $E[X]$ or for x in the tails of the distribution. For a random variable X_n, one general form of this approximation is

$$P\{X_n \geq x\} \approx 1 - \Phi(r_n) + \phi(r_n)(w_n^{-1} - r_n^{-1}), \qquad (6)$$

where r_n and w_n depend on the cumulant-generating function of X_n, its first two derivatives, and x. In certain cases, r_n takes the form of a likelihood ratio statistic and w_n takes the form of a Wald statistic. Under favorable conditions, the relative error of (3.2) is $O(n^{-3/2})$ for r_n bounded. The apparent simplicity of (6) and its excellent properties are probably the main reasons why the article of Lugannani and Rice (1980) is one of the most referenced in the statistical literature on tail probability approximations. This is interesting, as the article itself is generally difficult to follow and provides little useful insight into how (6) is obtained. In this regard, a very useful and important expository article is that of Daniels (1987), where an exceptionally clear discussion and derivation of (6) is given in the case of a sample mean. Quite general forms of (6), derived without any reference to probability or sample means, can also be found in the literature prior to the work of Lugannani and Rice (1980) (see, e.g., Bleistein and Handelsman 1975, sec. 9.4; Olver 1974, sec. 9). Some interesting statistical articles only partially representative of the diversity of applications of (6) and related formulas include (in order of publication) those of Robinson (1982), Daniels (1983), Fraser (1990), Barndorff-Nielsen (1991), DiCiccio and Martin (1991), Wood, Booth, and Butler (1993), Skates (1993), and Booth, Hall, and Wood (1994). More recently, Zheng (1998) combines (6) and bootstrap calibration in order to construct confidence sets for parameters of lattice distributions. Another recent article is that of Strawderman and Wells (1998), which derives single-saddlepoint approximations of the form (6) for the conditional distribution of the common odds ratio in k 2×2 tables. Their approximations also apply more generally to noncentral hypergeometric distributions and sums of non-iid Bernoulli random variables.

Skovgaard (1987) developed an approximation for conditional probabilities of the form $P\{\bar{Y} \geq y | \mathbf{X} = \mathbf{x}\}$ by combining (6) with the technique of double-saddlepoint approximation. Skovgaard's work represents a substantial generalization of the work of Barndorff-Nielsen and Cox (1979), as his approximation is for tail probabilities (i.e., instead of a conditional density) and allows for the marginal distributions of $\bar{Y}$ and $\mathbf{X}$ to be lattice, continuous, or a mix of the two. The form of this approximation is the same as (6), with r_n and w_n suitably defined. In part due to its relative simplicity, Skovgaard's work has been instrumental in developing tail probability approximations in the realm of conditional inference. Two useful papers here include those by Davison (1988), who showed how Skovgaard's approximations could be used for conditional inference in certain generalized linear models, and Kolassa and Tanner (1994), who showed how to combine Gibbs sampling and Skovgaard's approximation for carrying out conditional inference in generalized linear models.

We now turn to the important work of Barndorff-Nielsen (1980, 1983, 1986) and the rather substantial number of related articles that this work has generated on increasing our understanding of the properties of likelihood, with a particular focus on the connections between likelihood and higher-order asymptotics. Daniels (1958) observed that for n iid observations for an exponential family with canonical parameter θ, (5) could also be written as

$$\left(\frac{n}{2\pi}\right)^{1/2} [i(\hat{\theta})]^{-1/2} \frac{L(\theta | t(\mathbf{x}))}{L(\hat{\theta} | t(\mathbf{x}))}, \tag{7}$$

where $t(\mathbf{x})$ is sufficient for θ, $\hat{\theta}$ is the maximum likelihood estimator (MLE), $i(\hat{\theta})$ is the expected information matrix evaluated at $\theta = \hat{\theta}$, and $L(\theta | t(\mathbf{x}))$ is the likelihood function. Barndorff-Nielsen (1980) showed that (7), with $i(\hat{\theta})$ replaced by the observed information evaluated at $\theta = \hat{\theta}$, leads to either an exact or a very accurate approximation to the conditional distribution of the MLE given an ancillary statistic in transformation and exponential families. The resulting density approximation, generally referred to as the p^* formula, was explored further by Barndorff-Nielsen (1983), where such results are extended to more general parametric models. It is perhaps worth pointing out here that although the p^* approximation can be motivated without the use of Laplace or saddlepoint approximations (e.g., Barndorff-Nielsen 1988), strong connections to these methods of integral approximation are also undeniable (e.g., Barndorff-Nielsen 1990a; Fraser 1988). The likelihood vignette by Reid discusses the origin and properties of the p^* approximation from a likelihood perspective in more detail.

More directly useful for inference are tail probability approximations. As discussed earlier, the approximations of Lugannani and Rice (1980) and Skovgaard

(1987) are derived directly using saddlepoint methods and can be applied provided that the relevant cumulant-generating function is available or can be computed. However, asymptotic approximations for carrying out conditional inference for a single parameter in the presence of nuisance parameters that are likelihood based also exist. These approximations rest on the same considerations as those underlying the p^* formula and largely stem from Barndorff-Nielsen (1986), who developed the r^* approximation for a single component of a MLE based on a mean and variance–adjusted version of a signed log-likelihood ratio statistic. The resulting tail probability approximation takes the general form (6) and is computed via the "modified directed likelihood" (e.g., Barndorff-Nielsen and Cox 1994, sec. 6.6). This approximation has strong connections to various corrected versions of profile likelihood (see, e.g., Barndorff-Nielsen 1994). In particular, as shown by Pierce and Peters (1992), r^* involves two main corrections to the usual profile likelihood, one that adjusts for the presence of nuisance parameters and the other that attempts to make an adjustment for a lack of normality (i.e., as measured through the curvature, or, equivalently, the amount of information, about the parameter of interest in the profile log-likelihood). One difficulty with the original r^* approximation is the need to explicitly specify an ancillary statistic. This fact spurred considerable additional research in this area, and simpler versions of the r^* approximation that avoid this difficulty were given by Barndorff-Nielsen (1991, 1994); DiCiccio and Martin (1993) considered a related approximation with a more Bayesian flavor. As discussed by Reid (1996), in addition to being less accurate, the computations required for these alternative approximations are still quite difficult. This especially holds true outside canonically parameterized exponential and transformation families. For more extensive reviews of recent work see the review articles of Reid (1995, 1996). (For a summary of the many results in this area largely due to Barndorff-Nielsen, see Barndorff-Nielsen and Cox 1994, chap. 8.)

To this point, not much has been said about Laplace's method. Although applications of Laplace's method do appear sporadically in the statistics and applied probability literature prior to 1986, its current popularity can be traced to the important article of Tierney and Kadane (1986) on approximating posterior expectations. Tierney and Kadane (1986) convincingly demonstrated that Laplace's approximation holds significant practical value for Bayesians. In particular, they showed how to accurately approximate posterior expectations of the following general form:

$$E[w(\theta)|\text{data}] = \frac{\int e^{l_n(\theta)+g(\theta)}\pi(\theta)\,d\theta}{\int e^{l_n(\theta)}\pi(\theta)\,d\theta}, \tag{8}$$

where $l_n(\theta)$ is the log-likelihood function computed from data, $\pi(\theta)$ is the prior, and $w(\theta) = e^{g(\theta)}$ is a smooth positive function of θ. With suitable manipulation (and conditions), the above may be placed in a form to which the Laplace approximation of Section 2.1 can be separately applied to the numerator and denominator, producing an accurate approximation to (8). Follow-up articles by Kass, Tierney, and Kadane (1988) and Tierney, Kass, and Kadane (1989) further extended these results and also provided discussion on some practical issues involved in using these approximations. Aside from providing a useful method that continues to facilitate complicated Bayesian computations, the Tierney and Kadane article is important outside its Bayesian context, as it clearly renewed general interest in Laplace's method among statisticians working in many different areas of statistics (e.g., mixed-effects and measurement error models).

4. WHERE SHOULD WE GO FROM HERE?

There are many distinct directions in which the field of asymptotic approximations in statistics could proceed, but here I discuss only two: improving accessibility and applicability and encouraging fruitful combinations between computing power and asymptotic approximation.

4.1 Accessibility and Applicability

Casella, DiCiccio, and Wells (1995), in a discussion to a review article of Reid (1995) on conditional inference, remarked that a major failing of present literature on conditional inference is that it "has been developed by the cognoscenti for their use." Although this statement is rather strong (and obviously irked Reid, as seen in her rejoinder), it also contains some unfortunate truth. There is no doubt that the results being developed in the conditional inference school of asymptotics have led to profound new insight into the role and properties of likelihood. However, far less has been accomplished in the way of convincing the statistical community that such results are useful at interesting levels of generality. Of greatest use to practitioners are tail probability formulas, for these form the core of most inferential procedures. Such formulas are generally couched in the form (6), and at first sight appear exceedingly simple—that is, of course, until one delves further into the compact notation used in defining r_n and w_n. One then finds that computation of (6) is generally very complicated, involving, for example, derivatives of the log-likelihood function with respect to the MLE. More work in this area is needed before this statistically driven version of higher-order asymptotic theory attains some reasonable level of practical utility. Some useful recent progress is evident (see, e.g., Severini 1999).

In terms of expanding applicability, it is worth emphasizing a point that has been made a number of times in the past: that although saddlepoint and p^*-type approximations coincide in certain cases, they are also distinct entities. Saddlepoint approximations and related methods are purely mathematical tools designed to approximate integrals of a given form, such as (5). In contrast, p^*-type formulas have a deeper philosophical motivation rooted within basic tenets (and controversies) of statistical theory. The development of a purely statistical theory of higher-order asymptotic expansion based on the principles of likelihood and conditional inference is attractive. However, I would argue that despite some claims to the contrary, this necessarily limits general applicability. For example, such a theory does not cover much of the current research today on estimators derived from more general estimating equations. This fact does not seem to bother those working in the area of conditional inference, the justification being that such inferential methodologies are "at odds with the viewpoint of conditionality" (Barndorff-Nielsen and Cox 1994). But whether this truly represents a problem is entirely a function of one's inferential philosophy. Irrespective of existing dogma, an important fact to keep in mind is that the statistical models used in most real problems are no longer dominated by those amenable to purely likelihood-based analyses, and that more accurate inferences than those provided by first-order arguments are in general quite desirable, whatever the inferential context.

4.2 Computing and Asymptotics

Asymptotic approximations in statistics can be extremely complex. Efforts to keep such approximations manageable often result in nebulous error terms of the form O (something small). These $O(\cdot)$ error terms describe the "asymptotic order" of approximation and generally receive a misplaced level of trust. Significant discrepancies between two approximations claimed to be accurate to the same general asymptotic order (consider, e.g., single versus double saddlepoint approximations) often can be explained in terms of what is *not* captured by the leading term of their respective expansions, or, more precisely, what is hidden within the respective $O(\cdot)$ error terms. Although this fact is well known, it is often ignored because establishing sharp error bounds requires substantial knowledge regarding the global asymptotic behavior of the integrand.

In univariate problems, there is really no need to rely on asymptotic expansions at all, because most reasonable numerical quadrature routines can yield "exact" results to *user-controlled* levels of error. A distinct advantage of numerical quadrature is that one need not rely on a $O(\cdot)$ error term that represents some vague, hypothetical level of accuracy under a given set of assumptions that may or may not hold. However, a difficulty is that two different representations of the same exact integral can exhibit substantially different numerical behavior. Such difficulties are of especially significant concern in the computation of integrals with rapidly oscillating integrands, such as Fourier inversion integrals (see, e.g., Davis and Rabinowitz 1984). Few existing quadrature schemes seem to exploit the basic principles that underlie the theory of asymptotic expansion. For certain classes of integrands, preliminary transformations may create integrals amenable to asymptotic approximations. As many approximations can be viewed as "one-point" quadrature routines, such preliminary transformations should improve both the efficiency and the accuracy of numerical quadrature routines. For example, this basic idea underlies the results of Liu and Pierce (1994), who proposed a method for improving the efficiency of Gauss–Hermite quadrature.

Combining the principles of asymptotic expansion with numerical quadrature in the multivariate setting is also possible, though more problematic. Some work in this area has been done (see, e.g., DiCiccio, Kass, Raftery, and Wasserman 1997 for results relevant to the computation of Bayes factors, and Kolassa 1999 for progress in the area of combining Gibbs sampling and Skovgaard's approximation for computing conditional distribution functions in generalized linear models). Genz and Monahan (1998) devised stochastic integration rules for multivariate integrals of form $\int g(x)p(x)\,dx$, where p is either a multivariate normal or t density; adaptation of these efficient routines to Laplace-type integrals seems feasible and may significantly widen their scope of applicability. Arguably, with today's computing power, many lower-dimensional problems can be handled efficiently in such a manner. Gelman and Meng (1998) discussed interesting "path sampling" Monte Carlo methods for high-dimensional integration problems that originate from the statistical physics literature. These methods are designed to reduce high-dimensional integration problems to problems of much lower dimension. The results of DiCiccio et al. (1997), who considered combining Laplace's method with bridge sampling (a special case of path sampling), suggest that fruitful combinations of path sampling methods and asymptotic approximation may be possible.

REFERENCES

Ablowitz, M. J., and Fokas, A. S. (1997), *Complex Variables: Introduction and Applications*, Cambridge, U.K.: Cambridge University Press.

Bak, J., and Newman, D. J. (1996), *Complex Analysis* (2nd ed.), New York: Springer-Verlag.

Barndorff-Nielsen, O. E. (1980), "Conditionality Resolutions," *Biometrika*, 67, 293–310.

——— (1983), "On a Formula for the Distribution of the Maximum Likelihood Estimator," *Biometrika*, 70, 343–365.

——— (1986), "Inference on Full and Partial Parameters Based on Standardized Signed Log-Likelihood Ratio," *Biometrika*, 73, 307–322.

——— (1988), Discussion of "Saddlepoint Methods and Statistical Inference," by N. Reid, *Statistical Science*, 3, 288–289.

——— (1990a), "p^* and Laplace's Method," *REBRAPE (Brazilian Journal of Probability and Statistics)*, 4, 89–103.

——— (1991), "Modified Signed Log-Likelihood Ratio," *Biometrika*, 78, 557–563.

——— (1994), "Adjusted Versions of Profile Likelihood and Directed Likelihood, and Extended Likelihood," *Journal of the Royal Statistical Society*, Ser. B, 56, 125–140.

Barndorff-Nielsen, O. E., and Cox, D. R. (1979), "Edgeworth and Saddlepoint Approximations With Statistical Applications" (with discussion), *Journal of the Royal Statistical Society*, Ser. B, 41, 279–312.

——— (1994), *Inference and Asymptotics*, London: Chapman and Hall.

Bleistein, N., and Handelsman, R. A. (1975), *Asymptotic Expansions of Integrals*, New York: Dover.

Booth, J., Hall, P., and Wood, A. T. A. (1994), "On the Validity of Edgeworth and Saddlepoint Approximations," *Journal of Multivariate Analysis*, 51, 121–138.

Casella, G., DiCiccio, T. J., and Wells, M. T. (1995), Discussion of "The Roles of Conditioning in Inference," by N. Reid, *Statistical Science*, 10, 138–199.

Daniels, H. (1954), "Saddlepoint Approximations in Statistics," *Annals of Mathematical Statistics*, 25, 631–650.

———— (1958), Discussion of "The Regression Analysis of Binary Sequences" by D. R. Cox, *Journal of the Royal Statistical Society*, Ser. B, 20, 236–238.

———— (1983), "Saddlepoint Approximations for Estimating Equations," *Biometrika*, 70, 89–96.

———— (1987), "Tail Probability Approximations," *International Statistical Review*, 55, 37–48.

Davis, P., and Rabinowitz, P. (1984), *Methods of Numerical Integration* (2nd ed.), New York: Academic Press.

Davison, A. C. (1988), "Approximate Conditional Inference in Generalized Linear Models," *Journal of the Royal Statistical Society*, Ser. B, 50, 445–461.

de Bruijn, N. G. (1961), *Asymptotic Methods in Analysis* (2nd ed.), New York: Wiley.

Debye, P. (1909), "Näherungsformeln für die Zylinderfunktionen für Grosse Werte des Arguments und Unbeschränkt Veränderliche Werte des Index," *Math. Ann.*, 67, 535–558.

DiCiccio, T. J., Kass, R. E., Raftery, A., and Wasserman, L. (1997), "Computing Bayes Factors by Combining Simulation and Asymptotic Approximations," *Journal of the American Statistical Association*, 92, 903–915.

DiCiccio, T. J., and Martin, M. A. (1991), "Approximations of Marginal Tail Probabilities for a Class of Smooth Functions With Applications to Bayesian and Conditional Inference," *Biometrika*, 78, 891–902.

———— (1993), "Analytical Approximations to Conditional Distribution Functions," *Biometrika*, 80, 781–790.

Fraser, D. A. S. (1988), "Normed Likelihood as Saddlepoint Approximation," *Journal of Multivariate Analysis*, 27, 181–193.

———— (1990), "Tail Probabilities From Observed Likelihoods," *Biometrika*, 77, 65–76.

Fulks, W. (1960), "Asymptotics I: A Note on Laplace's Method," *American Mathematical Monthly*, 67, 880–882.

Fulks, W., and Sather, J. O. (1961), "Asymptotics II: Laplace's Method for Multiple Integrals," *Pacific Journal of Mathematics*, 11, 185–192.

Gelman, A., and Meng, X-L. (1998), "Simulating Normalizing Constants: From Importance Sampling to Bridge Sampling to Path Sampling" (with discussion), *Statistical Science*, 13, 163–185.

Genz, A., and Monahan, J. (1998), "Stochastic Integration Rules for Infinite Regions," *SIAM Journal on Scientific Computing*, 19, 426–439.

Jensen, J. L. (1995), *Saddlepoint Approximations*, Oxford, U.K.: Clarendon Press.

Kass, R. E., Tierney, L., and Kadane, J. B. (1988), "Asymptotics in Bayesian Computation," in *Bayesian Statistics 3*, eds. J. M. Bernardo, M. H. DeGroot, D. V. Lindley, and A. F. M. Smith, Oxford, U.K.: Clarendon Press.

Kolassa, J. (1999), "Convergence and Accuracy of Gibbs Sampling for Conditional Distributions in Generalized Linear Models," *The Annals of Statistics*, 27, 129–142.

Kolassa, J., and Tanner, M. A. (1994), "Approximate Conditional Inference in Exponential Families via the Gibbs Sampler," *Journal of the American Statistical Association*, 89, 697–702.

Liu, Q., and Pierce, D. (1994), "A Note on Gauss–Hermite Quadrature," *Biometrika*, 81, 624–629.

Lugannani, R., and Rice, S. O. (1980), "Saddlepoint Approximation for the Distribution of the Sum of

Independent Random Variables," *Advances in Applied Probability*, 12, 475–490.

Olver, F. W. J. (1974), *Asymptotics and Special Functions*, New York: Academic Press.

Pierce, D., and Peters, D. (1992), "Practical Use of Higher-Order Asymptotics for Multiparameter Exponential Families" (with discussion), *Journal of the Royal Statistical Society*, Ser. B, 54, 701–725.

Reid, N. (1995), "The Roles of Conditioning in Inference" (with discussion), *Statistical Science*, 10, 138–157.

———— (1996), "Likelihood and Higher-Order Approximations to Tail Areas: A Review and Annotated Bibliography," *Canadian Journal of Statistics*, 24, 263–278.

———— (2000), "Likelihood," *Journal of the American Statistical Association*, 95.

Riemann, B. (1863), "Sullo Svolgimento del Quoziente di due Serie Ipogeometriche in Frazione Continua Infinita," reprinted in *Complete Works*, New York: Dover, 1953.

Robinson, J. (1982), "Saddlepoint Approximations for Permutation Tests and Confidence Intervals," *Journal of the Royal Statistical Society*, Ser. B, 44, 91–101.

Severini, T. (1999), "An Empirical Adjustment to the Likelihood Ratio Statistic," *Biometrika*, 86, 235–248.

Skates, S. (1993), "On Secant Approximations to Cumulative Distribution Functions," *Biometrika*, 80, 223–235.

Skovgaard, I. (1987), "Saddlepoint Expansions for Conditional Distributions," *Journal of Applied Probability*, 24, 875–887.

Stigler, S. (1986), *The History of Statistics: The Measurement of Uncertainty Before 1900*, Cambridge, MA: Harvard University Press.

Strawderman, R. L., and Wells, M. T. (1998), "Approximately Exact Inference for the Common Odds Ratio in Several 2 × 2 Tables" (with discussion), *Journal of the American Statistical Association*, 93, 1294–1306.

Tierney, L., and Kadane, J. B. (1986), "Accurate Approximations for Posterior Moments and Marginal Densities," *Journal of the American Statistical Association*, 81, 1032–1038.

Tierney, L., Kass, R. E., and Kadane, J. B. (1989), "Fully Exponential Laplace Approximations to Expectations and Variances of Nonpositive Functions," *Journal of the American Statistical Association*, 84, 710–716.

Watson, G. N. (1944), *A Treatise on the Theory of Bessel Functions* (2nd ed.), London: Cambridge University Press.

Wood, A. T. A., Booth, J., and Butler, R. (1993), "Saddlepoint Approximations to the CDF of Some Statistics With Nonnormal Limit Distributions," *Journal of the American Statistical Association*, 88, 680–686.

Zheng, X. (1998), "Better Saddlepoint Confidence Intervals via Bootstrap Calibration," *Proceedings of the American Mathematical Society*, 126, 3669–3679.

Minimaxity

William E. Strawderman

1. INTRODUCTION

A minimax statistical decision procedure is, by definition, one that minimizes (over the class of procedures) the maximum (over the parameter space) of the risk function. Minimaxity is occasionally put forward as a principle roughly in the form that "one should use a minimax procedure in any statistical problem" (see, e.g., Berger 1985; Robert 1994). However, it seems to typically be the case that the principle is put forward mainly for the purpose of refuting it. In fact, there may be no statistician who actively supports the minimax principle as a prescription for action. Nonetheless, the study of minimaxity has been an important part of theoretical statistics for more than 50 years, has led to a deeper understanding of various areas of statistics, has provided useful new procedures, and as Brown (1994) argued, has been an organizing theme for many statistical developments. (See also the vignette on decision theory by Brown.)

2. MINIMAX ESTIMATION

There is no universally successful method for finding a minimax estimator (or procedure, more generally). Two methods have proven particularly helpful: the least favorable prior approach and the invariance approach.

2.1 The Least Favorable Prior Approach

The basis of the least favorable prior approach is the following result.

Proposition 1. If δ is a Bayes estimator with respect to a prior distribution π and if the Bayes risk $r(\pi)$ is equal to the supremum risk, $\sup_\theta R(\theta, \delta)$, then δ is minimax (and π is a least favorable prior).

A closely related result is the following.

Proposition 2. If π_n is a sequence of priors $n-1, 2, \ldots$ and δ is an estimator such that $\lim_{n \to \infty} r(\pi_n) = \sup_\theta R(\theta, \delta)$, then δ is a minimax estimator (and π_n is a

William E. Strawderman is Professor of Statistics, Department of Statistics, Rutgers University, New Brunswick, NJ 08903 (E-mail: straw@stat.rutgers.edu). This research was supported by National Science Foundation grant DMS-9704524.

least favorable sequence).

The term least favorable (sequence) refers to the fact that for any other prior, $\pi', r(\pi') \leq r(\pi)(r(\pi') \leq \lim_{n \to \infty} r(\pi_n))$. Hence a prior that is least favorable (to the statistician) gives the largest possible Bayes risk and is in this sense the most difficult prior distribution to deal with. (For proofs of Propositions 1 and 2, see Berger 1985, sec. 5.3, or Lehman and Casella 1998, sec. 5.1.)

An easy corollary of Proposition 1 is that a constant risk Bayes estimator is minimax. Here is a well-known example.

Example 1.   Let X have a binomial distribution with parameters n (known) and p (unknown). Consider estimation of the parameter p with squared error loss $L(p, \delta) = (\delta - p)^2$. A beta prior with parameters $\alpha = \beta = \sqrt{n}/2$ gives the Bayes estimator $(X + \sqrt{n}/2)/(n + \sqrt{n})$, which has constant risk $1/[4(1 + \sqrt{n})^2]$ and is hence minimax by Proposition 1.

In this example it would perhaps be difficult to guess that this particular prior would be least favorable. And, in fact, the minimax estimator is generally considered to not be particularly good. The maximum likelihood estimator (MLE) and uniformly minimum variance unbiased (UMVU) estimator, $X/n = \hat{p}$, has a risk $([p(1-p)]/n)$ that (for moderate or large n) is much better in a neighborhood of $p = 0$ or $p = 1$ and only slightly worse in a neighborhood of $p = \frac{1}{2}$.

In fact, if we change the loss to $L(\delta, p) = (\delta - p)^2/p(1-p)$, then the estimator $\hat{p} = X/n$ has constant risk $(= 1/n)$ and is Bayes with respect to the uniform [beta $(1, 1)$] prior. Hence $\hat{p}$ is minimax for this natural loss. The uniform prior might well be considered a more natural guess as a least favorable prior. This example illustrates that minimaxity considerations can be, and often are, quite sensitive to the choice of loss function.

Another interesting feature of this example is that although the minimax estimator for each of the losses is unique, the least favorable prior in each case is not unique. For example, for squared error loss, any prior whose first $n + 1$ moments match that of the beta $(\sqrt{n}/2, \sqrt{n}/2)$ prior will give rise to the same Bayes minimax estimator.

Proposition 2 is useful in finding minimax estimators in certain situations where there is no proper least favorable prior. Perhaps the best-known example of this phenomenon is estimation of the mean of a normal distribution. We discuss this in the following example.

Example 2.   Let X have a normal distribution with mean θ and variance 1. We wish to estimate θ with squared error loss $L(\theta, \delta) = (\delta - \theta)^2$. The more general case of a sample of size n from a normal distribution with known variance, σ^2, can be reduced to the foregoing by sufficiency (and change of scale). Of course, X has constant risk equal to 1, and it is perhaps natural to think that X can be shown to be minimax by Proposition 1. However, an easy and well-known result states that an unbiased estimator, δ, can be Bayes with respect to π under squared error loss if and only if the integral with respect to the prior π of the risk of δ is equal to 0 (see, e.g., Bickel and Mallows 1988). As X has constant risk equal to 1, the integrated risk with respect to any prior will be 1 and not 0. Hence X cannot be Bayes with respect to any (proper) prior, and Proposition 1 cannot be used to deduce minimaxity. Intuitively, it seems that a least favorable prior would be uniformly distributed on $(-\infty, \infty)$, but this measure would not be a proper (integrable) prior. Proposition 2 comes to the rescue here in the following way. Let the prior, π_n, be a normal distribution with mean 0 and variance n. The Bayes estimator $[n/(n+1)]X$ has Bayes risk, $r(\pi_n)$, equal to $n/(n+1)$, which approaches $R(\theta, X) = \sup_\theta R(\theta, X)$. Hence X is minimax, and the sequence π_n is least favorable. The sequence of priors, π_n, is becoming more and more spread out and in a sense may be said to be approaching a uniform distribution on $(-\infty, \infty)$.

It is worth noting that "the uniform distribution on $(-\infty, \infty)$" (i.e., Lebesgue measure), may be formally treated as a "prior." The usual formal calculations result in a proper posterior distribution—in fact, the normal distribution with mean X and variance 1. Hence X is in this sense a "Bayes" estimator with respect to the (improper) uniform prior distribution on $(-\infty, \infty)$.

The minimax estimator in this example is unique, although the least favorable sequence is not. Interestingly, up to constant multiples, the "uniform" prior is the unique generalized prior with X as a formal Bayes estimator.

2.2   The Invariance Approach

Location, scale, and location-scale families are examples of so-called group families to which invariance methods may be applicable. Under mild conditions (e.g., transitivity and local compactness) on the group of transformations (on the parameter space), a minimum risk equivariant (MRE) procedure may be found as a generalized or formal Bayes procedure with respect to the right-invariant (generalized) prior distribution. If, further, the group is amenable (see, e.g., Bondar and Milnes 1981), then the MRE is also minimax. This so-called Hunt–Stein theorem relies in part on a version of Proposition 2. Berger (1985) gave an accessible version of the development of the MRE as a generalized Bayes estimator with respect to the right-invariant measure, and Bondar and Milnes (1981) discussed a development of the Hunt–Stein theorem. Lehmann (1986) provided for a proof of the Hunt–Stein theorem for hypothesis testing, and Lehmann and Casella (1998, p. 421) gave a discussion and further references. Among the most important examples is the following.

Example 3.   Suppose that $\mathbf{X} = (X_1, X_2, \ldots, X_n)$ has density $f(X_1 - \theta, X_2 - \theta, \ldots, X_n - \theta)$, $-\infty < \theta < \infty$, with respect to Lebesgue measure on R^n. We wish to estimate θ with squared error loss $L(\theta, \delta) = (\delta - \theta)^2$. Here the distribution of $\mathbf{X}$ is a location family, and the problem is invariant under the location group. An equivariant estimator, δ, is one such that $\delta(X_1+a, X_2+a, \ldots, X_n+a) = \delta(X_1, X_2, \ldots, X_n)+a$ for all $\mathbf{X}, a$. The group $\bar{G}$ acting on the parameter space is the one-dimensional location group. It is both transitive and locally compact. Hence the MRE is given by the generalized Bayes estimator with respect to the right-invariant measure, which in this case is Lebesgue measure. The estimator is called Pitman's estimator. The group is also amenable, and hence Pitman's estimator is minimax. Pitman's estimator is given in this case by

$$\delta(\mathbf{X}) = \frac{\int_{-\infty}^{\infty} \theta f(X_1 - \theta, X_2 - \theta, \ldots, X_n - \theta) \, d\theta}{\int_{-\infty}^{\infty} f(X_1 - \theta, X_2 - \theta, \ldots, X_n - \theta) \, d\theta}.$$

2.3   Additional Methods, Examples, and Comments

Minimax procedures exist under general conditions. Proposition 1 may be used quite generally in finite-parameter and finite-action problems. Game theoretic techniques and linear programming techniques are occasionally useful. If in a particular problem with loss $L(\theta, \delta)$, we have an admissible procedure, δ, with positive risk $R(\theta, \delta)$, then δ will be an admissible procedure with constant risk for the problem with loss $L(\theta, \delta)/R(\theta, \delta)$. Because constant-risk admissible procedures are easily seen to be minimax, δ is minimax in the new problem. This again indicates the sensitivity of minimaxity to choice of loss function.

Minimaxity may also be affected by restrictions on the parameter space. Consider the following alteration of Example 2.

Example 2 (continued).   Suppose we know that $\theta \geq 0$. In this example X remains a minimax estimator, but it is not unique. Two alternative minimax estimators are $\max(X, 0)$ and the generalized Bayes estimator with respect to the uniform prior on $[0, \infty)$. There are, in fact, many generalized priors giving rise to minimax procedures.

If we further restrict θ to lie in the compact interval $[a, b]$, then X is no longer minimax, because the estimator that truncates X at a and at $b, \min[\max(X, a), b]$, has a risk that is everywhere strictly less than 1, the risk of X. Compactness of $[a, b]$ and continuity of all risk functions imply that the supremum risk is also strictly less than 1. The minimax procedure in this case is unique and is Bayes with respect to a unique prior supported on a finite set of points. If the interval $[a, b]$ is sufficiently small $(b - a < 1.05)$, then the prior puts mass $\frac{1}{2}$ at a and b. The number of support points of the prior grows with $b - a$ in a way that still is not well understood (see Casella and Strawderman 1981 for more details). Bickel (1981) found the weak limit of the sequence of least favorable priors, renormalized to the interval $[0, 1]$.

3. MULTIPARAMETER ESTIMATION

One of the most important developments in decision theory was the discovery by Stein (1956) that the usual estimator (i.e., the vector of sample means) of the mean vector of a multivariate normal population is inadmissible under sum of squared error loss. Because the usual estimator is minimax in this setting (independent copies of the sequence of priors in Example 2 and Proposition 2 work), any improvement is also minimax. As Brown (1994) pointed out, it is not clear whether admissibility or minimaxity is the main focus or the beneficiary of the discovery, although it is clear that research in both of these decision theory foci received a boost and greatly benefited from Stein's fundamental and monumental discovery. The following example illustrates some of the developments.

Example 4.   Let $\mathbf{X} = (X_1, \ldots, X_p)$ have a p-variate normal distribution with mean vector $\boldsymbol{\theta} = (\theta_1, \ldots, \theta_p)$ and covariance matrix $\sigma^2 \mathbf{I}$ where $\mathbf{I}$ is the $p \times p$ identity matrix. Let the loss be $L(\theta, \delta) = \sum (\delta_i - \theta_i)^2 = \|\delta - \theta\|^2$. Baranchik (1970) showed that an estimator of the form $(1 - \{[ar(\mathbf{X}'\mathbf{X})\sigma^2]/\mathbf{X}'\mathbf{X}\})\mathbf{X}$ dominates the estimator $\mathbf{X}$ provided that $0 < a \leq 2(p-2)$ and $r(\mathbf{X}'\mathbf{X})$ is nondecreasing and bounded between 0 and 1. The case $r(\cdot) \equiv 1$ is the result of James and Stein (1961). An enormous literature developed in several directions. Efron and Morris (1971, 1972, 1973) developed a variety of estimators of the foregoing form in this and related settings and studied them from the perspective of empirical Bayes estimation (see the empirical Bayes vignette by Carlin and Louis). Strawderman (1971) found proper Bayes minimax estimators of Baranchik's form for five and more dimensions. Berger (1976), Bhattacharya (1966), Bock (1975), Efron and Morris (1971, 1972, 1973), Strawderman (1978), and others studied the problem when the covariance was not a multiple of the identity or loss was not $\|\delta - \theta\|^2$. Berger (1975), Brandwein and Strawderman (1978, 1980), and Strawderman (1974) found explicit improvements for distributions other than the normal and losses other than quadratic. Hwang and Casella (1982) and others found improved confidence sets centered at James–Stein-type estimators. It seems plausible that the development of robust Bayes procedures and hierarchical Bayes procedures owes much to Stein's discovery. Much of the development of such Stein-type or shrinkage-type procedures moved away from questions of minimaxity to questions of empirical, hierarchical, or robust Bayes estimation, but a great deal of the impetus for these studies lies in the fact that the James–Stein estimator not only beats the "usual" estimator everywhere, but it wins by a large margin in a relatively small neighborhood of the origin. Indeed, if the covariance matrix is the identity, then the risk at $\boldsymbol{\theta} = \mathbf{0}$ of the James–Stein estimator under $L(\theta, \delta) = \|\delta - \theta\|^2$ is 2, compared to p, the constant risk of the usual estimator. This amounts to a nearly 100% improvement for large p. Hence if one has reasonably good prior information, then large gains are possible at no cost.

The Stein phenomenon has been demonstrated to persist in virtually all multiparameter problems (with noncompact parameter space). Berger (1980), Casella and Strawderman (1994), Clevenson and Zidek (1975), and Hwang (1982) have explored a variety of other distributional settings.

4. MINIMAX TESTS

Minimaxity ideas can be useful in a variety of hypothesis testing settings. Consider testing $H : \theta \in \Omega_H$ versus $K : \theta \in \Omega_K$. The following is an example of how least favorable distributions can lead to a uniformly most powerful (UMP) test in a situation where one might not be expected to exist.

Example 5. Suppose that $X_1, \ldots, X_n$ are iid $N(\theta, \sigma^2)$ with both θ and σ^2 unknown. Suppose that we wish to test $H : \sigma^2 \leq \sigma_0^2$ versus $K : \sigma^2 > \sigma_0^2$. General theory tells us that there is a uniformly most powerful unbiased (UMPU) test for this problem, but we typically do not expect a UMP test. Lehmann (1986, sec. 3.9) showed that for this problem (and *not* for testing $H : \sigma^2 \geq \sigma_0^2$ versus $K : \sigma^2 < \sigma_0^2$), one can find, for any (θ_1, σ_1^2) in K, a prior distribution, π, over H such that the Neyman–Pearson level-α test, ϕ_π, for testing $H' : f(X) = \int P_{\theta, \sigma^2}(X) \, d\pi(\theta, \sigma^2)$ versus $K' : (\theta_1, \sigma_1^2)$ is level α for H versus K and does not depend on (θ_1, σ_1^2). A Proposition 1-type result implies that the resulting test is UMP for H versus K. The UMP test is, of course, the test that rejects if $\sum (X_i - \bar{X})^2 > \sigma_0^2 \chi_{n-1,\alpha}^2$ where $\chi_{n-1,\alpha}^2$ is the upper 100α percentile of a chi-squared distribution with $n - 1$ degrees of freedom.

One may in a similar manner replace both H and K by prior distributions and use the Neyman–Pearson lemma to find a best level-α test of the reduced hypotheses. This device can be most helpful when H and K are not contiguous (e.g., when there is an indifference zone Ω_I). Huber's theory of robust tests is based on such a device (see, e.g., Huber 1981; Lehmann 1986).

5. ASYMPTOTIC MINIMAXITY AND OTHER CONSIDERATIONS

Minimaxity considerations have played a role in asymptotics and optimal design of experiments. I briefly visit these areas in this section. LeCam (1953) developed a theory of local asymptotic minimaxity (LAM) and showed the MLE to be LAM under weak conditions.

Pinsker (1980) developed exact asymptotic minimax rates and constants for a class of (nonparametric) function estimation problems. He showed that in certain cases the asymptotic minimax risk among linear procedures coincides with the asymptotic minimax risk. Donoho and Liu (1991) demonstrated that under weak conditions, the minimax linear risk is not greater than 1.25 times that for all estimators. Minimaxity has also had an important role to play in the theory of optimal experimental design, essentially beginning with the work of Kiefer and Wolfowitz (1959).

6. IS MINIMAXITY Y2K COMPATIBLE?

It is interesting to speculate on the role of minimaxity in the post-Y2K era. Actually, it is even necessary, because the editor directed that I do so! Here are some possibilities.

Bayesians have spent considerable effort on developing "objective" (nonsubjective) priors—also referred to as default or reference priors (Berger and Bernardo 1992; Bernardo 1979; Jeffreys 1961)—or probability matching priors (Datta and Ghosh 1995). (Also see the Bayes vignette by Berger.) Recent attention has largely focused on the multiparameter case, as Jeffreys's prior seems well accepted in the single-parameter case. It may be worthwhile to search for default priors from a risk-based (i.e., minimax) perspective. For example, one might search for a parameterization for which the Fisher information matrix is a constant multiple of the identity (when such exists) and take as a default prior a bounded superharmonic function on the boundary between proper and improper priors. Such a prior in the normal (identity covariance) case would give admissible minimax estimators. Such procedures might well have good Bayesian and frequentist properties over fairly broad classes of problems.

Componentwise thresholding in wavelet estimation problems is justified in part on the basis of asymptotic minimaxity considerations. It is possible that a combination of thresholding and Stein-type estimation might allow improved estimation in certain problems. For example, the Stein estimation might reasonably be applied to the larger-scale components, and thresholding applied to those with smaller scales.

In any event, it seems likely that minimaxity as an idea, an ideal, an organizing theme, or as a useful touchstone will continue to play an important role in statistical developments well into the new era.

REFERENCES

Baranchik, A. J. (1970), "A Family of Minimax Estimators of the Mean of a Multivariate Normal Distribution," *Annals of Mathematical Statistics*, 41, 642–645.

Berger, J. (1975), "Minimax Estimation of Location Vectors for a Wide Class of Densities," *The Annals of Statistics*, 3, 1318–1328.

——— (1976), "Admissible Minimax Estimation of a Multivariate Normal Mean With Arbitrary Quadratic Loss," *The Annals of Statistics*, 4, 223–226.

——— (1980), "Improving on Inadmissible Estimators in Continuous Exponential Families With Applications to Simultaneous Estimation of Gamma Scale Parameters," *The Annals of Statistics*, 8, 545–571.

——— (1985), *Statistical Decision Theory and Bayesian Analysis* (2nd ed.), New York: Springer.

Berger, J. O., and Bernardo, J. M. (1992), "Ordered Group Reference Priors With Application to Multinomial Probabilities," *Biometrika*, 79, 25–37.

Bernardo, J. M. (1979), "Reference Posterior Distributions for Bayesian Inference," *Journal of the Royal Statistical Society*, Ser. B, 41, 113–147.

Bhattacharya, P. K. (1966), "Estimating the Mean of a Multivariate Normal Population With General Quadratic Loss Function," *Annals of Mathematical Statistics*, 37, 1819–1824.

Bickel, P. J. (1981), "Minimax Estimation of the Mean of a Normal Distribution When the Parameter Space is Restricted," *The Annals of Statistics*, 9, 1301–1309.

Bickel, P. J., and Mallows, C. (1988), "A Note on Unbiased Bayes Estimation," *American Statistician*, 42, 132–134.

Bock, M. E. (1975), "Minimax Estimators of the Mean of a Multivariate Normal Distribution," *The Annals of Statistics*, 3, 209–218.

Bondar, J. V., and Milnes, P. (1981), "Amenability: A Survey for Statistical Applications of Hunt–Stein and Related Conditions on Groups," *Z. Wahrsch. Verw. Geb.*, 57, 103–128.

Brandwein, A. C., and Strawderman, W. E. (1978), "Minimax Estimation of Location Parameters for Spherically Symmetric Unimodal Distributions Under Quadratic Loss," *The Annals of Statistics*, 6, 377–416.

——— (1980), "Minimax Estimation of Location Parameters for Spherically Symmetric Distributions With Concave Loss," *The Annals of Statistics*, 8, 279–284.

Brown, L. D. (1994), "Minimaxity, More or Less," in *Statistical Decision Theory and Related Topics, V*, eds. S. S. Gupta and J. O. Berger, New York: Springer.

Casella, G., and Strawderman, W. E. (1981), "Estimating a Bounded Normal Mean," *The Annals of Statistics*, 9, 870–878.

——— (1994), "On Estimating Several Binomial N's," *Sankhya*, 56, 115–120.

Clevenson, M. L., and Zidek, J. (1975), "Simultaneous Estimation of the Mean of Independent Poisson Laws," *Journal of the American Statistical Association*, 70, 698–705.

Datta, G. S., and Ghosh, J. K. (1995), "On Priors Providing Frequentist Validity for Bayesian Inference," *Biometrika*, 82, 37–46.

Donoho, D. L., and Liu, R. C. (1991), "Geometrizing Rates of Convergence, III," *The Annals of Statistics*, 19, 668–701.

Efron, B., and Morris, C. N. (1971), "Limiting the Risk of Bayes and Empirical Bayes Estimators—Part I: The Bayes Case," *Journal of the American Statistical Association*, 66, 807–815.

——— (1972), "Empirical Bayes on Vector Observations—An Extension of Stein's Method," *Biometrika*, 59, 335–347.

——— (1973), "Stein's Estimation Rule and Its Competitors—An Empirical Bayes Approach," *Journal of the American Statistical Association*, 68, 117–130.

Huber, P. J. (1981), *Robust Statistics*, New York: Wiley.

Hwang, J. T. (1982), "Semi-Tail Upper Bounds on the Class of Admissible Estimators in Discrete Exponential Families With Applications to Poisson and Binomial Cases," *The Annals of Statistics*, 10, 1137–1147.

Hwang, J. T., and Casella, G. (1982), "Minimax Confidence Sets for the Mean of a Multivariate Normal Distribution," *The Annals of Statistics*, 10, 868–881.

James, W., and Stein, C. (1961), "Estimation With Quadratic Loss," in *Proc. Fourth Berkeley Symp. Math. Statist. Prob.*, 1, Berkeley, CA: University of California Press, pp. 311–319.

Jeffreys, H. (1961), *The Theory of Probability*, Oxford, U.K.: Oxford University Press.

Kiefer, J. C., and Wolfowitz, J. (1959), "Optimum Designs in Regression Problems," *Annals of Mathematical Statistics*, 30, 271–294.

LeCam, L. (1953), "On Some Asymptotic Properties of Maximum Likelihood Estimates and Related Bayes Estimates," Univ. of Calif. Publ. in Statist., 1, 277–330.

Lehmann, E. L. (1986), *Testing Statistical Hypotheses* (2nd ed.), New York: Springer.

Lehmann, E. L., and Casella, G. (1998), *Theory of Point Estimation* (2nd ed.), New York: Springer.

Pinsker, M. S. (1980), "Optimal Filtering of Square Integral Signals in Gaussian White Noise" (in Russian), *Problemy Peredachi Inf.*, 16(2), 52–68; English trans., *Problems Inf. Transmission*, 120–133.

Robert, C. P. (1994), *The Bayesian Choice: A Decision-Theoretic Motivation*, New York: Springer.

Stein, C. (1956), "Inadmissibility of the Usual Estimator for the Mean of a Multivariate Distribution," in *Proc. Third Berkeley Symp. Math. Statist. Prob.*, 1, Berkeley, CA: University of California Press, pp. 197–206.

Strawderman, W. E. (1971), "Proper Bayes Minimax Estimators of the Multivariate Normal Mean," *Annals of Mathematical Statistics*, 42, 385–388.

——— (1974), "Minimax Estimation of Location Parameters for Certain Spherically Symmetric Distributions," *Journal of Multivariate Analysis*, 4, 255–264.

——— (1978), "Minimax Adaptive Generalized Ridge Regression Estimators," *Journal of the American Statistical Association*, 73, 623–627.

Afterword

George Casella

I hope that you have done what I just did—sat down and read this entire collection of vignettes. If you haven't, please treat yourself and do it right now. Make it a vacation reading project . . . teach a special topics class that reads through the vignettes . . . but please don't miss them!

These thought-provoking pieces have many common themes, and only through reading the collection in its entirety will the themes become clear. I'm sure that I did not see all of them, but a few themes that emerge are as follows:

- Large datasets. We have to develop all sorts of tools to handle massive datasets. This means not just data mining, but also computational and inferential tools.

- High-dimensional/nonparametric models. The simple parametric model may not yet be dead, but it is dying. Hierarchical, robust, semiparametric and nonparametric, high-dimensional parametric, measurement error, and more are what we are going to be doing. (Check out Efron's discussion of bootstrapping a data-mining excursion.)

- Accessible computing. Many of the vignettes call for more accessible computing, whether it be MCMC, robust methods, nonparametrics, or other. We now need the generation beyond S-PLUS—more user friendly and able to easily do very sophisticated analyses.

- Bayes/frequentist/who cares? The empirical Bayesian is not a Bayesian. The Bayesian who uses MCMC needs frequentist methods. Maybe we are finally learning that we are *statisticians*, and should use whatever are the best available methods to solve the problem at hand. Which brings me to . . .

- Theory/applied/why differentiate? Good statistics does not exist in a vacuum. Throughout these vignettes, we constantly see that good (and interesting and even elegant) theory arises from trying to solve real problems. So there is no distinction between the applied and the theoretical statistician. We all must know both ends of our field!

My colleague Marty Wells made an interesting observation about the theory/applied pseudodistinction. He said that the applied statistician, who never knows what problem will be presented, has to be prepared to use almost any branch of mathematics or statistics to provide an answer. (In recent years, to solve real problems, I have seen the use of complex analysis, functional analysis, differential equation theory, Markov chain and martingale theory, abstract algebra, and nonlinear optimization theory, to name a few.) Compared to the theoretical statistician, the applied statistician may need an as broad, or an even broader, understanding of mathematics and theoretical statistics. So as theory and applications become more intertwined, thorough knowledge of each becomes even more crucial.

Author Index

Subject Index

A

abortion, 48

abstract algebra, 493

accelerated failure time models, 319

accelerated moment release, 211

accelerometer, 204

acceptance rate, 306

accept–reject algorithms, 304

accumulated prediction error criterion, 233

accurate interval estimates, 346

active drug v. placebo, 370

adaptive

    algorithms, 309

    arithmetic code, 232

    estimation, 203

    rejection sampling, 344

    tests, 174

    wavelet estimation, 298

additive Gaussian white noise channel, 235

additive model, 7, 35, 335, 336, 416

additive mortality hypothesis, 29

additive separable models, 336

admissibility, 295, 379

admixture, 259

ADPCM speech coder, 234

advanced importance sampling techniques, 309

advanced manufacturing, 197, 264–270

AECM algorithms, 404

affine equivariant estimators, 412

affine transformations, 361

age structure, 190

age-graded intracohort changes, 192

aggregated packet process, 218

aging, 194

AIDS, 8, 326, 45, 327, 328

air pollution, 22, 76, 77, 78, 80

AIRS air pollution database, 77

Akaike Information Criterion (AIC), 29, 212, 352, 353, 355, 457–458

algebraic coding, 235

algebraic functions, use in statistics, 380

aligned rank tests, 359

allele frequencies, 260

allele sharing among affected siblings, 260

allele types, 402

alleles, 259

allelic association, 259, 260, 261

Alzheimer's, 261

ambient pollution, 79

American Political Science Association, 154

American Society of Composers, Authors, & Publishers (ASCAP), 181

American Stock Exchange, 106

amino acid sequence, 92

amino acids, 91, 93

Amstat News, 273

analysis of choice, 150

analysis of duration data, 150

analysis of variance (ANOVA), 37, 323, 335, 381

analytic approximations, 346

analytical chemistry, 258

ancillarity, 434–436

ancillarity paradoxes, 440

ancillary statistic, 433, 478

animal abundance from line-transects, 462

animal behavior in earthquakes, 211

animal breeding, 1, 2, 34–39

animal testing, 256